Fundamentals of Geothermal Heat Pump Systems

Louis Lamarche

Fundamentals of Geothermal Heat Pump Systems

Design and Application

 Springer

Louis Lamarche
Mechanical Engineering
École de Technologie Supérieure
Montréal, QC, Canada

ISBN 978-3-031-32175-7   ISBN 978-3-031-32176-4   (eBook)
https://doi.org/10.1007/978-3-031-32176-4

This Springer imprint is published by the registered company Springer Nature Switzerland AG
The registered company address is: Gewerbestrasse 11, 6330 Cham, Switzerland

Paper in this product is recyclable.

To Natalie, Julien, Frédérique, and my regretted friend Christian.

Preface

As many other academic books, this one originates from lecture notes that were used for several years in a graduate course that I teach at École de Technologie Supérieure, the university where I work. Teaching also a course in solar thermal systems, I was fortunate to rely on several classical academic references in which examples and sample problems were available to accompany the students. In geoexchange systems or ground source heat pumps in general, several books were recently written but were mostly oriented toward practical engineers or sometimes a portrayal of recent researches in the field. Although some were very interesting, few examples or problems were available. I hope this contribution may help filling some gap in this interesting field. Starting this project, many more challenges than I expected have come up; some of which were the choice of the subject matter covered in the book, the order in which they are presented and so on. I made the choice to include a wider range of topics than I usually deal with during a semester to make it more complete and useful to students that want to start some research projects in that field. Some mathematical aspects were associated with starred sections and can be skipped without impairing the rest of the book. Another difficult choice was about the notation used. Being more accustomed to write brief articles, it is not rare that I used different notations in different papers associated with the specific subject treated. In a book, the consistency of the notation is of prime importance. Not only that the same property should not change notation in the same book (k and λ for example for the conductivity), but ideally, the same letter should not be used for too many variables. Although this is quite impossible with 26 letters (plus fortunately the Greek ones), but it should, I think, be restricted as much as possible. For example, I don't mind using the letter "d" for a diameter and distance between two points (at least they share the same unit) but using the letter "Q" for heat and flow rate didn't seem logical to me. One of my main concern was about the symbol "h" for enthalpy, heat transfer coefficient, and head losses in pipes. In some heat transfer books, the letter "i" is sometimes used for enthalpy, but it did feel awkward to me; so I arbitrary borrowed the Plank's constant letter $\hbar$ for the heat transfer coefficient, hoping the physicians will forgive me. A classical trick to overcome the problem is to use subscripts and superscripts and it was used in this book, but too much of them makes the reading too confusing. So, the arbitrary choices I made were done to make a compromise between those approaches. I also point out that I use the symbol "log" for the natural logarithm and, if necessary, "$\log_{10}$" for logarithm in the base 10. This choice comes from one of my former mathematics teacher saying always that the only logarithm worth to talk about is the natural logarithm, the others being unnecessary amusing artifices. This convention is the same as the one used in many common languages like matlab and python.

Talking about units, the system of units (SI versus IP system) was also an important choice to make. Since my early years as an engineering student, I always was a fervent supporter of the SI system (with a decimal point of course and not that stupid comma), and I never thought that 40 years later, we would still talk about inches and BTU. But of course, living in North America, we are dealing with data in IP system on a daily basis. Nevertheless, I made the deliberate choice to use the IP system as little as possible, hoping that any engineer can apply a rule of three if needed, the hope being to share a common system 1 day. Saying that, several problems will use the IP system especially when dealing with manufacturers data sheet and American standards.

As in all engineering fields, the use of software or programming tools is nowadays very accessible to everyone, many of them becoming free and open source. Although the use of graphical methods is sometimes useful to have a quick idea of a quantity, solving, what used to be complex before, is now quite easy. In this book, I made the deliberate choice to solve all problems and examples with the python computer language. This is a very powerful open source interpreted language, easy to learn, and with tremendous amount of tutorial and examples available. I am always impressed when I can nowadays solve more complex problems in one afternoon than during my whole PhD period. Saying that, I am more a practical engineer than a software programmer, so I hope that pure programmers will forgive me for not doing things sometimes the "pythonesque" way. For example, I had a lot of reluctance to write np.cos or np.sin everywhere. I finally tried to follow the guidance rule of programming, but I still wonder who would be so stupid to write a new module with a function name "sqrt" that would not be the square root of a number. All the functions I used are available and can be modified if needed by anyone.

As a final remark, I must apologize for all researchers that I have forgotten to cite; I unfortunately do not have the time to read all papers in the field and these possible oversights were not deliberate.

Downloads

Sample codes that solve all examples in the book can be downloaded from the author's GitHub repository at https://github.com/LouisLamarche/Fundamentals-of-Geothermal-Heat-Pump-Systems. This repository contains numerous sample computer codes and function libraries that anyone can use to solve many problems related to ground-source heat pump systems. The programs that solve the problems at the end of the chapters can be obtained from the author on demand.

Montréal, QC, Canada   Louis Lamarche
June 2020

- Alors, que devrais-je faire plus tard ?
- Curieux
- Mais... ce n'est pas un métier
- Pas encore ...

–François Truffaut, *Jules et Jim*

Chaque génération, sans doute, se croit vouée à refaire le monde. La mienne sait pourtant qu'elle ne le refera pas. Mais sa tâche est peut-être plus grande. Elle consiste à empêcher que le monde se défasse.
–Albert Camus, *speech at the Nobel Banquet, Stockholm, Dec. 10, 1957*

There is no sin so great as ignorance. Remember this.

–Rudyard Kypling, *Kim*

If we want things to stay as they are, things will have to change.
–Giuseppe Tomasi di Lampedusa, *The Leopard*

Contents

1.1 Introduction

Geothermal (from *geo*, earth; *thermos*, hot) energy is a large field that deals essentially with all applications which exchange energy with the earth. This subject is thus very large, and this book will focus mainly on one part of geothermal energy: the heat exchange with geothermal heat pumps, usually called ground-source heat pumps (GSHP), ground-coupled heat pumps (GCHP), or geo-exchange systems. In this first chapter however, some broad notions related to all aspects of geothermal applications will be summarized.

1.2 The Earth Structure

The earth's center temperature is very high, in the order of 5500 °C, and it can easily be understood that the earth will serve mainly as a heat supplier. However, as it will be seen later, it can also be used for cooling applications. Even though the earth's center is very hot, all geothermal applications are done at a relatively shallow depth compared with the earth's radius. The earth's structure is divided in several regions.

Other criteria are also used to characterize the earth structure. For example, the *lithosphere* (0–170 km) includes the crust as well as the solid part of the outer mantle. The rest of the outer mantle, which is in liquid phase, forms the *asthenosphere* (Fig. 1.1).

Since all geothermal applications are not more than few kilometers deep, they are all situated essentially in the crust region. However, the energy coming from the earth is influenced by the deeper regions. The typical temperature ranges vary from 0 to 300 °C. The very large thermal energy coming from the center of the earth comes in part from the thermal energy created at the beginnning of the earth and also from the radioactive reactions that take place in the inner parts of it. The overall mean of the earth heat flux that is observed is in the order of 0.05 W/m^2. This value seems rather small, knowing, for example, that the solar flux is in the order of 1000 W/m^2 and it might be worth arguing if this small energy flux is worthwhile to harvest. However, since it is present for millions of years, a large amount of this energy is stored in the soil and can be exploited. Being stored in a given volume, the concept of *renewable energy* may be argued. Before trying to answer these questions, let us try to evaluate the order of magnitude of the temperatures involved. In Table 1.1, it is seen that the earth's crust is approximately 35 km depth, where the highest temperature is around 1100 °C. This gives a mean gradient of the order of 0.03 °C/m. At a depth of 1 km, which is already quite deep, temperatures in the order of 30–40 °C would be observed, which is small if we want to use it as a heating medium. These last two statements seem to give little interest in geothermal applications. However, it should be remembered that these are mean values. Fortunately (or unfortunately for some), the earth's underground is far from being homogeneous, and many regions will experience much more important thermal loads and temperature gradients [1]. This energy is partly due to movement of tectonic plates where fractures often occur due to tension-compression phenomena. These faults bring by convection hot magma from the inner parts of the earth, which offer very large thermal energy potential. The movements are also responsible for other phenomena like volcanic eruptions and earthquakes. It is not a surprise that most high-temperature geothermal applications are in active geological regions like the *ring of fire*.

In these regions, temperature gradients of the order of 0.15 °C/m can be observed. In the case where water supply is present, we can, in certain conditions, obtain a *geothermal reservoir*. Depending of the temperature ranges, different applications can be defined (Table 1.2):

Whatever the temperature values, the geothermal applications can be separated between:

- Electricity production applications.
- Direct applications.

Fig. 1.1 (**a**) Earth structure. (**b**) Earth temperatures

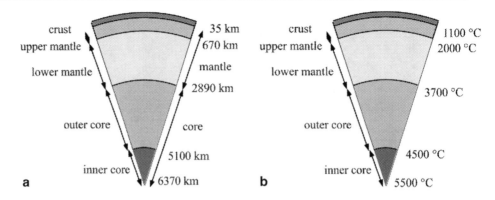

Table 1.1 Earth structure

Layer	Depth (km)	Temperature (°C)
Crust	0–35	10–1100
Outer mantle	35–670	1100–2000
Inner mantle	670–2890	2000–3700
Outer core	2890–5100	3700–4500
Inner core	5100–6370	4500–5500

Table 1.2 Temperature range for geothermal applications

High temperature	>150 °C	Electricity production
Medium temperature	90–150 °C	Electricity production, direct thermal heating
Low temperature	60–90 °C	Direct thermal heating
Very low temperature	0–50 °C	Heating and cooling with heat pumps

1.3   Geothermal Electricity Production

1.3.1   Dry-Steam Power Plants

Direct applications were already used by Romans, but electricity production is relatively recent. The first geothermal power plant was exploited in Lardorello in Italy [2]. When the thermal reservoir is at very high temperature, it is possible to extract saturated (or superheated) steam that can be directly used to produce electricity through a steam turbine. In 1904, Piero Ginori Conti was the first to link a steam turbine to a generator and was able to produce enough electricity to turn on five light bulbs. Fifteen years later, one MWe of electricity was produced. Nowadays, the plant produces nearly 795 MWe of electric power [3]. The Geysers, located in California, is the largest dry-steam installation in the United States, with an installed capacity of 725 MWe [4]. At Lardorello, and in few other specific places, the temperature and pressure are such that saturated steam is already present. The low-pressure steam is then condensed and re-injected in the ground through a re-injection well. This water was previously rejected at the surface, but this could cause environmental problems due to the chemical composition of the groundwater and also could adversely affect the continuous production of the plant. A schematic of the process is illustrated in Figs. 1.2 and 1.3:

Example 1.1 A dry-steam reservoir provides saturated vapor at 2 Mpa to be used for electricity generation. Expansion through the well-head valve reduces the pressure to 700 kPa. The isentropic efficiency of the turbine is 80%, and the condenser pressure is 20 kPa. If the flow rate is 25 kg/s, find the power produced by the plant.

Solution The steam properties are calculated with the CoolProp[a] library, and the following values are found:

	T	p	h	s	x
	°C	kPa	kJ/kg	kJ/kg-K	
1	212.38	2000	2798.29	6.3390	1
2	179.54	700	2798.29	6.7869	–
3s	60	20	2235.63	6.7869	0.84
3	60	20			
4	60	20	251.42	0.8320	0

To find the other properties, classical analysis is followed. The enthalpy of point 3 is found from the efficiency:

$$h_3 = h_2 - \eta(h_2 - h_{3s}) = 2348.16 \text{ kJ/kg}$$

$$x_3 = \frac{2348.16 - 251.42}{2608.94 - 251.42} = 0.89$$

$$\dot{W} = 25(2798.29 - 2348.16)/1000.0 = 11.25 \text{ MW}$$

■

[a]http://www.coolprop.org/.

Turbine efficiency is often dependent on the quality at the exit of the turbine, which in turn depends on the efficiency. This is known as the *Baumann rule* [5]:

$$\eta = \eta^* \frac{1 + x_3}{2} \tag{1.1}$$

where η^* is called the dry turbine efficiency. From the definition of the quality and the turbine efficiency, the following

Fig. 1.2 Dry-steam schematic

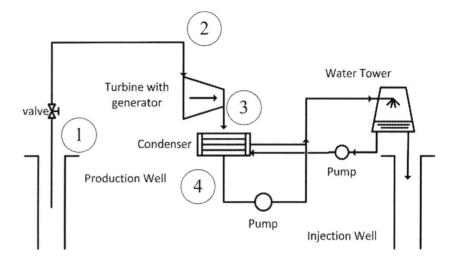

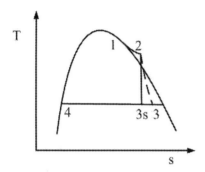

Fig. 1.3 Dry-steam cycle

expression is found:

$$h_3 = \frac{h_2 - \eta^*/2(h_2 - h_{3s})(1 - h_4/h_{fg})}{1 + \eta^*/2(h_2 - h_{3s})/h_{fg}} \tag{1.2}$$

Example 1.2 Repeat the previous example where the dry turbine efficiency is 84% and the real efficiency follows the Baumann rule.

Solution The properties in the table of the previous example remain unchanged. The enthalpy of point 3 is now:

$$h_3 = \frac{2798.29 - 0.42(2798.29 - 2235.63)(1 - 251.42/2357.52)}{1 + 0.42(2798.29 - 2235.63)/2357.52}$$
$$= 2351.46 \text{ kJ/kg}$$

$$\eta = \frac{2798.29 - 2351.46}{2798.29 - 2235.63} = 0.794$$

$$x_3 = \frac{2351.46 - 251.42}{2357.52} = 0.89$$

(continued)

Baumann's rule can be verified:

$$\eta = 0.42 \cdot 1.89 = 0.794$$

Finally:

$$\dot{W} = 25(2798.29 - 2351.46)/1000.0 = 11.17 \text{ MW}$$

∎

Pressure reduction through valves or friction in the well produces irreversibilities, which reduces the potential work available. This maximum possible work can be evaluated using the *exergy efficiency*, which is defined as the actual work divided by the exergy variation. In the case of an adiabatic turbine, it is given by DiPippo [2]:

$$\eta_{ex} = \frac{\dot{W}}{\dot{m}(h - h_o) + T_o(s - s_o)} \tag{1.3}$$

where T_o, s_o, h_o are associated with ambient conditions.

Example 1.3 In Example 1.2, two exergy efficiencies can be defined. If the reservoir conditions are used:

$$h = 2798.29 \text{ KJ/kg} \quad, \quad s = 6.339 \text{ KJ/kg-K}$$

$$\eta_{ex} = 0.489$$

The exergetic efficiency can also be evaluated using the conditions at the entrance of the turbine. In that case, we have:

$$h = 2798.29 \text{ KJ/kg} \quad, \quad s = 6.7869 \text{ KJ/kg-K}$$

$$\eta_{ex} = 0.573$$

(continued)

Fig. 1.4 Flash power plant schematic

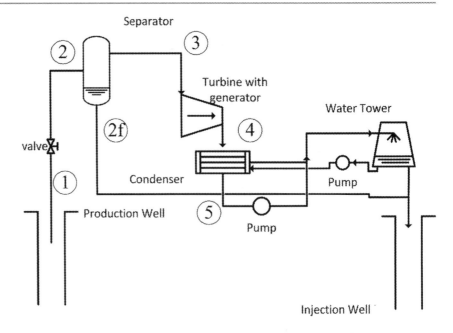

One way to get some of this energy is to have a second low-pressure turbine stage. It is called a double flash cycle. These plants are more efficient but of course are more expansive. More than two stages are also possible [6].

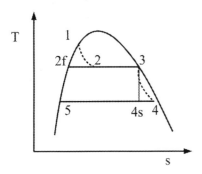

Fig. 1.5 Flash cylcle

Example 1.4 A reservoir provides water at 4 Mpa and 200 °C. The water flashes at 100 kPa through the well-head valve, and the vapor produced goes through a turbine, whose dry turbine efficiency is 85%. The condenser pressure is 20 kPa, and the flow rate of water is 25 kg/s. Find the power produced by the plant.

Solution The enthalpy of the pressurized water is:

$$h_1 = 853.26 \text{ KJ/kg}$$

At the exit of the well-head valve, the enthalpy is the same. At 100 kPa, we have:

$$h_{2f} = 417.50 \text{ KJ/kg}, h_3 = h_{2g} = 2674.95 \text{ KJ/kg}$$

$$x_2 = \frac{853.26 - 417.50}{2674.95 - 417.50} = 0.193$$

$$s_3 = 7.359 \text{ KJ/kg-K} = s_4$$

$$h_{4s} = 2426.21 \text{ KJ/kg}$$

$$h_5 = 251.42 \text{ KJ/kg}$$

From (1.2),

$$h_4 = \frac{2674.95 - 0.425(2674.95 - 2426.21)(1 - 251.42/2357.52)}{1 + 0.425(2674.95 - 2426.21)/2357.52}$$
$$= 2469.76 \text{ kJ/kg}$$

Example 1.3 (continued)
The latter one is associated with irreversibility in the plant, whereas the first one takes into account the ones in the well also.   ∎

1.3.2 Flash Power Plants

It is however rare to have the special conditions that produce dry-steam at the surface. Most of the time, hot water at high pressure is present. It is then possible to use the fact that part of this water will evaporate (flash) when the pressure decreases. The vapor phase is then recovered to rotate the turbine, and the remaining liquid is re-injected back into the ground. We have a so-called flash cycle (Fig. 1.4).

The following diagram (Fig. 1.5) illustrates this:

One disadvantage of flashing systems is that part of the high-energy fluid is directly returned to the re-injection well.

(continued)

Fig. 1.6 Binary plant schematic

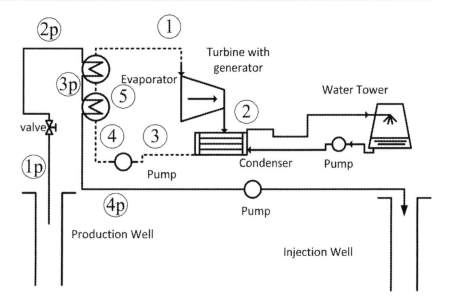

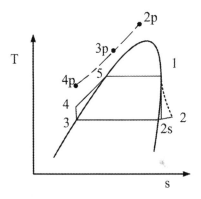

Fig. 1.7 Binary plant cycle

Example 1.4 (continued)
Finally:

$$\dot{W} = x_3 \dot{m}(h_3 - h_4) = 1 \text{ MW}$$

As it is seen, for similar conditions than the dry-steam plant, the output work is much less, since only part of the flow is vaporized. ∎

1.3.3 Binary Plants

Heat sources with temperatures lower than those found in direct steam generation but high enough to have a significant energy potential are often present in nature. These fluids, at temperatures around or below 150 °C, can be used in binary cycles involving a secondary fluid whose condensation temperature is lower. The general principle is illustrated in Fig. 1.6:

Hydrocarbons are often used as secondary fluids since their critical temperature is lower than the one of water. The power cycle is often referred to as *Organic Rankine cycle*, (ORC). Electricity generation at temperatures as low as 74 °C was obtained in Chena Hot Springs in Alaska [7]. Three 200 kW ORC power plants were installed and run since 2006 at a cost of 5¢ per kWh as opposed to a production cost of 30 ¢ previously obtained with diesel plant. R134a was used as the working fluid in that case (Fig. 1.7).

Example 1.5 Find the work delivered by a binary plant using iso-pentane ($i - C_5H_{12}$). The following conditions are met: Iso-pentane enters the turbine as saturated vapor:

$$T_{evap} = 150\,°C$$

and leaves the condenser as saturated liquid with:

$$T_{cond} = 40\,°C$$

$$\eta_{turb} = 0.85 \quad , \quad \eta_{pump} = 0.80$$

$$\dot{m}_{isopentane} = 12 \text{ kg/s}$$

$$\dot{m}_{water} = 80 \text{ kg/s}$$

$$T_{water,in}(\text{point 2p}) = 160\,°C$$

$$P_{water,in}(\text{point 2p}) = 1 \text{ Mpa}$$

(continued)

Example 1.5 (continued)

Solution The different properties associated with the secondary fluid are easily found using CoolProp. The results are given in the following table:

	T	p	h	s	x
	$^\circ$C	kPa	kJ/kg	kJ/kg-K	
1	150	1868	526.64	1.367	1
2s	77.4	151.5	431.46	1.367	–
2		151.5			
3	40	151.5	28.30	0.0919	0
4s		1868	31.16	0.0919	
4		1868			
5	150	1868	330.02	0.9022	0

A particularity of hydrocarbons is that they have a positive slope for the saturated vapor boundary on the $T - s$ diagram. This makes the exit vapor superheated at the exit of the turbine, whereas a water turbine would have wet steam at the exit if the entrance is saturated. From the efficiency of the turbine, we have:

$$h_2 = h_1 - \eta_t(h_1 - h_{2s}) = 447.74 \text{ KJ/kg-K}$$

The isentropic value at the exit of an ideal pump can be found from the table of compressible liquid, which is available assuming constant entropy. If this table is not available, it is usually found from the classical relation:

$$h_{4s} \approx h_3 + v_3(p_4 - p_3) = 31.16 \text{ KJ/kg}$$

where $v_3 = 0.00167$ m^3/kg. The difference between both approaches is very small. The real pump output can then be found from the pump efficiency:

$$h_4 = h_3 + (h_{4s} - h_3)/\eta_{pump} = 31.87 \text{ KJ/kg-K}$$

From the results:

$$\dot{W}_{turb} = \dot{m}(h_1 - h_2) = 970.84 \text{ kW}$$

$$\dot{W}_{pump} = \dot{m}(h_4 - h_3) = 42.88 \text{ kW}$$

$$\dot{W}_{net} = \dot{m}(h_4 - h_3) = 927.96 \text{ kW}$$

$$q_{in} = \dot{m}(h_1 - h_4) = 5937.24 \text{ kW}$$

$$\eta_{ther} = \frac{\dot{W}_{net}}{q_{in}} = 0.156$$

Primary Fluid Analysis

In order to have saturated vapor of iso-pentane at 150 $^\circ$C at the entrance of turbine, water at higher temperature should be provided and exchange heat through the heat exchanger. Heat exchanger analysis can be

done using LMTD method or NTU-ϵ methods [8] (see Chap. 2). Different approaches are used for the sensible heat gain from point 4 to the the saturated liquid (point 5) and for the constant-temperature latent heat gain. For the latent portion, the heat exchange needed is:

$$q_1 = \dot{m}(h_1 - h_5) = 2359.88 \text{ kW}$$

The water at the exit of the latent part becomes:

$$T_{3p} = T_{2p} - \frac{q_1}{\dot{m}_{water}C_p} = 153 \,^\circ\text{C}$$

This temperature should be higher than the evaporator temperature (150 $^\circ$C). Here it is 3 degrees higher; this temperature difference is called the *pinched temperature*, and it has of course to be positive. We can verify that if the water flow rate would have been smaller, this temperature would have been negative, which is impossible. In pinched temperature analysis [9], this temperature is often the input to the analysis. The exit temperature of the pre-heater becomes:

$$q_2 = \dot{m}(h_5 - h_4) = q_{in} - q_1 = 3577.71 \text{ kW}$$

$$T_{4p} = T_{3p} - \frac{q_2}{\dot{m}_{water}C_p} = 142.28 \,^\circ\text{C}$$

This temperature has also to be greater than the iso-pentane temperature at the entrance of the heat exchanger, which is the case since from the table, we have that:

$$T_4 = 40.93 \,^\circ\text{C}$$

As expected, this temperature is near the temperature upstream of the pump. The exergy efficiency of the plant can be computed knowing that the maximum work from the pressurized water would be:

$$h = 2798.29 \text{ KJ/kg} \quad, \quad s = 6.339 \text{ KJ/kg-K}$$

$$\eta_{ex} = 0.489$$

The exergetic efficiency can also be evaluated using the conditions at the entrance of the turbine. In that case we have:

$$h = 675.70 \text{ KJ/kg} \quad, \quad s = 1.942 \text{ KJ/kg-K}$$

$$h_o = 104.83 \text{ KJ/kg} \quad, \quad s = 0.367 \text{ KJ/kg-K}$$

$$\eta_{ex} = 0.12$$

∎

(continued)

Fig. 1.8 Kalina cycle schematic

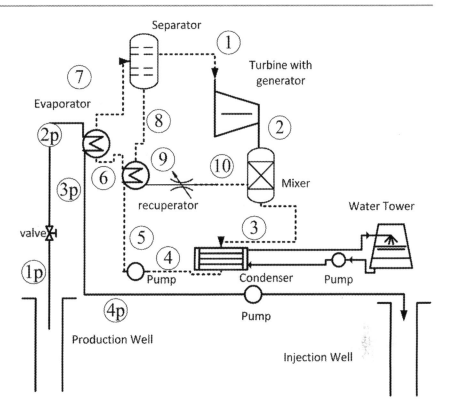

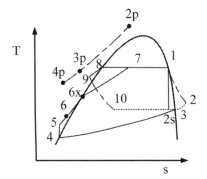

Fig. 1.9 Kalina cycle

Use of recuperators and preheaters can be added to the basic cycle to improve the efficiency of the whole cycle.

1.3.4 Kalina Cycle

A *Kalina cycle* [10] is a cycle used in a binary plant where the secondary fluid used is a mixture, often water-ammonia where the temperature glide reduces the irreversibilies associated with the heat exchangers. Several Kalina cycles are described in the literature [11]; a simple version of it (called KCS-11) is represented in Figs. 1.8 and 1.9:

Example 1.6 Find the efficiency of a Kalina cycle where the following conditions are met: The hot water coming from the ground is at the same temperature (160 °C) and flow rate (80 kg/s) as the previous example. The ammonia mass fraction is 0.8 at the entrance of the separator. At the inlet of the turbine, the vapor is saturated and:

$$T_{in,turbine} = 150\,°C$$

$$p_{in,turbine} = 3\ MPa$$

The solution leaves the condenser as saturated liquid with:

$$T_{out,cond} = 40\,°C$$

$$\eta_{turb} = 1\ ,\ \eta_{pump} = 1$$

Solution Classical cycle analysis using a pure substance is most often done with some basic assumptions like a prescribed state at the inlet of the turbine and/or at the exit of the condenser. These assumptions avoid the modeling of the external heat exchangers. Similar assumptions are made here, but the use of a zeotropic mixture complicates some of these assumptions. For a pure substance, imposing the saturated temperature and pressure would be overestimated; this is not the

(continued)

Example 1.6 (continued)

case with a mixture since the mass fraction at the exit separator is unknown and will depend on the vapor quality at its inlet.

When the working fluid is a binary mixture, one has to give, or find, the ratio of the mass of one mixture with respect to the total mass (mass fraction). The symbol x is often used, but since this symbol was already used for the quality of liquid-vapor mixture, the symbol y will be used instead.

– inlet of the separator (point 7)

$$p_5 = p_6 = p_7 = p_1 = p_8 = p_9 = p_H = 3 \text{ MPa}$$

$$p_2 = p_3 = p_4 = p_L = ?$$

$$y_3 = y_4 = y_5 = y_6 = y_7 = 0.8$$

$$y_7 = 0.8 \ , \quad p_7 = 3 \text{ Mpa} \ , \quad T_7 = 423.15 \text{ K}$$
$$\Rightarrow x_7 = 0.879, \ h_7 = 1816 \text{ kJ/kg}$$

The exit of the separator can be found:

$$p_1 = 3 \text{ Mpa} \ , \quad T_1 = 423.15 \text{ K} \ , \quad x_1 = 1 \text{ (dew point)}$$
$$\Rightarrow y_1 = 0.865 \ , \ s_1 = 6.22 \text{ kJ/kg-K} \ , \ h_1 = 1992 \text{ kJ/kg}$$

$$p_8 = 3 \text{ Mpa} \ , \quad T_8 = 423.15 \text{ K} \ , \quad x_8 = 0 \text{ (bubble point)}$$
$$\Rightarrow y_8 = 0.329 \ , \ h_8 = 548 \text{ kJ/kg}$$

The flow rate fraction in the recuperator is found either by taking the liquid portion at the entrance of the separator:

$$f = \frac{\dot{m}_8}{\dot{m}_{total}} = 1 - x_7 = 0.121$$

or by doing a mass balance on the binary mixture and on one component.

$$f = \frac{\dot{m}_8}{\dot{m}_{total}} = \frac{y_7 - y_1}{y_8 - y_1} = \frac{0.8 - 0.865}{0.329 - 0.865} = 0.121$$

The low pressure is not given since the temperature and quality are imposed at the exit of the condenser. It can be however obtained from mixture properties:

$$y_4 = 0.8 \ , \quad T_4 = 313.15 \text{ K} \ , \quad x_4 = 0 \text{ (bubble point)}$$
$$\Rightarrow p_4 = 1.18 \text{ Mpa}, \ s_4 = 1.70 \text{ kJ/kg-K}, \ h_4 = 310.9 \text{ kJ/kg}$$

$$p_2 = p_3 = p_4 = p_L = 1.18 \text{ MPa}$$

The exit of the pump can be found from:

$$y_5 = 0.8 \ , \quad p_5 = 3 \text{ mPa} \ , \quad s_5 = s_4$$
$$\Rightarrow T_5 = 313.5 \text{ K} \ , \ h_5 = 313.5 \text{ kJ/kg}$$

The internal recuperator impact depends on the efficiency of the heat exchanger. Defining:

$$h_{6m} = h(T = T8, \ p = p_6, \ y = y_6) = 1823 \text{ kJ/kg}$$

$$h_{9m} = h(T = T5, \ p = p_9, \ y = y_9) = 55.7 \text{ kJ/kg}$$

$$q_{max} = \min((h6m - h5), \ f(h8 - h9m)) = 59.7 \text{ kJ/kg}$$

$$q_{rec} = \epsilon q_{max} = 47.8 \text{ kJ/kg}$$

$$h_6 = h_5 + q_{rec} = 361.3 \text{ kJ/kg}$$

$$h_9 = h_8 - q_{rec} = 154.1 \text{ kJ/kg}$$

$$h_{10} = h_9 \text{ isenthalpic valve}$$

Exit of the turbine (point 2s):

$$p_2 = 1.18 \text{ Mpa} \ , \quad s_2 = s_1 \ , \quad y_2 = 0.865$$
$$\Rightarrow h_{2s} = 1830 \text{ kJ/kg}$$

$$h_2 = h_1 - \eta(h_1 - h_{2s}) = h_{2s}$$

$$\Rightarrow T_2 = 384.3 \text{ K}$$

Mixer:

$$h_3 = f h_{10} + (1 - f)h_2 = 1627 \text{ kJ/kg},$$

$$T_3 = 381.6 \text{ K} = 108.4 \,^{\circ}\text{C}$$

The total work per unit mass rate:

$$\dot{w} = (1 - f)(h1 - h2) - (h5 - h4) = 139 \text{ kJ/kg}$$

The inlet heat:

$$q_{in} = (h7 - h6) = 1455 \text{ kJ/kg}$$

$$\eta_{ther} = \frac{139}{1455} = 9.5\%$$

With a 12 kg/s flow rate, the total work is:

$$\dot{W} = 1665 \text{ kW}$$

Primary Fluid Analysis

The temperature glide associated with the mixture evaporation makes the pinching process less important. It is still possible to split the evaporator in two sections: one associated with the liquid phase and one to the "latent" region. The working fluid will reach the bubble line at the temperature:

$$T_{6x} = T(p = p_6, \ y = y_6, \ x = 0) = 352 \text{ K} = 78.9 \,^{\circ}\text{C}$$

(continued)   (continued)

Example 1.6 (continued)

$$h_{6x} = 512.6 \text{ kJ/kg}$$

$$q_{latent} = 12(1816 - 512.6) = 15645 \text{ kW}$$

$$T3p = 160 - \frac{15645}{10 \cdot 4.19} = 113\,°C > T_{6x}$$

$$q_{sensible} = 12(512.6 - 361.3) = 1815 \text{ kW}$$

$$T3p = 113 - \frac{1815}{10 \cdot 4.19} = 108\,°C > T_6$$

Comments *It is hard to compare directly the Kalina cycles to ORC cycles since the operating temperature ranges may differ although such comparisons are often done in the literature (see, for example, [12]), but this is outside the scope of this book. Here the same conditions are used, where in general "optimum conditions" are looked upon for each fluid. Without a direct comparison with the previous example, a comment can be made on the condenser temperature. In both examples, the cold water temperatures were not given. It was assumed that the cold side was sufficient to remove the heat required. In the case of a pure substance, the imposed condenser of 40 °C meant that the outlet cold water temperature should be lower than 40 °C (say 38 °C for a 2 K pinch temperature), which means that it should be lower than that at the inlet. In the case of a condenser with a temperature glide, one of the advantage is to have lower temperature difference, and a better (although not exact) comparison that could be done is to assume that the outlet condenser temperature would be few degrees more than the cooling water inlet temperature. In such a case, the efficiency obtained would be higher.*

∎

One of the first Kalina-plant was introduced at Unterhaching, Germany [13], but was recently been shut down due to technical problems, often associated to new technology. A plant that until recently was still in operation is in Husavik, Iceland [14].

1.3.5 Enhanced Geothermal Systems

Regardless of the types of cycles described in the previous section, they all require the presence of underground water at relatively high temperature. The presence of this water requires very specific conditions for the presence of a *geothermal reservoir*. These conditions are unfortunately not present in many places in the world. The conditions for a geothermal reservoir are:

- A heating source (magma, etc.)
- A underground water reserve.
- An impermeable rock region underneath the water reserve
- A porous or fractured rock region above the water reserve to allow the heated water to percolate from and to the surface.

A new approach was recently attempted to allow the use of deep geothermal energy in places where there would be an interesting thermal potential but where there is no reservoir. These are called *enhanced geothermal systems* (EGS) or *hot dry rocks* (HDR) systems. The principle is to take advantage of the presence of hot rocks with great thermal energy potential and create an artificial geothermal reservoir by fracturing the rock and injecting water which, when heated, will be pumped as in a standard plant. The first EGS project that was developed was the Fenton Hill project [15] in New Mexico in the early 1970s. Wells of up to 3 km were drilled, and although the concept of artificially created fractured reservoir was proven, the project was dismantled by 2000 due to lack of funding. Several EGS projects have been initiated, but few are in operation. The Soultz-sous-Forets site in France is one of the best-known EGS project [16].

1.3.6 Geopressure Reservoirs

The water pressured in geothermal reservoirs is often associated to hydrostatic pressure (piezometric head). In some special ground formation, overburden stress on some water reservoir increases substantially; these conditions are called *geopressured reservoirs* and were observed along the coastline of the Gulf of Mexico [17].

1.3.7 Geothermal Electricity Production Statistics

Geothermal power plants have increased significantly in the past years as shown in Fig. 1.10 [18]:

In can be remarked that the total production in 2014 was larger than 73 TeraWatts-Hours (TWh) (or 263 PJoules). Although this number looks rather large, it represents approximately 0.046% of the total energy consumed in the world [19] (Fig. 1.11).

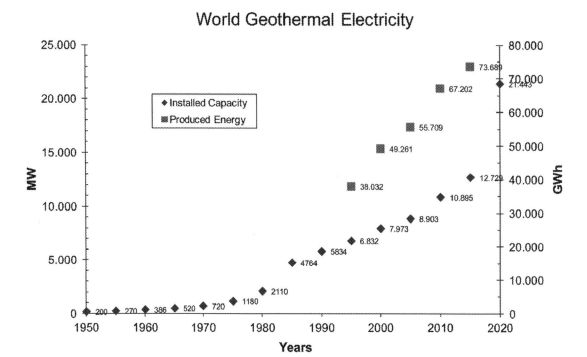

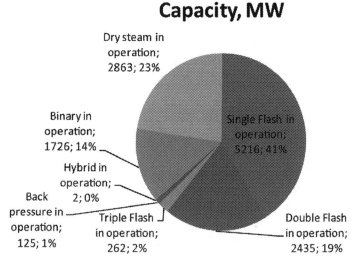

Fig. 1.10 Geothermal electricity generation [18] (Reprint with permission)

Fig. 1.11 Electricity generation
[18] (Reprint with permission)

A better comparison would be to look only at electricity generation (Table 1.3). Table 1.4 compares the electricity generated by geothermal plants with other types of technology [19].

1.4   Geothermal Direct Applications

Direct applications imply that thermal energy is directly used for different purposes, mainly heating, but in some cases cooling applications. In the presence of a source at moderately warm temperature (≈45–60 °C), one can use directly the source of heat for heating buildings. In the case where the temperatures are too low, the geothermal source can be used to pre-heat water or air. A special, but very important, case is the direct heating and cooling of buildings using heat pumps. This is the main application which will be discussed in this book. Meanwhile, we can quickly look at some other direct applications.

1.4.1   Direct Applications Statistics

As in the case of high-temperature applications, a review of the direct geothermal applications is performed every five years. In the last report in 2021 [20], the evolution of geothermal direct applications is summarized as shown in Fig. 1.12:

As seen in the figure, almost 1 ExaJoules of heat was generated from geothermal sources. This can be compared to the total production of heat worldwide [19] of 13.8 ExaJoules for 7% of the total heat produced.

The different sectors associated with direct applications changed over the years. In the last 15 years, geothermal heat pumps became the largest sector of applications for direct geothermal uses (Fig. 1.13).

In countries like Sweden, geothermal (or ground-source) heat pumps became very popular and contribute largely as the heating source for the country. In countries like Iceland, the availability of water at higher temperature allows the use of direct heating without the use of heat pumps most of the time. Almost 90% of the heating is provided by the hot water. Almost 100% of Reykjavik is heated by a district heating system, of which energy comes from geothermal hot water. Many other applications, like greenhouse heating, snow melting of sidewalks and routes, fish farming, and bathing, are typical applications of geothermal hot water in Iceland [21].

Table 1.3 World Electricity Generation (2021) [19]

Source	Generation (TWh)	Pourcentage
Coal	9707	40.8
Oil	1024	4.3
Gaz	5154	21.7
Total (fossil)	15885	66.8
Nuclear	2536	10.65
Hydro	3894	16.35
Renewables (other hydro)	1489	6.3
Total	23804	100

Table 1.4 Electricity Generation from Renewables (2021) [19]

Source	Generation (TWh)	Pourcentage
Hydro	4327	15
Biomass	746	3
Wind	1870	7
Geothermal	97	0.34
Solar PV	1003	4
Concentrating solar	15	0.05
Marine	1	0.004
Total	8060	28

1.5 Direct Applications Using Heat Pumps

As noted previously, this book will be devoted mostly on ground-source heat pump (GSHP) applications, so most of the following chapters will look into this specific subject. In this introductory chapter, the different configurations will briefly be discussed.

The principle of heat pumps will be addressed in Chap. 3. Simply stated, a heat pump transfers heat from a cold region to a warmer one. Following the second principle of thermodynamics, work (most of the time electric) has to be done to realize that. Heat pumps can extract the thermal energy from different sources like air in air source heat pumps (ASHP). As the world suggests, ground-source heat pumps take out the energy from the ground. However, different configurations can be used to accomplish the same task. GSHP can be categorized in different ways, a possible one is:

Fig. 1.12 Direct applications statistics [20] (Reprint with permission)

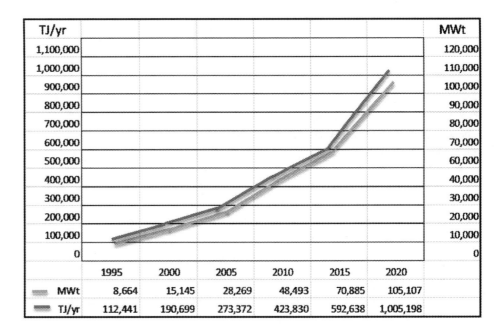

	1995	2000	2005	2010	2015	2020
MWt	8,664	15,145	28,269	48,493	70,885	105,107
TJ/yr	112,441	190,699	273,372	423,830	592,638	1,005,198

Fig. 1.13 Direct applications
[20] (Reprint with permission)

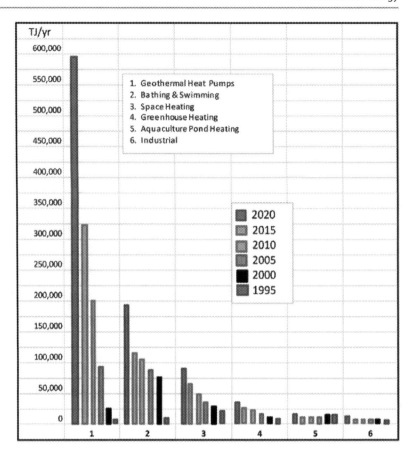

1. Groundwater heat exchange systems
   (a) open-loop water systems.
   (b) Standing-column systems.
2. Ground-coupled heat pumps (GCHP)
   (a) Vertical boreholes systems.
   (b) Horizontal boreholes systems.
   (c) Pile systems.
   (d) Direct-exchange (DX) systems.
3. Surface water heat pumps (SWHP)

Ground water systems exchange heat with water aquifers. In open-loop systems, the water from the aquifer is pumped directly into the heat pump or most often through a heat exchanger and returned in the ground in a secondary well. These systems are very efficient but are restricted to area where a water aquifer is available and where local legislations allow them. Standing column systems are kind of a mixture between open-loop and secondary-loop systems; water exchanges heat by conduction with the adjacent ground, and when it becomes too hot or too cold, a bleed of a portion of water is started, which is similar to open-loop behavior [22, 23]. Surface water systems exchange heat with waters in lake or ponds using a secondary fluid through a heat exchanger. Again these systems can be a very economic alternative, but often, environmental regulations preclude their use. The most common geoexchange systems transfer heat by conduction

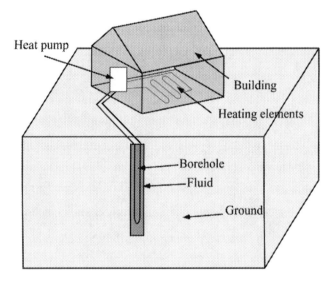

Fig. 1.14 Secondary loop vertical GSHP system

with the surrounding ground either directly (DX-GCHP) or via a secondary fluid (SL-GCHP), often water or water with antifreeze, in vertical, horizontal, or inclined boreholes. In DX systems [24, 25], the refrigerant goes directly into the ground, and no intermediate heat exchanger is needed. The boreholes are mostly made in cooper. This improves heat exchange with the ground and limits the total length of

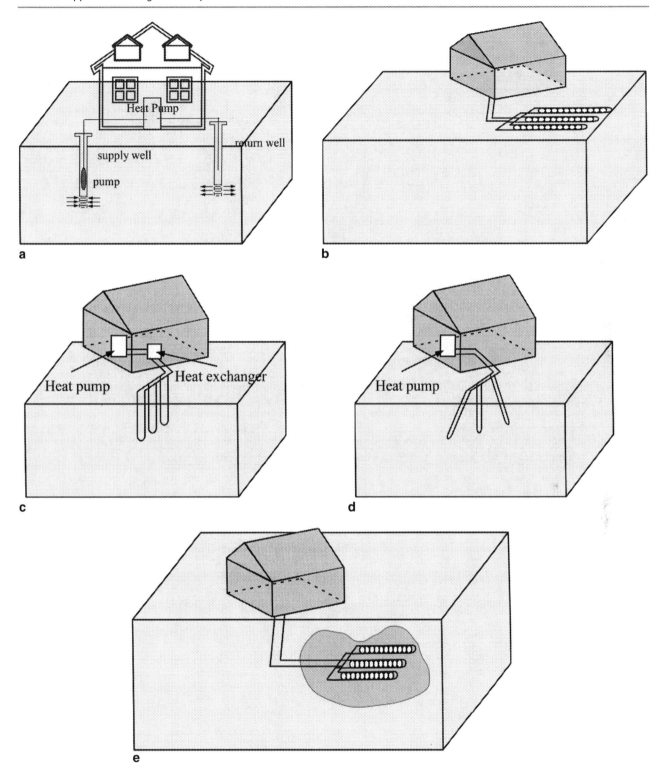

Fig. 1.15 (**a**) Open-loop system. (**b**) Horizontal loop. (**c**) Vertical loop. (**d**) DX system. (**e**) Surface water system

boreholes. However, the market share of these systems is very limited, partly due to the fear that they might be less reliable. So, the most common configuration used in geoexchange applications is a heat pump linked to a heat exchanger where a secondary fluid exchanges heat in the ground in one or several boreholes. In each borehole, a U-tube, or several U-tubes, made in high-density plastic, is separated from the ground by a sealed grout. A schematic of such system is shown in Fig. 1.14. GSHP systems are efficient but complex energy systems that have interactions between all elements shown on the figure, and a good knowledge of how these components interact is important. A brief discussion on all of these subsystems will be given in this book, although most of the focus will be drawn on the interaction between the ground, the borehole(s), and the heat pump (Fig. 1.15).

References

1. Watson, A.: Geothermal Engineering. Springer, Berlin (2016)
2. DiPippo, R.: Geothermal Power Plants—Principles, Applications, Case Studies and Environmental Impact, 4th edn. Butterworth-Heinemann, Oxford (2016)
3. Parri, R., Lazzeri, F.: 19—Larderello: 100years of geothermal power plant evolution in Italy. In: DiPippo, R. (ed.) Geothermal Power Generation, pp. 537–590. Woodhead Publishing (2016)
4. http://geysers.com. Accessed 28 May 2019
5. Baumann, K.: Some recent developments in large steam turbine practice. J. Inst. Electr. Eng. **59**(302), 565–623 (1921)
6. Rotokawa II/Nga Awa Purua Geothermal Power Plant. https://www.renewable-technology.com/projects/geothermal-power-plant/ Accessed 31 May 2022
7. Chena Geothermal Power Plant Project Final Report Prepared for the Alaska Energy Authority. https://geothermalcommunities.eu/assets/elearning/7.13.FinalProjectReport_ChenaPowerGeothermalPlant.pdf. Accessed 28 May 2019
8. Bergman, T.L., Incropera, F.P.: Fundamentals of Heat and Mass Transfer, 7th edn. Wiley, New York (2011)
9. Kemp, I.C.: Pinch Analysis and Process Integration: A User Guide on Process Integration for the Efficient Use of Energy. Elsevier, Amsterdam (2011)
10. Kalina, A.I.: Combined-cycle system with novel bottoming cycle. J. Eng. Gas Turbines Power **106**(4), 737–742 (1984)
11. Zhang, X., He, M., Zhang, Y.: A review of research on the Kalina cycle. Renewable Sustain. Energy Rev. **16**(7), 5309–5318 (2012)
12. Bombarda, P., Invernizzi C.M., Pietra, C.: Heat recovery from Diesel engines: a thermodynamic comparison between Kalina and ORC cycles. Appl. Thermal Eng. **30**(2), 212–219 (2010). https://doi.org/10.1016/j.applthermaleng.2009.08.006
13. Richter, B.: Geothermal energy plant unterhaching, Germany. In: Proceedings World Geothermal Congress (2010)
14. Ogriseck, S.: Integration of Kalina cycle in a combined heat and power plant, a case study. Appl. Thermal Eng. **29**(14–15), 2843–2848 (2009)
15. Brown, D.W. et al.: Mining the Earth's Heat: Hot Dry Rock Geothermal Energy. Springer, Berlin (2012)
16. MIT interdisciplinary panel (J.F. Tester Chairman). The Future of Geothermal Energy. Impact of Enhanced Geothermal Systems (EGS) on the United States in the 21st Century. Technical report. MIT, 2006, 372 pp. http://geothermal.inel.gov/publications/future_of_geothermal_energy.pdf
17. Green, B.D., Nix, R.G.: Geothermal—The Energy Under Our Feet: Geothermal Resource Estimates for the United States. Technical report. National Renewable Energy Lab.(NREL), Golden, CO (United States)
18. Bertani, R.: Geothermal power generation in the world 2010 2014 update report. Geothermics **60**, 31–43 (2016)
19. IEA: World Energy Outlook 2022. Technical report. https://iea.blob.core.windows.net/assets/830fe099-5530-48f2-a7c1-11f35d510983/WorldEnergyOutlook2022.pdf
20. Lund, J.W., Toth, A.N.: Direct utilization of geothermal energy 2020 worldwide review. Geothermics **90**, 101915 (2021). ISSN: 0375-6505. https://doi.org/10.1016/j.geothermics.2020.101915
21. Ragnarsson, A.: Geothermal development in Iceland 2010–2014. In: Proceedings World Geothermal Congress, 2015 (2015)
22. Deng, Z., Rees, S.J., Spitler, J.D.: A model for annual simulation of standing column well ground heat exchangers. HVAC&R Res. **11**(4), 637–655 (2005). https://doi.org/10.1080/10789669.2005.10391159
23. Nguyen, A., Beaudry, G., Pasquier, P.: Experimental assessment of a standing column well performance in cold climates. Energy Build. **226**, 110391 (2020). https://doi.org/10.1016/j.enbuild.2020.110391
24. Ndiaye, D.: Reliability and performance of direct-expansion groundcoupled heat pump systems: issues and possible solutions. Renewable Sustain. Energy Rev. **66**, 802–814 (2016). ISSN: 1364-0321
25. Fannou, J.-L. et al.: Experimental analysis of a direct expansion geothermal heat pump in heating mode. Energy Build. **75**, 290–300 (2014). https://doi.org/10.1016/j.enbuild.2014.02.026

2.1 Introduction

Focus will be given in this book about direct use of geothermal resources for heating and cooling buildings using heat pumps. In order to design such systems, heat transfer between a calorimetric fluid and the ground as well as energy that earth-coupled systems must deliver to buildings must be evaluated. Both aspects require some basic knowledge on heat transfer principles that most engineers may already have, but a quick reminder will be done in this chapter. In this book, several concepts that will help the ground-source heat pump engineer to work together with building engineers are introduced. Several good books [1, 2] can be consulted for more details.

2.2 Heat Transfer Basics

Heat transfer mechanisms involve energy transfer from a hot source to a cold sink using three modes: *conduction*, *convection*, and *radiation*.

2.2.1 Heat Transfer by Conduction

Heat transfer by *conduction* occurs in solids and in inert fluids, which is rare, since most fluids will move and produce convection. Heat conduction is governed by *Fourier's law*:

$$\vec{q}''_{cond} = -k\nabla T \qquad (2.1)$$

The temperature field is found from the *heat equation* given by:

$$\frac{\partial}{\partial x}\left(k\frac{\partial T}{\partial x}\right) + \frac{\partial}{\partial y}\left(k\frac{\partial T}{\partial y}\right) + \frac{\partial}{\partial z}\left(k\frac{\partial T}{\partial z}\right) + \dot{q} - \rho C_p \frac{\partial T}{\partial t} = 0 \qquad (2.2)$$

where k [W/m-K] is the conductivity of the medium analyzed, a very important property in heat transfer analysis. Through out this book, several solutions of this equation will be presented associated with different geometries. In this introduction, only certain simple solutions associated with simple geometries under steady-state conditions are presented. In such cases, the final results can be expressed as thermal resistances:

$$q_{cond}[\text{W}] = \frac{\Delta T}{R_{cond}} = UA\Delta T \qquad (2.3)$$

UA [W/K] is the conductance, inverse of the resistance (R [K/W]), associated with the problem. Heat transfer analysis is often done on a unit area basis (Watts per meter square). In that case, the heat transfer flux is given by:

$$q''_{cond}[\text{W/m}^2] = \frac{q}{A} = \frac{\Delta T}{R''_{cond}} = U\Delta T \qquad (2.4)$$

where:

$$R''[\text{K-m}^2/\text{W}] = R\,A \qquad (2.5)$$

Finally, analysis per unit meter is also a common approach especially when dealing with radial geometry:

$$q'_{cond}[\text{W/m}] = \frac{q}{L} = \frac{\Delta T}{R'_{cond}} = UP\Delta T \qquad (2.6)$$

where:

$$R'[\text{K-m/W}] = R\,L \qquad \text{and} \qquad P: \text{perimeter} \qquad (2.7)$$

2.2.1.1 Wall Conduction Resistance

In the case of a plane geometry, representing either a wall, window, isolation, etc., resistance is:

$$R_{wall} = \frac{L}{kA} \qquad (2.8)$$

$$R''_{wall} = \frac{L}{k} \tag{2.9}$$

$$U_{wall} = \frac{k}{L} \tag{2.10}$$

where L is the wall thickness.

2.2.1.2 Radial Conduction Resistance

In the case of a radial annulus, representing, for example, isolation around a cylinder pipe, the thermal resistance is:[1]

$$R_{radial} = \frac{\log(r_o/r_i)}{2\pi k L} \tag{2.11}$$

$$R'_{radial} = \frac{\log(r_o/r_i)}{2\pi k} \tag{2.12}$$

with L being the length of the annular cylinder.

2.2.1.3 Spherical Conduction Resistance

In the case of a spherical annulus, representing, for example, isolation around a sphere, the thermal resistance is:

$$R_{sphere} = \frac{1}{4\pi k}\left(\frac{1}{r_i} - \frac{1}{r_o}\right) \tag{2.13}$$

2.2.2   Heat Transfer by Radiation

Heat transfer by radiation is a very complex physical phenomenon where the thermal energy is transported by electromagnetic waves at the speed of light. Radiation is the only heat transfer mode that can exist in vacuum, but it can also be present between radiative surfaces having different temperatures separated by either vacuum or transparent gas. Diatomic gas like oxygen, nitrogen, helium, and dry air are basically transparent for thermal radiation. Radiative surfaces are either in solid, liquid, or gaseous state if the gas is sensitive to radiation (e.g., water vapor). Thermal radiation is associated mostly with the infrared spectrum. Radiation from very high temperature sources (like the sun) has radiation with shorter wavelengths (less than $3\,\mu m$), which has different radiation characteristics, but for most of the applications, thermal radiation is associated with large wavelength behavior.

A radiative surface emits radiative heat at a rate given by the *Stephan-Boltzman* relation:

$$E(W/m^2) = \epsilon \sigma T_s^4 \tag{2.14}$$

where:

- σ: Stephan-Boltzman constant = 5.67×10^{-8} W/m^2-K^4
- ϵ: emissivity of the surface ($\epsilon \leq 1$)

A perfect emitting surface is called a *black body* or *black surface* and has its emissivity equal to one ($E = E_b$). In some applications, the spectral behavior of the surface is important; in that case, the spectral emissivity is needed. The relation between both emissivities is:

$$\epsilon = \frac{\int_0^\infty \epsilon_\lambda E_{\lambda,b} d\lambda}{E_b}$$

If the spectral emissivity (ϵ_λ) is constant, it is then equal to the total emissivity (ϵ). If not, the total emissivity is a weighted average of the spectral one. At a radiative surface, incident radiation is also present and called *irradiation*, $G(W/m^2)$. At the surface, this irradiation can be reflected, absorbed, or transmitted:

$$G_{inc} = \rho G_{inc} + \alpha G_{inc} + \tau G_{inc} = (\rho + \alpha + \tau)G_{inc} \tag{2.15}$$

where it is seen that:

$$\rho + \alpha + \tau = 1 \tag{2.16}$$

At an *opaque surface*, the transmissivity is zero, and:

$$\rho + \alpha = 1 \Rightarrow \rho = 1 - \alpha \tag{2.17}$$

From Kirchoff law [3], it can be shown that for most surfaces:

$$\epsilon_\lambda = \alpha_\lambda \tag{2.18}$$

From that, it is easy to see that if both are constants, one also have that:

$$\epsilon = \alpha \tag{2.19}$$

Such a behavior is called a *gray surface*. In practice, even though the spectral emissivity is not constant, if it is "almost constant" in the range of the emissive and irradiation spectrum, the surface can be assumed to be "gray." This is almost always done for opaque diffuse surfaces except when dealing with sun radiation. Once the emissive power (E) and the irradiation (G) are known, the global radiative heat transfer leaving a surface can be found. For an opaque surface, it is given as:

$$q_{rad} = A(E - \alpha G) = A(\epsilon E_b - \alpha G) \tag{2.20}$$

where for a gray surface ($\alpha = \epsilon$).

Analytic expressions to evaluate the thermal heat transfer are in general quite complex especially if it involves many surfaces at different temperatures. If we limit ourselves to the case of thermal exchange between two gray surfaces, some

---

[1] log represents the natural logarithm. $\log_{10}$ will be used for logarithm to the base 10.

simple formula can be used to evaluate the radiative heat transfer. All these expressions can be expressed as:

$$q_{rad} = K_{12}\sigma A_1(T_1^4 - T_2^4) \qquad (2.21)$$

where:

- σ: Stephan-Boltzman constant $= 5.67 \times 10^{-8}$ W/m^2-K^4
- A_1: area of the smaller surface where radiation takes place
- K_{12}: radiative parameter associated with a given geometry

2.2.2.1 Large Enclosure

Many radiative application involves a small flat or convex surface (A_1) surrounded by a large enclosure ($A_2 \gg A_1$). In this case, we have:

$$K_{12} = \epsilon_1$$
$$q_{rad} = \epsilon_1 \sigma A_1(T_1^4 - T_2^4) \qquad (2.22)$$

where ϵ_1 is the *emissivity* of surface 1 and ϵ_2 the emissivity of the large surface, which does not influence the results.

2.2.2.2 Parallel Planes

Radiation between parallel planes ($A_1 = A_2$) is observed, for example, in double paned windows. In that case, we have:

$$K_{12} = \frac{1}{\frac{1}{\epsilon_1} + \frac{1}{\epsilon_2} - 1} \qquad (2.23)$$

2.2.2.3 Concentric Cylinders

$$K_{12} = \frac{1}{\frac{1}{\epsilon_1} + \frac{1-\epsilon_2}{\epsilon_2}\frac{r_1}{r_2}} \qquad (2.24)$$

2.2.2.4 Concentric Spheres

$$K_{12} = \frac{1}{\frac{1}{\epsilon_1} + \frac{1-\epsilon_2}{\epsilon_2}\left(\frac{r_1}{r_2}\right)^2} \qquad (2.25)$$

Heat transfer by radiation involves expressions with temperature at power 4 (temperatures should then be in Kelvin) and as such cannot be expressed as a temperature difference divided by a radiative resistance. However, it is possible to rewrite Eq. (2.21) as:

$$q_{rad} = \underbrace{K_{12}\sigma A_1(T_1^2 + T_2^2)(T_1 + T_2)}_{\hbar_{rad} A}(T_1 - T_2) \qquad (2.26)$$

Using this linearized expression, a radiative resistance can be defined:

$$R_{rad} = \frac{1}{\hbar_{rad} A} \qquad , \qquad R''_{rad} = \frac{1}{\hbar_{rad}} \qquad (2.27)$$

It must however be remembered that this resistance is non-linear since it depends on the surface temperature, which, in turn, depends on the heat rate, so an iterative solution is sometimes needed to solve the problems.

2.2.3 Heat Transfer by Convection

Heat transfer by *convection* is observed at the interface between a solid (or liquid) surface and a moving fluid. It involves complex exchange phenomenon, which are simplified using a global exchange coefficient, which is often found from experimental correlations. In such a case, the convection heat transfer can be expressed as:[2]

$$q_{conv} = \hbar_{conv} A\Delta T \qquad (2.28)$$

and so:

$$R_{conv} = \frac{1}{\hbar_{conv} A} \qquad (2.29)$$

$$R''_{conv} = \frac{1}{\hbar_{conv}} \qquad (2.30)$$

$$R'_{conv} = \frac{1}{\hbar_{conv} P} \qquad (2.31)$$

Convection coefficients $\hbar$ are mostly always evaluated with empirical expressions, which depend on the geometry studied and thermo-fluids properties. These expressions are given with respect to a dimensionless parameter associated with the convection coefficient called *Nusselt Number*. This dimensionless convection coefficient is defined as:

$$Nu_{L_c} = \frac{\hbar_{conv} L_c}{k_f} \qquad (2.32)$$

where L_c is a characteristic length associated with a given geometry, length of a flat plate or diameter of a cylinder, etc. and k_f the fluid conductivity. Since convection coefficients are not uniform over surfaces, one often differentiates between local Nusselt numbers giving local coefficients and average Nusselt numbers, leading to average convection coefficients. In the latter case, a bar is often put above the coefficient and over the Nusselt number. In this book, only average values will be used, and no over bars will be written for clarity. Convection coefficients can be found for *forced convection*, where the fluid movement is due to pressure gradients and when the fluid velocity or flow rate is normally known. *Natural convection* is when the fluid movement is due to buoyancy effects due to temperature differences. A large

---

[2]The symbol h is normally used for the heat transfer coefficient, but here the symbol $\hbar$ will be used not to confuse it with the enthalpy.

number of expressions for Nusselt numbers are found in the literature [4].

2.2.3.1 Forced Convection-External Flows

Nusselt correlations associated with forced convection are expressed with respect to two major dimensionless numbers: the *Reynolds number* and the *Prandl number*. The Reynolds number is defined as:

$$Re_{L_c} = \frac{u_f L_c}{v_f} = \frac{\rho u_f L_c}{\mu_f} \tag{2.33}$$

Again L_c [m] is a typical characteristic length, u_f [m/s] is a reference fluid velocity, v_f [m^2/s] is the kinematic viscosity, and μ_f [Pa-s] is the dynamic viscosity of the fluid. The Prandtl number defined as:

$$Pr = \frac{v_f}{\alpha_f} \tag{2.34}$$

represents the dimensionless ratio between momentum and thermal *diffusivity*. Examples of Nusselt correlations useful in practice are for:

Flat Plate

This case can be applied to ceiling, walls, windows, etc. to flat plates are useful to evaluate convection coefficient for flows around roofs. For the fully laminar case, one have [3]:[3]

$$Nu_L(laminar) = 0.664 Re_L^{1/2} Pr^{1/3}, \qquad Re_L < 500000 \tag{2.35}$$

For higher Reynolds numbers (>500000), the flow is mixed:

$$Nu_L(mixed) = (0.037 Re_L^{4/5} - A) Pr^{1/3} \tag{2.36}$$

$$A = 0.037 Re_c^{4/5} - 0.664 Re_c^{1/2} \tag{2.37}$$

The boundary layer is assumed to remain laminar until the Reynolds number reaches a critical Reynolds number, which is in the order of 500000 for smooth surfaces. For rough surfaces, the transition can occur before 500000. In the case where the turbulent regime is tripped at the beginning of the plate (fully turbulent), the latter expression is still valid with ($A = 0$):

$$Nu_L(turbulent) = 0.037 Re_L^{4/5} Pr^{1/3} \tag{2.38}$$

Unless specified, the fluid properties are evaluated at the *film temperature*:

$$T_f = \frac{T_\infty + T_s}{2} \tag{2.39}$$

---

[3]Nusselt numbers are sometimes associated with local or mean (over the surface) values. In the latter case, an over-bar is sometimes used. In this introduction, only mean values are discussed, and no over-bar is used.

Cross Flow Around a Cylinder

Several correlations were proposed; Churchill Bernstein [5] suggested:

$$Nu_D = 0.3 + \frac{0.62 Re_D^{1/2} Pr^{1/3}}{(1 + (0.4/Pr)^{2/3})^{1/4}} \left[1 + \left(\frac{Re_D}{282000} \right)^{5/8} \right]^{4/5} \tag{2.40}$$

Cross Flow Around a Sphere

A similar correlation was proposed by Whitaker [6] for the sphere:

$$Nu_D = 2 + (0.4 Re_D^{1/2} + 0.06 Re_D^{2/3}) Pr^{0.4} \left(\frac{\mu}{\mu_s} \right)^{1/4} \tag{2.41}$$

In this expression, all properties are evalated at the fluid temperature except μ_s evaluated at the surface temperature. This ratio is often close to one in practice.

2.2.3.2 Forced Convection-Internal Flows
Circular Pipes

Heat transfer in pipes is very important in GSHP applications. For circular pipes, in the case where the flow is fully developed, we have [3]:

$$Nu_d = 3.66 \quad Re_D < 2300 \text{ laminar, uniform temperature}$$

$$Nu_d = 4.36 \quad Re_d < 2300 \text{ laminar, uniform heat flux} \tag{2.42}$$

$$Nu_d = 0.023 Re_d^{4/5} Pr^n \quad (n = 0.4, T_b > T_f, n = 0.3, T_b < T_f)$$

$$Re_d > 10000 \text{ turbulent, Dittus-Boelter} \tag{2.43a}$$

$$Nu_d = \frac{f/8(Re_d - 1000) Pr}{1 + 12.7(f/8)^{1/2}(Pr^{2/3} - 1)} \tag{2.43b}$$

$$Re_D > 2300 \text{ turbulent, Gnielinski}$$

For the fully turbulent case $Re_d > 10000$, two equations are proposed. The Gnielinski [7] is preferred in the transition region ($2300 < Re_D < 10000$). In the Gnielinski formula, f is called the friction factor and can be calculated with the Petukhov relation [8]:

$$f = (0.79 \log(Re_d) - 1.64)^{-2} \tag{2.44}$$

when the pipes are smooth. For rough pipes, the Moody diagram or the Colebrook formula can be used.

$$1/\sqrt{f} = -2 \log_{10} \left(\frac{\epsilon/d}{3.7} + \frac{2.51}{Re_D \sqrt{f}} \right) \tag{2.45}$$

Gnielinski [9, 10] proposed a modified version of their correlation.

$$Nu_{corr} = Nu \left[1 + \left(\frac{d}{L} \right)^{2/3} \right] K \qquad (2.46)$$

The correction factor K for liquids was given by:

$$K = \left(\frac{Pr}{Pr_w} \right)^{0.11} \qquad (2.47)$$

In GSHP applications, the ratio d/L is quite small, and the variation of Prandtl number in the typical range of temperatures encountered is also small, so these two factors will be ignored here for simplicity. However, the same author suggested also to use a corrected factor in the transition region ($2300 < Re < 10000$). Since they suggested that this correction avoid discontinuity, which might be harmful during a design process, its suggestion will be followed. The value suggested for the region ($2300 < Re < 4000$):

$$Nu_{trans} = (1 - \gamma)Nu_{lam} + \gamma Nu_{turb,4000} \qquad (2.48)$$

with:

$$\gamma = \frac{Re - 2300}{4000 - 2300} \qquad (2.49)$$

The value of 10000 was proposed in [9], whereas in [10], he suggested 4000. The latter value will be used in this book. In [9], Gnielinski also proposed a modified value that takes into account the entry length in laminar regime. It is well known [3] that the entry length has a greater impact in the laminar case, but even in that case, it will be neglected in our calculations. It must be remembered that the aspect ratio diameter-length in geothermal applications is quite small, and also, neglecting the entry length will always give a lower heat transfer coefficient, which will lead to more conservative designs.

Annular Pipes: Laminar

The flow in the annular part of concentric tubes is important in counterflow heat exchangers and in coaxial boreholes applications. Several correlations were proposed in the literature, and they all depend on the *hydraulic diameter* as the reference length. The hydraulic diameter is defined for non-circular pipes as:

$$d_h = \frac{4A_c}{P}$$

which, in the case of the annulus pipe, gives:

$$d_h = d_o - d_i = d_o \left(1 - \frac{d_i}{d_o} \right) = d_o(1 - a) \qquad (2.50)$$

with:

$$a = \frac{d_i}{d_o} \qquad (2.51)$$

the diameters ratio. In the case where the flow is laminar, several correlations were proposed depending if the heat flux is uniform (UHF) at the boundaries or the wall temperature is uniform (UWT). In geothermal applications, it is often neither the case, but only the UWT case will be presented, since even if the wall temperature is not uniform, the ground temperature is and the temperature profile inside a borehole follows more a profile similar to the UWT case in practice. In [9], the following expressions are suggested in the laminar regime in the fully developed region. All expressions are corrections of the circular case. Three cases are presented:

BC-1, Inner Surface Heated (Outer Surface Isolated)

This case is often assumed in counter-flow heat exchangers.

$$Nu_{i,d_h} = 3.66 + 1.2a^{-0.8} \qquad (2.52)$$

BC-2, Outer Surface Heated (Inner Surface Isolated)

This case would occur in ideal coaxial borehole where no heat exchange is assumed between the inner portion and the annulus.

$$Nu_{o,d_h} = 3.66 + 1.2a^{0.5} \qquad (2.53)$$

BC-3, Both Surfaces Heated at Equal Temperature

$$Nu_{i,o,d_h} = 3.66 + \left[4 - \frac{0.102}{a + 0.02} \right] a^{0.04} \qquad (2.54)$$

These relations are also corrected in [9] to take into account the entry length, but as explained in the previous section, this correction will not be used in our calculations.

Annular Pipes: Turbulent

The turbulent regime is often more important to enhance heat transfer. Again several authors (see, e.g., [11]) have presented experimental correlations. One of the best-known study was given by Kays and Lung [12], where the authors presented results at different diameters ratio. Their results are given in tables and can be found in [4]. In [13], Petukhov and Roizen proposed the following relation that matches the Kays and Lung results to ±5%:

$$\begin{aligned} \frac{Nu_i}{Nu_{pipe}} &= 0.86\zeta a^{-0.16} \\ \frac{Nu_o}{Nu_{pipe}} &= 1 - 0.14a^{0.6} \end{aligned} \qquad (2.55)$$

with Nu_{pipe}, the Nusselt number for a circular pipe, and:

$$\begin{aligned} \zeta &= 1 + 7.5 \left(\frac{1 - 5a}{a Re_{dh}} \right)^{0.6} &, a < 0.2 \\ \zeta &= 1 &, a > 0.2 \end{aligned} \qquad (2.56)$$

The same authors proposed their own correlation for the circular pipe for air, but any other correlations can be used. These correlations were found with uniform heat flux imposed at a given surface and the other one being isolated. Corrections factors involving the flux ratio are proposed in [13], but in practice, it is not easy to use since this ratio is often unknown. Instead, a weighted average proposed in [4] can be used to evaluate a mean Nusselt number:

$$Nu_m = \frac{Nu_o + aNu_i}{1 + a} \qquad (2.57)$$

Incropera and Dewitt [3] suggest that the correction for the diameter ratio is small and the formula used with the circular pipe (2.43b) with the Reynolds based on the hydraulic diameter can be valid. Gnielinski [9, 14] proposed the following correlation for fully turbulent regime ($Re > 10000$):

$$Nu_{d_h} = \frac{f_a/8 Re_{d_h} Pr}{k_1 + 12.7(f_a/8)^{1/2}(Pr^{2/3} - 1)}\left[1 + \left(\frac{d_h}{L}\right)^{2/3}\right] F_{ann} K \qquad (2.58)$$

with:

$$k_1 = 1.07 + \frac{900}{Re_{d_h}} - \frac{0.63}{1 + 10Pr} \qquad (2.59)$$

The friction factor suggested in annulus region differs than for the circular pipe and is given in [9] by:

$$f_a = (1.8 \log_{10}(Re^*) - 1.5)^{-2}$$

To be consistent with the previous definition of friction factor, it will be expressed as a function of natural logarithm:

$$f_a = (0.781 \log(Re^*) - 1.5)^{-2} \qquad (2.60)$$

with the modified Reynolds number:

$$Re^* = Re_{d_h} \frac{(1 + a^2)\log(a) + 1 - a^2}{(1 - a)^2 \log(a)} \qquad (2.61)$$

The correction factor K was the same as for the circular case (2.47), whereas the F factor is a correction associated with the way the heat flows through the annular pipes. As in the laminar case, these expressions are given:

BC-1

$$F_{ann,I} = 0.75a^{-0.17} \qquad (2.62)$$

BC-2

$$F_{ann,II} = 0.9 - 0.15a^{0.6} \qquad (2.63)$$

BC-3

$$F_{ann,III} = \frac{a0.75a^{-0.17} + 0.9 - 0.15a^{0.6}}{1 + a} \qquad (2.64)$$

As pointed out earlier, the correction factor K for geothermal applications is quite small and will be omitted in our applications. In [15], the same author suggested instead to use a

similar correlation than with circular pipes as:

$$Nu_{d_h} = \frac{f_a/8(Re_{d_h} - 1000)Pr}{1 + 12.7(f_a/8)^{1/2}(Pr^{2/3} - 1)}\left[1 + \left(\frac{d_h}{L}\right)^{2/3}\right] F_{ann} K \qquad (2.65)$$

with the friction factor f_a given again by (2.60) and the F_{ann} given previously. In [15], the author gave factors only for BC-1 and BC-2, strangely stating that no corrections factors are available for BC-3. Again, for part of the transition region ($2300 < Re < 4000$), the same linear interpolation will be used:

$$Nu_{trans} = (1 - \gamma)Nu_{lam} + \gamma Nu_{turb,4000}$$

Example 2.1 Water flows through the annulus portion of a coaxial borehole. The flow rate is 0.5 kg/s, the inner diameter is 2 cm and the outer diameter 4 cm. Evaluate the heat transfer coefficient assuming BC-I. Evaluate the properties at $T_{ref} = 300$K.

Solution The solution is given by:
    Properties of water at 300 K.

$$\rho = 997 \text{ kg/m}^3 \quad , \quad \mu = 8.54e - 4 \text{ Pa-s}, \quad Pr = 5.86$$
$$C_p = 4.18 \text{ kJ/kg-K} \quad , \quad k = 0.609 \text{ W/m-K}$$

$$d_h = d_o - d_i = 0.02 \text{ m}, \quad a = 0.5, \quad A_c = \pi(d_o^2 - d_i^2)/4 = 9.4 \text{ m}^2$$

$$u = \frac{\dot{m}}{\rho A_c} = 0.53 \text{ m/s}$$

$$Re_{d_h} = \frac{\rho u d_h}{\mu} = 12428$$

With the flow being turbulent, four different relations will be compared:

- The classical relation for circular ducts as proposed in Incropera and Dewitt (Eq. (2.43b))
- The correction suggested by Petukhov (Eq. (2.55))
- The two relations suggested by Gnielinski (Eqs. (2.58, (2.65)).

BC-I : Heat transfer through the inner tube, outer tube isolated. Using (2.43b):

$$f = (0.79 \log(12428) - 1.64)^{-2} = 0.02964$$

$$Nu_{d_h} = \frac{f/8(Re_{d_h} - 1000)Pr}{1 + 12.7(f/8)^{1/2}(Pr^{2/3} - 1)} = 90.4$$

(continued)

Example 2.1 (continued)

$$\hbar = \frac{90.4 \cdot 0.609}{0.02} = 2755 \text{ W/m}^2\text{-K}$$

Using the Petukhov correction:

$$Nu_{d_h} = 0.86a^{-0.16}90.4 = 86.9$$

$$\hbar = 2647 \text{ W/m}^2\text{-K}$$

Using Eq. (2.58):

$$Re^* = Re_{d_h}\frac{(1+a^2)\log(a) + 1 - a^2}{(1-a)^2\log(a)} = 8350$$

$$f_a = f_a = (0.7817\log(Re^*) - 1.5)^{-2} = 0.032$$

$$F_{ann} = 0.75 * a^{-0.17} = 0.84$$

Since the total length of the exchanger is not given, it is logical to neglect the factor $(1 + d/L)$, the same for the K factor.

$$k_1 = 1.07 + \frac{900}{Re_{d_h}} - \frac{0.63}{1+10Pr} = 1.13$$

$$Nu_{d_h} = \frac{f_a/8 Re_{d_h} Pr}{k_1 + 12.7(f_a/8)^{1/2}(Pr^{2/3} - 1)}F_{ann} = 81$$

$$\hbar = \frac{81 \cdot 0.609}{0.02} = 2471 \text{ W/m}^2\text{-K}$$

and finally with Eq. (2.65):

$$Nu_{d_h} = \frac{f_a/8(Re_{d_h} - 1000)Pr}{1 + 12.7(f_a/8)^{1/2}(Pr^{2/3} - 1)}F_{ann} = 84$$

$$\hbar = \frac{84 \cdot 0.609}{0.02} = 2567 \text{ W/m}^2\text{-K}$$

The largest difference between the largest and smallest value is approximately 10%, which is smaller than the order of precision often expected with convection coefficients ($\approx$20%). The most important conclusion is maybe that the use of the circular correlation seems to slightly overestimate the heat transfer. It was not possible to evaluate the correction factor due to the temperature effect since the temperatures were not given. Let's assume that the water, which was expected to be at a mean temperature of 300 K, exchange heat with a ground at 10 °C and that the borehole temperature is at a mean temperature ($\approx$18 °C(291 K)). The correction factor would then be:

$$K = \left(\frac{5.86}{Pr(283K)}\right)^{0.11} \approx 0.98$$

which again is less than the precision expected. Again since the length of the borehole, it was not possible to evaluate its effect on the correction factor, but if we assume a length of 100 meters, the correction factor would be:

$$\left[1 + \left(\frac{d_h}{L}\right)^{2/3}\right] = \left[1 + \left(\frac{0.02}{100}\right)^{2/3}\right] = 1.0034$$

which is negligible. Similar conclusions would be observed with the other boundary conditions.

Comments *Heat transfer through annular pipes are more complex than in the case of circular pipes. Although not exhaustive, several expressions were compared in this example, and it was found that they all fall within approximately 5%. Using the correlation for circular pipe as suggested by Incropera and Dewitt gives an overestimation of more than 10% (at least in this example).* ∎

Helicoidal Pipes

Helicoidal pipes are often used in heat exchangers and in coiled ground heat exchangers. A schematic of a coiled tube is shown in Fig. 2.1.

Numerous studies have been performed on these configurations, and reviews can be found in [16–18]. Without going into too many details, two general remarks are generally reported associated with the centrifugal forces in a curved pipe:

- They have a stabilizing effect on the laminar flow, retarding the transition to the turbulent regime.

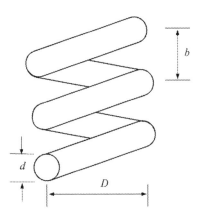

Fig. 2.1 Helicoidal tube

(continued)

- They cause secondary flows increasing the heat transfer and the pressure drop, especially in the laminar regime.

Higher critical Reynolds numbers expressions were suggested by many authors [18]. Only two will be given, one suggested by Srinivasan et al. [19]:

$$Re_{d,cr-hel} = 2100\left[1 + 12\lambda^{1/2}\right] \quad (2.66)$$

and one by Schmidt [20]:

$$Re_{d,cr-hel} = 2300\left[1 + 8.6\lambda^{0.45}\right] \quad (2.67)$$

with:

$$\lambda = \frac{d}{D} \quad (2.68)$$

It has been suggested by Truesdell and Adler [21] to take into account the effect of the pitch by replacing the coil diameter by a corrected diameter defined as:

$$D_c = D\left[1 + \left(\frac{b}{\pi D}\right)^2\right] \quad (2.69)$$

But for typical applications, this correction is small and will be neglected. The extended laminar regime will experience also a larger heat transfer. Nusselt correlations are often given in terms of the *Dean number*:

$$De = Re\sqrt{\lambda} \quad (2.70)$$

where again, the corrected coil diameter can be used in the case of the more severe cases. Secondary flows in coiled tubes result also in nonuniform local heat transfer, and authors are sometimes interested in peripherally local or averaged Nusselt numbers [16]. Correlations for constant heat flux as well as constant wall temperature are suggested in the literature. For the latter case, a correlation was suggested by Manlapaz and Churchill [22], also proposed in the book of Incropera and DeWitt [3].

$$Nu_d = \left[\left(3.66 + \frac{4.343}{x_1}\right)^3 + 1.158\left(\frac{De}{x_2}\right)^{3/2}\right]^{1/3}\left(\frac{\mu}{\mu_s}\right)^{0.14} \quad (2.71)$$

$$x_1 = \left(1 + \frac{957}{De^2 Pr}\right)^2 \quad, \quad x_2 = 1 + \frac{0.477}{Pr}$$

A correlation suggested by Gnielinski [9] is:

$$Nu_d = 3.66 + 0.08\left[1 + 0.8\lambda^{0.9}\right]Re_d^m Pr^{1/3}\left(\frac{Pr}{Pr_s}\right)^{0.14} \quad (2.72)$$

$$m = 0.5 + 0.2903\lambda^{0.194}$$

It should be pointed out that they used an unusual definition of the Reynolds number:

$$Re = \frac{\dot{m}d}{\mu}$$

Example 2.2 Evaluate the heat transfer coefficient in a helicoidal coil having 50 cm of diameter, 10 cm of pitch. The pipe has a 2.5 cm diameter, and 0.1 kg/s of water is flowing inside. Evaluate the properties at 300 K and 290 K for the wall temperature.

Solution The solution is straightforward. With the properties at 300 K being the same as the previous example, at 290 K:

$$\mu_s = 1.08e - 3 \text{ Pa-s} \quad, \quad Pr_s = 7.66$$

The Reynolds number is:

$$Re_d = \frac{4\dot{m}}{\pi d\mu} = 5965$$

which should be turbulent for straight pipes but less than Eq. (2.66) (7734) or Eq. (2.67) (7437), which indicates laminar regime. The Dean number is:

$$De_d = Re_d\sqrt{\lambda} = 1333$$

Using Eq. (2.71):

$$x_1 \approx 1, x_2 = 1.081$$

$$Nu_d = \left[\left(3.66 + \frac{4.343}{1}\right)^3 + 1.158\left(\frac{1217}{1.081}\right)^{3/2}\right]^{1/3}$$

$$\times \left(\frac{8.5}{10.8}\right)^{0.14} = 35.8$$

$$\hbar = \frac{35.8 \cdot 0.609}{0.025} = 872 \text{ W/m}^2\text{-K}$$

Using the correlation suggested by Gnielinski and keeping the same Reynolds number gives:

$$Nu_d = 3.66 + 0.08\left[1 + 0.8 \cdot 0.05^{0.9}(5965)^{0.662}5.86^{1/3}\right]$$

$$\times \left(\frac{5.86}{7.66}\right)^{0.14} = 50$$

$$\hbar = \frac{50 \cdot 0.609}{0.025} = 1219 \text{ W/m}^2\text{-K}$$

(continued)

Example 2.2 (continued)
Using their modified form of Reynolds number, the final value would be in the order of 1050 W/m²-K, a little bit closer than the first correlation. Equation (2.71) will be used in this book. ∎

As previously stated, turbulent flow will be delayed in a curved coil. A value up to 22000 is sometimes suggested [9] for the fully turbulent flow. In such a case, Incropera and DeWitt [3] suggest to use correlations for straight pipes. Gnielinski [9] proposed the following correlation:

$$Nu_{d_h} = \frac{f_c/8 Re_d Pr}{1 + 12.7(f_c/8)^{1/2}(Pr^{2/3} - 1)} \left(\frac{Pr}{Pr_s}\right)^{0.14} \quad (2.73)$$

where a specific friction coefficient for turbulent coiled tubes proposed by Mishra and Gupta [23]:[4]

$$f_c = \left(f_{sp} + 0.03\sqrt{\lambda}\right)\left(\frac{\mu_s}{\mu}\right)^{0.27} \quad (2.74)$$

where f_{sp} is the friction factor for straight pipes. In [9], the Blasius expression for smooth pipes was proposed:

$$f_{sp} = \frac{0.3164}{Re_d^{0.25}} \quad (2.75)$$

but, of course, the Petukhov expression (2.44) can also be used or the Colebrook relation for non-smooth pipes. Since this expression is strictly valid for $Re_d > 22000$, the usual linear interpolation between laminar and fully turbulent expression is suggested by Gnielinski [9]:

$$Nu_{trans} = (1 - \gamma)Nu_{lam} + \gamma Nu_{turb,22000}$$

with:

$$\gamma = \frac{Re - Re_{cr}}{22000 - Re_{cr}}$$

Example 2.3 Redo the last example for a flow rate of 0.3 kg/s of water.

Solution The only difference will now be that the Reynolds number will be 3 times larger:

$$Re_d = 3 \cdot 5965 = 17895$$

(continued)

Being in the transition region, different final values can be evaluated. Following Incropera and Dewitt's [3] suggestion, the expression for straight pipes can be used:

$$f = (0.79 \log(Re) - 1.64)^{-2} = 0.0269$$

$$Nu_{d,turb1} = \frac{f/8(Re_d - 1000)Pr}{1 + 12.7(f/8)^{1/2}(Pr^{2/3} - 1)} = 125$$

$$\hbar_{circ,pipe} = \frac{125 \cdot 0.609}{0.025} = 3054 \text{ W/m}^2\text{-K}$$

Using the correlation suggested by Gnielinski [9]:

$$f_c = (0.0269 + 0.03\sqrt{0.05})\left(\frac{10.8}{8.5}\right)^{0.27} = 0.0358$$

$$Nu_{d,turb2} = \frac{f_c/8(Re_d)Pr}{1 + 12.7(f_c/8)^{1/2}(Pr^{2/3} - 1)}\left(\frac{5.86}{7.66}\right)^{0.14} = 155$$

where the heat transfer augmentation associated with the curved coil is clearly seen although it is sometimes neglected in turbulent regime. Since the flow is not fully turbulent, the linear interpolation can be used, and again, several choices can be used for the critical Reynolds number as well as for the laminar Nusselt number. Using only Eqs. (2.66) and (2.71) gives:

$$Re_{cr} = 2100(1+12\sqrt{0.05}) = 7734 \quad, \quad De_{cr}^* = 1729 \quad, \quad x_1 \approx 1$$

$$Nu_{lam_{Re=7734}} = \left[\left(3.66 + \frac{4.343}{1}\right)^3 + 1.158\left(\frac{1729}{1.081}\right)^{3/2}\right]^{1/3}$$

$$\left(\frac{8.5}{10.5}\right)^{0.14} = 40$$

$$\gamma = \frac{17895 - 7734}{22000 - 7734} = 0.71$$

$$Nu_d = 0.29 \cdot 40 + 0.71 \cdot 155 = 122$$

$$\hbar = \frac{122 \cdot 0.609}{0.025} = 2983 \text{ W/m}^2\text{-K}$$

Comments *It is surprising to observe that taking an average value between the laminar expression and the fully turbulent one, the final value of the Nusselt number is similar to the one used for straight pipes. In the fully turbulent regime where no average value is suggested, the heat transfer enhancement using Eq. (2.73) rather Eq. (2.43b) is still observed. It should be remembered that, as it is the case for all problems involving convection coefficients, the accuracy is very limited and the general guideline is to use conservative values for design purposes and,*

---

[4]In Mishra and Gupta paper, the Fanning friction coefficient C_f is given, so the expression is divided by 4.

(continued)

Example 2.3 (continued)
   if possible, more precise expressions for simulation
   applications. ∎

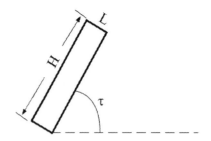

Fig. 2.2 Rectangular cavity

2.2.3.3 Free Convection
Although, in GSHP applications, free convection is rarer, classical correlations will be stated for completeness. In free convection, the Reynolds number is not involved since no reference velocity is known. Instead the *Grashof number* defined as:

$$Gr_{L_c} = \frac{g\beta|T_s - T_\infty|L_c^3}{\nu^2} \tag{2.76}$$

is one of the major parameter involved. Another closely related parameter is the *Rayleigh number*:

$$Ra_{L_c} = Gr_{L_c}Pr = \frac{g\beta|T_s - T_\infty|L_c^3}{\nu\alpha} \tag{2.77}$$

Vertical Plates
The following expression was proposed by Churchill and Chu [24]:

$$Nu_L = \left(0.825 + \frac{0.387Ra_L^{1/6}}{\left[1 + (0.492/Pr)^{9/16}\right]^{8/27}}\right)^2 \tag{2.78}$$

where L is the height of the plate. The correlation is assumed valid for the heated (fluid going up) or cooled plate (fluid going down).

Horizontal Plates
Natural convection on horizontal plates depends if the fluid is heated or cooled. In the case where the plate is hot and the plate is facing up or if the plate is cold and facing down, Lloyd and Moran [25] suggested:

$$\begin{aligned} Nu_{L_c} &= 0.54Ra_{L_c}^{1/4} &, Ra_{L_c} < 10^7 \\ Nu_{L_c} &= 0.15Ra_{L_c}^{1/3} &, Ra_{L_c} > 10^7 \end{aligned} \tag{2.79}$$

In the case where the plate is hot and the plate is facing down or if the plate is cold and facing up, Radziemska and Lewandowski [26] suggested:

$$Nu_{L_c} = 0.52Ra_{L_c}^{1/5} \tag{2.80}$$

In both cases, the characteristic length suggested for the horizontal plate is given by:

$$L_c = \frac{A_s}{P} \tag{2.81}$$

Horizontal Cylinder
The correlation suggested by Churchill and Chu [27] is:

$$Nu_D = \left(0.60 + \frac{0.387Ra_D^{1/6}}{\left[1 + (0.559/Pr)^{9/16}\right]^{8/27}}\right)^2 \tag{2.82}$$

Sphere
The correlation suggested by Churchill [28] is:

$$Nu_D = 2 + \frac{0.589Ra_D^{1/4}}{\left[1 + (0.469/Pr)^{9/16}\right]^{4/9}} \tag{2.83}$$

Rectangular Cavities
A vertical cavity of height H and width L is shown in Fig. 2.2.

Applications involving vertical cavities include double glazed windows, while inclined cavities are studied in solar panel applications. For small Rayleigh numbers, no convection occurs, and the heat transfer is primary conduction (and radiation if it is a gas). The Rayleigh number above, whose convection occurs, is called the *critical Rayleigh number*. For vertical cavities, Catton [29] suggested the following correlations:

$$\begin{aligned} Nu_L &= 0.18\left(\frac{PrRa_L}{0.2 + Pr}\right)^{0.29} &, 1 < \tfrac{H}{L} < 2 \\ Nu_L &= 0.22\left(\frac{PrRa_L}{0.2 + Pr}\right)^{0.28}\left(\frac{H}{L}\right)^{-1/4} &, 2 < \tfrac{H}{L} < 10 \end{aligned} \tag{2.84}$$

Suggestions for larger aspect ratios are given in [30].

$$Nu_L = 0.42Ra_L^{1/4}Pr^{0.012}\left(\frac{H}{L}\right)^{-0.3} \quad, \quad 10 < \frac{H}{L} < 40 \tag{2.85}$$

For vertical plates, the critical Rayleigh number is in the order of 10^3. For inclined cavities, Hollands et al. [31] suggest:

$$Nu_L = 1 + 1.44\left[1 - \frac{1708}{Ra_L\cos\tau}\right]^{\bullet}\left[1 - \frac{1708(\sin 1.8\tau)^{1.6}}{Ra_L\cos\tau}\right]$$

$$+ \left[\left(\frac{Ra_L\cos\tau}{5830}\right)^{1/3} - 1\right]^{\bullet} \quad, \quad \frac{H}{L} > 12 \tag{2.86}$$

with the convention:

$$[x]^\bullet = \begin{cases} x & , \ x > 0 \\ 0 & , \ x < 0 \end{cases} \tag{2.87}$$

For inclined numbers, the critical Rayleigh number is given by:

$$Ra_{L_{cr}} = \frac{1708}{\cos \tau} \tag{2.88}$$

2.3 Heat Exchangers

Most of this book will be devoted to ground heat exchangers with emphasis on the particularity of these exchangers compared with classical heat exchangers. A brief review of classical heat exchangers will be helpful, since such heat exchangers are often integral parts of GSHP systems. Heat exchangers can be built with different configurations: parallel, counterflow concentric tubes, shell-and-tube, cross-flow, and plate heat exchangers. Whatever the configuration, a schematic of a heat exchanger is shown on Fig. 2.3. A hot fluid delivers heat to a cold fluid. Each side is sometimes called source side and load side, but in cooling applications, it is not so clear what means source and load, so each side will be labeled as hot ($_h$) and cold ($_c$). Two classical approaches are used to analyze heat exchangers: the logarithm temperature difference method and the NTU-ϵ method. Only the second one will be discussed in this introduction. If no phase change occurs, the heat exchanged by the fluid is given by:

$$q_h = \underbrace{(\dot{m}C_p)_h}_{C_h}(T_{h,i} - T_{h,o}) \quad , \quad q_c = \underbrace{(\dot{m}C_p)_c}_{C_c}(T_{c,o} - T_{c,i})$$
$$\tag{2.89}$$

The product of the heat capacity and the flow rate is given the name C to simplify the writing. In heat exchanger analysis, those two heat transfers are assumed to be the same ($q_h = q_c = q$).

The efficiency of a heat exchanger is defined as:

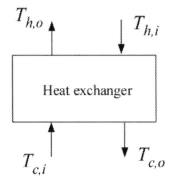

Fig. 2.3 Heat exchanger schematic

$$\epsilon = \frac{q}{q_{max}} \tag{2.90}$$

where q_{max}, as the name suggests, is the *maximum heat exchange* in the heat exchanger for the given conditions. It can be shown that it is given by:

$$q_{max} = C_{min}(T_{h,i} - T_{c,i}) \tag{2.91}$$

$$C_{min} = \min(C_c, C_h) \tag{2.92}$$

The efficiency can be expressed as a function of two dimensionless parameters:

$$\epsilon = f(NTU, C_r) \tag{2.93}$$

where:

$$C_r = \frac{C_{min}}{C_{max}} \quad , \quad NTU = \frac{UA}{C_{min}} \tag{2.94}$$

NTU is the *Number of Thermal Units* and is associated with the capacity of the heat exchanger to transfer heat. The *overall heat transfer coefficient* (UA) is associated with the total thermal resistance between the hot and cold fluid.

$$UA = \frac{1}{R_{tot}} \tag{2.95}$$

where the total resistance R_{tot} comprises the convection resistance associated with the hot fluid, the cold fluid flow, as well as the conduction resistance associated with the solid between both streams, resistance which is often small. In some cases, dirt (or dust) can accumulate, and a possible extra *fooling resistance* can be present [3]:

$$R_{tot} = R_{conv,h} + R_{fouling,h} + R_{cond} + R_{fouling,c} + R_{conv,c} \tag{2.96}$$

where again several of these resistances are sometimes neglected in practice. Expressions that relate the number of thermal units and the efficiency of the heat exchanger can be found in Table 2.1 taken from [3].

Typical problems associated with heat exchangers can be split into two types: performance analysis and design problems. In the first type of problems, the geometry of the heat exchanger is known. In the NTU-ϵ method, the following steps are then followed:

1. The total resistance between the hot and cold fluid is evaluated.
2. q_{max}, C_r, and NTU are calculated.
3. The efficiency is found from Table 2.1.
4. The heat transfer and exit temperatures are then evaluated.

In design problems, the same expressions are used in a reverse order:

Table 2.1 Heat exchanger efficiencies

Configuration	Efficiency	
Parallel flow	$\epsilon = \frac{1-\exp(-NTU(1+C_r))}{1+C_r}$	
Counter flow	$\epsilon = \frac{1-\exp(-NTU(1-C_r))}{1-C_r\exp(-NTU(1-C_r))}$	$C_r < 1$
	$\epsilon = \frac{NTU}{1+NTU}$	$C_r = 1$
Shell-and-tube		
n shell pass	$\epsilon = \left[\left(\frac{1-\epsilon_1 C_r}{1-\epsilon_1}\right)^n - 1\right]\left[\left(\frac{1-\epsilon_1 C_r}{1-\epsilon_1}\right)^n - C_r\right]^{-1}$	$C_r < 1$
	$\epsilon = \frac{n\epsilon_1}{1+\epsilon_1(n-1)}$	$C_r = 1$
1 shell pass	$\epsilon_1 = 2\left[1 + C_r + G\frac{1+\exp(-NTU_1\cdot G)}{1-\exp(-NTU_1\cdot G)}\right]^{-1}$	$G = \sqrt{1+C_r^2}$
	NTU_1, Number of thermal units for one shell	
Cross-flow		
Both fluids unmixed	$\epsilon = 1 - \exp\left(\frac{1}{C_r}NTU^{0.22}\left[\exp(-C_r NTU^{0.78}) - 1\right]\right)$	
C_{max} mixed	$\epsilon = \frac{1}{C_r}(1 - \exp(-C_r(1 - \exp(-NTU))))$	
C_{min} mixed	$\epsilon = 1 - \exp\left(\frac{1}{C_r}(1 - \exp(-C_r NTU))\right)$	
All exchangers		
$C_r = 0$	$\epsilon = 1 - \exp(-NTU)$	

1. The heat transfer is evaluated.
2. q_{max}, C_r are calculated.
3. The efficiency is found from its definition.
4. The NTU is found from Table 2.2.
5. The total resistance between the hot and cold fluid is evaluated.
6. The requested geometric parameter is then found to reach the expected resistance.

Example 2.4 Cold water is heated in a shell-and-tube heat exchanger from 15 to 45 °C. The mass flow rate of water is 2.5 kg/s. The hot fluid is superheated vapor at 170 °C and atmospheric pressure. The water flows through six thin tubes of 40 mm of diameter in two shells with two passes. The vapor flows through the shell side at a flow rate of 3 kg/s. Estimate the total length of tubes needed to heat the water. The heat transfer coefficient for vapor is estimated as 375 W/m²-K.

Solution The $NTU - \epsilon$ method is straightforward. This is an example of a design problem, where some parameter of the heat exchanger geometry, here the length, is searched. We first evaluate the efficiency needed. From CoolProp, we can evaluate the specific heat of water and vapor:

$$C_{p,c} = 4.18 \text{ KJ/kg-K} \quad, \quad C_{p,h} = 1.98 \text{ KJ/kg-K}$$

$$C_c = 4.18 \cdot 2.5 = 10.45 \text{ kW/K} = C_{max}$$

$$C_h = 1.98 \cdot 3.0 = 5.95 \text{ kW/K} = C_{min}$$

(continued)

$$q = C_c(T_{co} - T_{ci}) = 313.5 \text{ kW}$$

$$T_{ho} = 170 - \frac{313.5}{5.95} = 117.3 \,°C$$

At atmospheric pressure, the vapor is still superheated.

$$C_r = \frac{5.95}{10.45} = 0.569 \quad, \quad q_{max} = 5.95(170-15) = 922 \text{ kW}$$

$$\epsilon = \frac{313.5}{922} = 0.34$$

The number of thermal units is found from Table 2.2:

$$G = \sqrt{1 + C_r^2} = 1.15$$

$$F = \left(\frac{\epsilon C_r - 1}{\epsilon - 1}\right)^{1/2} = 1.105$$

$$\epsilon_1 = \frac{F-1}{F - C_r} = 0.197$$

$$E = \frac{2/\epsilon_1 - (1 + C_r)}{G} = 7.48$$

$$NTU_1 = -\frac{1}{G}\log\left(\frac{E-1}{E+1}\right) = 0.234$$

$$NTU = 2 \cdot 0.234 = 0.468$$

$$R_{tot} = \frac{1}{0.468 \cdot 5.95} = 0.359 \text{ K/kW}$$

The total resistance consists here of two convective resistances, one on the hot side and one on the cold side

(continued)

Table 2.2 Heat exchanger NTU

Configuration	NTU	
Parallel flow	$NTU = -\frac{\log[1-\epsilon(1+C_r)]}{1+Cr}$	
Counter flow	$NTU = \frac{1}{Cr-1}\log\left[\frac{\epsilon-1}{\epsilon C_r-1}\right]$	$C_r < 1$
	$NTU = \frac{\epsilon}{1-\epsilon}$	$C_r = 1$
Shell-and-tube		
	$NTU_1 = -\frac{1}{G}\log\left[\frac{E-1}{E+1}\right]$	$E = \frac{2/\epsilon_1-(1+C_r)}{G}$
n shell pass	$\epsilon_1 = \frac{F-1}{F-C_r}$, $F = \left(\frac{\epsilon C_r-1}{\epsilon-1}\right)^{1/n}$	$C_r < 1$
	$\epsilon_1 = \frac{\epsilon}{n-\epsilon(n-1)}$	$C_r = 1$
1 shell pass	$\epsilon_1 = \epsilon$	
Cross flow		
Both fluids unmixed	$NTU = \epsilon^{-1}(\epsilon, Cr)$	
C_{max} mixed	$NTU - \log\left(1 + \left(\frac{1}{C_r}\right)\log(1-\epsilon C_r)\right)$	
C_{min} mixed	$NTU - \left(\frac{1}{C_r}\right)\log(C_r\log(1-\epsilon)+1)$	
All exchangers		
$C_r = 0$	$NTU = -\log(1-\epsilon)$	

Example 2.4 (continued)
and a conductive resistance:

$$R_{tot} = \frac{1}{\hbar_c A_c} + R_{cond} + \frac{1}{\hbar_h A_h}$$

Since the tube is *thin*, the surface area on both sides is equal, and the conductive resistance is neglected:

$$R_{tot} = \frac{1}{L}\left(\frac{1}{\pi D}\left[\frac{1}{\hbar_c} + \frac{1}{\hbar_h}\right]\right) = \frac{R'_{tot}}{L}$$

If the heat transfer coefficient on the shell side is given, we can calculate the one in the tubes. From CoolProp, the water properties at a mean temperature of 303 K are:

$$\mu = 7.97e-4 \text{ Pa-s} \quad , \quad Pr = 5.42$$
$$C_p = 4180 \text{ J/kg-K} \quad , \quad k = 0.614 \text{ W/m-K}$$

$$Re_d = \frac{4 \cdot 2.5 \cdot 0.04}{6\pi \, 7.97e-4} = 16636$$

Using, for example, the Dittus-Boelter relation (Eq. (2.43b)):

$$Nu_d = 0.023 \cdot 16636^{4/5} 5.42^{0.4} = 108$$

$$\hbar_c = \frac{108 \cdot 0.614}{0.04} = 1654 \text{ W/m}^2\text{-K}$$

(continued)

$$R'_{tot} = \frac{1000}{\pi 0.04}\left(\frac{1}{1654} + \frac{1}{375}\right) = 4.34 \text{ K-m/kW}$$

$$L_{tot} = \frac{4.34}{0.359} = 12 \text{ m}$$

Since we have two shells, two passes, and six tubes, the shell length would be:

$$L = \frac{12}{24} = 0.5 \text{ m}$$

Comments *The vapor was treated here as an ideal gas with a constant specific heat. We could have used the vapor properties and evaluate the heat as the difference between the enthalpy times the flow rate.* ∎

Example 2.5 Some fins are added to the shell side in order to increase by a factor of two the heat transfer area on that side. The overall efficiency of the fin array is 0.75, and it is assumed that the convection coefficients don't change. Evaluate the new heat transfer obtained. We can assume that the properties and the heat transfer coefficient do not change.

(continued)

Example 2.5 (continued)

Solution It is now a performance problem. The new NTU is easily obtained:

$$A_c = \pi D L_{tot} = 9.1 \text{ m}^2$$

$$R_{tot} = 1000 \left(\frac{1}{1654 \cdot 9.1} + \frac{1}{0.75 \cdot 375 \cdot 18.2} \right) = 0.262 \text{ K/kW}$$

$$NTU = \frac{1}{0.262 \cdot 5.95} = 0.642$$

$$NTU_1 = \frac{0.642}{2} = 0.321$$

$$\epsilon_1 = 2 \left[1 + C_r + G \frac{1 + \exp(-NTU_1 \cdot G)}{1 - \exp(-NTU_1 \cdot G)} \right]^{-1} = 0.251$$

$$Z = \frac{1 - \epsilon_1 C_r}{1 - \epsilon_1} = 1.15$$

$$\epsilon = \frac{Z^2 - 1}{Z^2 - C_r} = 0.422$$

$$q = 0.422 \cdot 922 = 390 \text{ kW}$$

$$T_{co} = 15 + \frac{390}{10.5} = 52.3 \,^\circ\text{C}$$

$$T_{ho} = 170 - \frac{390}{5.95} = 104.5 \,^\circ\text{C} \qquad \blacksquare$$

Although the fluid heat capacity varies with the temperature, this variation often neglected in the concept of C_{min}, C_{max} where an average value evaluated at the mean temperature is calculated as long as the fluid does not change state. If the variation is important, an equivalent heat capacity can be found as:

$$C_p = \frac{h_i - h_o}{T_i - T_o} \tag{2.97}$$

In the case of some special fluids like transcritical fluids of some mixtures, the concept of average heat capacity is not trivial. A general form of (2.91) can be expressed as:

$$q_{max} = \min \left[\dot{m}_h (h_h(T_{h,i}) - h_h(T_{c,i})), \dot{m}_c (h_c(T_{h,i}) - h_c(T_{c,i})) \right] \tag{2.98}$$

In the case when one fluid change phase (condenser or evaporator), the heat exchange is given as:

$$q = \dot{m} \Delta h \tag{2.99}$$

In the special case where a fluid goes from saturated vapor to saturated liquid (or the contrary), it is given as:

$$q = \dot{m} \Delta h_{fg} \tag{2.100}$$

where h_{fg} is the latent heat between the two phases and $\dot{m}$ the amount of fluid that changes phase. Since the heat exchange is at constant temperature,[5] it represents an infinite heat capacity ($C \to \infty, C_r = 0$).

Example 2.6 Repeat the previous example where the area on the shell side is multiplied by four instead of two. The heat transfer coefficient when the vapor condenses is assumed to be ten times larger.

Solution Redoing exactly the same analysis as in the previous example would give a vapor temperature of 83 °C, which does not make sense since at atmospheric pressure, the vapor condenses into liquid at 100 °C. Many problems involving phase change are simplified assuming that the fluid goes from saturated vapor to saturated liquid (in a condenser) or the contrary in an evaporator. Here, the problem is more complicated since there is a sensible portion (from superheated to saturated) and a latent part. This is usually done by separating the heat exchanger in two parts (or three if it goes into compressed liquid). The first part is considered as a design problem where the area needed to bring the vapor to saturated state (A_a) is analyzed. Using the remaining area ($A_b = A - A_a$), a performance problem is done to evaluate the state at the exit of the heat exchanger:

Part a:[a]

$$q_a = C_h(T_{hi} - T_{sat}) = 416 \text{ kW}$$

The design problem is to evaluate the total area (or length) needed to provide the needed heat.

$$\epsilon_a = \frac{q_a}{q_{max,a}}$$

$$q_{max,a} = C_{min}(170 - T_{ci,a}) = 5.95(170 - T_{ci,a})$$

The problem now is that the entrance liquid temperature ($T_{ci,a}$) is unknown and an iterative procedure is needed. Let's assume that $L_a = 0.75L = 9$ m, $L_b = 0.25L = 3$ m. The entrance water temperature in part a will be the outlet water temperature of part b:

$$R_{tot,b} = \frac{1000}{\pi \, 0.04 \cdot 3} \left(\frac{1}{1654} + \frac{1}{0.75 \cdot 4 \cdot 3750} \right) = 0.304 \text{ K/kW}$$

---

[a]Since the saturated state is reached, a better approach would be to evaluate the heat as $q_a = \dot{m}_h(h_i - h_{sat})$.

(continued)

---

[5]For a pure substance.

Example 2.6 (continued)

In the condenser, the specific heat of the vapor is infinite and $C_{min} = C_c = 10.45$ kW/K.

$$q_{max,b} = 10.45(100 - 15) = 888 \text{ kW}$$

$$NTU_b = \frac{1}{10.45 \cdot 0.304} = 0.314$$

$$\epsilon_b = 1 - \exp(-0.314) = 0.269$$

$$q_b = 239 \text{ kW}$$

$$T_{co,b} = T_{ci,a} = 37.9\,°\text{C}$$

$$q_{max,a} = 5.95(170 - 37.9) = 785 \text{ kW}$$

$$\epsilon_a = \frac{416}{785} = 0.53$$

$$NTU_a = 938$$

$$R_{tot,a} = 0.179 \text{ K/kW}$$

$$R'_{tot,a} = \frac{1000}{\pi 0.04}\left(\frac{1}{1654} + \frac{1}{0.75 \cdot 4 \cdot 375}\right) = 1.98 \text{ K-m/kW}$$

$$L_a = \frac{1.98}{0.179} = 11 \text{ m} , L_b = 1 \text{ m}$$

And another iteration is performed. At convergence, the results are:

$$L_a = 10 \text{ m} , L_b = 2 \text{ m}$$

$$q_b = 166 \text{ kW}$$

$$q = 582 \text{ kW}$$

$$T_{co,b} = T_{ci,a} = 30.8\,°\text{C}$$

$$T_{co} = 30.8 + \frac{416}{10.45} = 70.7\,°\text{C}$$

$$h_{sat,v} = 2675 \text{ kJ/kg-K}$$

$$h_{sat,l} = 419 \text{ kJ/kg-K}$$

$$\epsilon_a = 0.5 \ , \ \epsilon_b = 0.19$$

$$h_o = 2675 - 166/3 = 2620 \text{ kJ/kg-K}$$

$$x_o = \frac{2620 - 419}{2675 - 419} = 0.97$$

The vapor is still almost saturated at the exit of the exchanger. ∎

Example 2.7 Repeat the last problem where the unknown hot fluid flow rate exits as a saturated liquid.

Solution Such phase change problems are classical, but often, only the latent part is treated. In such a case, in a performance problem like this one, the total area is known, and the solution is quite simple. Here, it is more complicated since we have two parts. The solution is still iterative and follows similar steps as before. Again assuming, for example, that $L_b = 3$ m, we have:

$$\epsilon_b = 0.269$$

$$q_b = 239 \text{ kW}$$

$$T_{co,b} = T_{ci,a} = 37.9\,°\text{C}$$

$$h_{fg} = 2675 - 419 = 2256 \text{ kJ/kg-K}$$

$$\dot{m}_b = \frac{239}{2256} = 0.11 \text{ kg/s}$$

Part a:

$$C_h = 0.11 \cdot 1.98 = 0.21 \text{ kW/K} = C_{min}$$

$$q_{max,a} = 0.21(170 - 37.9) = 27.8 \text{ kW}$$

$$q_a = 0.21(170 - 100) = 14.7 \text{ kW}$$

$$\epsilon_a = 0.53$$

$$C_r = 0.02$$

$$NTU_a = 0.76$$

$$R_{tot,a} = 6.257 \text{ K/kW}$$

$$L_a = \frac{1.98}{6.257} = 0.32 \text{ m} , L_b = 11.68 \text{ m}$$

After convergence, the result is:

$$L_a = 1.34 \text{ m} , L_b = 10.76 \text{ m}$$

$$T_{co,b} = T_{ci,a} = 72.16\,°\text{C}$$

$$\dot{m}_h = 0.265 \text{ kg/s}$$

$$q_a = 0.265 \cdot 1.98(170 - 100) = 36.7 \text{ kW}$$

$$q = 634 \text{ kW}$$

$$T_{co} = 72.16 + \frac{36.7}{10.45} = 75.7\,°\text{C}$$

∎

In many applications, the state at the exit is controlled, either by a steam trap in a condenser or an expansion device in an evaporator. In such a case, the flow rate is unknown.

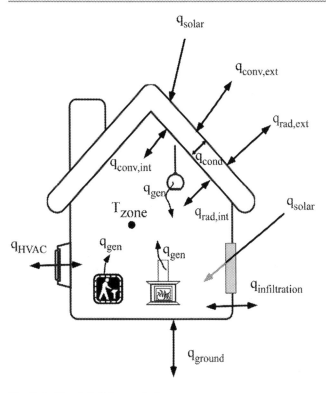

Fig. 2.4 Simple building analysis

2.4 Building Loads

Basic heat transfer calculations involving conduction, radiation, and convection will be done thoroughly this book. In the rest of this chapter, they will be applied for the evaluation of building loads. As stated before, only simple approaches to this complex subject will be covered. A simple building is shown in Fig. 2.4

Knowledge of the thermal behavior of the building can be found by performing an energy balance on the building. In a simple case, if the whole building is treated as a single zone having a single temperature, the balance would give:

$$\sum q_{in} + \sum q_{internal} - \sum q_{out} = \dot{E}_{zone} = m\, C_p \frac{d\, T_{zone}}{dt}$$
$$(2.101)$$

In the case of a steady-state analysis, the right-hand term becomes zero:

$$\sum q_{in} + \sum q_{internal} - \sum q_{out} = 0 \qquad (2.102)$$

Internal heat sources are here represented by a separated term but can be included in the input gains (q_{in}). In Fig. 2.4, some examples of potential heat gains and losses are shown. Most of the numerical values will be evaluated with the concept of thermal resistance introduced in the previous sections. Some basic examples may help illustrate the concepts:

Example 2.8 A 12 m^2 composite wall is composed of a concrete (k $= 2.1$ W/m-K) thickness of 24 cm and an isolation (k $= 0.04$ W/m-K) of 16 cm thickness. The outside temperature is $-10\,^\circ$C and is 22 $^\circ$C inside. The outside convection coefficient is 25 W/m^{2}K and 8 W/m^{2}K inside. If radiation is neglected, find:

(a) The total resistance
(b) The global coefficient $U\,A$
(c) The thermal loss from the inside to the outside

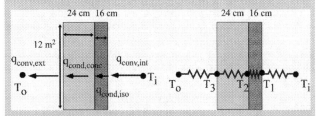

Solution All heat transfer mechanisms can be represented by simple resistances:

$$R_{conv,o} = \frac{1}{25 \cdot 12} = 0.0033 \text{ K/W}$$

$$R_{conv,i} = \frac{1}{8 \cdot 12} = 0.0104 \text{ K/W}$$

$$R_{cond,c} = \frac{0.24}{2.1 \cdot 12} = 0.0095 \text{ K/W}$$

$$R_{cond,i} = \frac{0.16}{0.04 \cdot 12} = 0.333 \text{ K/W}$$

$$R_{tot} = 0.3566 \text{ K/W}$$

$$U\,A = 2.804 \text{ W/K}$$

$$q = 2.804(22 + 10) = 89.73 \text{ W}$$

Although not requested, the intermediate temperatures can be easily calculated:

$$T_1 = 22 - 89.73 \cdot 0.0104 = 21.06\,^\circ\text{C}$$

$$T_2 = 21.06 - 89.73 \cdot 0.333 = -8.84\,^\circ\text{C}$$

$$T_3 = -8.84 - 89.73 \cdot 0.0.0095 = -9.7\,^\circ\text{C}$$

Comments *As expected, the thermal resistance from the isolation is the highest one. As long as we have*

(continued)

temperature differences, temperatures can be left in Celsius. In other cases, they should be transformed in Kelvin. ∎

Example 2.9 In the previous example, the exterior wall ($\epsilon = 0.9$) exchanges heat with the surroundings by radiation. The exterior wall also absorbs 80% of the heat coming from the sun (750 W/m^2) when it is present. Find the heat losses when:

(a) There is no sun $T_{surr} = -10\,°C$.
(b) There is no sun $T_{surr} = -20\,°C$.
(c) Sun is present $T_{surr} = -10\,°C$.

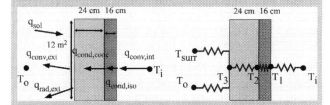

Solution (a) Several approaches can be taken to solve the problem. In the first two cases, the sun radiation is not taken into account. If we want to use the concept of radiative resistance, the surface temperature is needed, which is not known before the problem is solved. One possibility is to guess the outdoor temperature. In this problem, a possible guess would be to assume that this temperature does not change significantly from the previous example:

$$T_{so} \approx -9.7\,°C \approx 263.45K$$

$$T_{surr} \approx -10.0\,°C \approx 263.15K$$

Since the exterior surface exchanges radiation with the surroundings, which represent a very large enclosure, the radiative coefficient is given by (with all temperatures in Kelvin):

$$\hbar_{rad} = \epsilon\sigma(T_{so} + T_{surr})(T_{so}^2 + T_{surr}^2) = 3.73 \text{ W/m}^2\text{ K}$$

In the case of plane geometries (wall, windows, ceilings, etc.), the surface area where the heat is transferred is often not known, and the analysis is often done on a unit area basis. Here, the area is known, and it is not mandatory to work it this way, but it can still be done. This approach cannot be done however in the

case where the area changes like in radial geometries.

$$R''_{rad} = \frac{1}{\hbar_{rad}} = 0.268 \text{ m}^2\text{ K/W}$$

$$R''_{conv,o} = \frac{1}{\hbar_{conv,o}} = 0.04 \text{ m}^2\text{K/W}$$

$$R''_{conv,i} = \frac{1}{\hbar_{conv,i}} = 0.125 \text{ m}^2\text{ K/W}$$

$$R''_{cond,c} = \frac{L_c}{k_c} = 0.114 \text{ m}^2\text{K/W}$$

$$R''_{cond,i} = \frac{L_i}{k_i} = 4 \text{ m}^2\text{ K/W}$$

Since the temperature of the surroundings is assumed to be the same temperature than the outside air temperature, the external resistances are in parallel and:

$$R''_{ext} = \frac{R''_{conv,o}R''_{rad}}{R''_{conv,o}+R''_{rad}} = \frac{1}{h_{conv,o}+h_{rad}} = 0.0348 \text{ m}^2\text{K/W}$$

$$R''_{tot} = 0.0348 + 0.114 + 4 + 0.125 = 4.274 \text{ m}^2\text{K/W}$$

$$q'' = \frac{32}{4.274} = 7.49 \text{ W/m}^2$$

$$q = 12 \cdot 7.487 = 89.84 \text{ W}$$

Normally from this result, the new surface temperature should be recalculated and the calculation redone. However, it is seen that, since the heat loss has not changed a lot, the final result would not change significantly. The reason for that is that the radiative mode is much less important than the convective mode in this problem. Without going into details, the final result, after convergence, would change at the 4th decimal. The initial guess was here pretty good since we did a similar example before. In any cases, we could have solve directly the nonlinear problem coming from the following surface energy balance:

$$q''_{in} = q''_{out} = q''_{conv,o} + q''_{rad}$$

$$\frac{T_i - T_3}{R''_{cond,c} + R''_{cond,i} + R''_{conv,i}} = h_{conv,o}(T_3 - T_o) + \epsilon\sigma(T_3^4 - T_{surr}^4)$$

The solution of this nonlinear equation would give:

$$T_3 = 263.41K = -9.74\,°C \quad , \qquad q'' = 7.49 \text{ W/m}^2$$

Again the value is almost the same since our first guess was very good.

(continued)   (continued)

Example 2.9 (continued)

(b) Here the problem is similar, and the only difference is that the surrounding temperature differs from the air temperature. In that case, it is not possible to compute an equivalent external resistance. The external surface temperature becomes an unknown that has to be solved, either using the approximate radiative resistance approach or the exact heat balance. In the first case, a first guess might be to choose the same radiative loss coefficient as before ($\hbar_{rad} \approx 3.73$ W/m²-K) or recompute it with the new surrounding temperature assuming the same surface temperature:

$$h_{rad} = \epsilon\sigma(263.45+253.14)(263.45^2+253.14^2) = 3.52 \text{ W/m}^2 \text{ K}$$

$$q''_{in} = q''_{out} = q''_{conv,o} + q''_{rad}$$

$$\frac{T_i - T_3}{R''_{cond,c} + R''_{cond,i} + R''_{conv,i}} = h_{conv,o}(T_3 - T_o) + h_{rad}(T_3 - T_{surr})$$

$$\frac{22 - T_3}{4.24} = 25(T_3 + 10) + 3.52(T_3 + 20)$$

$$T_3 = -10.96\,^\circ\text{C}$$

$$q''_{in} = q''_{out} = 7.77 \text{ W/m}^2$$

Solving the same nonlinear balance than in (a) except changing the surroundings temperature would give:

$$T_3 = -10.95\,^\circ\text{C}$$

$$q''_{in} = q''_{out} = 7.77 \text{ W/m}^2$$

again a very similar result since the initial guess of the surface temperature was close to the real one.

(c) In the last case, the heat coming from the sun that is absorbed by the wall should be taken into account in the surface balance. The new balance becomes:

$$q''_{in} = q''_{out}$$

$$q''_{solar} + q''_{wall} = q''_{conv,o} + q''_{rad}$$

$$0.8 \cdot 750 + \frac{22 - T_3}{4.24} = 25(T_3 + 10) + 3.73(T_3 + 10)$$

$$T_3 = 10.98\,^\circ\text{C}$$

$$q_{loss} = \frac{22 - 10.98}{4.24} = 2.60 \text{ W/m}^2$$

It is seen now that the surface temperature, due to solar radiation, is quite different than what was assumed to evaluate the radiation coefficient, and another iteration would be necessary, or one can solve the nonlinear

relation:

$$0.8 \cdot 750 + \frac{295.15 - T_3}{4.24} = 25(T_3 - 263.15) + \epsilon\sigma(T_3^4 - 263.15^4)$$

which would give:

$$T_3 = 283.8K = 10.65\,^\circ\text{C}$$

$$q''_{loss} = 2.68 \text{ W/m}^2 = 516.33 + 86.35 - 600 \text{ W/m}^2$$

∎

2.4.1 Solar Gains

2.4.1.1 Opaque Walls

In the last example, comparing cases (a) and (c) showed the impact of *solar gain* on the opaque wall. This gain can be easily obtained by subtracting the total losses of (a) and (c):

$$\Delta q''_{solar} = 7.49 - 2.68 = 4.81 \text{ W/m}^2$$

This value is near what can be expected using the relation given in [32]:

$$\Delta q''_{solar} = \frac{U\alpha G_{sol}}{\hbar_{out}} \tag{2.103}$$

Taking from part (a) of the last example $U = 0.233$ W/m²-K and $\hbar_{out} = \hbar_{conv} + \hbar_{rad} = 28.73$ W/m²-K. One would find:

$$\Delta q''_{solar} = \frac{0.233 \cdot 600}{28.73} = 4.89 \text{ W/m}^2$$

The difference comes from the nonlinear effect of the infrared radiation. Keeping the radiative heat transfer coefficient constant in case (a) and (c), the two results would be equal. It is observed that it is much less than the incident absorbed radiation since most the heat is reflected back in the environment. The concept of *sol-air temperature* ($T_{sol-air}$) is also associated with the impact of solar radiation on walls. It is defined as an "equivalent outside air temperature" that would result in the same heat losses. For example, in the case where both air and surrounding temperature are the same (cases (a) and (c) in the last example), the heat balance at the external surface was given by:

$$\alpha G_{inc} + q''_{out} = \hbar_{conv,o}(T_{s,o} - T_o) + \epsilon\sigma(T_{s,o}^4 - T_{surr}^4)$$

Assuming $T_o = T_{surr}$ and using the concept of radiative exchange coefficient:

$$\alpha G_{inc} + q''_{out} = (\hbar_{conv,o} + \hbar_{rad,o})(T_{s,o} - T_o) = \hbar_o(T_{s,o} - T_o)$$

(continued)

The idea of sol-air temperature is to define a pseudo temperature that will include the effect of the absorbed radiation:

$$q''_{out} = \hbar_o(T_{s,o} - T_o) - \alpha G_{inc} = \hbar_o(T_{s,o} - T_{sol-air})$$

which gives:

$$T_{sol-air} = T_o + \frac{\alpha G_{inc}}{\hbar_o} \qquad (2.104)$$

which is, for the example:

$$T_{sol-air} = -10 + \frac{600}{28.73} = 10.88\,^\circ C$$

When the surrounding temperature is different, an equivalent temperature can also be defined:

$$q''_{out} = \hbar_{conv,o}(T_{s,o} - T_o) + \hbar_{rad,o}(T_{s,o} - T_{surr}) - \alpha G_{inc}$$

$$= \hbar_o(T_{s,o} - T_{sol-air})$$

$$T_{sol-air} = \frac{\alpha G_{inc}}{\hbar_o} + \frac{\hbar_{conv,o} T_o + \hbar_{rad,o} T_{surr}}{\hbar_o}$$

or:

$$T_{sol-air} = T_o + \frac{\alpha G_{inc}}{\hbar_o} - \frac{\hbar_{rad,o}(T_o - T_{surr})}{\hbar_o} \qquad (2.105)$$

In practice, the surrounding temperature is most of the time lower than the air temperature, since the radiation is done with lower sky temperature, especially with horizontal surfaces, while part of the radiation exchange with vertical surfaces is done with the ground.

2.4.1.2 Glazing Surfaces

In an opaque surface, no radiation is transmitted through it.[6] For glazing surfaces, this is not the case, and the analysis is more complex. First of all, many glazing surfaces are not *gray*, which means that their behavior depends on the wavelength (selective surface). While a simple glass window might be almost transparent to solar radiation (short wavelengths), it is often almost opaque to long wavelengths, a phenomenon called *greenhouse effect*. The solar gain of glazing materials will then be influenced by the transmitted solar radiation but also by the absorbed radiation of the window. The situation is even more complex since many windows have double- or triple-pane glazing. Let's illustrate this with an example with a single-pane window:

Example 2.10 A similar example than in the previous one will be analyzed with a single pane window. The window is 5 mm thick and has a conductivity of 1.0 W/m-K. The window has a 1 m width and a 2 m height. Both surfaces have an emissivity of 0.84. The outside temperature is $-10\,^\circ C$ and the inside temperature $22\,^\circ C$. The outside wind is 9 m/s and flows in the transverse direction of the window. Free convection is observed inside the house. Evaluate the heat losses when:

(a) There is no sun $T_{surr} = -10\,^\circ C$.
(b) Sun is present $T_{surr} = -10\,^\circ C$ (750 W/m^2 normal direction).

The solar characteristics associated with the solar radiation are:

$$\tau = 0.83, \rho = 0.08, \alpha = 0.09$$

Solution (a) Without the sun, the solution is quite similar to the previous problems, the difference is that the convection coefficients are not given and will be evaluated. The air properties as well as the radiative coefficients need the surface temperature. Expecting a small conductive resistance, it is assumed that the surface temperature on the outside and inside surfaces is almost the same (Hyp $\approx 0\,^\circ C$). The external properties will then be evaluated at $T \approx -5\,^\circ C$ and the internal ones at $T \approx 11\,^\circ C$:

$$v_o = 1.29e-5 \text{ m}^2/\text{s} \quad, \quad Pr_o = 0.71 \quad, \quad k_o = 0.024 \text{ W/m-K}$$
$$v_i = 1.43e-5 \text{ m}^2/\text{s} \quad, \quad Pr_i = 0.71 \quad, \quad k_i = 0.025 \text{ W/m-K}$$

External coefficients (based on a 1 m width):

$$Re = \frac{9 \cdot 1}{1.29e-5} = 700000$$

From (2.36):

$$Nu = (0.037 Re^{0.8} - 871) Pr^{1/3} = 787 = \frac{\hbar_{conv,o} \cdot 1}{0.024}$$

$$\hbar_{conv,o} = 18.87 \text{ W/m}^2\text{-K}$$

(continued)

---

[6]We should maybe say thermal radiation since for some special wavelengths (microwave, X-rays, etc.), some radiation can be transmitted.

Example 2.10 (continued)

The approximative radiative coefficient is given by:

$$\hbar_{rad,o} = \epsilon_o \sigma (T_{so} + T_{surr})(T_{so}^2 + T_{surr}^2) = 3.67 \text{ W/m}^2 \text{ K}$$

$$\hbar_o = 18.87 + 3.67 = 22.54 \text{ W/m}^2\text{-K}$$

Internal coefficients (based on a 2 m height):

with $\beta = \dfrac{1}{T_{fi}} = \dfrac{1}{284} = 0.0035 K^{-1}$, $\alpha = \dfrac{\nu}{Pr} = 2.0e-5 \text{ m}^2\text{/s}$

$$Ra = \frac{9.8 \cdot 0.0035(22 - 0)2^3}{1.43e - 5 \cdot 2.0e - 5} = 2.1e10$$

$$Nu = \left(0.825 + \frac{0.387 Ra_L^{1/6}}{\left[1 + (0.492/Pr)^{9/16}\right]^{8/27}}\right)^2 = 319$$

$$\hbar_{conv,i} = 4.02 \text{ W/m}^2\text{-K}$$

Since the emissivity of the internal surface is given, the radiation at the internal surface will be taken into account contrary to the previous examples. This heat exchange will be done with the internal surroundings (internal walls and ceiling) and not the air, but here it will be assumed that they are at the same temperature:

$$\hbar_{rad,i} = \epsilon_o \sigma (T_{s1} + T_{inf,i})(T_{si}^2 + T_{inf,i}^2) = 4.38 \text{ W/m}^2 \text{ K}$$

$$\hbar_i = 4.02 + 4.38 = 8.40 \text{ W/m}^2\text{-K}$$

$$R_{tot}'' = \frac{1}{22.54} + \frac{0.005}{1.0} + \frac{1}{8.40} = 0.044 + 0.005 + 0.119$$

$$= 0.168 \text{ m}^2\text{-K/W}$$

It is observed that the conduction resistance is small and is almost always neglected but will be kept for now.

$$U = \frac{1}{0.168} = 5.94 \text{ W/m}^2\text{-K}$$

$$q'' = 5.99 \cdot 32 = 190.1 \text{ W/m}^2$$

The surface temperatures can be found:

$$T_{si} = 22 - 190.1 \cdot 0.119 = -0.62 \,°\text{C},$$

$$T_{so} = -0.62 - 190.1 \cdot 0.005 = -1.57 \,°\text{C}$$

Since the initial temperature hypothesis is very near the values found, an iterative solution would give almost the same result. Actually it gives a heat loss of 189.8 W/m², which is almost the same thing.

(continued)

b) In the presence of the sun, two new phenomena will be observed; first the transmitted solar flux is assumed to be absorbed by the internal enclosure, which represents the major part of the solar gain, but part of the energy absorbed by the window will be transmitted inside as for the opaque wall. As a first approximation, we can keep the same coefficients as in the first case. An energy balance on the window will give:

$$\alpha G_{inc} + \frac{T_i - T_{so}}{R_{cond}'' + R_{int}''} = \frac{T_{so} - T_o}{R_{out}''}$$

$$0.09 \cdot 750 + \frac{22 - T_{so}}{0.124} = \frac{T_{so} + 10}{0.044}$$

$$\Rightarrow T_{so} = 0.63 \,°\text{C} \ , \ T_{si} = 1.50 \,°\text{C}$$

Again, those two values are often equal when the conduction resistance is neglected. The heat losses from the window is now:

$$q_{out}'' = \frac{22 - 1.50}{0.119} = 172.3 \text{ W/m}^2$$

$$\Delta q_{losses}'' = 190.1 - 172.3 = 17.8 \text{ W/m}^2$$

The total heat gain of the window is then:

$$\Delta q_{solar}'' = \Delta q_{losses}'' + \tau G_i nc = 17.8 + 622.5 = 640.3 \text{ W/m}^2$$

Again, the solution was found neglecting the nonlinear coupling of the radiation transfer. Solving, the nonlinear energy balance would result in a very similar final result since the surface temperatures do not change a lot. The final results would be:

$$T_{so} = 0.61 \,°\text{C} \ , \ T_{si} = 1.47 \,°\text{C} \ , \ q_{out}'' = 173.2 \text{ W/m}^2$$

$$\Delta q_{solar}'' = 639 \text{ W/m}^2$$

∎

It is seen, in the previous examples that the solar gains are much higher in a glazing surface than in an opaque surface, which is quite normal. This is obviously an advantage in cold climate but a handicap in summer in hot climate. The solar gain is given normally as a relative portion of the incoming radiation. The *solar heat gain coefficient* (SHGC) is defined as:

$$SHGC = \frac{\Delta q_{solar}''}{G_{inc}} \tag{2.106}$$

In the previous example, it would give a value of SHGC = 0.854. This value was found by evaluating the heat exchanges with and without sun. It is shown in [32] that, for the single-pane window, it is given by:

$$SHGC = \tau + \alpha \frac{U}{h_o} \qquad (2.107)$$

In the previous example, this would give:

$$SHGC = 0.83 + 0.09 \frac{5.94}{22.54} = 0.854$$

Again, the two values are exactly the same if the exchange coefficients are kept constants but may differ a little bit if the nonlinear expressions are solved. For the double-pane window, it is [32]:

$$SHGC = \tau + \alpha_o \frac{U}{h_o} + \alpha_i U \left(\frac{1}{h_o} + \frac{1}{h_s} \right) \qquad (2.108)$$

where α_o, α_i are the abortion coefficient for the outer and inner pane, respectively, and h_s the overall exchange coefficient (convection+radiation) between the two panes. In the previous example, the exchange coefficients were evaluated using prescribed relations. In practice, ASHRAE fundamentals [33] propose design values for typical calculations. Those values are:

$$\hbar_i \approx 8.3 \ \text{W/m}^2\text{-K} \quad , \quad \hbar_0 \approx 28.0 \ \text{W/m}^2\text{-K}$$

It is observed that the values found in our example are in the same order of magnitude. Finally, in practice, solar radiation is split in three components: direct (or beam) radiation, diffuse, and ground reflected solar radiation. Moreover, the beam radiation is highly dependent on the angle of incidence (assumed normal in the previous example). Optical coefficients are often given in terms diffuse and beam radiation, the latter given in function of the angle of incidence. Finally, the last example did not take into account the edge effects due to the non-negligible heat transfer by the frame. Many of these topics are covered in the literature [33]. Doing correctly heat balances is the most important objective of these examples.

2.4.2 Infiltration Loads

Buildings can lose or gain energy through walls by conduction and also by mass transfer through walls or windows. In winter, for example, warm inside air can go out of the building, and by mass conservation, cold outside air will be drawn in. This new air has also to be heated. Knowing the total amount of air, the heat losses will be:

$$q_{loss} = \dot{m} C_p (T_{in} - T_{out}) \qquad (2.109)$$

The flow rate depends on several factors like the age of the building, the tightness of the envelope, the outside wind, etc. ASHRAE standard [1] gives more details on how to evaluate these loads. In modern houses, the envelope is so airtight that a ventilation system is used to exchange fresh air with the outside. However, the inside air give some of its thermal energy to the outside air stream through a heat recovery ventilator (HRV).

Example 2.11 A 72 m³ building has infiltration, which represents approximately 0.5 Vol/hour. If the outside temperature is 30 °C and the inside temperature 22 °C, what is the infiltration load associated with the process?

Solution Taking the air properties at a mean temperature of 300 K, we have:

$$\rho = 1.177 \ \text{kg/m}^3, \ C_p = 1006 \ \text{J/kg-K}$$

The infiltration flow is then:

$$\dot{m} = \frac{0.5 \cdot 72\rho}{3600} = 0.0118 \ \text{kg/s}$$

$$q = 0.0118 \cdot 1006(22 - 30) = -94.76 \ \text{W}$$

The negative sign is associated here as a negative heating load (heat gain). It can, of course, be interpreted as a positive cooling load of 94.76 W. Different sign conventions can be used as it will be discussed later on. ∎

Knowing all conduction and infiltration loads associated to a building, we can calculate the total losses as:

$$q_{tot} = \underbrace{\left(\sum_i (\dot{m} C_p)_i + \sum_j (UA)_j \right)}_{K_{tot}} \Delta T \qquad (2.110)$$

2.4.3 Latent Loads

The foregoing examples dealt with loads that were proportional to a temperature difference. These loads are called sensible loads. Latent loads represent a process of energy exchange where phase change occurs at constant temperature. The most frequent example in buildings is the vaporization or condensation of water vapor. Sources of latent heat come mainly from two major sources:

- The release of water vapor by the occupants.
- The supply (or removal) of water vapor by the infiltration or ventilation of buildings.

From either source, the latent loads can be evaluated by:

$$q = \dot{m}h_{fg} \tag{2.111}$$

where h_{fg} represents the enthalpy variation between the liquid phase to the vapor phase. In practice, latent loads are only important in cooling mode. The energy that is needed to dehumidify the air has to be provided by the cooling system. In heating mode, the energy needed to humidify the air is provided by other systems.

Example 2.12 If a person emits approximately 0.001 m^3 of vapor each day, what mean heat load does it represent? Assume the temperature of the air at 25 °C.

Solution At 298 K, the energy needed to change water from liquid to vapor is:

$$h_{fg} = 2442 \text{ KJ/kg}$$

The mean rate of evaporation is:

$$\dot{m} = \frac{0.001 \cdot 1000}{24 \cdot 3600} = 1.157e - 5 \text{ kg/s}$$

from (2.111) we get:

$$q_{cool,lat} = 28.27 \text{ W}$$

Since in this example water is vaporized, this number represents a heating gain (a cooling load). If the heating loads are considered as positive and cooling loads as negative, as the convention in this book, a negative sign should be added.

$$q_{lat} = -28.27 \text{ W}$$

∎

In the case of air infiltration, the amount of water evaporated is the difference between the amount of water coming in from the outside minus the one inside.

$$q = \dot{m}_{air}(w_i - w_o)h_{fg} \tag{2.112}$$

where w is the absolute humidity. It can be found from the relative humidity from a classical thermodynamic relation:

$$w = \frac{0.622 p_w}{p_{atm} - p_w} \tag{2.113}$$

with:

$$p_w = \phi p_{sat,w} \tag{2.114}$$

where ϕ is the relative humidity, p_w the partial pressure of water, $p_{s,w}$ the vapor saturation pressure, and p_{atm} the atmospheric pressure.

Example 2.13 What is the latent load associated with the infiltration in example 2.11 if the relative humidity is 30% inside and 80% outside?

Solution From the water tables, we have that the vapor pressures are:

$$p_{sat}|_{T=30C} = 4.247 \text{ kPa}, \ p_{sat}|_{T=22C} = 2.645 \text{ kPa}$$

and:

$$h_{fg}|_{T=30C} = 2439 \text{ kJ/kg}$$

From (2.113), we find:

$$w_i = 0.0049 \quad , \quad w_o = 0.0216$$

From (2.112), the absolute value of the cooling latent load is:

$$q = 0.0118(0.0049 - 0.0216)2439 \cdot 1000 = -480.3 \text{ W}$$

Comments *The properties of humid air can also be evaluated using the humid air properties of Cool-Prop, and the difference will be small.*

∎

2.4.4 Dynamic Analysis

As mentioned earlier, building loads estimation is a very broad field, but some knowledge of the basic concepts is important, in particular the notion of energy balance and heat transfer modes. All of the previous examples were associated with steady-state analysis, which is valid in many cases when the time scales are higher than the time constants associated with the thermal component analyzed. A typical case where this will not be the case is the thermal behavior of the ground, which will be studied in details in the next chapters. Even in building loads evaluations, the dynamic analysis is of primary importance especially for cooling loads [2] and non-residential buildings [33]. In such a case, simulation programs are used most of the time.

2.4.5 Capacity Sizing

Once all the loads are evaluated, sizing of the equipment can be done using typical design conditions for heating and cooling. Usually, these conditions consider the coldest day of the year for heating and hottest day of the year for cooling. In theory, it is possible that the worst day does not coincide with these particular days due to internal gains. However, to have this precise knowledge, building simulations is mandatory.

2.5 Annual Energy Estimation

The calculation of the loads for the design day allows the choice of heating and air conditioning systems. To estimate the annual consumption of energy, evaluations of the energy demand over discrete period of times are needed over a typical year. Thus, we would like to have the heat losses (or gains):

- every month
- or every day
- or every hour

Obviously one can easily switch from hourly loads to daily, monthly, etc., but not the contrary. In typical applications of building simulations, annual energy simulations are useful to estimate energy efficiency measures or to evaluate the energetic performance of a building. In geo-exchange applications, it is also important for system design, as it will be seen in later chapters.

Many simulation software are now available to evaluate the energy profile of a building. The non-exhaustive following list names a few:

- DOE2
- Energy + (https://energyplus.net/)
- Transys (http://www.trnsys.com/)
- ESP-R (https://www.esru.strath.ac.uk/applications/esp-r/)
- etc.

Several interfaces have also been developed to facilitate the use of these calculation kernels. In this book, it will be assumed most of the time that these calculations were done, and the available information will be used to design and analyze geothermal systems. It is however useful to have simpler tools that can be used by hand to have a general idea of the energy profile of the building. Classical methods to do so are *Degree-day method* and *BIN method*.

2.5.1 Degree-Day Method

The basis of the degree-day method is quite simple. For example, in heating mode, it is assumed that above a given arbitrary temperature (T_{base}), no heating is necessary. This temperature is lower than the building set point temperature due to internal gains. Knowing this temperature, it is possible to define the number of degree days in a certain period of time as:

$$DD = (T_{base} - T_o)^\bullet N_j = (T_{base} - T_o)^\bullet \frac{N_h}{24} \qquad (2.115)$$

where

$$x^\bullet = \begin{cases} x & \text{for } x > 0 \\ 0 & \text{for } x < 0 \end{cases} \qquad (2.116)$$

T_o is the outside temperature
N_d is the number of days where T_o is the outside temperature
N_h is the number of hours where T_o is the outside temperature

Several choices are possible for T_{base}, but $18\,^\circ$C was often chosen in the past. In the recent handbook [33], ASHRAE suggests to use the balance method to evaluate the base temperature, which becomes the *balance temperature*. This temperature represents the outside temperature when the internal gains compensate the heating losses (or gains).

$$K_{tot}(T_{bal} - T_i) + \sum_i q_{gains,i} = 0$$

$$T_{bal} = T_i - \frac{\sum_i q_{gains,i}}{K_{tot}} \qquad (2.117)$$

T_i is often chosen as $22\,^\circ$C in heating mode and $25\,^\circ$C in cooling mode. Knowing the number of degree days in a given period, the consumption over this period can be estimated as:

$$Q = \frac{K_{tot}DD}{\eta} \text{ W-day} = \frac{K_{tot}24DD}{\eta} \text{ Whr} = \frac{K_{tot}24 \cdot 3600DD}{\eta} \text{ J} \qquad (2.118)$$

η is the efficiency of the heating or cooling equipment. When the base temperature is fixed at $18\,^\circ$C, a correction factor was included in this formula but not with the balance temperature.

Example 2.14 Text file, "Temp_Mtl.txt," gives the dry bulb temperature for a typical year in Montreal, Canada, each hour of the year. In a typical building with $K_{tot} = 31$ W/K, find the heating energy required for every month of the year using the degree-day

(continued)

Example 2.14 (continued)
method. The internal gains are 225 W and the set point temperature 22 °C. Assume $\eta = 100\%$ (electric heating).

Solution The balance temperature can easily be calculated with (2.117):

$$T_{bal} = 22 - \frac{225}{31} = 14.74\,°C$$

From the 8760 data values of temperature, it is easy to evaluate the number of degree-hours, keeping only the values of temperature below T_{bal}. Doing so for every month, the following table is obtained:

Month	DD	Q_h(KWh)
January	761	566.65
February	675	502.87
March	540	402.31
April	255	190.45
May	98	73.40
June	21	15.72
July	6	4.7
August	8	6.1
September	59	44.0
October	204	152.41
November	363	270.43
December	667	496.79
Total	3664	2725.25

Comments *Even though, temperatures below the equilibrium temperature occur in summer, the internal temperature will probably be acceptable during these months, and no heat is normally needed during that time.* ∎

2.5.2   BIN Method

In the BIN method, the number of occurrences that the outdoor temperature is between a given interval (bin) is counted and tabulated. In the table, the time of the day is often separated in bins in order to separate occupied and unoccupied periods. Doing so, it is possible to take into account different balance temperatures depending on the period. The fact

that not only temperature difference, like in the degree-day method, but absolute temperature is known, the analyst can take into account the variation of performance of the systems with outdoor temperature. Knowing these values, the energy consumption can be evaluated using:

$$Q_{bin} = \frac{K_{tot,bin}N_{bin}(T_{bal,bin} - T_{o,bin})^{\bullet}}{\eta_{bin}}\ \text{Whr} \qquad (2.119)$$

$$Q_{total} = \sum Q_{bin} \qquad (2.120)$$

Example 2.15 Using the same data file than the previous example, produce a BIN table using four °C intervals. Separate the bins in two groups: group A corresponds to unoccupied hours, between midnight and 9:00 AM, and group B between 9 and 17 hours corresponding to occupied hours.

Temperature bin	$N_{b,A}$	$N_{b,B}$
-27:-23	4	22
-23:-19	13	117
-19:-15	34	265
-15:-11	28	405
-11:-7	49	434
-7:-3	65	692
-3:1	70	741
1:5	70	892
5:9	85	739
9:13	77	885
13:17	103	955
17:21	82	910
21:25	44	686
25:29	6	239
29:33	0	48
Total	730	8030

(b) Assuming that the loss coefficient and that the heat gains are the same as in the previous example, evaluate the heat loss in each of the bins calculated in the previous example.

Calculation is quite simple using (2.119). The only problem is how to treat the bin [13:17]. Since the middle temperature is 15.5 °C, it is higher than the equilibrium, and a simple approach would be to consider zeros number of heating hours for that bin.

(continued)

Example 2.15 (continued)

Temperature bin	$\Delta T^\bullet$	$Q_{bin,A}$ kWhr	$Q_{bin,B}$ kWhr	
$-27: -23$	39.7	4.9	27.1	
$-23: -19$	35.7	14.4	129.6	
$-19: -15$	31.7	33.5	260.8	
$-15: -11$	27.7	24.1	348.3	
$-11: -7$	23.7	36.1	319.4	
$-7: -3$	19.7	39.8	423.5	
$-3:1$	15.7	34.2	361.6	
$1:5$	11.7	25.5	324.7	
$5:9$	7.7	20.4	177.4	
$9:13$	3.7	8.9	102.7	
$13:17$	0.0	0.0	0.0	
$17:21$	0.0	0.0	0.0	
$21:25$	0.0	0.0	0.0	
$25:29$	0.0	0.0	0.0	
$29:33$	0.0	0.0	0.0	
Total		241.7	2475.0	2716.73

A more precise approach would be to use a smaller bin around T_{bal}, and it will be kept this way for this simple example. Of course, keeping the same heating gains and balance temperatures for all bins gives an energy consumption very similar to the degree-day method. Again the advantage of the bin is to have different possible values for the different bins either due to the occupation schedule or to the effect of the outside temperature on the efficiency of the systems.

Comments *In this simple example, the weekends were not taken into account. Since the same parameters were kept for the occupied and unoccupied bins, it is not important. In the case where it is important, all the occurrences in the weekend days should be put in group A. The simplest way to do this is to multiply the number in group B by 5/7 and add the rest in group A.*

∎

2.6 Hourly Loads

In order to design and size a geo-exchange system, the building loads have to be known. Often, it will be assumed that this part of the analysis is already done. These loads have

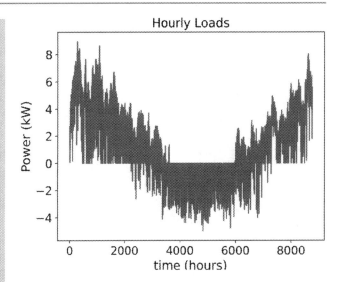

Fig. 2.5 Heating and cooling loads

however to be processed adequately. As a first example, let's assume that a building simulation was done and the building loads on an hourly basis are represented in Fig. 2.5:

As seen in the figure, the heating loads here are assumed to be positive in heating mode and negative in cooling mode. This is an arbitrary choice and, even though this will be the convention that will be followed in this book, it is not always the case, especially in ground-coupled applications. Another way to illustrate and process these data is to separate heating and cooling loads and treat them as positive values (Fig. 2.6):

Regardless of the convention used, it is important that they follow the rules associated with the tools used to analyze the systems. From hourly loads, several important information can be drawn:

(a) From the maximum data values, capacity of the systems can be assessed.
(b) In some applications, aggregated loads can be useful. In ground-coupled applications, monthly loads and block loads on several hours are of interest. This can be easily be generated from hourly loads.

Figure 2.7 shows the mean demand from the same data.

Interesting information can be drawn if we plot the number of occurrences that the power needed is above a certain value. An example of such demand curve is shown in Fig. 2.8.

This information is interesting since it gives a visual idea of how often the full capacity of the load is needed and it may help the designer choose a relevant capacity installation for its system.

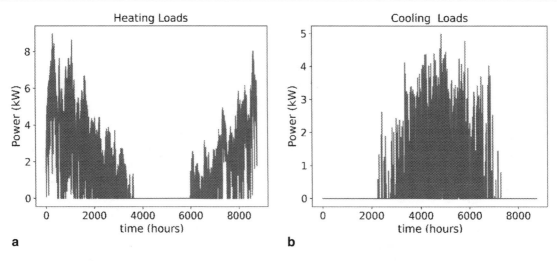

Fig. 2.6 (**a**) Heating loads. (**b**) Cooling loads

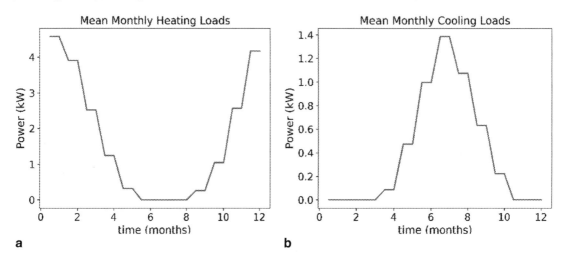

Fig. 2.7 (**a**) Monthly average heating loads. (**b**) Monthly average cooling loads

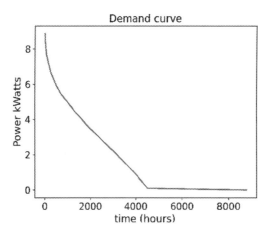

Fig. 2.8 Demand curve

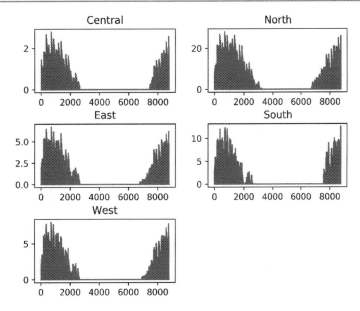

Fig. 2.9 Five zones heating loads

2.7 Multizone Buildings

A five-zone simple building was modeled in a building simulation software, and the hourly loads are shown on Fig. 2.9:

From the tabulated values,[7] it is easy to evaluate the peak values that the heating system must deliver. These are given in the following table:

Central	North	East	South	West	Total
2.8 kW	28.2 kW	6.7 kW	12.5 kW	8.0 kW	58.2 kW

The maximum loads associated with each zone are necessary to choose the heat pumps needed to cover the loads. For the ground heat exchanger and to evaluate the pumping energy, it is often needed to account for the diversity of the loads whether they all peak at the same time of the year or not. Sfeir et al. [34] and Kavanaugh Rafferty [35] define diversified design block loads (DBL) as:

$$DBL_{heating_j} = \sum_{i=1}^{n} q_{heating_{i,j}} \qquad (2.121)$$

with n the number of zones and j the number of hours (8760) [34] or number of hourly blocks (four- or six-hour-block period [35]). In the example shown above, the maximum diversified block load defined as:

$$DBL_{heating_{max}} = \max(DBL_{heating_j}) \qquad (2.122)$$

is given by:

$$DBL_{heating_{max}} = 57 \text{ kW}$$

for one-hour period, which is less than the sum of the five peak loads.

2.8 Chapter Summary

In this chapter, some basic knowledge on heat transfer was given, which will be used throughout this book. Building loads and heat transfer fundamentals were given to help the ground-coupled system engineer to interface correctly with mechanical engineers that may be more familiar with these specific subjects. Once the building loads are known, they often have to be processed to fit with a specific format needed for the future analysis.

Problems

Pr. 2.1 The cross-sectional view of a ground coaxial heat exchanger is shown in Fig. Pr_2.1. A flow rate of 0.4 l/s of a water-glycol solution passes through the annulus and center portion of the heat exchanger.

(a) Using the following properties, evaluate the heat transfer coefficients for the inner region as well as for the annulus section (Assume BC-3).

$$k = 0.5 \text{ W/m-K} \quad , \quad \rho = 1020 \text{ kg/m}^3 \quad ,$$
$$\mu = 0.0025 \text{ Pa-s} \quad , \quad Pr = 20$$

---

[7]From the file: "Multi_heating_loads.txt".

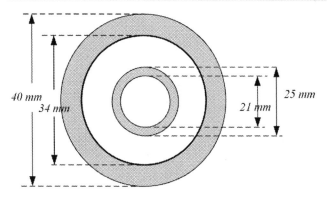

Fig. Pr_2.1 Cross section of coaxial pipe

Answer, $h_{inn} = 934$ W/m²-K ,$h_{ann} = 2121$ W/m²-K (with Eq. (2.58))

(b) The inner pipe is made of plastic ($k = 0.4$ W/m-K), whereas the outer pipe is made of a high-conductivity plastic ($k = 0.8$ W/m-K). Find the unit-length resistance between the inner and annulus region (R'_{12}) as well as the outer resistance (R'_1)/
   Answer, $R'_{12} = 0.092$ m-K/W, $R'_1 = 0.037$ m-K/W

Pr. 2.2 A cross-flow heat exchanger is formed using a 10-m-long coaxial tube. The inner diameter is 6 cm and the outer diameter 10 cm. The inner tube is thin, and the outer tube is well isolated. A hot ($T_{hi} = 40$ °C) mixture of water and glycol passes though the inner portion of the tube with a 0.5 kg/s flow rate, while 1 kg/s of cold water ($T_{ci} = 7$ °C) goes through the inner part. The properties are given by:

$$k_h = 0.50 \text{ W/m-K} \quad , \quad \rho_h = 1020 \text{ kg/m}^3 \quad ,$$

$$\mu_h = 0.0025 \text{ Pa-s} \quad , \quad Pr_h = 20$$

$$k_c = 0.57 \text{ W/m-K} \quad , \quad \rho_c = 1000 \text{ kg/m}^3 \quad ,$$

$$\mu_c = 0.0014 \text{ Pa-s} \quad , \quad Pr_c = 10$$

(a) Find the heat exchanged and the outlet temperatures.
   Answer, $q = 13785$ W,$T_{co} = 10.3$ °C,$T_{ho} = 32.9$ °C
(b) What would be the answer if the hot mixture passes through the annular section?

Answer, $q = 3854$ W,$T_{co} = 7.9$ °C,$T_{ho} = 38.0$ °C

Pr. 2.3 A building has a loss coefficient of $K = 1000$ W/K. During working hours (9 to 5), the set point temperature is 22 °C, and the heat gains are 6000 W. During the week-end and the night, the heat gains are 3000 W, and the setpoint is

18 °C. The dry-bulb temperatures during a typical year are given in the file "Temperature.txt". Using the BIN method with a 3 K bin interval, evaluate the energy needed to heat the building during a typical year. The building is heated by an air-air heat pump having a constant COP of 3, but that is stopped if the outside temperature is below −10 °C (electrical resistance (COP = 1)) is then used.
   Answer, E(thermal) = 65688 kWh, E(electric) = 26755 kWh

Pr. 2.4 The file "loads_building1.txt" gives the hourly heating loads (kW) in the first column and the cooling loads in the second column. Evaluate the monthly energy consumption:

(a) Using the real number of hours for each month.
(b) Assuming 12 months of 730 hours.

Answer

	Heating (a) (kWh)	Cooling (a) (kWh)	Heating (b) (kWh)	Cooling (b)) (kWh)
Jan	3408	0	3398	0
Feb	2629	0	2748	0
Mar	1877	0	1848	0
Apr	896	63	839	71
May	243	352	220	357
Jun	0	718	0	728
Jul	0	1031	0	1038
Aug	0	799	0	769
Sep	190	456	203	455
Oct	781	165	782	165
Nov	1849	0	1890	0
Dec	3101	0	3047	0

Pr. 2.5 A composite wall is made of two wooden parts separated by a small aperture where a vacuum has been made. The wall separated cold outside air ($T_o = -10$ °C) from inside air ($T_i = 20$ °C). Both external facades are subjected to free convection. The outside wall also exchanges heat with the surroundings by radiation ($T_{sky} = -20$ °C, $\epsilon_4 = 0.85$). The radiation on the internal side is neglected. The emissivities of both sides of the aperture are 0.8 ($\epsilon_2 = \epsilon_3 = 0.8$). The other parameters are (Fig. Pr_2.2):

$$H = W = 2 \text{ m} \quad , \quad L_a = L_b = 5 \text{ cm}$$

$$k_a = 0.75 \text{ W/m-K} \quad , \quad k_b = 0.25 \text{ W/m-K}$$

• Estimate the heat loss of the wall.
• Redo the calculation if a low emissivity coating is applied to the aperture surfaces($\epsilon_2 = \epsilon_3 = 0.1$).

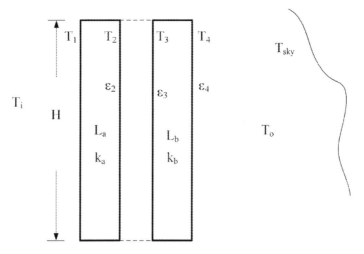

Fig. Pr_2.2 Cross section of composite wall

- The vacuum is broken, and some convection is present inside the aperture with a convection coefficient of $\hbar_i = 1.5$ W/m^2-K ($q''_{conv} = \hbar_i (T_2 - T_3)$). What are the losses with the high emissivity?
- Redo with the low emissivity.

Answer (a) 133 W, (b) 29 W, (c) 148 W (d) 106.6 W

References

1. Spitler, J.D.: American Society of Heating, Refrigerating and Air-Conditioning Engineers (2014). ISBN: 978-1-936504-76-3
2. McQuiston, F.C., Parker, J.D., Spitler, J.D. (2004). Heating, Ventilating, and Air Conditioning: Analysis and Design. Wiley, New York (2004)
3. Bergman, T.L., Incropera, F.P.: Fundamentals of Heat and Mass Transfer, 7th edn. Wiley, New York (2011)
4. Rohsenow, W.M., Hartnett, J.P., Cho, Y.I.: Handbook of Heat Transfer. 3nd edn. McGraw-Hill, New York (1998). ISBN: 007053554X
5. Churchill, S.W., Bernstein, M.: A correlating equation for forced convection from gases and liquids to a circular cylinder in cross-flow. J. Heat Transf. **99**(2), 300–306 (1977)
6. Whitaker, S.: Forced convection heat transfer correlations for flow in pipes, past at plates, single cylinders, single spheres, and for flow in packed beds and tube bundles. AIChE J. **18**(2), 361–371 (1972)
7. Gnielinski, V.: New equations for heat and mass transfer in turbulent pipe and channel flow. Int. Chem. Eng. **16**(2), 359–368 (1976)
8. Petukhov, B.S.: Heat Transfer and Friction in Turbulent Pipe Flow with Variable Physical Properties, vol. 6, pp. 503–564. Elsevier, Amsterdam (1970). https://doi.org/10.1016/S0065-2717(08)70153-9
9. VDI Heat Atlas. VDI-Buch. Springer-Verlag, Berlin Heidelberg (2010). ISBN: 9783540778776
10. Gnielinski, V.: On heat transfer in tubes. Int. J. Heat Mass Transf. **63**, 134–140 (2013). https://doi.org/10.1016/j.ijheatmasstransfer.2013.04.015
11. Dirker, J., Meyer, J.P.: Convective heat transfer coefficients in concentric annuli. Heat Transf. Eng. **26**(2), 38–44 (2005)
12. Kays, W.M., Leung, E.Y.: Heat transfer in annular passages—hydrodynamically developed turbulent flow with arbitrarily prescribed heat flux. Int. J. Heat Mass Transf. **6**(7), 537–557 (1963)
13. Petukhov, B.S., Ginzburg, I.P., Kasperovich, A.S. (eds.): Heat and Mass Transfer, Volume I: Convective Heat Exchange in a Homogeneous Medium, vol. 1842. Heat and Mass Transfer (1967)
14. Gnielinski, V.: Heat transfer coefficients for turbulent flow in concentric annular ducts. Heat Transf. Eng. **30**(6), 431–436 (2009)
15. Gnielinski, V.: Turbulent heat transfer in annular spaces, a new comprehensive correlation. Heat Transf. Eng. **36**(9), 787–789 (2015)
16. Ghobadi, M., Muzychka, Y.S.: A review of heat transfer and pressure drop correlations for laminar flow in curved circular ducts. Heat Transf. Eng. **37**(10), 815–839 (2016). https://doi.org/10.1080/01457632.2015.1089735
17. Naphon, S., Wongwises, P.: A review of flow and heat transfer characteristics in curved tubes. Renewable Sustain. Energy Rev. **10**(5), 463–490 (2006)
18. Vashisth, V., Kumar, S., Nigam, K.D.P.: A review on the potential applications of curved geometries in process industry. Ind. Eng. Chem. Res. **47**(10), 3291–3337 (2008)
19. Srinivasan, P.S.: Pressure drop and heat transfer in coils. Chem. Eng. **218**, CE113–CE119 (1968)
20. Schmidt, E.F.: Wärmeübergang und druckverlust in rohrschlangen. Chemie Ingenieur Technik **39**(13), 781–789 (1967)
21. Truesdell, L.C. Jr., Adler, R.J.: Numerical treatment of fully developed laminar flow in helically coiled tubes. AIChE J. **16**(6), 1010–1015 (1970)
22. Manlapaz, R.L., Churchill, S.W.: Fully developed laminar convection convection from a helical coil. Chem. Eng. Commun. **9**(1–6), 185–200 (1981). https://doi.org/10.1080/00986448108911023
23. Mishra, P., Gupta, S.N.: Momentum transfer in curved pipes. 1. Newtonian fluids. Ind. Eng. Chem. Process Design Dev. **18**(1), 130–137 (1979)
24. Churchill, S.W., Chu, H.H.S.: Correlating equations for laminar and turbulent free convection from a vertical plate. Int. J. Heat Mass Transf. **18**(11), 1323–1329 (1975)
25. Lloyd, J.R., Moran, W.R.: Natural convection adjacent to horizontal surface of various planforms. J. Heat Transf. **96**(4), 443–447 (1974)
26. Radziemska, E., Lewandowski, W.M.: Heat transfer by natural convection from an isothermal downward-facing round plate in unlimited space. Appl. Energy **68**(4), 347–366 (2001)

27. Churchill, S.W., Chu, H.H.S.: Correlating equations for laminar and turbulent free convection from a horizontal cylinder. Int. J. Heat Mass Transf. **18**(9), 1049–1053 (1975)

28. Churchill, S.W.: Comprehensive, theoretically based, correlating equations for free convection from isothermal spheres. Chem. Eng. Commun. **24**(4–6), 339–352 (1983)

29. Catton, I.: Natural convection in enclosures. In: Proceedings of the Sixth International Heat Transfer Conference, vol. 6, pp. 13–31 (1978)

30. MacGregor, R.K., Emery, A.F.: Free convection through vertical plane layers moderate and high Prandtl number fluids. J. Heat Transf. **91**(3), 391–401 (1969)

31. Hollands, K.G.T. et al.: Free convective heat transfer across inclined air layers. J. Heat Transf. **98**(2), 189–193 (1976)

32. Reddy, J.F., Kreider, T.A., Curtiss, P., Rabl, A.: Heating and cooling of buildings: principles and practice of energy efficient design, 3rd edn. Mechanical and Aerospace Engineering. CRC Press. Boca Raton, Taylor & Francis Group (2017). ISBN: 9781315362915

33. American Society of Heating, Refrigerating and Air-Conditioning Engineers. ASHRAE Handbook, Fundamentals. The Society, New York (2017)

34. Sfeir, A. et al.: A methodology to evaluate pumping energy consumption in GCHP systems. ASHRAE Trans. **111**(1) (2005)

35. Kavanaugh, S., Rafferty, K.: Geothermal heating and cooling: design of ground-source heat pump systems. ASHRAE Atlanta (2014)

3.1 Thermodynamic Cycles of Heat Pumps

The two main elements of a ground-source heat pump are the ground and the heat pump itself. Although whole books are devoted solely to the subject of heat pumps, it is worthwhile to introduce several concepts about them. An ideal heat pump is a thermal system that transfer heat from a cold area to a warmer one using the perfect refrigeration cycle. It has four primary processes (Fig. 3.1):

1. 1–2: A cold fluid is compressed isentropically at high pressure and high-temperature vapor.
2. 2–3: The hot vapor condenses as it loses heat to the surroundings.
3. 3–4: The high-pressure liquid expands in an isenthalpic expansion device and becomes a cold mixture.
4. 4–1: The liquid evaporates as it pumps heat from the surroundings.

The cycle can be presented in a thermodynamic diagram, either T-s or p-h diagram (Fig. 3.2):

When a heat pump is used to heat a building, the useful energy is the heat released in the condenser. The performance of the heat pump is then given by the ratio between this energy and the amount of electric energy given at the compressor:[1]

$$COP_{heating} = \frac{q_{out}}{\dot{W}} = \frac{q_{out}}{q_{out} - q_{in}} \qquad (3.1)$$

In cooling mode, the useful heat is the one pumped at the evaporator:

$$COP_{cooling} = \frac{q_{in}}{\dot{W}} = \frac{q_{in}}{q_{out} - q_{in}} \qquad (3.2)$$

---

[1]All quantities, in this chapter are assumed to be positive. In most thermodynamic books, heat released by the system is negative.

In this book, SI units will be used most of the time. However, many manufacturers come from North-America, where the IP (inch-pound) system is often used. In this system, the same dimensionless definition of the COP is used in heating mode, but a dimensional performance index is used for cooling applications, called energy efficiency ration (EER) defined as:

$$EER = \frac{q_{in}(\text{BTU/hr})}{\dot{W}(\text{W})} = 3.415\, COP_{cooling} \qquad (3.3)$$

Sometimes, different ratios are used for large chillers as they define performance indices as the amount of work that has to be given divided by the heat that is taken out at the evaporator. They are defined as:

$$kW/ton = \frac{\dot{W}(kW)}{\dot{q}_{in}(ton)} = \frac{3.52}{COP_{cooling}} \qquad (3.4)$$

$$hp/ton = \frac{\dot{W}(hp)}{\dot{q}_{in}(ton)} = \frac{4.71}{COP_{cooling}} \qquad (3.5)$$

Example 3.1 An ideal heat pump, in cooling mode, extracts heat at 180 kPa and rejects heat at 1 Mpa. Find:

(a) The ideal COP of heat pump.
(b) The flow rate needed to pump 5 kW.

Solution The solution is quite straightforward using CoolProp to find the properties of the four different points of the ideal cycle:

	T	p	h	s	x
	°C	kPa	kJ/kg	kJ/kg-K	
1	−12.71	180	391.02	1.735	1
2	46.20	1000	426.76	1.735	–
3	39.39	1000	255.50		0
4	−12.71	180	255.50		0.348

(continued)

L. Lamarche, *Fundamentals of Geothermal Heat Pump Systems*,
https://doi.org/10.1007/978-3-031-32176-4_3

Fig. 3.1 Heat pump schematic

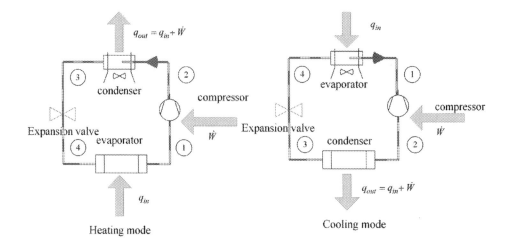

Heating mode   Cooling mode

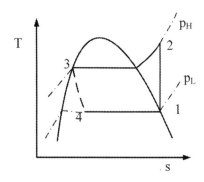

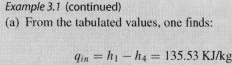

Fig. 3.2 Refrigeration ideal cycle

Fig. 3.3 Carnot cycle

Example 3.1 (continued)

(a) From the tabulated values, one finds:

$$q_{in} = h_1 - h_4 = 135.53 \text{ KJ/kg}$$

$$w_{in} = h_2 - h_1 = 35.74 \text{ KJ/kg}$$

$$COP = \frac{135.53}{35.74} = 3.79$$

(b) Since $\dot{q}_{in} = 5$ kW,

$$\dot{m} = \frac{5}{135.53} = 0.037 \text{ kg/s} \qquad \blacksquare$$

1. 1–2: isentropic compression
2. 2–3: isotherm heat rejection
3. 3–4: isentropic expansion
4. 4–1: isotherm heat input

It is shown in thermodynamics that the COP associated with the Carnot cycle is given by:

$$COP_{heating} = \frac{T_H}{T_H - T_L} = \frac{1}{1 - T_L/T_H} \qquad (3.6)$$

$$COP_{cooling} = \frac{T_L}{T_H - T_L} = \frac{1}{T_H/T_L - 1} \qquad (3.7)$$

where, of course, the temperatures are all in Kelvin. It is easily seen that both COP tend to infinity when the lower temperature tends to the higher temperature.

3.1.1 Carnot Cycle

Although refrigeration systems do not follow the Carnot cyle, this "ideal cycle" is interesting since it gives an upper limit above which no other cycle is possible. The inverse Carnot cycle is defined as four ideal processes (Fig. 3.3):

Example 3.2 A building has an inside temperature of 22°C. Assume that the heat can be released by a Carnot heat pump at the same temperature and that the heat can be pumped ideally at the outside temperature. Find the ideal Carnot COP in the case where:

(continued)

Fig. 3.4 Refrigeration real cycle

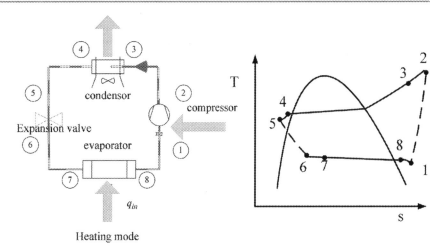

Heating mode

(continued)

Example 3.2 (continued)
(a) The outside air temperature is −10°C (air-air heat pump)
(b) The outside ground temperature is 10°C (geothermal heat pump)

Solution Application of 3.6 gives:

$$COP_{ASHP} = \frac{295.15}{32} = 9.2, \quad COP_{GSHP} = \frac{295.15}{12} = 24.6$$

Of course, real heat pumps will have lower COP, but in theory, since the upper limit of the GSHP is much higher than the ASHP, it should be expected that the real COP will still be higher. Unfortunately, if the conception of the system is not done properly, this might not always be the case. ∎

3.1.2 Real Cycle

In practice a real heat pump follows a cycle that is not ideal. Real characteristics include:

- 1–2 Non-isentropic compression, which can also be non-adiabatic.
- 2–3 Possible pressure and heat losses in the discharge lines.
- 3–4 Condensation with pressure losses and non-saturated liquid at the exit of the condenser.
- 4–5 Possible pressure and heat losses in the lines.
- 5–6 Possible non-adiabatic expansion.
- 6–7 Possible pressure and heat losses in the lines.

- 7–8 Evaporation with pressure losses and non-saturated vapor at the exit of the evaporator.
- 8–1 Possible pressure and heat losses in the aspiration lines (Fig. 3.4).

In order to analyze a real cycle, measurement of the thermodynamic states should be provided, or typical values can be assumed from experience.

Example 3.3 Let's redo Example 3.1 in a real cycle assuming the following data:

- compressor is adiabatic with an isentropic efficiency of 0.7.
- pressure and heat losses in the discharge line are negligible.
- Losses of 0.05 bar in the condenser and the evaporator.
- Pressure of 1 Mpa is at the entrance of the condenser and 180 kPa at the entrance of the evaporator.
- Sub-cooling of 3°C at the exit of the condenser.
- Negligible losses between the condenser and the expansion valve.
- Adiabatic valve.
- Negligible losses between the expansion valve and the evaporator.
- Superheat of 4°C at the exit of the evaporator.
- Gain of 1°C and loss of 0.05 bar in the aspiration line.

Solution From the assumptions made, the real cycle can be described by the following figure:

(continued)

Example 3.3 (continued)

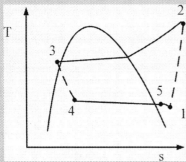

Several known points are available from the assumptions made:

	T	p	h	s	x
	°C	kPa	kJ/kg	kJ/kg-K	
1		170			–
2		1000			–
3		995			–
4	−12.71	180			–
5		175			–

Point 5 can be found knowing the saturation temperature at pressure 175 kPa. From CoolProp, it is found to be: $T_{sat}(175\,\text{kPa}) = -13.41°C$. The temperature at point 5 is then $-13.41 + 4 = -9.41°C$ and $-8.41°C$ at point 1. Knowing the pressure ($175 - 5 = 170\,\text{kPa}$), the enthalpy and entropy at point 1 can be found:

$$h_1 = 394.44 \text{ kJ/kg} \quad , \quad s_1 = 1.754 \text{ kJ/kg K}$$

Enthalpy at point 2, is given by:

$$h_{2s} = h(p=1\text{ Mpa}, s=1.754 \text{ kJ/kg K}) = 432.98 \text{ kJ/kg}$$

$$h_2 = h_1 + \frac{h_{2s} - h_1}{\eta_s} = 449.28 \text{ kJ /kg}$$

$$T_2 = T(p = 1 \text{ Mpa}, h = 449.28 \text{ kJ/kg }) = 340.55 \text{ K}$$

Saturation temperature at 995 kPa being 39.2°C, we have $T_3 = 39.2 - 3 = 36.2°C$. We then have:

$$h_3 = h(p=995\text{kpa}, T = 309.35 \text{ K}) = 250.76 \text{ kJ/kg} = h_4$$

The final Table becomes:

(continued)

	T	p	h	s	x
	°C	kPa	kJ/kg	kJ/kg-K	
1	−8.41	170	394.44	1.754	–
2	67.4	1000	449.29	1.803	–
3	36.20	995	250.76		–
4	−12.71	180	250.76	1.196	0.325
5	−9.41	175	393.95		–

$$COP = \frac{393.95 - 250.76}{449.29 - 394.44} = 2.63$$

$$\dot{m} = \frac{5}{143.19} = 0.035 \text{ kg/s}$$

Comments   *The effect on the COP is easily shown in this example. Heat pumped in the aspiration line could have been added to useful heat if the heat removed would be in the physical volume where the cooling load is needed.*

∎

3.2 Practical Aspects of Heat Pumps

3.2.1 ISO 13256

Although it is important to understand the thermodynamic principles of a heat pump, a GSHP designer will have, in practice, to select real heat pumps available on the market and will decide its choice on the loads of the system and from the performance data given by the manufacturer. Interpretation of these data is of primary importance in practical situations. Most of ground-source heat pumps will exchange heat with the ground using a secondary fluid, which is water or water with antifreeze. They are then considered as water-source heat pumps as opposed to air-source heat pumps. For the building side, the heat that is provided in winter and pumped in summer can be delivered by air (water-to-air heat pumps) or water (water-to-water heat pumps) or both (water for heating and fan-coil for cooling). Practical information useful for system designs are available from manufacturers and are dictated by AHRI/ASHRAE ISO standards: 13256-1 for water-to air heat pumps and 13256-2 for water-to-water heat pump [1,2]. These standards specify nominal conditions under which the heat pumps must be tested (Table 3.1). Performance data at off-nominal conditions should also be provided to correct the performances in real applications. In

Table 3.1 Entering Water Temperature (EWT), ISO-13256

	WLHP	GWHP	GCHP	GCHP-PL
EWT (Cooling)	30°C (86°F)	15°C (59°F)	25°C (77°F)	20°C (68°F)
EWT (Heating)	20 °C (68 °F)	10 °C (50 °F)	0 °C (32 °F)	5°C (41 °F)

Table 3.2 Load side Conditions, ISO-13256

Water-to-Water	Water-to-Air	
Interior water entering temperature	Dry bulb interior air temperature	Wet bulb interior air temperature
12 °C (53.6 °F) (Cooling)	27 °C (80.6 °F) (Cooling)	19 °C (66.2 °F) (Cooling)
40 °C (104 °F) (Heating)	20 °C (68 °F) (Heating)	

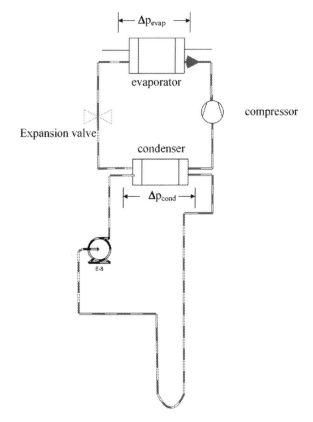

Fig. 3.5 ISO 13256 pressure work

ISO 13256 standard, the total work associated with a specific heat pump must include the work done by the compressor and also the work needed to transport the secondary fluid through the evaporator and condenser heat exchangers.

Looking at Fig. 3.5, the total *corrected work* is defined as:

$$\dot{W}_{corr} = \dot{W}_{comp} + \dot{W}_{evap} + \dot{W}_{cond} + \dot{W}_{control} \qquad (3.8)$$

with:

$$\dot{W}_{evap} = \frac{\dot{V}_{evap}\Delta P_{evap}}{\eta_{cond}} \qquad (3.9)$$

$$\dot{W}_{cond} = \frac{\dot{V}_{cond}\Delta P_{cond}}{\eta_{evap}} \qquad (3.10)$$

which implies that the work needed to transport the fluid outside the heat pump, that is, in the ground loop for GCHP, for example, and the work needed to move the fluid inside the building on the load side are not included. The total capacity given is also corrected to represent only the part given by the

heat pump (see [1]. As brine mixtures are also used instead of water, corrections factors are also provided to take care of the different properties of these fluids. Please note also that many manufacturers give their specifications in IP units, and conversion factors are then applied if needed.

Water-loop heat pumps (WLHP) refer to heat pump where the water (or brine) is connected to a heat supply (boiler or waste heat) in winter and cooling tower in summer. The source temperature range is expected to be between 15°C and 40°C. Groundwater heat pumps (GWHP) are used in open systems where the temperature is normally between 5°C and 25°C. Finally Ground-coupled heat pumps (GCHP) refer to secondary loop systems where part-load conditions (-PL) are also supplied. Their temperature range is between −5°C and 40°C. On the load side, the standard specify the following nominal conditions: ISO-13256 specifications for a typical heat pump are given in Appendix A (Figs. A.1 and A.2). When used in real applications, off-nominal conditions occur, and corrections factors need to be provided. Manufacturers provide either correction factors or look-up table to provide the sufficient information. Figures A.3 and A.4 are examples of correction factors for air flow rate correction. Corrections factors for water-to-air heat pumps include correction factors for (Table 3.2):

- water inlet temperature
- air flow different from nominal values
- air dry bulb and wet-bulb temperature
- water flow rate different from nominal values

Corrections for air temperature different than the one given in the standard are given in Figs. A.5, A.6, A.5, and A.6. Finally, the composition of the brine also plays a major part in the performance of a particular heat pump, and example of correction factors is given in Figs. A.9 and A.10.

One of the most important factors influencing the performance of the heat pump is the temperature of the source brine. The effect of the brine temperature and the flow rates are often given by a look-up table by the manufacturer. Examples for water-to-air heat pump are shown in Figs. A.11, A.12, A.13, A.14, and A.15 and in Figs. A.17, A.18, A.19, and A.20 for water-to-water heat pumps. In the case where such information is not available, it is possible to find the impact of the EWT from ISO-13256 specifications by interpolating the nominal values from the table given (Fig. A.1). This table does not however give correction factors for water flow rate and sensible cooling capacity. Approximation correction factors have been proposed by Kavanaugh and Raferty [3] to estimate those corrections factors. For the impact of the flow rate, they suggested to use:

$F_{tc} = 0.0281x_{fr} + 0.9715$ Total cooling capacity
$F_{hc} = -0.0962x_{fr}^2 + 0.2878x_{fr} + 0.8085$ Heating capacity
$F_{pc} = 0.1102x_{fr}^2 - 0.3157x_{fr} + 1.206$ Power cooling mode
$F_{ph} = -0.0154x_{fr}^2 + 0.067x_{fr} + 0.9483$ Power heating mode
$$(3.11)$$

with:
$$x_{fr} = \frac{\dot{m}}{\dot{m}_{nominal}} \qquad (3.12)$$

Example 3.4 Evaluate the performance (total capacity, power, and COP) of the heat pump NV036 (Appendix A) at full load in the following conditions using only the ISO-13256 specifications:

- Cooling mode
- Entering water temperature of the water is 30°C.
- Water flow rate is 6 gpm.
- Nominal water flow rate is 8 gpm.
- Dry-bulb air temperature 25°C.
- Relative humidity of 40%
- Air flow rate is 1150 cfm.

Solution In imperial units, the initial data are:

$$EWT{=}86°F \;, \; AETdb{=}77°F \;, \; AETwb{=}61.2°F$$

From Fig. A.1, the uncorrected performance data are:

$$TC = 32 \text{ kBTU/hr} \;, \; EER = 18.0 \;, \; \dot{W} = 1.77 \text{ kW}$$

The first correction from the EWT is not needed here since the EWT is nominal for WLHP. The second correction factor is for the air flow rate: that can be interpolated from Fig. A.3 with($x_{cfm} = \frac{1150}{1300} = 69.2\%$):

$$F_{tc_{cfm}} = 0.977 \;, \; F_{sc_{cfm}} = 0.919 \;, \; F_{pow_{cfm}} = 0.978$$

The correction factor for the air temp Fig. A.7:

$$F_{tc_{wb}} = 0.925 \;, \; F_{sc_{db,wb}} = 1.11 \;, \; F_{pow_{wb}} = 0.994 \;,$$

The correction for the water flow rate is not given but can be estimated from 3.11:
$$F_{tc_{gpm}} = F_{sc_{gpm}} = 0.0281 \cdot 0.75 + 0.9715 = 0.992$$

(continued)

$$F_{pow_{gpm}} = 0.1102 \cdot 0.75^2 - 0.3157 \cdot 0.75 + 1.206 = 1.03$$
$$TC_{corr} = 32 \cdot 0.977 \cdot 0.925 \cdot 0.992 = 28.7 \text{ kBTU/hr}$$
$$\dot{W}_{corr} = 1.77 \cdot 0.978 \cdot 0.994 \cdot 1.03 = 1.79 \text{ kW}$$
$$EER_{corr} = 16.1$$

∎

Water-to-water heat pumps also have performance data that depend on the source parameters, as well as the load parameters. Instead of correction factors, manufacturers often present their data in a form of a table look-up involving the four parameters. Examples are shown in Figs. A.17 and A.18.

Example 3.5 Redo the same problem not using the AHRI/ISO-13256 table but the performance data provided by the manufacturer (Fig. A.11)

Solution Since the table is given for several values of the source side, interpolation replaces correction factors for the source EWT source and load flow rate. From the lookup table, interpolate the performance data at the correct gpm, cfm, and EWT:

$$TC = 36.2 \text{ kBTU/hr} \;, \; EER = 16.9 , \; \dot{W} = 2.15 \text{ kW}$$

The correction factors for the air temperature do not change:

$$F_{tc_{wb}} = 0.925 \;, \; F_{pow_{wb}} = 0.994 \;,$$

The final performance are:

$$TC_{corr} = 36.2 \cdot 0.925 = 33.5 \text{ kBTU/hr}$$
$$\dot{W}_{corr} = 2.15 \cdot 0.994 \cdot 0.99 = 2.145 \text{ kW}$$
$$EER_{corr} = 15.6$$

Comments *The results differ from the previous estimations. A possible reason is how the fan and pump power are treated. In the ISO norm, it is clear what should be included, but it may not be the same for the performance data given by the manufacturer. In any case, the qualitative impact of the different parameters are the same, and conservative approaches should be used in case of doubt.*

∎

Whether the performances data are taken from the ISO norm or the manufacturer data, it should be remembered that they do not include the energy needed to move the fluid in the ground loop neither inside the building, and their influence should also be taken into account (see Chap. 7).

3.3 Refrigerants

A very important aspect of geothermal heat pumps is the refrigerant used in the heat pump. Indeed, these systems are very efficient, and incentives should be initiated to increase their use in order to limit our energy footprint as well as our impact on climate change. However, although the reduction in energy consumption can reduce the CO_2 emissions especially in the case where electricity is generated by fossil fuels, the potential leaks of refrigerant may lead to a negative impact on climate change.

Refrigeration compression systems were first introduced in the nineteenth century where the refrigerant fluids used then were ammonia (NH_3), carbon dioxide (CO_2), and sulfur dioxide (SO_2). Around 1930, chlorofluorocarbons (CFC) were introduced. Due to their negative impact on the ozone layer, they were replaced by hydrochlorofluorocarbons ($HCFC$), which have a lower impact on the ozone layer. Both are now replaced since the Montreal protocol [4] and replaced by hydrofluorocarbons (HFC), which have zero impact on the ozone layer. However, these refrigerants do have a large impact on global warming potential (GWP), and the replacement of these refrigerants is also under consideration as part of the Kigali amendment [5]. Hydrofluoroolefins (HFO) are new (fourth-generation) refrigerants that have a very low global warming potential and are considered as possible replacements of HFCs but are slightly flammable. GWP is a relative index which gives the equivalent impact of 1 kg of fluid with respect with 1 kg of CO_2. Table 3.3 shows some values for some common refrigerants:

Besides their impact on climate change, refrigerants are also categorized based on their toxicity and flammability. A letter A is assigned to nontoxic fluids, whereas letter B refers to a toxic refrigerant. A number ranging from 1 (non-flammable), 2 (mildly flammable, burning velocity >10 cm/sec), 2L (lightly flammable, burning velocity <10 cm/sec), to 3 (highly flammable) is also assigned to a refrigerant as shown in Table 3.3.

Refrigerant notation follows the ISO 817 standard and ANSI/ASHRAE 34-2019 [6, 7]. All refrigerants start with a letter R (Refrigerant) followed by two to four numbers: R-WXYZ. Z is the number of fluor atoms, Y is the number of hydrogen +1, X is the number of carbon −1, and finally, W, which is often absent, is the number of carbon double-bonds (HFO). Refrigerants with only two numbers, R50-,

are methane (CH_4), and by-products only have one carbon atom. Serie 600 is associated with hydrocarbons (butane, isobutane, etc.). Special hydrocarbons like methane (R-50), ethane (R-170), and propane (R-290) follow however the general rule. Inorganic components are assigned to series 700 like ammonia (R-717), water (R-718), and carbon dioxide (R-744). Most of the new refrigerants are mixtures, and their properties when they change phases are different than a pure substance. Since the composition of the mixture changes with the phase change, the saturation temperature also changes, which is a different behavior than for pure substances. This behavior is called a temperature glide. These fluids are called zeotropic (or non-azeotropic) mixtures, and the R400 family is dedicated to these refrigerants (R-407, R410, etc.). Azeotropic mixtures are fluids where the temperature glide is small, and their behavior during phase change is similar to pure substances. Series 500 is associated with these mixtures. R410a is an almost azeotropic since its associated glide is very small, but it is still in the zeotropic family [8].

3.3.1 Total Equivalent Warming Potential

As pointed out previously, heat pumps are considered environment-friendly since their energy footprint is much less than the usual heating systems. However, since the refrigerant used may have a potential negative impact on climate change, care should be taken when one tries to evaluate its environmental outcome. Of course, this impact will occur only if there is a potential leak of refrigerant, which unfortunately can happen in practice. A quantitative index has been introduced to evaluate this potential impact, and it is called *Total Equivalent Warming Impact* (TEWI) [9]. It includes the direct impact (DGWP) coming from potential leak of refrigerant and indirect impact (IDGWP), which depends on the electricity used. It is defined as:

$$TEWI = \underbrace{GWP(L \cdot n) + m(1 - \alpha))}_{DGWP} + \underbrace{n \cdot E \cdot \beta}_{IDGWP} \quad (3.13)$$

where

GWP : Global warming potential index
L : Annual quantity of refrigerant leak during a year (kg/y)
n : Life cycle of the equipment (y)
m : Quantity of refrigerant in the heat pump at installation (kg)
α : Recovery rate of refrigerant at the end of its working life (0–1)
E : Annual energy consumption (kWh/y)
β : Carbon factor associated with electricity generation (kg CO_2/kWh)

Table 3.3 ODP and GWP of some refrigerants

Refrigerant number	Chemical name			ODP	GWP	Safety group
12	Dichlorofluoromethane	CFC		1	8100	A1
22	Chlorofluoromethane	HCFC		0.055	1500	A1
32	Difluoromethane	HFC		0.0	675	A2L
134a	1.1.1.2.tetrafluoroethane	HFC		0	1300	A1
1234ze	1.3.3.3-tetrafluoropropene	HFO		0	0	A2L
1234yf	2.3.3.3.tetrafluoropropene	HF0		0	4	A2L
290	Propane	Hydrocarbon		0	≈0	A3
407c	R32/125/134a	HFC		0	1530	A1
410a	R32/125	HFC		0	1730	A1
717	Ammonia	Inorganic		0	0	B2
718	Water	Inorganic		0	0	A1
744	Carbon dioxide	Inorganic		0	1	A1

Table 3.4 Electricity generation carbon index

	Electricity source					
	Coal	Oil	Gaz	Solar PV	Hydro	Nuclear
Carbon Factor (kg CO_2/MWh)	1140	960	580	80	2	2

Table 3.5 Fossil fuel carbon index

Coal	Oil	Natural Gaz
390 kg/MWh	295 kg/MWh	220 kg/MWh

This index gives the total effect of greenhouse gas (GHG) emissions taking one kg of CO_2 as basic unit. Carbon emission factors depend on the energy source to generate electricity. Typical values are given in the following values (Table 3.4):

These values are approximate values since they depend on the type of coal or oil but are in a typical range. Since electrical utilities use different sources of energy, mean carbon factor is often associated with different regions or countries. Since power plants evolve every year, this index is changing over the years. In order to compare the impact of the use of heat pumps, it is important to have an idea on the production of equivalent mass of CO_2 produced by fossil fuel combustion.

Again, these are typical values since the composition of fossil fuels varies. Implication of refrigerant usage on GHG emissions is shown in the following example:

Example 3.6 A residential house needs 25,000 kWh of heating during a typical winter. Compute the TEWI in the following cases:

(a) The house is heated by an oil furnace (80% efficiency)

(b) The house is heated by a gas furnace (80% efficiency)

(continued)

(c) The house is heated by electric baseboards (100% efficiency)

(d) The house is heated by an air-air heat pump (COP = 2.5), which has 4 kg of refrigerant 410a

(e) The house is heated by a SL-GSHP heat pump (COP = 3.5), which has 4 kg of refrigerant 410a

(f) The house is heated by a DX-GSHP heat pump (COP = 3.5), which has 6.5 kg of refrigerant

In order to evaluate the DGWP, assume a 3% leak per year and that the total amount of refrigerant is recuperated at the end of the life cycle. Compute the TEWI considering that electricity is generated by:

1. hydro
2. coal
3. oil
4. gas

Solution Evaluation of the GHG emissions for fossil fuel heating is straightforward. For gas furnace, we have that the total amount of energy needed is:

$$E = \frac{25000}{0.8} = 31250 \text{ kWh}$$

And from Table 3.5, we have:

$$GHG_{gaz\_furnace} = 6875 \text{ kg } CO_2/\text{year}$$

Similarly, for oil furnace:

$$GHG_{oil\_furnace} = 9218.75 \text{ kg } CO_2/\text{year}$$

If the house is equipped with electric baseboards, the heat demand will be 25,000 kWh, and the GHG foot-

(continued)

Example 3.6 (continued)

print will depend on how the electricity is done:

$$GHG_{electric}$$

$$= \begin{cases} 0.002 \cdot 25000 = 50 \text{ kg } CO_2/\text{year} & - hydro \\ 1.14 \cdot 25000 = 28500 \text{ kg } CO_2/\text{year} & - coal \\ 0.96 \cdot 25000 = 24000 \text{ kg } CO_2/\text{year} & - oil \\ 0.58 \cdot 25000 = 15000 \text{ kg } CO_2/\text{year} & - gaz \end{cases}$$

If a heat pump is used, two factors must be taken into account: the direct impact depends on the potential leak. This value is of course unknown, and in this example, we chose an arbitrary value of 3%. For the air-air heat pump and the secondary-loop geothermal heat pump, it is assumed that the refrigerant charge is the same. For the DX system, the charge is larger since the ground loop is filled with refrigerant.

$$DGWP_1 = 4.0 \cdot 0.03 \cdot 1730 = 207.6 \text{ kg } CO_2/\text{year}$$

$$DGWP_2 = 6.5 \cdot 0.03 \cdot 1730 = 337.35 \text{ kg } CO_2/\text{year}$$

The indirect factor will depend on the electricity used, which is assumed to be less for GSHP than for air-air heat pump:

$$E_1 = \frac{25000}{2.5} = 10000 \text{ kWh/year}$$

$$E_2 = \frac{25000}{3.5} = 7142.85 \text{ kWh/year}$$

For electricity generated by coal, the air-air heat pump TEWI will be:

$$DGWP = 207.6, IDGWP = 1.14 \cdot 10000 = 11400$$

$$TEWI = 11607.6 \text{ kg } CO_2/\text{year}$$

The same calculation is done for all other cases, and the final results are summarized in the table:

Electricity	Heating type					
					SL-	DX-
Source	Gaz	Oil	Electric	AA-HP	GSHP	GSHP
Hydro	6875	9218.8	50	227.6	221.9	351.63
Coal	6875	9218.8	28,500	11,607.6	8350.4	8480.2
Oil	6875	9218.8	24,000	9807.6	7064.7	7194.5
Gas	6875	9218.8	14,500	6006.6	4350.4	4480.2

From this simple example, it is easy to see that electric heating without the use of a heat pump is a

(continued)

stupid choice, except in the case where the electricity is produced from renewables. Even so, this choice would not be relevant if electricity could be exchanged easily from jurisdiction based on GHG emissions. For example, imagine that a consumer change its electric baseboards for a gas furnace in a region where the electricity comes from hydro. From a climate change point of view, it may not be a good idea since the potential increase in GHG emissions would be an increase of:

$$6875 - 50 \approx 6825 \text{ kg } CO_2/\text{year}$$

If however, the electricity saved is sold to a region where the electricity is made mostly by gas, the reduction from electricity generation would be:

$$0.58 \cdot 25000 = 14500 \text{ kg } CO_2/\text{year}$$

That is, a net reduction of 7675 kg CO_2/year. The reduction would of course be more important if the electricity comes from coal. If the consumer uses instead a SL-GSHP, the effect would be even better:

$$\text{Increase} \approx 222 - 50 = 172 \text{ kg } CO_2/\text{year}$$

$$\text{Decrease} \approx 0.58 \cdot (25000 - 7142.85) \approx 10357 \text{ kg } CO_2/\text{year}$$

hence a net gain of 10,185 kg CO_2/year. From an environmental point of view, the choice is trivial. Care should be taken however since the direct impact factor is highly dependent on the potential leak assumption. If, for example, DX-GSHP loses entirely its refrigerant content, the DGWP factor for that year would be:

$$6.5 \cdot 1730 = 11245 \text{ kg } CO_2$$

If the consumer uses this technology in a country where the electricity is generated by gas, hoping that it might save approximately 2400 kg CO_2 each year, in just one leak, the benefit of almost 5 years of operation is lost. It is then important to continuously improve heat pump technology using natural refrigerants with lower GWP index. ∎

Problems

Pr. 3.1 A heat pump using R134a has the following cycle:

- Pressure losses are neglected.
- Evaporator pressure is 180 kPa.

Fig. Pr_3.1 Two heat pumps connections

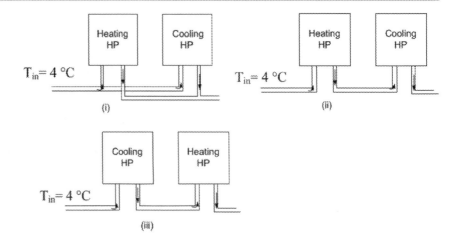

- Condenser pressure is 1 MPa.
- Liquid is saturated at the exit of the condenser.
- There is a 5 K superheat at the exit of the evaporator.
- The isentropic compressor efficiency is 80%.

Using CoolProp, find the COP's of the heat pump.
Answer: COP(heating) = 4.05, COP(cooling) = 3.05

Pr. 3.2 Propylene-glycol (20% volume) at 4°C (ρ = 1024 kg/m³,C_p = 3.94 kJ/kg-K) is used at the source inlet of two heat pumps. One of the heat pumps is in cooling mode and the other one in heating mode. Both capacities (load side) are 30 kW. The flow rate in each heat pump is 1 l/s. The COP is given by:

$$COP_{cl} = 8 - 0.18T_{in} + 0.001T_{in}^2$$
$$COP_{ch} = 3.5 + 0.06T_{in} - 0.0005T_{in}^2$$

where all temperatures are in Celsius. Assume that the fluid properties are constants.

(a) Compare the electric power for the three configurations shown (Fig. Pr_3.1):
(b) What is the exit temperature in each case?

Answer: (a) (i) 12.15 kW, (ii) 11.67 kW, (iii) 11.31 kW
(b) (i) 5.5 °C, (ii) 6.9 °C, (iii) 6.8 °C

Pr. 3.3 Due to internal gains, an internal zone requires cooling, while another one has a heating demand at the same time. It is possible to use external air to have "free cooling," but a better solution is to use a heat pump to recover the heat lost. The simplified situation is shown in Fig. Pr_3.2, where T_2 is the temperature of the internal zone. The rejected heat is used to heat zone 1. If the heat is not sufficient, an auxiliary heater is used; if the heat is too large, the excess heat is rejected in a water loop. If we have:

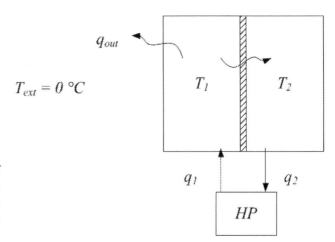

Fig. Pr_3.2 Two zones heat exchange

- $T_{ext} = 0$ °C.
- $q_{ext} = (UA)_1(T_1 - T_{ext})$, with$(UA)_1 = 600$ W/K.
- $COP_{heating} = 3$
- Internal gains (zone 2) = 12 kW.
- $(UA)_{1,2} = 400$ W/K
- $T_1 = T_2 = 22$ °C.

Find the excess heat or the auxiliary heat needed.
Answer: Auxiliary heat = 4.8 kW

Pr. 3.4 In practice, the heat exchange in a heat pump depends on the temperature. From a manufacturer datasheet, the cooling and heating capacity of a typical heat pump are given as:

$$Q_{evap}(kW) = 11.02 - 0.033T_1 - 0.164T_2 + 0.012T_1T_2$$
$$-4.86e - 05T_2^2$$

$$q_{cond}(kW) = 12.38 - 0.04T_1 - 0.136T_2 + 0.012T_1T_2$$

$$-4.43e - 05T_2^2$$

with both temperatures in Celsius. Redo the last problem solving for the two temperatures (T_1, T_2) giving that no auxiliary heat is used.

Answer: $T_1 = 23°C$, $T_2 = 21.4°C$

Pr. 3.5 In real life, the heat pump has to be located somewhere, and the heat must be transferred from the heat pump to the rooms. Knowing that the heat pump used in the last example is a water-to-water heat pump and that the heat is exchanged in a water-to-air heat exchanger in each room, redo the last problem knowing now that the temperatures given in the previous problem are now the water inlet temperature:

$$q_{evap}(kW) = 11.02 - 0.033T_{in,cond} - 0.164T_{in,evap}$$
$$+0.012T_{in,cond}T_{in,evap} - 4.86 \times 10^{-5}T_{in,evap}^2$$

$$q_{cond}(kW) = 12.38 - 0.04T_{in,cond} - 0.136T_{in,evap}$$
$$+0.012T_{in,cond}T_{in,evap} - 4.43 \times 10^{-5}T_{in,evap}^2$$

with the following data:

$$\dot{m}_{water} = 0.5 \text{ kg/s} \quad , \quad Cp_{water} = 4200 \text{ J/kg-K}$$
$$\dot{m}_{air} = 0.72 \text{ kg/s} \quad , \quad Cp_{air} = 1007 \text{ J/kg-K}$$
$$\epsilon = 0.9$$

the same for both zones.

Answer: $T_1 = 22.42°C$, $T_2 = 20.95°C$, $T_{in,cond} = 32.9°C$, $T_{in,evap} = 11.5°C$

Pr. 3.6 Water is used in a supermarket to cool the refrigeration systems. The heat that is extracted can be exhausted outside in cooling towers, but part of it can be recovered to preheat the domestic hot water (DHW). You want to use a CO_2 EcoCute water-to-water heat pump from the Mayekawa company [10] and a heat exchanger, but you are not sure what would be the best configuration. The heat pump delivers hot water at 65°C or 90°C. For the 65°C setting, the performance chart of the heat pump can be modelled by the following expressions:

$$CAP(kW) = 66 - 0.52T_L + 2.74T_S - 0.019T_LT_S$$
$$-0.023T_ST_S = f_1(T_S, T_L)$$

$$COP = 3.78 - 0.042T_L + 0.076T_S - 0.00086T_LT_S$$
$$-0.00018T_ST_S = f_2(T_S, T_L)$$

where T_L, the load inlet temperature, and T_S, the source inlet temperature, are in Celsius. The flow rate of the heated water by the heat pump is adjusted in order to have the prescribed outlet temperature (65°C in this case) according to the giving capacity. You hesitate between the following configurations (Fig. Pr_3.3):

(a) In the first scenario, the hot fluid passes through the heat pump before the heat exchanger, so the heat pump COP will be higher. The cold-side temperature is the same at the entrance of the heat pump and the heat exchanger.

(b) In the second case, the hot fluid passes through the heat exchanger before the heat pump, so the heat exchanger temperature difference will be higher. The cold-side temperature is the same at the entrance of the heat pump and the heat exchanger.

(c) In the third case, the hot fluid passes through the heat exchanger before the heat pump. The total cold stream

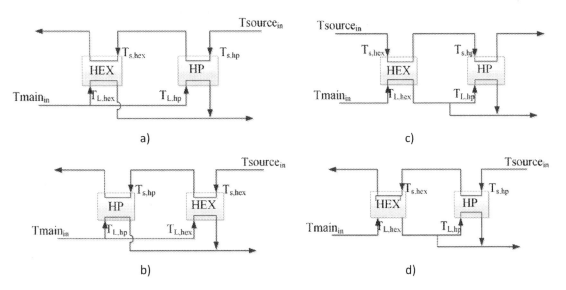

Fig. Pr_3.3 Four scenarios for heat recovery

passes through the cold side of the heat exchanger. In such a case, the flow rate difference can modify the heat exchanger efficiency, but this effect will be neglected here.

(d) In the last case, the hot fluid passes through the heat pump before the heat exchanger, and the total cold stream passes through the cold side of the heat exchanger.

The following conditions are given:

$$T_{source_{in}} = 30\,°C\,, \qquad T_{main_{in}} = 10\,°C$$

$$\dot{V}_{source} = 200\ \text{l/min}\,, \qquad \dot{V}_{load} = 70\ \text{l/min}$$

We assume that the cooling fluid is also water ($C_p = 4.18\,\text{kJ/kg-K}$) and that the heat exchanger efficiency is 0.8.

- What is the configuration giving the largest heat recovery?
- What is the system COP neglecting pump power?

Answer: scenario 3
      case (a): q_{tot} = 142 kW, COP = 6.37
      case (b): q_{tot} = 158 kW, COP = 7.06
      case (c): q_{tot} = 172 kW, COP = 7.13
      case (d): q_{tot} = 154 kW, COP = 6.63

Pr. 3.7 A heat pump using R410A as refrigerant is used in a cooling application where a "desuperheater" is used to preheat domestic water with the superheated vapor at the exit of the compressor. The heat pump is modeled with an almost ideal refrigeration cycle, assuming no pressure losses in the refrigerant circuit, where we have:

- 2 K of subcooling at the end of the condenser.
- 3 K of superheat at the exit of the evaporator.
- The compressor has a 0.6 isentropic efficiency.

The heat pump has a 3 tons cooling capacity (10,541 W), a low pressure of 1 Mpa, and a high pressure of 2 Mpa. The desuperheater is an helicoidal counterflow annulus heat exchanger where the water flows inside the annular region. The geometric features of the heat exchanger are:

- Inner diameter of annulus: 16 mm
- Outer diameter of annulus: 21 mm
- Negligible tube thickness
- 8 turns with a 35 mm pitch
- Coil diameter: 18 cm

The overall heat transfer coefficient (on the water side) between the water and the refrigerant has been measured experimentally and is given by:

$$U_o = 110 Re_w^{0.22} \qquad \text{W/m2-k}$$

where Re_w is the water Reynolds number. The water flows in the annulus region with a 0.2 kg/s mass flow rate and a 20°C inlet temperature. Evaluate the heat gained by the water in these conditions.

Answer: NTU=0.68, ϵ=0.465, q_{surc}=6243 W, $T_{water,out}$ = 27.5°C

References

1. Payne, W.V., Domanski, P.A.: A Comparison of Rating Water-source Heat Pumps Using ARI Standard 320 and ISO Standard 13256-1. Technical report. 2001
2. Kim, J., et al.: Verification study of a GSHP system Manufacturer data based modeling. In: Renewable Energy, vol. 54 (2013), pp. 55–62
3. Kavanaugh, S., Rafferty, K.: Ground-Source Heat Pumps: Design of Geothermal Systems for Commercial and Institutional Buildings. ASHRAE, Atlanta (1997)
4. United Nations: Montreal Protocol. https://ozone.unep.org/treaties/1987-montreal-protocol-substances-deplete-ozone-layer. [Online; accessed 26 July-2022]
5. United Nations: Kigali Amendment. https://treaties.un.org/Pages/ViewDetails.aspx?src=IND&mtdsg_no=XXVII-2-f&chapter=27&clang=_en. [Online; accessed 26 July-2022]
6. Calm, J.M., Hourahan, G.C: Physical, safety, and environmental data for current and alternative refrigerants. In: Proceedings of 23rd International Congress of Refrigeration (ICR2011), Prague, Czech Re-public, August 2011, pp. 21–26 (2011)
7. American Society of Heating, Refrigerating and Air-Conditioning Engineers. ANSI/ASHRAE Standard 34-2019, Designation and Safety Classification of Refrigerants. The Society, New York (2019)
8. J Steven Brown PhD, P.E.: HFOs: new, low global warming potential refrigerants. In: ASHRAE Journal, vol. 51(8), p. 22 (2009)
9. Makhnatch, P., Khodabandeh, R.: The role of environmental metrics (GWP, TEWI, LCCP) in the selection of low GWP refrigerant. In: Energy Procedia, vol. 61 (2014). International Conference on Applied Energy (ICAE2014), pp. 2460–2463. https://doi.org/10.1016/j.egypro.2014.12.023
10. MAYEKAWA, Specification CatalogTechnical Datal. https://www.master.ca/documents/Water_Source_EcoCute_Application_Manual_Ver.1.pdf [Online; accessed 13-October-2019]

4.1 Introduction

As already pointed out, the principle of a ground-source heat pump, in heating mode, is to extract part of the heat from the ground. For a GSHP with a secondary loop, the heat is first transferred from ground to a secondary heat carrier fluid and then from the fluid to the refrigerant. The thermal problem illustrated in Fig. 4.1 seems relatively complex. A fluid exchanges heat with a potentially non-homogeneous soil in a three-dimensional geometry. It is now possible to solve this kind of problems with numerical tools easily affordable, but for design purposes, it is of interest to have simpler analytical tools to quickly have a good order of magnitude of what is needed. For this, the general practice is to separate the thermal problem into three distinct parts:

1. **First, the heat transfer between the ground and the borehole wall (r_b) is analyzed.**
2. Then, how this heat is transferred to the fluid inside the well through the grout is looked over.
3. Finally, we examine how the heat is transported by the fluid to the heat pump.

The first step will be analyzed in this chapter, the last two in the next chapter and the resulting sizing method based on combining the three steps will be given in Chap. 6. The goal of this chapter will then be to analyze the classical methods used to find the borehole temperature (T_b) knowing the heat transfer rate in the ground.

4.2 Classical Models to Calculate the Mean Borehole Temperature

At the periphery of a typical borehole, the temperature can vary with the azimuth angle, but this dependence is small in the ground and is neglected most of time. The thermal problem, for a single borehole, is then modeled by the axisymmetric heat equation:

$$\frac{\partial}{\partial r}\left(k\frac{\partial T}{\partial r}\right) + \frac{k}{r}\frac{\partial T}{\partial r} + \frac{\partial}{\partial z}\left(k\frac{\partial T}{\partial z}\right) = \rho C_p \frac{\partial T}{\partial t} \quad (4.1)$$

$$r_b < r < \infty$$
$$0 < z < \infty$$

with the following boundary and initial conditions:

I.C	$T(r, z, 0) = T_i(r, z) = T_o(z)$	(4.2a)
B.C.1	$T(r, 0, t) = T_s(r, t)$	(4.2b)
B.C.2	$T(\infty, z, t) = T(r, \infty, t) = T_o(z)$	(4.2c)
B.C.3	$-k\frac{\partial T}{\partial r}\big|_{r=r_b} = f(z, t)$	(4.2d)

In practice, the initial temperature distribution is assumed to be independent of the radial location but can vary with the depth. For now, it will be considered uniform to simplify the notation. The last boundary condition will be specified later on. Since in most applications typical boreholes are very slim, the axial effects are often neglected in geothermal applications. After several years, it may however be important [1]. In addition, the variation of the thermal conductivity is most of the time neglected. In that case the last relation can be simplified to:

$$\frac{\partial}{\partial r}\left(\frac{\partial T}{\partial r}\right) + \frac{1}{r}\frac{\partial T}{\partial r} = \frac{1}{\alpha}\frac{\partial T}{\partial t} \quad (4.3)$$

with the boundary conditions:

I.C.	$T(r, 0) = T_i(r) = T_o$	(4.4a)
B.C.1	$T(\infty, t) = T_o$	(4.4b)
B.C.2	$-k\frac{\partial T}{\partial r}\big|_{r=r_b} = f(t)$	(4.4c)

© The Author(s), under exclusive license to Springer Nature Switzerland AG 2023
L. Lamarche, *Fundamentals of Geothermal Heat Pump Systems*,
https://doi.org/10.1007/978-3-031-32176-4_4

Fig. 4.1 Typical Borehole
configuration

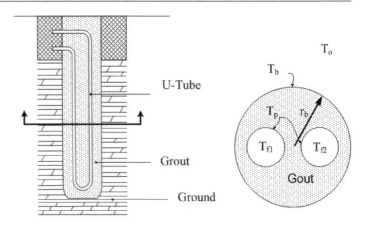

4.2.1   Infinite Cylindrical Source (ICS)

The solution of the last problem requires a boundary condition at the borehole wall. In practice, the heat exchange can be quite complex, but here, a constant, the heat flux is assumed for $t > 0$ and known at $r = r_b$ (heat flux step).

$$2\pi r_b \, k \frac{\partial T}{\partial r}\bigg|_{r=r_b} = q_b'(t) = q_o' \, u(t) = \begin{cases} = 0, \; t < 0 \\ = q_o', \; t \geq 0 \end{cases} \quad (4.5)$$

where, by convention, the heat flux is positive when it is extracted from the ground (heating mode) (see Chap. 2). The solution of (4.3) subjected to a heat pulse, called *thermal step response*, can be found in many good textbooks on heat transfer [2, 3]. It is however given normally in terms of dimensionless variables. Using the following variables:

$$\tilde{r} = r/r_b \quad (4.6)$$

$$\tilde{t} = \alpha t/r_b^2 \equiv Fo \quad (4.7)$$

where the non-dimensional time is given a specific name in heat transfer, called the *Fourier number*. The solution is found to be: [2]:

$$T(\tilde{r}, Fo) - T_o = \frac{-q_o'}{k}$$

$$\times \underbrace{\frac{1}{\pi^2} \int_o^\infty \frac{e^{-\xi^2 Fo} - 1}{\xi^2(J_1^2(\xi) + Y_1^2(\xi))} \left[J_o(\tilde{r}\xi) Y_1(\xi) - J_1(\xi) Y_o(\tilde{r}\xi) \right] d\xi}_{G_r(\tilde{r}, Fo)}$$

$$(4.8)$$

This solution allows us to evaluate the temperature with respect to time and radial position. Usually, it is evaluated at $r = r_b$ where the temperature T_b is the temperature that is of interest:

$$T_b(Fo) - T_o = \frac{-q_o'}{k} \frac{2}{\pi^3} \underbrace{\int_o^\infty \frac{1 - e^{-\xi^2 Fo}}{\xi^3(J_1^2(\xi) + Y_1^2(\xi))} d\xi}_{G(Fo)} \quad (4.9)$$

knowing that:

$$J_o(\xi) Y_1(\xi) - J_1(\xi) Y_o(\xi) = \frac{-2}{\pi \xi} \quad (4.10)$$

This equation is often referred to as the *G-function*, a capital g not to be confused with g, lowercase, a formalism used by Eskilson that will be studied later on. This is a dimensionless factor associated with the thermal resistance of the ground. The evaluation of the last expressions is not easy to perform due to the oscillatory behavior of Bessel's functions. Correlations are often used to evaluate them although modern computers can now easily solve the exact integral. For the *G-function*, evaluated at $\tilde{r} = 1$, the Cooper relation is often used:

$$G(Fo)$$

$$= \begin{cases} \frac{Fo^{1/2}}{2\pi} \Big(C_a + C_o Fo^{1/2} + C_1 Fo + C_2 Fo^{3/2} + \\ \qquad\qquad C_3 Fo^2 + C_4 Fo^{5/2} + C_5 Fo^3 \Big), \\ \qquad Fo \leq 6.124633 \\[2mm] \frac{(2z(8Fo(1 + 2Fo) - 1 - 3z) + 16Fo + \pi^2 + 3)}{128\pi Fo^2}, \\ \qquad Fo > 6.124633 \end{cases}$$

$$(4.11)$$

with:

$C_a = 1.128379, C_o = -0.5, C_1 = 0.2756227,$

$C_2 = -0.1499385, C_3 = 0.0617932, C_4 = -0.01508767,$

$C_5 = 0.001566857$

$$z = \log\left(\frac{4Fo}{\exp(\gamma)}\right)$$

$\gamma = 0.577215665$, Euler Constant

Analog expressions were suggested by Fossa [4] and Bernier [5] for $\tilde{r} \neq 1$.

4.2.2 Infinite Line Source (ILS)

Before looking at how the last equations can be used to solve practical problems, another solution will be presented that is often used to solve almost the same thermal problem. It is called the *infinite line source* (ILS). This solution is given in the classical book of Carslaw and Jaeger [2]. It assumes that the heat, instead of being released at $r = r_b$, is injected at the center of the borehole. The final solution is given by:

$$T(\tilde{r}, Fo) - T_o = \frac{-q'_o}{4\pi k} \int_{\tilde{r}^2/4Fo}^{\infty} \frac{e^{-u}}{u} du = \frac{-q'_o}{4\pi k} E_1(\tilde{r}^2/(4Fo))$$

$$= \frac{-q'_o}{k} \underbrace{\frac{1}{4\pi} E_1(\tilde{r}^2/(4Fo))}_{G_{ils_r}(\tilde{r}, Fo)}$$

(4.12)

where E_1 is called the *exponential integral function*. As before, the case where $\tilde{r} = 1$ is more relevant and is given by:

$$T(\tilde{r}, Fo) - T_o = \frac{-q'_o}{4\pi k} E_1(1/(4Fo)) = \frac{-q'_o}{k} \underbrace{\frac{1}{4\pi} E_1(1/(4Fo))}_{G_{ils}(Fo)}$$

(4.13)

For Fourier numbers larger than 4, the expression can be simplified with:

$$G_{ils}(Fo) \approx \frac{1}{4\pi}(\log(4Fo) - \gamma)$$ (4.14)

where γ is called the *Euler-Mascheroni* constant or sometimes the *Euler constant*, whose value is 0.577215665...It can be seen on Fig. 4.2 that for Fourier numbers larger than 10, the three solutions give very similar results. For small Fourier numbers, (4.14) cannot be used since it gives non-physical negative values.

For low Fourier numbers, the two solutions differ, and the impact of this difference will be discussed later on. For now, it suffices to see that the two solutions are very similar for most time of interest. One advantage of the infinite line source versus the cylindrical source line is that it is a function of a single dependent similarity variable and not two. One can easily use it regardless of the spatial location. Indeed:

$$G_{ils_r}(\tilde{r}, Fo) = \frac{1}{4\pi} E_1(\tilde{r}^2/(4Fo)) = \frac{1}{4\pi} E_1(1/(4Fo^*))$$

$$= G_{ils}(Fo^*)$$

with:

$$Fo^* = Fo/\tilde{r}^2$$

This property would be useful if, for example, only a graphical solution of the function would be available at $\tilde{r} = 1$ and

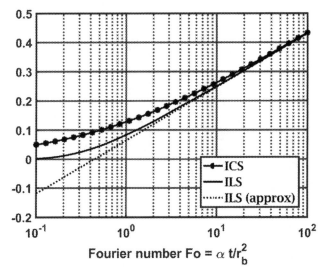

Fig. 4.2 Comparison between ICS and ILS models

solution at other locations would be needed. Nowadays, computer solutions are used, and this is less relevant. From (4.12), the heat flux can be computed from Fourier law:

$$q''(\tilde{r}, Fo) = -k \frac{\partial T}{\partial r} = \frac{-q'_o}{2\pi r_b \tilde{r}} e^{-\tilde{r}^2/(4Fo)}$$

It is seen that the heat rate per unit length at the borehole wall is:

$$q'_b(Fo) = -q'_o e^{-1/(4Fo)}$$ (4.15)

Unlike the infinite cylindrical source, the heat flux is not imposed at the periphery of the borehole, but the difference is less than 5 % for $Fo > 5$ and less than 2 % for $Fo > 10$. It must however be remembered that in a real borehole, the heat is released at the heat pump and to the ground through the heat transfer fluid. It is not imposed at the periphery neither than the center, so in both cases, assumptions are made. There is a fundamental difference between the ICS and ILS: the first one solves a classical partial differential equation with standard boundary conditions, whereas the latter does not satisfy any boundary conditions at the borehole wall; it is rather a solution based on superposition of elementary solutions. It is "mathematically" valid inside the borehole but "physically" valid only for $r \geq r_b$.

The solution of the infinite line source is often given with a different, but equivalent, notation. Kavanaugh and Rafferty [6], Bose et al. [7] and Ingersoll [8] use them with the following notation:

$$T(\tilde{r}, Fo) - T_o = \frac{-q'_o}{4\pi k} E_1(\tilde{r}/(4Fo)) = \frac{-q'_o}{k} \frac{1}{2\pi} I(X)$$

(4.16)

with:

$$X = \frac{r}{2\sqrt{\alpha t}}$$

It can be seen that:

$$I(X) = 2\pi G_{ils}(Fo') \qquad \text{with} \qquad Fo' = \frac{1}{4X^2}$$

Example 4.1 Evaluate the temperature variation from a heat pulse of -40 W/m after 250 days at a distance of 2 meters in a ground where the diffusivity is 0.1 m^2/day and the conductivity is 2.5 W/m-K. The borehole radius is 10 cm. Calculate the temperature variation using:

- the ILS
- the ICS

Solution Using, for example, the last notation:

$$X = \frac{r}{2\sqrt{\alpha t}} = 0.2$$

$$I(0.2) = 1.34, \quad \Rightarrow \Delta T = \frac{40 \cdot 1.34}{2.5 \cdot 2 \cdot \pi} = 3.414°C$$

which is equivalent to:

$$Fo = \frac{\alpha t}{r_b^2} = 25000, \quad \tilde{r} = 200$$

$$G_{ils_r}(25000, 200) = 0.2134, \Rightarrow \Delta T = \frac{40 \cdot 0.2134}{2.5}$$

$$= 3.414°C$$

The ground temperature would then be 3.41 degrees higher than the undisturbed temperature.

As pointed out previously, the ILS does not need two arguments. If a single table was available for $\tilde{r} = 1$, the solution could still be found knowing that:

$$Fo^* = \frac{Fo}{\tilde{r}} = 6.25 = \frac{1}{4X^2}$$

$$G_{ils}(6.25) = 0.2134, \qquad \Rightarrow \Delta T = 3.414°C$$

Actually, for the ILS, the borehole radius is not important and could have been omitted.

(b) For the ICS, the solution is not self-similar, so the solution needs the two arguments. Several regression expressions are available in the literature for several values of $\tilde{r}$ but not to our knowledge for $\tilde{r} = 200$. For this solution, the exact integral (4.8) was solved:

(continued)

$$G_{ics}(250000, 200) = 0.2133, \quad \Rightarrow \Delta T = 3.412°C$$

More significant numbers were kept to illustrate that the solution is different, but in practice, it gives essentially the same values, at least for the given conditions.
∎

4.2.3 *Source Solutions

Before looking at other possible solutions, it is worthwhile to look in deeper details on how the ILS solution was found. It comes from the superposition of the solution for an *instantaneous point source* at a given point in space (ξ, η, ζ) and at $t = 0$. This solution (also known as *Green's function* in unconfined space) [2] is given by:

$$T(x, y, z, t) - T_o = -Q\, F_i(t, x, y, z, \xi, \eta, \zeta)$$

where Q represents the total quantity of heat (in Joules) drawn during the pulse, and

$$F_i(t, x, y, z, \xi, \eta, \zeta) = \frac{1}{8\rho C_p (\pi \alpha t)^{3/2}} e^{-\frac{d^2}{4\alpha t}} \qquad (4.17)$$

will be associated with the *instantaneous point source* solution where

$$d^2 = (x - \xi)^2 + (y - \eta)^2 + (z - \zeta)^2$$

The linearity of the heat equation allows us to generate many different solutions by superposing the last expression in time and space. For example, the thermal response of a *continuous point source* is found by Carslaw and Jaeger [2] letting the point source run continuously from 0 to time t:

$$T(x, y, z, t) - T_o = -\int_0^t q(\tau)\, F_i(t - \tau, x, y, z, \xi, \eta, \zeta)d\tau$$

and $q[W]d\tau = dQ[J]$, the total heat during a period of time $d\tau$. This expression can be simplified with:

$$u = \frac{d}{2\sqrt{\alpha(t - \tau)}}, \quad du = \frac{d\, d\tau}{4\sqrt{\alpha}(t - \tau)^{3/2}}$$

$$T(x, y, z, t) - T_o = -\frac{1}{4\pi k\, d}\frac{2}{\sqrt{\pi}}\int_{d/2\sqrt{\alpha t}}^{\infty} q(\tau)e^{-u^2}du$$

(4.18)

When the heat load is constant, it becomes:

$$T(x, y, z, t) - T_o = -q \, F_c(t, x, y, z, \xi, \eta, \zeta)$$

$$F_c(t, x, y, z, \xi, \eta, \zeta) = \frac{1}{4\pi k \, d} \text{erfc} \left(\frac{d}{2\sqrt{\alpha t}} \right) \quad (4.19)$$

which will be associated with the *continuous point source* solution.

Another possibility is to superpose the point source in space and not in time. The *instantaneous line source* is the superposition of point source along a line parallel to the z-axis passing by the point ξ, η.

$$T(x, y, z, t) - T_o = - \int_{\infty}^{\infty} Q'(\zeta) \, F_i(t, x, y, z, \xi, \eta, \zeta) d\zeta$$

and $Q'[J/m]d\zeta = dQ[J]$, the total heat in a line element. For a uniform distribution of $Q'(\zeta) = Q'_o = $ constant:

$$T(x, y, z, t) - T_o = -\frac{Q'_o}{8\rho C_p(\pi \alpha \, t)^{3/2}} e^{-r^2/(4\alpha t)}$$

$$\times \int_{-\infty}^{\infty} e^{-(z-\zeta)^2/(4\alpha t)} d\zeta$$

with $r^2 = (x - \xi)^2 + (y - \eta)^2$. Using:

$$v = \frac{z - \zeta}{2\sqrt{\alpha t}}, \quad dv = \frac{-d\zeta}{2\sqrt{\alpha t}}$$

one finds:

$$T(x, y, z, t) - T_o = -\frac{Q'_o}{8\pi k \, t} e^{-r^2/(4\alpha t)} \left[\frac{2}{\sqrt{\pi}} \int_{-\infty}^{\infty} e^{-v^2} dv \right]$$

The last bracket is just 2erfc (0) and finally:

$$T(x, y, z, t) - T_o = -Q'_o \, F_l(t, x, y, \xi, \eta) \quad (4.20)$$

with:

$$F_l(t, x, y, \xi, \eta) = \frac{1}{4\pi k \, t} e^{-r^2/(4\alpha t)} \quad (4.21)$$

which will be associated with the *instantaneous infinite line source* solution.

The infinite line source solution (ILS), which is nothing else than a *continuous infinite line source*, is then easily obtained in two different ways: either by superposing the continuous point source (Eq. (4.19)) distributed along an axis parallel to the z-axis:

$$T(x, y, t) - T_o = - \int_{\infty}^{\infty} q'(\zeta) F_c(t, x, y, z, \xi, \eta, \zeta) d\zeta$$

or superposing in time the instantaneous infinite line source (Eq. (4.21)).

$$T(x, y, t) - T_o = - \int_0^t q'(\zeta) F_l(t - \tau, x, y, \xi, \eta) d\tau$$

It is found that it is easier with the latter approach.

$$T(x, y, t) - T_o = -\frac{1}{4\pi k} \int_0^t \frac{q'(\tau) e^{-r^2/(4\alpha(t-\tau))}}{t - \tau} d\tau \quad (4.22)$$

In the case where the heat transfer is constant, $q'(\tau) = q'_o$, and with:

$$u = \frac{r^2}{4\alpha(t - \tau)}, \quad \frac{du}{u} = \frac{d\tau}{t - \tau}$$

$$T(r, t) - T_o = -\frac{q'_o}{4\pi k} \int_{r^2/4\alpha t}^{\infty} \frac{e^{-u}}{u} du \quad (4.23)$$

$$= -\frac{q'_o}{4\pi k} E_1(r^2/4\alpha t) = -\frac{q'_o}{4\pi k} E_1(\tilde{r}^2/4Fo)$$

4.2.4 Finite Line Source (FLS)

The ILS and ICS solutions assumed that the heat flow is purely radial. When real boreholes are analyzed, the solution is assumed valid for all axial locations. In practice, due to the large aspect ratio, it is a good approximation for times that can go up to several years. However, in some cases, when the time of interest is long or when the boreholes are short, or both, these axial effects can become important [1]. A solution that is often used to take into account the axial effect is called the *finite line source* (FLS). It consists of using again the superposition of point sources, as described in the previous section but on a finite length of the borehole. Thus, the following solution:

$$T(r, z, t) - T_o = \int_0^t$$

$$\int_0^H \frac{-q'(\tau)}{8\rho C_p(\pi\alpha(t - \tau))^{3/2}} e^{-(r^2+(z-\zeta)^2)/(4\alpha(t-\tau))} d\zeta d\tau$$

being a superposition of the point source will satisfy the basic axisymmetric differential equation (4.1) and the boundary conditions ((4.2a) and (4.2b)). However, when axial variations are taking into account, the boundary condition at the air-ground boundary (4.2c) has also to be satisfied. In practice, the temperature on the ground surface will vary with time over a typical year. However, since the major part of vertical boreholes is in a region where the temperature is not affected by the air temperature variations, it can be assumed that the surface temperature remains equal to the undisturbed ground temperature T_o (or $T - T_o = 0$). A classical approach that can easily force the temperature variation to stay at 0 is the *method of images*. This approach is often applied when solutions of point sources or, like here, line sources, which are just a superposition of point sources, are looked for. In order to satisfy the homogeneous Dirichlet boundary condition at the surface, it suffices to add an image of the source with the same amplitude but with opposite sign in the fictitious domain $z > 0$. Moreover, the borehole will not necessarily start at $z = 0$. If the initial depth is called z_o, the final solution satisfying the surface boundary condition will be:

$$T(r, z, t) - T_o = \int_0^t \int_{z_o}^{H+z_o} \frac{-q'(\zeta, \tau)}{8\rho C_p (\pi \alpha (t - \tau))^{3/2}}$$

$$\left(e^{-(r^2+(z-\zeta)^2)/(4\alpha(t-\tau))} - e^{-(r^2+(z+\zeta)^2)/(4\alpha(t-\tau))} \right) d\zeta \, d\tau$$
(4.24)

When the heat flux is constant, the step response can be found by integrating with respect to time:

$$T(r, z, t) - T_o$$
(4.25)
$$= \int_{z_o}^{H+z_o} \frac{-q_o'(\zeta)}{4\pi k} \left(\frac{\text{erfc}\,(\frac{d^+}{\sqrt{4\alpha t}})}{d^+} - \frac{\text{erfc}\,(\frac{d^-}{\sqrt{4\alpha t}})}{d^-} \right) d\zeta$$

with $d^+ = \sqrt{r^2 + (z - \zeta)^2}, d^- = \sqrt{r^2 + (z + \zeta)^2}$. As before, the solution is often evaluated at $r = r_b$. It becomes:

$$T_b(z, t) - T_o$$
(4.26)
$$= \int_{z_o}^{H+z_o} \frac{-q_o'(\zeta)}{4\pi k} \left(\frac{\text{erfc}\,(\frac{d_b^+}{\sqrt{4\alpha t}})}{d_b^+} - \frac{\text{erfc}\,(\frac{d_b^-}{\sqrt{4\alpha t}})}{d_b^-} \right) d\zeta$$

As usual, it is better to work with dimensionless variables. Using the original reference variables ((4.6) and (4.7)), the final solution is:

$$T_b(\tilde{z}, Fo) - T_o$$
(4.27)
$$= \int_{\tilde{z}_o}^{\tilde{H}+\tilde{z}_o} \frac{-q_o'(\zeta)}{4\pi k} \left(\frac{\text{erfc}\,(\frac{\tilde{d}_b^+}{\sqrt{4Fo}})}{\tilde{d}_b^+} - \frac{\text{erfc}\,(\frac{\tilde{d}_b^-}{\sqrt{4Fo}})}{\tilde{d}_b^-} \right) d\zeta$$

with $\tilde{d}_b^+ = \sqrt{1 + (\tilde{z} - \zeta)^2}, \tilde{d}_b^- = \sqrt{1 + (\tilde{z} + \zeta)^2}$. This relation can be used to evaluate a new response factor (G-function) based on the FLS model. However, the last expression depends now on the axial variable z. The classical

approach for sizing boreholes (see Chap. 6) requires a reference *borehole temperature*. Since the last equation gives a temperature profile, Zeng et al. [9] suggested to use the temperature at the middle of the borehole as the reference temperature. They also assume that the heat flux along the borehole was uniform. These assumptions will then give:

$$T_{b_m}(Fo) - T_o$$
(4.28)
$$= -\frac{q_o'}{k} \frac{1}{4\pi} \underbrace{\int_{\tilde{z}_o}^{\tilde{H}+\tilde{z}_o} \left(\frac{\text{erfc}\,(\frac{\tilde{d}_{b,m}^+}{\sqrt{4Fo}})}{\tilde{d}_{b,m}^+} - \frac{\text{erfc}\,(\frac{\tilde{d}_{b,m}^-}{\sqrt{4Fo}})}{\tilde{d}_{b,m}^-} \right) d\zeta}_{G_{FLS,m}}$$

with $\tilde{d}_{b,m}^+ = \sqrt{1 + (\tilde{z}_o + \tilde{H}/2 - \zeta)^2}, \tilde{d}_{b,m}^- = \sqrt{1 + (\tilde{z}_o + \tilde{H}/2 + \zeta)^2}$.

4.2.4.1 Eskilson's Formalism

In the FLS context, two reference lengths are available: the borehole radius and the borehole height. In their pioneering work, Eskilson and Claesson [10] evaluated several response factors associated with different borehole arrangements (bore fields). They chose the borehole height as their reference length. Moreover, for a reason that will be discussed later on, they also chose a different characteristic time as reference:[1]

$$\bar{t} = \frac{t}{t_c} \qquad \text{where} \qquad t_c = \frac{H^2}{9\alpha} \qquad (4.29)$$

They called their response factors: *g-functions*. In the context of a single borehole, this function plays the same role as the other models previously discussed. The FLS model previously described can be formulated using that formalism. With:

$$\bar{t} = (3\bar{r}_b)^2 Fo \qquad \text{where} \qquad \bar{r}_b = \frac{r_b}{H}, \qquad \bar{z} = \frac{z}{H} \qquad (4.30)$$

The last relation will then be:

$$T_b(\bar{t}, \bar{z}) - T_o$$
$$= \int_{\bar{z}_o}^{1+\bar{z}_o} \frac{-q_o'(\zeta)}{4\pi k} \left(\frac{\text{erfc}\,(\frac{3\bar{d}_b^+}{\sqrt{4\bar{t}}})}{\bar{d}_b^+} - \frac{\text{erfc}\,(\frac{3\bar{d}_b^-}{\sqrt{4\bar{t}}})}{\bar{d}_b^-} \right) d\zeta$$
(4.31)

with:

$$\bar{d}_b^+ = \sqrt{\bar{r}_b^2 + (\bar{z} - \zeta)^2}, \bar{d}_b^- = \sqrt{\bar{r}_b^2 + (\bar{z} + \zeta)^2} \qquad (4.32)$$

---

[1] The overbar variables are dimensionless variables associated with Eskilson's characteristic variables.

$$T_{b,m}(\bar{t}) - T_o = T_b(\bar{z} + 0.5, \bar{t}) - T_o = \int_{\bar{z}_o}^{1+\bar{z}_o} \frac{-q'_o(\xi)}{4\pi k}$$

$$\times \left(\frac{\text{erfc}\left(\frac{3\bar{d}^+_{b,m}}{\sqrt{4\bar{t}}}\right)}{\bar{d}^+_{b,m}} - \frac{\text{erfc}\left(\frac{3\bar{d}^-_{b,m}}{\sqrt{4\bar{t}}}\right)}{\bar{d}^-_{b,m}} \right) d\xi \quad (4.33)$$

with $\quad \bar{d}^+_{b,m} = \sqrt{\bar{r}_b^2 + (\bar{z}_o + 0.5 - \zeta)^2}, \bar{d}^-_{b,m} = \sqrt{\bar{r}_b^2 + (\bar{z}_o + 0.5 + \zeta)^2}$. Again, if the heat flux is assumed uniform along the borehole, the last expression can be simplified to:

$$T_{b,m}(\bar{t}) - T_o = \frac{-q'_o}{2\pi k} \, {}_1g_m(\bar{t}, \bar{r}_b, \bar{z}_o)$$

with:

$${}_1g_m(\bar{t}, \bar{r}_b, \bar{z}_o) = \frac{1}{2} \int_{\bar{z}_o}^{1+\bar{z}_o} \left(\frac{\text{erfc}\left(\frac{3\bar{d}^+_{b,m}}{\sqrt{4\bar{t}}}\right)}{\bar{d}^+_{b,m}} - \frac{\text{erfc}\left(\frac{3\bar{d}^-_{b,m}}{\sqrt{4\bar{t}}}\right)}{\bar{d}^-_{b,m}} \right) d\zeta$$

$$(4.34)$$

The function $\,{}_1g_m(\bar{t}, \bar{r}_b, \bar{z}_o)$ will be called *g-function* for a single borehole based on the middle temperature. This formalism is, of course, equivalent to the one previously adopted (4.29) where:

$$G_{FLS,m}(Fo, \tilde{H}_b, \tilde{z}_o) = \frac{1}{2\pi} \, {}_1g_m(\bar{t}, \bar{r}_b, \bar{z}_o)\big|_{\bar{t}=(3\bar{r})^2 Fo} \quad (4.35)$$

A comparison between the FLS mode and the two other models (ICS and ILS) is given in the Fig. 4.3:

It can be remarked that, for small Fourier numbers, the difference between the ICS, the ILS, and the new FLS is the same as previously discussed. The ILS and the FLS look identical, which confirm that axial effects are negligible. Just to give us an idea, for a typical soil, $\alpha \approx 0.004$ m^2/hr, $r_b \approx 7.5$ cm, the maximum Fourier number shown on the figure (10,000) corresponds approximately to 560 days, which is small. To see the importance of the axial effects, we must talk normally in years. The same comparison for several years is shown in Fig. 4.4:

It is seen that from moderate Fourier numbers, the two lines ICS and ILS collapse, but the line corresponding to the FLS reaches a steady state due to the end effects neglected in the other models. On the same figure, the upper axis represents the dimensionless time chosen by Eskilson, and the reason for this particular choice is easily seen. It is at $\bar{t} \approx 1$ that the transition from the unsteady to the steady-state regime is reached, and this is the reason why they chose this value. This characteristic time is the time around which the steady regime occurs.

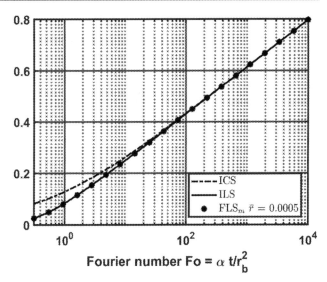

Fig. 4.3 Comparison between ICS, ILS and FLS Models, Low Fourier numbers

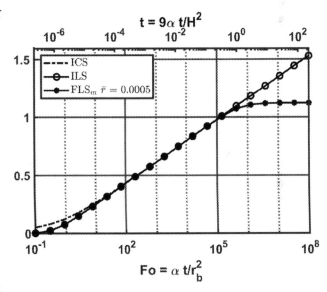

Fig. 4.4 Comparison between ICS, ILS and FLS Models, Large Fourier numbers

4.2.5 FLS Based on the Mean Temperature

The different models already discussed can bring some confusions if care is not taken. Part of it is due to the notation used (*G-function, g-function*), the different characteristic times used, as well as a factor 2π, which appears in some models and not in others. Although it would be interesting to standardize these notations, it is important to sort them out if we want to properly understand the literature. So as not to be mistaken, the context is important. If the relation suggested by Zeng et al. (4.29) is compared to the original *g-function* suggested by Eskilson, for a single borehole, small changes will be noticed. The reason for that is that the solution given by Eskilson did not come from an analytical model but from a numerical solution where the borehole temperature was assumed to be uniform at the borehole wall. Zeng et al. verified that, while keeping the uniform flux hypothesis, the difference between Eskilson and the FLS would be smaller, at least for a single borehole, if the mean borehole temperature instead of the temperature at the middle of the borehole is used as reference temperature. Using the relation given in the last section and with the change of variable $\zeta = \zeta - \bar{z}_o, \eta = \eta - \bar{z}_o$, we find:

$$T_b(\bar{t}) - T_o = \frac{-q'_o}{2\pi k} \, _1g(\bar{t}, \bar{r}_b, \bar{z}_o) \tag{4.36}$$

with:

$$_1g(\bar{t}, \bar{r}_b, \bar{z}_o) = \frac{1}{2} \int_{\bar{z}_o}^{1+\bar{z}_o} d\eta \int_{\bar{z}_o}^{1+\bar{z}_o}$$

$$\times \left(\frac{\text{erfc}(\gamma \bar{d}_b^+)}{\bar{d}_b^+} - \frac{\text{erfc}(\gamma \bar{d}_b^-)}{\bar{d}_b^-} \right) d\zeta \tag{4.37}$$

where:

$$\gamma = \frac{3}{2\sqrt{\bar{t}}} \tag{4.38}$$

$$\bar{d}_b^+ = \sqrt{\bar{r}_b^2 + (\eta - \zeta)^2}, \bar{d}_b^- = \sqrt{\bar{r}_b^2 + (2\bar{z}_o + \eta + \zeta)^2} \tag{4.39}$$

Zeng found that the complexity of the double integral made its use less practical. Lamarche and Beauchamp [11] modified the last expression for the case where $\bar{z}_o = 0$, in order to simplify it to a single integral. The generalization for the case where $\bar{z}_o \neq 0$ was done by Costes and Peysson [12]. The final expression for this case is given by:

$$_1g(\bar{t}, \bar{r}_b, \bar{z}_o) = \int_{\bar{r}_b}^{\sqrt{\bar{r}_b^2+1}} \frac{\text{erfc}(\gamma \zeta)}{\sqrt{\zeta^2 - \bar{r}_b^2}} d\zeta - D_A$$

$$-(1 + \bar{z}_o) \int_{\sqrt{\bar{r}_b^2+(1+2\bar{z})^2}}^{\sqrt{\bar{r}_b^2+4(1+\bar{z})^2}} \frac{\text{erfc}(\gamma \zeta)}{\sqrt{\zeta^2 - \bar{r}_b^2}} d\zeta$$

$$+\bar{z}_o \int_{\sqrt{\bar{r}_b^2+4\bar{z}_o^2}}^{\sqrt{\bar{r}_b^2+(1+2\bar{z}_o)^2}} \frac{\text{erfc}(\gamma \zeta)}{\sqrt{\zeta^2 - \bar{r}_b^2}} d\zeta - D_B \tag{4.40}$$

where:

$$D_A = \sqrt{\bar{r}_b^2 + 1} \, \text{erfc}(\gamma \sqrt{\bar{r}_b^2 + 1}) - \bar{r}_b \, \text{erfc}(\gamma \bar{r}_b)$$

$$-\frac{e^{-\gamma^2(\bar{r}_b^2+1)} - e^{-\gamma^2 \bar{r}_b^2}}{\gamma \sqrt{\pi}} \tag{4.41}$$

$$D_B = \sqrt{\bar{r}_b^2 + (1 + 2\bar{z}_o)^2} \, \text{erfc}(\gamma \sqrt{\bar{r}_b^2 + (1 + 2\bar{z}_o)^2})$$

$$-0.5 \left(\sqrt{\bar{r}_b^2 + 4\bar{z}_o^2} \, \text{erfc}(\gamma \sqrt{\bar{r}_b^2 + 4\bar{z}_o^2}) \right.$$

$$\left. +\sqrt{\bar{r}_b^2 + 4(1 + \bar{z}_o)^2} \, \text{erfc}(\gamma \sqrt{\bar{r}_b^2 + 4(1 + \bar{z}_o)^2}) \right)$$

$$-\frac{e^{-\gamma^2(\bar{r}_b^2+(1+2\bar{z}_o)^2)} - 0.5 \left(e^{-\gamma^2(\bar{r}_b^2+4\bar{z}_o^2)} + e^{-\gamma^2(\bar{r}_b^2+4(1+\bar{z}_o)^2)} \right)}{\gamma \sqrt{\pi}} \tag{4.42}$$

With a different approach, Claesson and Javed [13] proposed later an alternative simpler expression:

$$_1g(\bar{t}, \bar{r}_b, \bar{z}_o) = \int_{\gamma}^{\infty} e^{-\bar{r}_b^2 \zeta^2}$$

$$\frac{2\,\text{ierf}(\zeta) + 2\,\text{ierf}((1 + 2\bar{z}_o)\zeta) - \text{ierf}((2 + 2\bar{z}_o)\zeta) - \text{ierf}(2\bar{z}_o\zeta)}{2\zeta^2} d\zeta \tag{4.43}$$

where:

$$\text{ierf}(x) = \int_0^x \text{erf} x = x \, \text{erf}(x) - \frac{1}{\sqrt{\pi}} \left(1 - \exp(-x^2)\right) \tag{4.44}$$

Although this expression is associated with the work of Claesson and Javed, the same result was given earlier by Bandos et al. [14] for the case where $\bar{z}_o = 0$. $_1g$ will be called *g-function* for a single borehole based on the mean temperature, the only one that will be used from now on. Those last two expressions are the solution of the same mathematical problem and give the same value. A comparison between the different g-functions for a single borehole is shown in Fig. 4.5 for $\bar{r}_b = 0.0005, \bar{z}_o = 0.04$, the original parameters chosen by Eskilson:

Since the FLS solution based on the mean temperature is more precise, it will be the only one that we will used when the FLS solution is used unless it is explicitly specified. The FLS model is often used in the literature to generate the *g-functions* of Eskilson and is then often used with the dimensionless variables used by him as it was done in this section. However, nothing prevents to use the former formalism also:

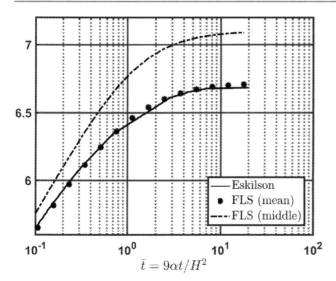

Fig. 4.5 Comparison between FLS Model based on the Mean and Middle Temperatures

$$G_{FLS}(Fo, \tilde{H}_b, \bar{z}_o) = \frac{1}{2\pi} \left. {_1g}(\bar{t}, \bar{r}_b, \bar{z}_o)\right|_{\bar{t}=(3\bar{r})^2 Fo} \qquad (4.45)$$

In such a case, the default values of Eskilson ($\bar{r} = 0.0005$, $\bar{z}_o = 0.04$) would become:

$$\tilde{H}_b = \frac{H}{r_b} = \frac{1}{\bar{r}} = 2000, \quad \bar{z}_o = \frac{z_o}{r_b} = \frac{z_o}{H}\frac{H}{r_b} = 80$$

4.2.6 *Finite Line Source, Revisited

As engineers, the application of proven formulas is often sufficient to do the job. Sometimes, it might be interesting to gain insight on how they were derived, especially if the same method can be used in other applications. In this section, the FLS model as presented by Bandos et al. [14] and Claesson [13] will be briefly discussed for the case where $\bar{z}_o = 0$. The logical step presented in the previous section was to integrate first in time to find the elementary solution of a continuous source and then integrate in space. The clever trick used by Bandos and Claesson was to integrate in space first. Let's rewrite Eq. (4.24) for the case of constant heat flux:

$$T(r, z, t) - T_o = \frac{-q'_o}{8\rho C_p(\pi\alpha)^{3/2}} \int_0^t \int_0^H$$

$$\frac{1}{(t-\tau)^{3/2}} \left(e^{-(r^2+(z-\zeta)^2)/(4\alpha(t-\tau))} \right.$$

$$\left. - e^{-(r^2+(z+\zeta)^2)/(4\alpha(t-\tau))} \right) d\zeta d\tau$$

In the previous section, the time integration was performed, and afterward, the mean value along the borehole was evaluated. Let's do the average first:

$$T_b(t) - T_o = \frac{-q'_o}{8\rho C_p(\pi\alpha)^{3/2}H} \int_0^t \frac{e^{-r_b^2/(4\alpha(t-\tau))}d\tau}{(t-\tau)^{3/2}} \int_0^H d\eta$$

$$\int_0^H \left(e^{-(\eta-\zeta)^2/(4\alpha(t-\tau))} - e^{-(\eta+\zeta)^2/(4\alpha(t-\tau))} \right) d\zeta$$

Using the following change of variables:

$$s = \frac{1}{\sqrt{4\alpha(t-\tau)}}, \quad u = s\eta, \quad v = s\zeta$$

$$T_b(t) - T_o = \frac{-q'_o}{4\pi k_s H} \frac{2}{\sqrt{\pi}} \int_{1/\sqrt{4\alpha t}}^\infty \frac{e^{-r_b^2 s^2}}{s^2} ds \int_0^{Hs} du \int_0^{Hs}$$

$$\left(e^{(u-v)^2} - e^{-(u+v)^2} \right) dv$$

The rest is trivial. For the inner integral, the first term is solved using $w = u - v$ and the second with $w = u + v$, leading to:

$$T_b(t) - T_o = \frac{-q'_o}{4\pi k_s H} \int_{1/\sqrt{4\alpha t}}^\infty \frac{e^{-r_b^2 s^2}}{s^2} ds \int_0^{Hs} 2\,\mathrm{erf}\,u$$

$$-\mathrm{erf}\,(u+Hs) - \mathrm{erf}\,(u-Hs)du$$

And direct integration:

$$T_b(t) - T_o = \frac{-q'_o}{4\pi k_s H} \int_{1/\sqrt{4\alpha t}}^\infty \frac{e^{-r_b^2 s^2}}{s^2} (4\,\mathrm{ierf}\,(Hs) - \mathrm{ierf}\,(2Hs))ds$$

Again, dimensionless variables are preferable. Using Eskilson's reference variables, the final result is:

$$T_b(\bar{t}) - T_o = \frac{-q'_o}{2\pi k_s} \int_{3/2\sqrt{\bar{t}}}^\infty \frac{e^{-\bar{r}_b^2\zeta^2}}{2\zeta^2} (4\,\mathrm{ierf}\,(\zeta) - \mathrm{ierf}\,(2\zeta))d\zeta$$

The same mathematical approach was often used in borehole analysis [15–17].

4.2.7 Models Summary

In the previous sections, classical models for the heat transfer in the ground produced by a single geothermal borehole were described. These are not the only one but the most frequently

used. All these models can be expressed with respect to
different time scales. In the case where the Fourier number
is based on the borehole radius and the dimensionless spatial
variables are normalized with the radius, the notation capital-
G will be used:

$$G_{ICS}(Fo, \tilde{r}), G_{ILS}(Fo, \tilde{r}), G_{FLS}(Fo, \tilde{H}, \tilde{z}_o)$$

$$\Delta T(Fo, \tilde{r}) = \frac{-q'}{k} G_{...}$$

Since the infinite cylindrical source solution evaluated at the
borehole radius was the first solution associated with the
letter capital-G, the notation $G(Fo)$ alone will be used for
the ICS solution:

$$G(Fo) = G_{ICS}(Fo, \tilde{r} = 1)$$

The long notation will still be used in some cases for clarity,
especially when comparisons are made between solutions
coming from different models. Lowercase-g formalism will
be used with the dimensionless time using Eskilson's characteristic time. Also, the height of the borehole is chosen as
the reference length for spatial variables. In practice, the only
analytical model that will be used for this formalism is the
FLS model.

$$\Delta T(\bar{t}, \bar{r}) = \frac{-q'}{2\pi k} g_{FLS}(\bar{t}, \bar{r}, \bar{z}_o)$$

4.3   Temporal Superposition

All models described previously assumed that the heat flow
is constant in time (*step response*). In practice, of course the
load profile varies over time. One can easily switch from one
to the other using the *Duhamel theorem* [3, 18]. This theorem
tells us that if $f(r, z, t)$ is the solution of a problem where
$q'(t) = u(t)$ is constant with respect to time, the general
solution when the excitation varies with time can be found
by the following expression:

$$\Delta T(r, z, t) = q'(0) f(x, z, t) + \int_0^t f(x, z, t-\tau) \frac{d\,q'(\tau)}{d\,\tau} d\tau \tag{4.46}$$

To solve, the analytical representation of the heat flow variation with time is needed. In practice, the analysis is simplified
by assuming the constant heat flux by small time intervals
(series of pulses). It is easy then to verify that the last relation
will become (Fig. 4.6):

$$\Delta T_b(t) = q'(0) f(t) + (q'(t_1) - q'(0)) f(t - t_1)$$
$$+ (q'(t_2) - q'(t_1)) f(t - t_2) + \ldots \tag{4.47}$$

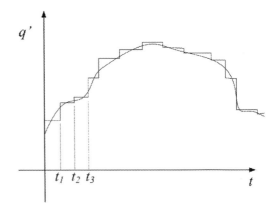

Fig. 4.6   Pulse representation of time-dependent heat flow

The same result can be interpreted as simply the *temporal
superposition* of the effect of several thermal pulses given the
linearity of the basic equation. The application of Duhamel's
theorem could be interesting if one wishes, for example, to
replace the approximation of piecewise constant heat distribution (see, e.g., [18]). In the previous expression, $f(t)$ can
be any response factor described in the previous section. So
if, for example, the infinite cylindrical source is used, the
result would be:

$$\Delta T_b(t) = -\frac{1}{k}(q'(0)G(Fo) + (q'(Fo_1) - q'(0))G(Fo - Fo_1)$$
$$+ (q'(Fo_2) - q'(Fo_1))G(Fo - Fo_2) + \ldots) \tag{4.48}$$

Example 4.2 In a 100-meter-long borehole, 1000
Watts of heat is extracted from the ground at $t = 0$,
1800 Watts at $t = 1$ hr, and 200 Watts is injected at $t = 3$ hr. The borehole radius is 7.5 cm. The undisturbed
ground temperature is 10°C. The soil conductivity is
2 W/m K, and its diffusivity $\alpha = 0.1$ m^2/day. What is
the borehole temperature after 5 hours if we use:

(a) the ICS
(b) the ILS
(c) the FLS

Solution   Fourier number based on the radius is:

$$Fo_1 = \frac{\alpha t}{r_b^2} = \frac{0.1 \cdot 1}{24 \cdot 0.075^2} = 0.7407, Fo_2 = 2.22, Fo = 3.704$$

$$G_{ICS}(3.704) = 0.198 \quad, \quad G_{ICS}(3.704 - 0.7407) = 0.1849$$

$$G_{ICS}(3.704 - 2.22) = 0.1469$$

(continued)

Example 4.2 (continued)

$$\Delta T_b(t) = -\frac{1}{2}(10 \cdot 0.198 + 8 \cdot 0.1849 - 20 \cdot 0.1469) = -0.26$$

$$T_{b,a} = 9.74°C$$

$$G_{ILS}(3.704) = 0.1739, G_{ILS}(3.704 - 0.7407) = 0.1574$$

$$G_{ILS}(3.704 - 2.22) = 0.1085$$

$$\Delta T_b(t) = -\frac{1}{2}(10 \cdot 0.1739 + 8 \cdot 0.1574 - 20 \cdot 0.1085) = -0.41$$

$$T_{b,b} = 9.59°C$$

For the FLS, the solution depends on two other parameters:

$$\tilde{H} = \frac{H}{r_b} = 1333.33$$

and $\tilde{z}_o = z_o/r_b$. Since this value is not given, the default value suggested by Eskilson $\bar{z}_o = z_o/H = 0.04$ is used. In practice, for small-time scale, this parameter does not influence the answer.

$$G_{FLS}(3.704) = 0.1737, G_{FLS}(3.704 - 0.7407) = 0.1573$$

$$G_{FLS}(3.704 - 2.22) = 0.1085$$

$$\Delta T_b(t) = -\frac{1}{2}(10 \cdot 0.1737 + 8 \cdot 0.1573 - 20 \cdot 0.1085)$$

$$= -0.41$$

$$T_{b,b} = 9.59°C$$

Since the time range is very small, there is a noticeable difference between the ICS and ILS but almost none between the ILS and the FLS.

As previously noted, the FLS solution is often used in the context of Eskilson formalism. In that case, the dimensionless time is:

$$\bar{t}_1 = \frac{9\alpha t}{H^2} = \frac{9 \cdot 0.1 \cdot 1}{24 \cdot 100^2} = 3.75e\text{-}6, \bar{t}_2$$

$$= 1.125e\text{-}5, \bar{t}_f = 1.875e\text{-}5$$

The $_1g - function$ evaluated for $\bar{r}_b = 0.00075$ and $\bar{z}_o = 0.04$ will give:

$$_1g(1.875e\text{-}5) = 1.091, \; _1g(1.875e\text{-}5 - 3.75e\text{-}6) = 0.9877$$

$$_1g(1.875e\text{-}5 - 1.125e\text{-}5) = 0.6813$$

(continued)

$$\Delta T_b(t) = -\frac{1}{2 \cdot 2\pi}(10 \cdot 1.091 + 8 \cdot 0.9877$$

$$-20 \cdot 0.6813) = -0.41$$

$$T_{b,c} = 9.59°C$$

which is of course the same thing.

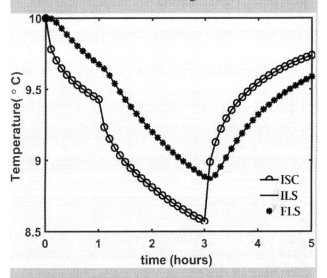

Comments *Redoing the same example replacing hours by days, the difference will be very small for all models.* ∎

4.4 Spatial Superposition

For now, the thermal response was found assuming a single borehole. Most of the time, there are more than one borehole. Of course, if the boreholes are far apart and if, at the entrance of the wells, the conditions are the same, we can consider that N-boreholes give N-times a single one. However if they are relatively close from each other, there may be interference effects that have to be taken into account. Due to the linearity of the thermal process, it is possible to evaluate the effect of several boreholes by summing the effect of all the ones taken separately.

Example 4.3 Three boreholes are distributed on a straight line as seen on the figure. The heat transfer per unit length associated with the boreholes are: $q_1' = 10$ W/m for borehole 1, $q_2' = -8$ W/m for borehole 2 and $q_3' = 12$ W/m for borehole 3.

(continued)

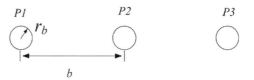

Example 4.3 (continued)

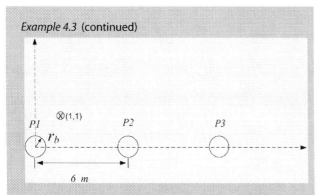

If the soil conductivity is 2.1 W/m-K and the volumetric heat capacity 2.2 MJ/m^3-K, find the temperature variation at point (1,1) after 10 days using the ILS.

Solution

$$\alpha_s = \frac{k_s}{\rho_s C_{p,s}} = \frac{2.1}{2.2e6} = 9.45\text{e-}7 \text{ m}^2/\text{s} = 0.0343 \text{ m}^2/\text{hr}$$

$$X_1 = \frac{d_1}{2\sqrt{\alpha_s t}} = \frac{1.41}{2\sqrt{0.00343 \cdot 240}} = 0.7786$$

$$I_1 = 0.224$$

$$\Delta T_1 = -0.1703$$

$$X_2 = \frac{d_2}{2\sqrt{\alpha_s t}} = \frac{5.1}{2\sqrt{0.00343 \cdot 240}} = 2.807$$

$$I_2 = 2.14\text{e-}5$$

$$\Delta T_2 = 1.3\text{e-}5$$

$$X_3 = \frac{d_3}{2\sqrt{\alpha_s t}} = \frac{11.04}{2\sqrt{0.00343 \cdot 240}} = 6.081$$

$$I_3 = 1.14\text{e-}18$$

$$\Delta T_3 = -1.05\text{e-}18$$

$$\Delta T = -0.170017$$

Comments   *The Ingersoll formalism was used in the example, but the $G_{ils}(Fo)$ would of course lead to the same result. It is noted that for the small time given (10 days), the effect of far boreholes are very small.*

■

Fig. 4.7 Superposition principle

quite logical for a single borehole, the presence of several boreholes makes the assumption less obvious. Indeed, it is noted on the figure that the effect of P2 on P1 will be larger on the right portion of the borehole wall than on the left one. Fortunately, in practice, the distance r_b is much smaller than the distance b, and the influence of P2 on P1 can be evaluated calculating only the temperature at the distance b regardless of the angular variation without a large error (Fig. 4.7).

Thus, the average temperature of the borehole one will be given by:

$$T_{b,1} - T_o = \Delta T_{1\rightarrow 1} + \Delta T_{2\rightarrow 1} + \Delta T_{3\rightarrow 1} \qquad (4.49)$$

Thus, the average temperature of the borehole one will be given by:

$$T_{b,1} - T_o = \frac{-1}{k_s}\left(q_1' G(Fo, 1) + q_2' G(Fo, \tilde{b}) + q_3' G(Fo, 2\tilde{b})\right) \qquad (4.50)$$

where as usual, G represents any models previously discussed. It can be seen in this example a symmetry between borehole 1 and 3. However, the borehole 2 will behave differently. For the overall response of geothermal field, we must make certain assumptions about the different behavior of these boreholes. Two simple approaches are possible: it can be assumed that the heat flux is the same in all boreholes or that the borehole temperature is the same for all boreholes. Although the latter seems more likely, the choice is more complex because it implies that the heat flow from the well 2 will be different than the others, and this difference will also change with time. For simplicity, the first assumption will be used for now. In this case, the temperature will vary between wells. Here because of the symmetry, boreholes 1 and 3 have the same temperature but not 2. The step response of the geothermal field is found by taking the average of the temperature:

$$\Delta \bar{T}_b = \frac{1}{3}(2\Delta T_{b,1} + \Delta T_{b,2})$$

The same principle of superposition can be used to evaluate the borehole mean temperature. Looking at a similar configuration than in the previous example, the temperature of the borehole 1, T_{b1}, will be influenced by the boreholes 1, 2, and 3. Whereas the axisymmetric approximation seemed

4.5   Eskilson's g-functions

As it was already pointed out, the concept of *g-functions* was initiated by the work of Eskilson and Claesson [10]. These functions were numerically calculated and put in a tabulated form into a family of response factors that were graphically available. Examples of typical plots that were included in Eskilson's work are shown in the following figures. On these figures, the values of $\bar{r}_b = 0.0005$ and $\bar{z}_o = 0.04$ are the default values proposed by Eskilson in his work. One of the problem with tabulated values is that it is difficult to evaluate the values for parameters that are different than the one used for the numerical evaluation. To evaluate the thermal response for different values of $\bar{r}_b$, Eskilson showed that the following formula can be used:

$$g(\bar{t}, \bar{r}_b^*, \bar{z}_o) = g(\bar{t}, \bar{r}_b, \bar{z}_o) - \log\left(\frac{r_b^*}{r_b}\right) \qquad (4.51)$$

However they do not suggest any correction factors for the variation of $\bar{z}_o$. It is possible to use the concept of spatial superposition to generate approximate expressions for *g-functions*. The idea was proposed by Lamarche and Beauchamp [11]. In order to take into account the axial effects, only the FLS model based on the mean temperature will be discussed in this section. Referring again to the 3 boreholes inline configuration previously discussed, its associated *g-function* would be:

$$\bar{T}_b - T_o = \frac{-q_o'}{2\pi k} g(\bar{t}, \bar{r}_b, \bar{z}_o)$$

$$g(\bar{t}, \bar{r}_b, \bar{z}_o) = \frac{1}{3}\left(3 \,_1g(\bar{t}, \bar{r}_b, \bar{z}_o) + 4 \,_1g(\bar{t}, \bar{b}, \bar{z}_o) + 2 \,_1g(\bar{t}, 2\bar{b}, \bar{z}_o)\right)$$

In that example, as well as in many examples, it was possible to take account of the symmetry to evaluate the thermal responses of complex geothermal fields, but in the most general case, it can still be put in the following form:

$$g(\bar{t}, \bar{r}_b, \bar{z}_o) = \frac{1}{N_b} \sum_{i=1}^{N_b} \left(\,_1g(\bar{t}, \bar{r}_b, \bar{z}_o) + \sum_{\substack{j=1 \\ j \neq i}}^{N_b} \,_1g(\bar{t}, \bar{d}_{ij}, \bar{z}_o) \right) \qquad (4.52)$$

or:

$$g(\bar{t}, \bar{r}_b, \bar{z}_o) = \,_1g(\bar{t}, \bar{r}_b, \bar{z}_o)$$

$$+ \frac{1}{N_b} \sum_{i=1}^{N_b} \left(\sum_{\substack{j=1 \\ j \neq i}}^{N_b} \,_1g(\bar{t}, \bar{d}_{ij}, \bar{z}_o) \right) \qquad (4.53)$$

with:

$$\bar{d}_{ij} = \sqrt{(\bar{x}_i - \bar{x}_j)^2 + (\bar{y}_i - \bar{y}_j)^2} \qquad (4.54)$$

This last relation implies that all boreholes are vertical and identical. Examples of *g-functions* based on FLS with identical heat flow in all boreholes are given in the following figures:

For small fields, the difference with the functions given by Eskilson is small. This difference is accentuated for larger geothermal fields. This difference comes from the fact that the numerical solutions of Eskilson make the assumption that the temperature of all the boreholes is the same and uniform along its length, whereas here, the heat flux is the same and uniform along the boreholes. Cimmino and Bernier [15, 19] proposed three types of conditions that we can imposed for complex fields:

1. Type I boundary condition: The heat flux is uniform along the borehole and identical for all boreholes (Solution given by (4.53))
2. Type II boundary condition: The heat flux is uniform along the borehole, and the mean temperature is the same for all boreholes
3. Type III boundary condition: The temperature is uniform and the same for all boreholes

The case discussed here corresponds to TYPE I. The others need a special mathematical treatment that will be discussed in Chap. 13.

Example 4.4 Find the mean borehole temperature variation in a 2 by 2 bore field if we have the following load conditions:

$$k_s = 2\,\text{W/m-K},\ q' = 3\ \text{W/m},\ r_b = 7.5\ \text{cm},\ z_o = 3.6\ \text{m}$$

$$H = 90\ \text{m}\ ,\ B = 6\ \text{m}\ ,\ \alpha_s = 0.1\text{m}^2/\text{s}\ ,\ t = 20\ \text{years}$$

Solution The analytic expression (4.53) can be used to find the solution. Due to the symmetry, the temperature increment can be calculated for only one borehole. We find:

$$t_c = \frac{90^2}{9 \cdot 0.1} = 9000\ \text{days}\ ,\quad \bar{t} = \frac{20 \cdot 365}{9000} = 0.81$$

$$\bar{r} = \frac{0.0075}{90} = 0.00083\ ,\quad \bar{z}_o = \frac{3.6}{90} = 0.04\ ,$$

$$\bar{b} = \frac{6}{90} = 0.067$$

$$g_{2x2} = \,_1g(\bar{t}, \bar{r}_b, \bar{z}_o) + 2 \,_1g(\bar{t}, \bar{b}, \bar{z}_o) + \,_1g(\bar{t}, \sqrt{2}\bar{b}, \bar{z}_o)$$

$$g_{2x2} = 5.879 + 2 \cdot 1.56 + 1.246 = 10.23$$

$$\Delta T = \frac{-3}{2\pi 2} 10.23 = -2.44°\text{C}$$

(continued)

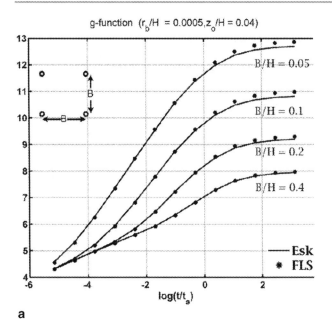

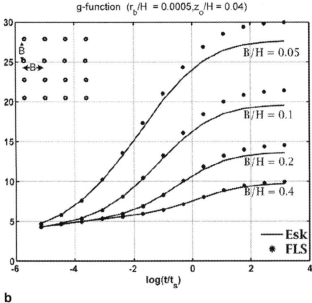

Fig. 4.8 (**a**)2×2 field g-function. (**b**) 4×4 field g-function

---

Example 4.4 (continued)

Assume now that we do not have access to an analytic expression but only to tabulated Eskilson's g-functions from a given library. Since these functions are given for the same $\bar{z}_o$, this parameter will not influence the result, and we will omit the parameter for clarity. However, the values of $\bar{b}$ and $\bar{r}$ are different. It is possible to find the answer from (4.51) and interpolation. For $\bar{b} = 0.05$, and $\log(0.81) = -0.21$, the tabulated value from Fig. 4.8 is approximately:

$$g(0.81, \bar{r} = 0.0005) \approx 11.54 \text{ for } \bar{b} = 0.05$$

$$g(0.81, \bar{r} = 0.0005) \approx 9.67 \text{ for } \bar{b} = 0.1$$

Using the correction for the radius:

$$g(0.81, 0.00083) = 11.54 - \log(0.00083/0.0005)$$

$$= 11.03 \text{ for } \bar{b} = 0.05$$

$$g(0.81, 0.00083) = 9.67 - \log(0.00083/0.0005)$$

$$= 9.16 \text{ for } \bar{b} = 0.1$$

Interpolating gives for $\bar{b} = 0.067$:

$$g(0.81, 0.00083)|_{\bar{b}=0.067} \approx 9.67 + \frac{11.03 - 9.16}{0.1 - 0.05}$$

$$(0.067 - 0.05) \approx 10.39$$

$$\Delta T = \frac{-3}{2\pi 2} 10.39 = -2.48°C \qquad \blacksquare$$

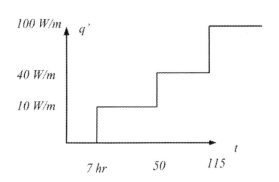

Fig. Pr_4.1 Three Pulses Heat exchange

4.6   Chapter Summary

Several analytical models to evaluate the borehole temperature were presented for vertical boreholes. Most of them are associated with a single borehole, whereas the *g-function* is associated with borehole arrangements (fields). Although not discussed, the concept of *g-function* can also be applied to inclined wells [20–22]. For now, no statements were given on the relative accuracy of those models, but some discussion can be found in the literature [23]. These models will be used in the next chapters to analyze GSHP systems.

Problems

Pr. 4.1 A geothermal borehole exchange heat given by the following pulses (Fig. Pr_4.1):

Fig. Pr_4.2 Four Pulses Heat exchange

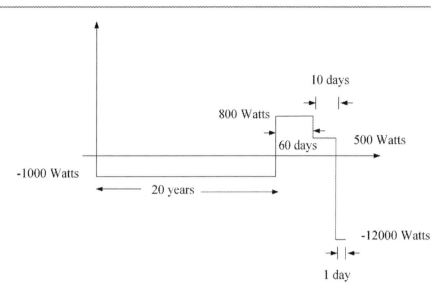

The diffusivity of the ground is 0.088 m²/day, and its conductivity is 2.25 W/m-K.

(a) What is the temperature variation at a distance of 145 mm of the origin after 150 hours using the approximated expression of the ILS (4.14)?
(b) What would be the same temperature variation if the first two pulses were aggregated?

Answer: (a) −10.87°C, *(b)* −10.66°C

Pr. 4.2 Two long vertical boreholes are separated by 3 m. At $t = 0$, 50 W/m of heat is injected in the ground in each borehole. Two temperature sensors are installed: one at the borehole radius (8 cm) of one borehole and another one at a distance of 620 mm from the same borehole on a line joining the centers of the two boreholes. After 48 hours, an increase of 6.9 K is measured at the borehole radius and 1 K at the sensor situated at 620 mm. Estimate what is the conductivity and the diffusivity of the soil using the infinite line source, thus assuming that the depth is not important.
Answer $k = 2.5$ W/m-K, $\alpha = 44.6$ cm²/hr

Pr. 4.3 Using the infinite cylindrical source, evaluate the borehole temperature at the end of the four pulses shown in Fig. Pr_4.2:

$$k_{soil} = 1.9 \text{ W/m-K}, \alpha_{soil} = 0.1 \text{ m}^2/\text{day}, T_o = 10\,°C, L = 200 \text{ m}$$

Answer, $T_b = 22.53\,°C$

(b) Redo the same thing if the second and third pulses are aggregated.
Answer, $T_b = 22.66\,°C$

Pr. *4.4 Show that the ILS can be evaluated by:

$$G_{ILS}(Fo) = \frac{1}{2\pi} \int_0^{\pi/2} \frac{\text{erfc}\left(1/(2\sqrt{Fo}\cos(x))\right)}{\cos(x)} dx$$

Pr. *4.5 Using the Duhamel theorem (Eq. (4.46)), show that the borehole temperature variation for a ramp input ($f(t) = tu(t)$) for the infinite line source is given by:

$$T(\tilde{r}, Fo) - T_o - = \frac{-1}{k}\left[Fo\left(G_{ils}(Fo) - \frac{1}{4\pi}E_2(1/(4Fo))\right)\right]$$

with:

$$E_2(x) = x \int_x^\infty \frac{e^{-t}}{t^2} dt$$

References

1. Marcotte, D., et al.: The importance of axial effects for borehole design of geothermal heat-pump systems. Renew. Energy **35**(4), 763–770 (2010). ISSN: 0960-1481
2. Carslaw, H.S., Jaeger, J.C.: Conduction of Heat in Solids, 2nd edn. (1959)
3. Hahn, D.W., Ozisik, M.N.: Heat Conduction, 3rd edn. Wiley, New York (2012)
4. Fossa, M.: Correct design of vertical borehole heat exchanger systems through the improvement of the ASHRAE method. Science and Technology for the Built Environment **23**(7), 1080–1089 (2017). https://doi.org/10.1080/23744731.2016.1208537
5. Bernier, M.: Ground-coupled heat pump system simulation. ASHRAE Trans. **106**(1), 605–616 (2001)
6. Kavanaugh, S., Rafferty, K.: Ground-Source Heat Pumps: Design of Geothermal Systems for Commercial and Institutional Buildings. ASHRAE, Atlanta (1997)
7. Bose, J.E., Parker, J.D., McQuiston, F.C.: Design data manual for closed-loop ground-coupled heat pump systems. In: Atlanta, Ga.: American Society of Heating, Refrigerating and Air-Conditioning Engineers (1985). ISBN: 0910110417
8. Ingersoll, L.R., Plass, H.J.: Theory of the ground pipe source for the heat pump. ASHVE Transactions **54**, 339–348 (1948)

9. Zeng, H.Y., Diao, N.R., Fang, Z.H.: A finite Line-Source Model for Boreholes in geothermal heat exchangers. Heat Transfer Asian Res. **31**(7), 558–567 (2002)

10. Eskilson, P.: Thermal analysis of heat extraction systems. PhD thesis. Lund University, Sweden (1987)

11. Lamarche, L., Beauchamp, B.: A new contribution to the finite line-source model for geothermal boreholes. Energ. Buildings **39**(2), 188–198 (2007)

12. Costes, V., Peysson, P.: Projet d'Initiation à la Recherche et au Développement-Capteurs géothermiques verticaux enterrés: validation expérimentale de nouveaux modèles déeveloppés dans l'environnement TRNSYS. In: Ecole Polytechnique, Montréal, Canada (2008)

13. Claesson, J., Javed, S.: An analytical method to calculate borehole fluid temperatures for time-scales from minutes to decades. ASHRAE Trans. **117**, 279–288 (2011)

14. Bandos, T.V., et al.: Finite line-source model for borehole heat exchangers: effect of vertical temperature variations. Geothermics **38**(2), 263–270 (2009)

15. Cimmino, M., Bernier, M., Adams, F.: A contribution towards the determination of g-functions using the finite line source. Appl. Therm. Eng. **51**(1–2), 401–412 (2013)

16. Lamarche, L.: g-function generation using a piecewise-inear profile applied to ground heat exchangers. Int. J. Heat Mass Transf. **115**, 354–360 (2017)

17. Lazzarotto, A.: A network-based methodology for the simulation of borehole heat storage systems. Renew. Energy **62**, 265–275 (2014)

18. Lamarche, L., Pasquier, P.: Higher order temporal scheme for ground heat exchanger analytical models. Geothermics **78**, 111–117 (2019). ISSN: 0375-6505. https://doi.org/10.1016/j.geothermics.2018.12.004

19. Cimmino, M., Bernier, M.: A semi-analytical method to generate g-functions for geothermal bore fields. Int. J. Heat Mass Transf. **70**(0), 641–650 (2014)

20. Lamarche, L.: Analytical g-function for inclined boreholes in ground-source heat pump systems. Geothermics **40**(4), 241–249 (2011)

21. Lazzarotto, A.: A methodology for the calculation of response functions for geothermal fields with arbitrarily oriented boreholes Part 1. Renew. Energy **86**, 1380–1393 (2016). https://doi.org/10.1016/j.renene.2015.09.056

22. Lazzarotto, A., Bjork, F.: A methodology for the calculation of response functions for geothermal fields with arbitrarily oriented boreholes, Part 2. Renew. Energy **86**, 1353–1361 (2016). https://doi.org/10.1016/j.renene.2015.09.057

23. Lamarche, L.: Short-time analysis of vertical boreholes, new analytic solutions and choice of equivalent radius. Int. J. Heat Mass Transf. **91**(12), 800–807 (2015). https://doi.org/10.1016/j.ijheatmasstransfer.2006.09.007

5.1 Borehole Resistance

As pointed out at the beginning of the last chapter, the borehole analysis is separated in three steps:

1. First the ground heat exchange is analyzed giving the mean borehole temperature T_b.
2. **From that, the average fluid temperature T_f is calculated**
3. **Lastly, the outlet fluid temperature is evaluated**

In the previous chapter, different analytical models were described to solve the first problem. In this chapter, the fluid temperature will be evaluated using the concept of *borehole resistance*.

In practice, the borehole capacitance is small and it is often neglected in borehole analysis. In that case, the borehole is modeled by a thermal resistance network. The case of a single U-tube is shown in Fig. 5.1. From the resistance network, it is possible to evaluate the heat transfered to the fluid:[1]

$$q_1' = \frac{T_b - T_{f1}}{R_1'} + \frac{T_{f2} - T_{f1}}{R_{12}'},$$

$$q_2' = \frac{T_b - T_{f2}}{R_2'} + \frac{T_{f1} - T_{f2}}{R_{12}'}$$

Assuming from symmetry that the two resistances are equal:

$$q_b' = q_1' + q_2' = \frac{T_b - (T_{f1} + T_{f2})/2}{R_1'/2} = \frac{T_b - T_f}{R_b'} \quad (5.1)$$

with:

$$T_f = \frac{T_{f1} + T_{f2}}{2} \quad (5.2)$$

---

[1] Remember that heat is positive entering the well.

is called the *mean fluid temperature* and:

$$R_b' = \frac{R_1'}{2} \quad (5.3)$$

the *borehole resistance*. The last equation can be reformulated as:

$$T_f = T_b - q_b' R_b' \quad (5.4)$$

Heat transfer from the fluid to the borehole radius is influenced by:

- conduction in the grout
- conduction in the plastic pipe
- convection through the fluid

The borehole resistance can then be expressed as:

$$R_b' = R_g' + \frac{\overbrace{R_{wall}' + R_{conv}'}^{R_p'}}{2} \quad (5.5)$$

where R_g' is the *grout resistance*. The last two expressions, associated with the *pipe resistance* R_p', are classical and can be found in any heat transfer textbook [1]:

$$R_{wall}' = \frac{\log(r_{po}/r_{pi})}{2\pi k_p}, \quad R_{conv}' = \frac{1}{\hbar_f 2\pi r_{pi}},$$

$$R_p' = R_{wall}' + R_{conv}' \quad (5.6)$$

$$R_b' = R_g' + \frac{R_p'}{2} \quad (5.7)$$

where k_p is the conductivity of the pipe material, often high-density polyethylene, which is in the order of 0.4 W/m-K. The grout resistance is more difficult to evaluate since it is associated with a nonstandard geometry. Since exact analytic expressions are not available, several researchers suggested expressions given either from experiment, numerical simulations, or analytic formulations. One of the first study made to evaluate this resistance was done by Paul [2] and Remund [3] at the South Dakota University. They realized several field measurements associated with three general cases as shown in Fig. 5.2.

The correlation thus obtained is given by the relation:

$$R'_{g,Paul} = \frac{1}{k_g S_g} \tag{5.8}$$

with the *shape factor*, S_g, given by:

$$S_g = \beta_o \left(\frac{r_b}{r_{po}} \right)^{\beta_1} \tag{5.9}$$

They considered three cases, called A, B, and C (see Fig. 5.2), where A is when the two pipes touch each other, case C is when the pipes are in contact with the borehole wall, and B is when the distance between the pipes is equal to the distance between the pipe and the wall. For the three cases, they suggested (Table 5.1):

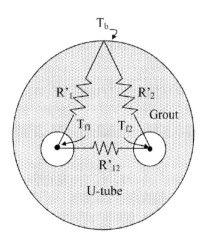

Fig. 5.1 Thermal resistance Delta-circuit

Even though the results of Paul and Remund are used as reference in several works, they have the disadvantage to be limited to three configurations. Sharqawy et al. [4] used exhaustive numerical simulations where they changed geometric parameters and thermal properties and proposed from their results the following correlation:

$$R'_{g,Shar} = \frac{1}{2\pi k_c} \left[\frac{-1.49x_c}{r_b} + 0.656 \log \left(\frac{r_b}{r_{po}} \right) + 0.436 \right] \tag{5.10}$$

Another expression was suggested by Hellstrom [5] called the *line-source method*. This analytic approach permits to evaluate the resistance expressions for different configurations, single U-tube, double U-tube, asymmetric configuration, etc. For the symmetric single U-tube, the solution is given by:

$$R'_{g,ls} = \frac{1}{4\pi k_g} \left(\log \left(\frac{r_b^2}{2x_c r_{po}} \right) - \sigma \log \left(1 - \left(\frac{x_c}{r_b} \right)^4 \right) \right) \tag{5.11}$$

with:

$$\sigma = \frac{k_g - k_s}{k_g + k_s}, \qquad \begin{array}{l} k_g : \text{ grout conductivity} \\ k_s : \text{ soil conductivity} \end{array} \tag{5.12}$$

Another well-known expression used to evaluate the borehole resistance is the *Multipole method* [6]. This is an iterative complex method, which evaluates the thermal resistance based on the order of the multipole. The order zero is the line-source method. Only the first-order multipole [5] will be given, but the full multipole method is available in the library.

Table 5.1 Paul-Remund Resistance parameters

Configuration	β_o	β_1
A	20.10	−0.9447
B	17.44	−0.6052
C	21.91	−0.3796

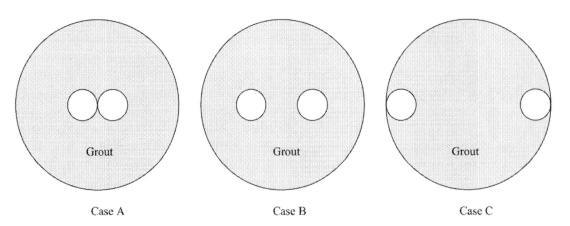

Fig. 5.2 Typical U-tube configurations

In the case of the multipole method, the grout and pipe resistance are not separated as suggested in expression (5.5). Instead, the total borehole resistance is given:

$$R'_{b,multi-1} = R'_{b,ls} - \frac{1}{4\pi k_g} \left[\frac{\left(1 - \frac{4\sigma x_c^4}{r_b^4 - x_c^4}\right)^2}{\lambda \left(\frac{2x_c}{r_{po}}\right)^2 + 1 + \frac{16\sigma r_b^4 x_c^4}{(r_b^4 - x_c^4)^2}} \right]$$
(5.13)

with:

$$\lambda = \frac{1 + 2\pi k_g R'_p}{1 - 2\pi k_g R'_p}$$
(5.14)

Example 5.1 A borehole of 7.5 cm of radius is filled with grout of conductivity 1.7 W/mK. The plastic pipe ($k_p = 0.4$ W/mK) has a 2.1 cm external radius and a 3.8 mm thickness. The center-to-center distance between the plastic pipe is 5 cm, and the flow rate is 0.4 l/s of water. If the soil conductivity is 2.5 W/mK, find the borehole resistance using:

(a) The Paul-Remund relation
(b) The Sharqawy relation
(c) The line-source relation
(d) The first-order multipole
(e) The multipole of order 10

Evaluate the properties of water at 30° C.

Solution The plastic resistance is:

$$R'_{cond} = \log\left(\frac{2.1/1.72}{2\pi 0.4}\right) = 0.0794 \text{ K-m/W}$$

From CoolProp, we have for water:

$$\mu = 8.0e - 4 \text{ Pa-s}, \ Pr = 5.44, \ k_f = 0.614 \text{ W/m-K},$$
$$\rho_f = 996 \text{ kg/m}^3$$

$$\dot{m} = 0.0004 \cdot 997 = 0.398 \text{ kg/s},$$

$$Re = \frac{4 \cdot 0.398}{\pi 0.344 \cdot 8e - 4} = 18432$$

Using Gnielinski's relation for turbulent flow (2.43b):

$$f = 0.0267, \quad Nu_D = 124.86,$$
$$h_f = 2229.2 \text{ W/m}^2\text{-K}$$

$$R'_{conv} = 0.004 \text{ K-m/W}$$
$$\Rightarrow R'_p = 0.083 \text{ K-m/W}$$

(continued)

(a) Paul's relation
A half spacing of $x_c = 0.025 = r_b/3$ is close to configuration B, so:

$$S_g = 17.44 (7.5/2.1))^{-0.6052} = 8.07$$

$$R'_{b,Paul} = 0.073 + 0.042 = 0.115 \text{ K-m/W}$$

(b) Sharkawi's relation gives:

$$R'g = \frac{1}{2\pi 1.7}\left(-1.49\frac{2.5}{7.5} + 0.656 \log\left(\frac{7.5}{2.1}\right) + 0.436\right)$$

$$= 0.0725 \text{ K-m/W}$$

$$R'_{b,Shar} = 0.0725 + 0.042 = 0.114 \text{ K-m/W}$$

(c) Line Source: Applying 5.11 gives:

$$\sigma = \frac{1.7 - 2.5}{1.7 + 2.5} = -0.19$$

$$R'g = \frac{1}{4\pi 1.7}\left[\log\left(\frac{7.5^2}{2 \cdot 2.5 \cdot 2.1}\right) + 0.19 \log\left(1 - \left(\frac{2.5}{7.5}\right)^4\right)\right]$$

$$= 0.07845 \text{ K-m/W}$$

$$R'_{b,ls} = 0.078 + 0.042 = 0.120 \text{ K-m/W}$$

(d) First-order multipole: Applying 5.13 gives:

$$\lambda = \frac{1 + 2\pi 1.7 \cdot 0.083}{1 - 2\pi 1.7 \cdot 0.083} = 17.638$$

$$R'_{b,multi-1} = 0.1202 - \frac{1}{4\pi 1.7}\left[\frac{\left(1 + \frac{4 \cdot 0.19 \cdot 2.5^4}{7.5^4 - 2.5^4}\right)^2}{\lambda \left(\frac{2 \cdot 2.5}{2.1}\right)^2 + 1 - \frac{16 \cdot 0.19 \cdot 7.5^4 \cdot 2.5^4}{(7.5^4 - 2.5^4)^2}}\right]$$

$$= 0.120 \text{ K-m/W}$$

(e) Tenth-order multipole:
No analytic expression exists, but using the function *Rb_multipole* gives:

$$R'_{b,multi-10} = 0.120 \text{ K-m/W}$$

Comments: *All methods give similar numerical values. Taking the tenth-order multipole as reference, the difference between method are 4.46 % (Paul-Remund), 4.8 % (Sharkawi), 0.19 % (line source), and 0.2 % (1st order multipole). Different configurations may however lead to larger discrepancies.*

(continued)

Example 5.1 (continued)

A full discussion on the origin of the differences is given in [7].

∎

Example 5.2 Compare the grout resistance computed with the:

- Paul method
- Sharquawy method
- line-source method

for cases

(a) A
(b) B
(c) C

for the following data:

$$k_s = 2.5 \text{ W/m-K}, k_g = 1.0 \text{ W/m-K}, r_b = 7.5 \text{ cm}$$

$$d_{po} = 33 \text{ mm}, d_{pi} = 27 \text{ mm}, R'_p = 0.08 \text{ K-m/W}$$

Solution

(a) For case A, Paul resistance is given by:

$$S_g = 20.10 \left(\frac{7.5}{1.65} \right)^{-0.94447} = 4.81$$

$$R'_{b,Paul} = \frac{1}{1.0 \cdot 4.81} + \frac{0.08}{2} = 0.248 \text{ K-m/W}$$

The Sharqawy relation gives (with $x_c \approx r_{po}$):

$$R'_{b,Shar} = \frac{1}{2\pi 1.0} \left(-1.49 \frac{1.65}{7.5} + 0.656 \log \left(\frac{7.5}{1.65} \right) + 0.436 \right)$$

$$+ 0.04 = 0.215 \text{ K-m/W}$$

Finally, for the line source, we have:

$$\sigma = \frac{1.0 - 2.5}{1.0 + 2.5} = -0.42$$

$$R'_{b,ls} = \frac{1}{4\pi 1.0} \left[\log \left(2 \frac{7.5 \cdot 7.5}{2 \cdot 1.65 \cdot 1.65} \right) \right.$$

$$\left. + 0.42 \log \left(1 - \left(\frac{1.65}{7.5} \right)^4 \right) \right] + 0.04 = 0.226 \text{ K-m/W}$$

(continued)

(b) Similar results are found for case where $x_c = r_b/3 = 2.5$ cm.

$$R'_{b,Paul} = 0.183 \text{ K-m/W}$$

$$R'_{b,Shar} = 0.188 \text{ K-m/W}$$

$$R'_{b,ls} = 0.192 \text{ K-m/W}$$

(c) Finally, for case C, $x_c = r_b - r_{po} = 2.5$ cm.

$$R'_{b,Paul} = 0.121 \text{ K-m/W}$$

$$R'_{b,Shar} = 0.083 \text{ K-m/W}$$

$$R'_{b,ls} = 0.109 \text{ K-m/W}$$

It can be seen, that except for case B, the difference is noticeable. Again, if we take the tenth-order multipole method as reference, the following differences will be observed:

Configuration	Paul-Remund	Sharkawi	Line source
A	12.8 %	2 %	2.7 %
B	3.2 %	0.5 %	1.5 %
C	13%	23%	2 %

∎

In general, the differences between the expressions for the evaluation of the borehole resistance can lead to large differences, which can influence the borehole sizing (see Chap. 6. It is usually accepted [7,8] that the multipole method is the most accurate.

5.1.1 Internal Resistance

In the evaluation of the borehole resistance, the short-circuit resistance R'_{12} shown in Fig. 5.1 was not taken into account. This resistance will have an effect however when the pipe-to-pipe interference is taken into account. Fewer expressions are suggested in the literature to evaluate it. This resistance is related to the *internal resistance*. The internal resistance is defined as the resistance associated with the thermal path between pipe 1 and pipe 2 assuming no heat flow into the ground (i.e., $q'_1 = -q'_2$). For the simple U-tube case, it is found by evaluating the equivalent resistance given by $R'_1 + R'_2$ in parallel with R'_{12} For the line-source method, assuming a symmetric configuration ($R'_1 = R'_2$), it is given by:

$$R'_{a,ls} = \frac{1}{\pi k_g} \left[\log\left(\frac{2x_c}{r_{po}}\right) + \sigma \log\left(\frac{1 + \left(\frac{x_c}{r_b}\right)^2}{1 - \left(\frac{x_c}{r_b}\right)^2} \right) \right] + 2R'_p$$

(5.15)

Hellstrom [5] also proposed an analytic expression for the first order multipole expression. Again it is expressed as a corrective term for the zero-th order (line-source) expression:

$$R'_{a,multi-1} = R'_{a,ls} - \frac{1}{\pi k_g} \left[\frac{\left(\frac{r_{po}}{2x_c}\right)^2 \left[1 + \frac{4\sigma x_c^2 r_b^2}{r_b^4 - x_c^4}\right]^2}{\lambda - \left(\frac{r_{po}}{2x_c}\right)^2 + \frac{2\sigma r_b^2 r_{po}^2 (r_b^4 + x_c^4)}{(r_b^4 - x_c^4)^2}} \right]$$

(5.16)

A last expression that has been suggested in the literature [9] is to use the shape factor associated with heat transfer between two cylinders in an infinite domain. This expression was used by Beier [10, 11] in their borehole modeling and is given by:

$$R'_{a,shape} = \frac{\cosh^{-1}\left(\frac{2x_c^2 - r_{po}^2}{r_{po}^2}\right)}{2\pi k_g} + 2R'_p$$

(5.17)

Example 5.3 With the same parameters as given in the last example, compute the internal resistance with:

- line source
- first-order multipole
- shape factor

Take $R'_p = 0.08$, and leave a 1 mm gap between the pipes for case A.

Solution

(a) For case A, the shape factor expression gives a null resistance if $x_c = r_{po}$, so we will let a small distance between the pipes:

$$x_c = r_{po} + 0.0005 = 0.017 \text{ m}$$

$$\sigma = \frac{1.0 - 2.5}{1.0 + 2.5} = -0.42$$

$$R'_{a,ls} = \frac{1}{\pi 1.0}$$

$$\times \left[\log\left(\frac{2 \cdot 1.7}{1.65}\right) - 0.42 \log\left(\frac{1 + \left(\frac{1.7}{7.5}\right)^2}{1 - \left(\frac{1.7}{7.5}\right)^2} \right) \right]$$

(continued)

$$+ 0.16 = 0.3761 \text{ K-m/W}$$

(b) With the first-order multipole correction:

$$R'_{a,multi-1} = 0.3761 - \frac{1}{\pi 1.0}$$

$$\left[\frac{\left(\frac{1.65}{2 \cdot 1.7}\right)^2 \left(1 - \frac{4 \cdot 0.42 \cdot 1.7^2 7.5^2}{7.5^4 - 1.7^4}\right)^2}{3.02 - \left(\frac{1.65}{2 \cdot 1.7}\right)^2 - \frac{2 \cdot 0.42 \cdot 7.5^2 1.65^2 (7.5^4 + 1.7^4)}{(7.5^4 - 1.7^4)^2}} \right]$$

$$= 0.3534 \text{ K-m/W}$$

(c) The shape-factor expression gives:

$$R'_{a,shape} = \frac{\cosh^{-1}\left(\frac{2*1.7^2 - 1.65^2}{1.65^2}\right)}{2\pi 1.0} + 0.16$$

$$= 0.2382 \text{ K-m/W}$$

The following table gives the values for the three cases and also the tenth-order multipole as reference:

Configuration	Line source	First-order multi	shape factor	Tenth-order multi
A	0.376	0.353	0.238	0.354
B	0.482	0.475	0.471	0.475
C	0.592	0.592	0.774	0.592

Comments: *As for the borehole resistance, B is the case where the difference is the smallest; case A shows a larger difference due in part, to the effect of the pipe resistance. Case C is influenced by the heat transfer through the ground.*

∎

5.1.2 *Line-Source Method

Sharqawy et al. found their results by imposing a uniform temperature at the borehole wall. In reality, this temperature varies, and this variation will influence the thermal behavior inside the borehole [7]. Methods that take these variations into account are the line-source method [5] and the multipole method [12], which were suggested by the Swedish researchers in the 1980s. The first one will be briefly explained.

The idea is similar to the line-source method discussed in the previous chapter but applied here to the steady-state regime. Each pipe inside the borehole is related to a heat source (or sink). The method assumes that the pipe dimension is small compared to the borehole, which is not always the case. The new thermal problem is:

$$
\begin{aligned}
\nabla T(r, \theta) &= \delta(\vec{r} - \vec{r}_o) & r < r_b \\
\nabla T(r, \theta) &= 0 & r > r_b
\end{aligned}
\tag{5.18}
$$

$$
T(r_b^-, \theta) = T(r_b^+, \theta)
$$

$$
-k_g \left.\frac{\partial T}{\partial r}\right|_{r=r_b^-} = -k_s \left.\frac{\partial T}{\partial r}\right|_{r=r_b^+}
\tag{5.19}
$$

It is well known that the thermal response of a single heat source in an infinite steady-state 2D domain (*unconfined steady-state Green's function*) is given by Carslaw and Jaeger [13]:

$$
T(r) = \frac{q'}{2\pi k} \log(\rho)
\tag{5.20}
$$

where:

$$
\rho = \sqrt{(x - x_n)^2 + (y - y_n)^2}
\tag{5.21}
$$

with (x_n, y_n) being the location of the source. The idea is to add to this particular solution a superposition of solutions, all of which satisfy the Laplace equation in order that the boundary conditions (5.19) are fulfilled. The calculations are lengthy, but the final results are given by Hellstrom [5]. He found that:

$$
T(\tilde{x}, \tilde{y}) = \frac{q'}{4\pi k_g} \left[\log((\tilde{x} - \tilde{x}_n)^2 + (\tilde{y} - \tilde{y}_n)^2) + \right.
$$
$$
\left. \sigma \log\left(\frac{(\tilde{x} - \tilde{x}'_n)^2 + (\tilde{y} - \tilde{y}'_n)^2}{\tilde{b}_n} \right) \right] \quad , \ r \le r_b
\tag{5.22}
$$

$$
T(\tilde{x}, \tilde{y}) = \frac{q'}{4\pi k_s} \left[(1 - \sigma) \log((\tilde{x} - \tilde{x}_n)^2 + (\tilde{y} - \tilde{y}_n)^2) + \right.
$$
$$
\left. \sigma \log(\tilde{x}^2 + \tilde{y}^2)) \right] \quad , \ r > r_b
\tag{5.23}
$$

where:

$$
(\tilde{x}_n, \tilde{y}_n) = \frac{(x_n, y_n)}{r_b} \quad , \quad \tilde{b}_n = \sqrt{\tilde{x}_n^2 + \tilde{y}_n^2} \quad , \quad \tilde{x}'_n = \frac{\tilde{x}_n}{\tilde{b}_n^2} \quad , \quad \tilde{y}'_n = \frac{\tilde{y}_n}{\tilde{b}_n^2}
\tag{5.24}
$$

The point $(\tilde{x}'_n, \tilde{y}'_n)$ is called the harmonic mirror point, which satisfies the properties:

$$
\sqrt{\tilde{x}_n'^2 + \tilde{y}_n'^2} \sqrt{\tilde{x}_n^2 + \tilde{y}_n^2} = 1
\tag{5.25}
$$

This temperature is given for an arbitrary constant level. The level chosen gives a zero mean borehole temperature. Indeed, (5.22) gives for $r = r_b$.

$$
T_b = \frac{q'}{4\pi k_s}(1 - \sigma) \log\left[(\cos(\theta) - \tilde{r}_n \cos(\alpha))^2 + (\sin(\theta) - \tilde{r}_n \sin(\alpha))^2 \right]
\tag{5.26}
$$

where:

$$
\tilde{x}_n = \tilde{r}_n \cos(\alpha), \qquad \tilde{y}_n = \tilde{r}_n \sin(\alpha)
$$

which shows that the borehole temperature varies with the angular position and is not uniform. However, we do have that the mean is zero:

$$
\bar{T}_b = \frac{q'}{4\pi k_s}(1 - \sigma) \int_0^\pi \log(1 + \tilde{r}_n^2 - 2\tilde{r}_n \cos(\theta - \alpha)) d\theta = 0
\tag{5.27}
$$

since:

$$
\tilde{r}_n = \frac{r_n}{r_b} < 1
$$

The effect of several pipes will add up, and temperature field inside the borehole will give:

$$
T(\tilde{x}, \tilde{y}) = \frac{1}{4\pi k_g} \sum_{n=1}^{N} q'_n \left[\log((\tilde{x} - \tilde{x}_n)^2 + (\tilde{y} - \tilde{y}_n)^2) + \right.
$$
$$
\left. \sigma \log\left(\frac{(\tilde{x} - \tilde{x}'_n)^2 + (\tilde{y} - \tilde{y}'_n)^2}{\tilde{b}_n} \right) \right]
\tag{5.28}
$$

In order to find the borehole resistances, the temperature is evaluated at the pipe perimeter. The line-source approach simplifies the calculation using a similar approach than the spatial superposition in Chap. 4. It evaluates the temperature level due to source n at $r = r_n$ but the effect of the other sources at $r = |r_n - r_m|$ that is at the center of the pipe. This assumption, which made a lot of sense for borehole interference, is more questionable here since the distances between pipes are small inside a typical borehole. Nevertheless, the results obtained are often quite good. The final result will give:

$$
T_b - T_{f,m} = \sum_{n=1}^{N} R'^O_{mn} q'_n
\tag{5.29}
$$

where:

$$
R'^O_{mm} = -\frac{1}{2\pi k_g} \left[\log(\tilde{r}_{pm}) + \sigma \log(1 - \tilde{b}_m^2) \right] + R_{pm}
\tag{5.30a}
$$

$$
R'^O_{mn} = -\frac{1}{2\pi k_g} \left[\log(\tilde{b}_{mn}) + \sigma \log(\tilde{b}'_{mn}) \right] \qquad m \ne n
\tag{5.30b}
$$

In order to find our borehole resistance as they were defined in the previous section, we must inverse these relations:

$$
q'_n = \sum_{n=1}^{N} K_{mn}(T_b - T_{f,m})
\tag{5.31}
$$

with $\mathbf{K} = (\mathbf{R}'^O)^{-1}$.

5.1.2.1 Single U-Tube Expressions

One of the main advantages of analytic approaches like the line-source method or the multipole approach is that any configurations can be calculated. However, we are most often interested in some special configurations; the most well-known is the single U-tube case. If we limit our analysis to the case where both legs are the same, $R'_{p1} = R'_{p2}$. In this case, it can be seen that:

$$R''^{o}_{11} = R''^{o}_{22} = -\frac{1}{2\pi k_g}\left(\log(\tilde{r}_{po}) + \sigma \log(1 - \tilde{x}_c^2)\right) + R'_p$$

(5.32)

$$R''^{o}_{12} = -\frac{1}{2\pi k_g}\left(\log(2\tilde{x}_c) + \sigma \log(1 + \tilde{x}_c^2)\right)$$

(5.33)

$$\begin{bmatrix} K_{11} & K_{12} \\ K_{12} & K_{11} \end{bmatrix} = \frac{1}{(R''^{o}_{11})^2 - (R''^{o}_{12})^2}\begin{bmatrix} R''^{o}_{11} & -R''^{o}_{12} \\ -R''^{o}_{12} & R''^{o}_{11} \end{bmatrix}$$

(5.34)

The heat flow system is then given by:

$$\begin{aligned} q'_1 &= K_{11}(T_b - T_1) + K_{12}(T_b - T_2) \\ &= (K_{11} + K_{12})(T_b - T_1) - K_{12}(T_2 - T_1) \\ &= \frac{(T_b - T_1)}{R'_1} + \frac{(T_2 - T_1)}{R'_{12}} \end{aligned}$$

(5.35)

R'_1 is sometimes written R'^{Δ}_1 associated with the shape of the network in the case of a single U-tube, but the Δ will be dropped here. From (5.34) we have:

$$R'_1 = \frac{1}{K_{11} + K_{12}} = \frac{(R''^{o}_{11})^2 - (R''^{o}_{12})^2}{R''^{o}_{11} - R''^{o}_{12}} = R''^{o}_{11} + R''^{o}_{12}$$

So:

$$R'_b = \frac{R'_1}{2} = -\frac{1}{4\pi k_g}\left(\log(2\tilde{x}_c\tilde{r}_{po}) + \sigma \log(1 - \tilde{x}_c^4)\right) + \frac{R'_p}{2}$$

(5.36)

The line-source formalism also permits the calculation of the interference resistance that is given by:

$$R'_{12} = -\frac{1}{K_{12}} = \frac{(R''^{o}_{11})^2 - (R''^{o}_{12})^2}{R''^{o}_{12}}$$

The closed form of the last expression is less compact. However, the *internal resistance* is simply given by:

$$R'_a = (R'_1 + R'_2) \, // \, R'_{12}$$

(5.37)

which in the usual case where $R'_1 = R'_2$, which gives:

$$R'_a = 2(R''^{o}_{11} - R''^{o}_{12})$$

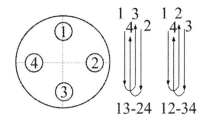

Fig. 5.3 Double-U-tubes parallel arrangements

$$= \frac{1}{\pi k_g}\left[\log\left(\frac{2\tilde{x}_c}{\tilde{r}_{po}}\right) + \sigma \log\left(\frac{1 + \tilde{x}_c^2}{1 - \tilde{x}_c^2}\right)\right] + 2R'_p$$

(5.38)

Another way to find R'_a is to set:

$$q'_1 = -q'_2$$

since, by assumption, the sum is zero. Doing so will gives the same result as before.

5.1.2.2 Double U-Tube Expressions

Another common configuration is the double U-tube configuration shown in Fig. 5.3.

With this configuration however, several cases are possible: A series configuration is possible where the exit of one U-tube is the inlet of the other one, a parallel configuration where the same fluid goes in two U-tubes, and independent circuits where two different fluids go inside separate U-tubes. We will only consider the parallel case here. However, even in this special case, we can see in Fig. 5.3 that three different cases can be expected. From the symmetry of the problem, we have from Eq. (5.30a):

$$R''^{o}_{11} = -\frac{1}{2\pi k_g}\left(\log(\tilde{r}_{po}) + \sigma \log(1 - \tilde{x}_c^2)\right) + R'_p$$

(5.39)

$$R''^{o}_{12} = -\frac{1}{2\pi k_g}\left(\log\left(\sqrt{2}\tilde{x}_c\right) + \frac{\sigma}{2} \log\left(1 + \tilde{x}_c^4\right)\right) = R''^{o}_{14}$$

(5.40)

$$R''^{o}_{13} = -\frac{1}{2\pi k_g}\left(\log\left(2\tilde{x}_c\right) + \sigma \log\left(1 + \tilde{x}_c^2\right)\right)$$

(5.41)

where all other resistances can be found from symmetry.

$$\begin{bmatrix} q'_1 \\ q'_2 \\ q'_3 \\ q'_4 \end{bmatrix} = \begin{bmatrix} K_{11} & K_{12} & K_{13} & K_{14} \\ K_{21} & K_{22} & K_{23} & K_{24} \\ K_{31} & K_{32} & K_{33} & K_{34} \\ K_{41} & K_{42} & K_{43} & K_{44} \end{bmatrix}\begin{bmatrix} T_b - T_1 \\ T_b - T_2 \\ T_b - T_3 \\ T_b - T_4 \end{bmatrix}$$

(5.42)

Rearranging and taking into account the symmetry:

$$q_1' = \frac{(T_b - T_1)}{R'_1} + \frac{(T_2 - T_1)}{R'_{12}} + \frac{(T_3 - T_1)}{R'_{13}} + \frac{(T_4 - T_1)}{R'_{12}}$$

$$q_2' = \frac{(T_1 - T_2)}{R'_{12}} + \frac{(T_b - T_2)}{R'_1} + \frac{(T_3 - T_2)}{R'_{12}} + \frac{(T_4 - T_2)}{R'_{13}}$$

$$q_3' = \frac{(T_1 - T_3)}{R'_{13}} + \frac{(T_2 - T_3)}{R'_{12}} + \frac{(T_b - T_3)}{R'_1} + \frac{(T_4 - T_3)}{R'_{12}}$$

$$q_4' = \frac{(T_1 - T_4)}{R'_{12}} + \frac{(T_2 - T_4)}{R'_{13}} + \frac{(T_3 - T_4)}{R'_{12}} + \frac{(T_b - T_4)}{R'_1}$$

$$(5.43)$$

From which it can be found that:

$$R'_1 = (K_{11} + K_{12} + K_{13} + K_{14})^{-1} = R''^o_{11} + 2R''^o_{12} + R''^o_{13}$$
$$(5.44)$$

$$R'_{12} = \frac{R''^o_{11}{}^2 + R''^o_{13}{}^2 + 2R''^o_{11}R''^o_{13} - 4R''^o_{12}{}^2}{R''^o_{12}}$$
$$(5.45)$$

$$R'_{13} = \frac{(R''^o_{11} - R''^o_{13})(R''^o_{11}{}^2 + R''^o_{13}{}^2 + 2R''^o_{11}R''^o_{13} - 4R''^o_{12}{}^2)}{R''^o_{13}{}^2 + R''^o_{11}R''^o_{13} - 2R''^o_{12}{}^2}$$
$$(5.46)$$

The borehole resistance is easy to evaluate since it relates the mean borehole temperature to the mean fluid temperature.

$$q_b' = q_1' + q_2' + q_3' + q_4' = \frac{T_b - T_f}{R'_b} = \frac{4(T_b - T_f)}{R'_1} \quad (5.47)$$

So that:

$$R'_b = \frac{R'_1}{4} = -\frac{1}{8\pi k_g}\left(\log\left(4\tilde{r}_{po}\tilde{x}_c^3\right) + \sigma\log\left(1 - \tilde{x}_c^8\right)\right) + \frac{R'_p}{4}$$
$$(5.48)$$

The internal resistance depends of the configuration: the following choices are possible:

Case 12-34

$$T_1 = T_3, T_2 = T_4 \qquad ; \qquad q_1' = q_3', q_2' = q_4'$$

Since the thermal circuit is more complicated, it is easier to find the internal resistance using the second approach, that is:

$$q_1' + q_3' = 2q_1' = -(q_2' + q_4') = -2q_2' \qquad (5.49)$$

$$q_1' = \frac{(T_b - T_1)}{R'_1} + \frac{2(T_2 - T_1)}{R'_{12}}$$
$$(5.50)$$

$$q_2' = -q_1' = \frac{(T_b - T_2)}{R'_1} + \frac{2(T_1 - T_2)}{R'_{12}}$$

Then:

$$2q_1' = q_a' = \frac{(T_2 - T_1)}{R'_a} = (T_2 - T_1)\left[\frac{1}{R'_1} + \frac{4}{R'_{12}}\right]$$

$$\frac{1}{R'_a} = \frac{1}{R'_1} + \frac{4}{R'_{12}} \qquad (5.51)$$

$$R'_a(12-34)$$
$$= \frac{1}{2\pi k_g}\left(\log\left(\frac{\tilde{x}_c}{\tilde{r}_{po}}\right) + \sigma\log\left(\frac{1 + \tilde{x}_c^4}{1 - \tilde{x}_c^4}\right)\right) + R'_p$$
$$(5.52)$$

Case 13-24

In this case:

$$T_1 = T_2, T_3 = T_4, \qquad q_1' = q_2' = -q_3' = -q_4'$$

$$q_1' + q_2' = 2q_1' = -2q_3' \qquad (5.53)$$

A similar procedure will give:

$$\frac{1}{R'_a} = \frac{1}{R'_1} + \frac{2}{R'_{12}} + \frac{2}{R'_{13}} \qquad (5.54)$$

$$R'_a(13-24) = \frac{1}{2\pi k_g}\left(\log\left(\frac{2\tilde{x}_c}{\tilde{r}_{po}}\right) + \sigma\log\left(\frac{1 + \tilde{x}_c^2}{1 - \tilde{x}_c^2}\right)\right) + R'_p$$
$$(5.55)$$

Case 12–43 will give similar results. Since this value is higher than the previous one, this is the one that will have less thermal interference and that is preferred. This is the value that is given in simulation software like DST [14] and EED [15], though in the latter case, the multipole method is used.

Example 5.4 Evaluate the borehole resistance and internal resistance (conf 13–24) using the same parameters as in Example 5.2 (case B) with a double U-tube.

Solution Solution is given by:

$$\sigma = \frac{1.0 - 2.5}{1.0 + 2.5} = -0.428$$

$$R'_{b,ls} = -\frac{1}{8\pi 1.0}\left[\log\left(4\cdot 0.33^3 \cdot 0.22\right)\right.$$

$$\left. -0.428\log\left(1 - 0.33^8\right)\right] + 0.02 = 0.156 \text{ K-m/W}$$

$$R'_{a,ls} = \frac{1}{2\pi 1.0}\left[\log\left(\frac{2\cdot 0.33}{0.22}\right)\right.$$

$$\left. -0.428\log\left(\frac{1 + 0.33^2}{1 - 0.33^2}\right)\right] + 0.08 = 0.241 \text{ K-m/W}$$

(continued)

Example 5.4 (continued)
Comments: *The borehole resistance of the double U-tube is logically smaller than the single U-tube, which brings some favorable arguments to those who promote their usage. However, the cost is more important, and only a deeper analysis can bring a more robust conclusion.* ∎

5.2 Outlet Temperature Calculation: Effective Resistance

As indicated in the introduction of this chapter, the goal of borehole modeling is to evaluate the outlet temperature of the field (the heat pump entering temperature (EWT)), as a function of the loads exchanged with the ground. In the previous sections, the mean borehole temperature was evaluated followed by the mean fluid temperature. In this last section, the evaluation of the outlet temperature will be considered.

5.2.1 Linear Approximation

In order to evaluate the outlet temperature, some assumptions have to be made. One common assumption is to consider that the temperature inside the borehole varies linearly (Fig. 5.4). Although not based on physical arguments, this distribution is often used in practice. In that case:

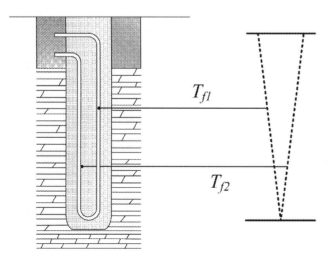

Fig. 5.4 Linear profile

$$T_{fx} = T_{fi} + \frac{T_{fo} - T_{fi}}{2H}x$$

where x is the axial distance from inlet of the borehole and the total length is twice the height of the borehole. At each section, for each pipe:

$$x_1 = x, \qquad x_2 = 2H - x$$

$$T_{f1}(x) = T_{fi} + \frac{T_{fo} - T_{fi}}{2H}x$$

$$T_{f2}(x) = T_{fi} + \frac{T_{fo} - T_{fi}}{2H}(2H - x)$$

$$\bar{T}_{fx} = \frac{T_{f1}(x) + T_{f2}(x)}{2} = \frac{T_{fi} + T_{fo}}{2} = T_f$$

So the *mean fluid temperature*, as defined in the previous section, is also the mean temperature at the inlet and outlet of the borehole. The outlet temperature can also be related to the heat transfer rate:[2]

$$q_b = \dot{m}C_p(T_{fo} - T_{fi})$$

Summing both relations gives:

$$T_{fo} = T_f + \frac{q_b}{2\dot{m}C_p} \tag{5.56}$$

$$T_{fi} = T_f - \frac{q_b}{2\dot{m}C_p} \tag{5.57}$$

From Eq. (5.4):

$$T_{fo} = T_b + q'_b\left(\frac{H}{2\dot{m}C_p} - R'_b\right) \tag{5.58}$$

In the case of several boreholes in parallel, this is still valid if $\dot{m}$ is the flow rate per borehole. We can also replace it by the total flow rate if H is replaced by L the total length. The last expression can be written:

$$T_{fo} = T_b - q'_b R'_b\left(1 - \frac{\gamma}{2}\right) \tag{5.59}$$

with:

$$\gamma = \frac{H}{\dot{m}C_p R'_b} \tag{5.60}$$

5.2.2 Exponential Distribution

The last assumption is valid if the heat flux is uniform, which is not the necessarily the case. A better assumption is

---

[2]Remember that the heat flux is positive when heat is extracted from the ground.

to take the ground far field temperature uniform, but since the ground resistance varies with time, it complicates the analysis. A simpler assumption is to assume that the borehole temperature is uniform although it may change with time. In that case, the solution is well known [1] to vary exponentially if interference effects between pipes are neglected [16]:

$$
\begin{aligned}
T_{fx} &= T_b + (T_{fi} - T_b) \exp\left(\frac{-x}{\dot{m}C_p R'_1}\right) \\
&= T_b + (T_{fi} - T_b) \exp\left(\frac{-x}{2\dot{m}C_p R'_b}\right)
\end{aligned}
\tag{5.61}
$$

A dimensionless temperature can be defined:

$$
\theta(x) = \frac{T_{fx} - T_b}{T_{fi} - T_b} = \exp\left(\frac{-x}{2\dot{m}C_p R'_b}\right)
\tag{5.62}
$$

where the outlet temperature (EWT) would be:

$$
\theta_o = \theta(2H) = \exp\left(\frac{-H}{\dot{m}C_p R'_b}\right) = \exp(-\gamma)
\tag{5.63}
$$

This temperature can be related to the total heat transfer per borehole:

$$
q_b = \dot{m}C_p(T_{fo} - T_{fi}) = \dot{m}C_p(T_{fo} - T_b - (T_{fi} - T_b))
$$

$$
q_b = \dot{m}C_p(T_{fo} - T_b)\left(\frac{\theta_o - 1}{\theta_o}\right)
\tag{5.64}
$$

The last expression can be written:

$$
T_{fo} = T_b - q'_b R'_b \left(\frac{\gamma \theta_o}{(1 - \theta_o)}\right)
\tag{5.65}
$$

For the exponential distribution, one has:

$$
T_{fo} = T_b - q'_b R'_b \left(\frac{\gamma \exp(-\gamma)}{(1 - \exp(-\gamma))}\right)
\tag{5.66}
$$

Taylor expansion of the last expression gives:

$$
T_{fo} \approx T_b - q'_b R'_b \left(1 - \frac{\gamma}{2} + \frac{\gamma^2}{12} + \dots\right)
$$

It can be observed that for small values of γ, both relations are similar. In Fig. 5.5, the outlet temperature is given for a borehole temperature $T_b = 10°C$, a heat transfer rate of $q'_b = 20$ W/m, and a borehole resistance of $R'_b = 0.08$ K-m/W, varying the parameter γ.

As can be observed, for small values of γ, the outlet temperature distributions are similar. At larger values of γ, not only do they start to diverge, but unrealistic temperatures are calculated, assuming a linear relation since temperatures larger than the borehole temperature are observed, which

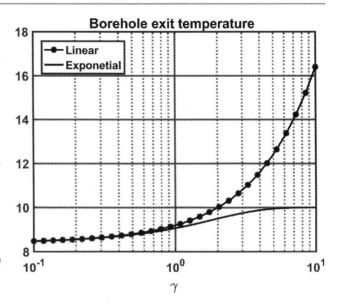

Fig. 5.5 Exit temperature using exponential and linear profiles

makes no sense. This is observed in the example for $\gamma > 2$. To have an idea of what this value represents, let's assume that $H = 100$ m, $R'_b = 0.08$ K-m/W, $C_p = 4.18$ KJ/kg-K, and the mass flow rate would be around 0.14 Kg/s for $\gamma = 2$. This is rather small, and in typical applications, the linear approximation is working quite well, but care should be taken especially for long boreholes with low flow rate.

It is interesting to relate the last expression to the inlet temperature. The inlet temperature is given by:

$$
T_{fi} = T_{fo} - \frac{q_b}{\dot{m}C_p}
$$

we find that:

$$
T_{fi} = T_b - q'_b R'_b \left(\frac{\gamma \theta_o}{1 - \theta_o} + \gamma\right)
$$

$$
T_{fi} = T_b - q'_b R'_b \left(\frac{\gamma}{1 - \theta_o}\right)
\tag{5.67}
$$

Summing (5.67)–(5.65) gives:

$$
\frac{T_{fi} + T_{fo}}{2} = T_f = T_b - q'_b R'_b \left(\frac{\gamma(1 + \theta_o)}{2(1 - \theta_o)}\right) = T_b - q'_b R'^*_b
\tag{5.68}
$$

with:

$$
R'^*_b = R'_b \left(\frac{\gamma(1 + \theta_o)}{2(1 - \theta_o)}\right)
\tag{5.69}
$$

Using (5.63), the last expression is given by:

$$
R'^\star_b = R'_b \frac{\gamma}{2}\left(\frac{1 + \exp(-\gamma)}{1 - \exp(\gamma)}\right)
$$

$$= R'_b \frac{\gamma}{2} \left(\frac{\exp(\gamma/2) + \exp(-\gamma/2)}{\exp(\gamma/2) - \exp(-\gamma/2)} \right)$$

$$R'^{\star}_b = R'_b \frac{\gamma}{2} \coth\left(\frac{\gamma}{2}\right) \qquad (5.70)$$

with Eqs. (5.69) and (5.70) equivalent, the latter, $R'^{\star}_b$, is specific to the exponential distribution, and the former, R'^{*}_b, is more general where the exit temperature θ_o can be different.

Indeed, it is interesting to see that the linear approximation would give:

$$\theta_{o,lin} = \frac{1 - \gamma/2}{1 + \gamma/2} \qquad (5.71)$$

Replacing this last expression in (5.69) leads to:

$$R'^{*}_b(linear) = R'_b \left(\frac{\gamma(1 + \frac{1-\gamma/2}{1+\gamma/2})}{2(1 - \frac{1-\gamma/2}{1+\gamma/2})} \right) = R'_b$$

An interesting analysis is also to compare the given relations with more classical ones found in heat transfer books. First, let's rewrite Eq. (5.68):

$$q'_b = \frac{T_b - T_f}{R'^{\star}_b} \qquad (5.72)$$

or, remembering that this is the heat transfer per unit length of borehole:

$$q_b = \frac{H(T_b - T_f)}{R'^{\star}_b} \qquad (5.73)$$

In internal flow, the heat flow extracted in a pipe of length $L = 2H$ is given by the classical expression:

$$q_b = U A \Delta T_{ln} = \frac{L \Delta T_{ln}}{R'_1} = \frac{H \Delta T_{ln}}{R'_b} \qquad (5.74)$$

with:

$$\Delta T_{ln} = \frac{(T_b - T_{fo}) - (T_b - T_{fi})}{\log\left(\frac{T_b - T_{fo}}{T_b - T_{fi}}\right)} \qquad (5.75)$$

It is easy to verify that both expressions are equivalent. From the definition of effective resistance, Eq. (5.73) can be rewritten:

$$q_b = \frac{2H(T_b - T_f)}{\gamma R'_b} \tanh(\gamma/2)$$

From the previous analysis, we have that:

$$\tanh(\gamma/2) = \frac{1 - \theta_o}{1 + \theta_o} = \frac{T_{fo} - T_{fi}}{2(T_b - T_f)}$$

which gives:

$$q_b = \frac{H(T_{fo} - T_{fi})}{\gamma R'_b} = \frac{-H((T_b - T_{fo}) - (T_b - T_{fi}))}{\gamma R'_b}$$

We also have:

$$\gamma = -\log(\theta_o)$$

From which:

$$q_b = \frac{-H((T_b - T_{fo}) - (T_b - T_{fi}))}{-\log\left(\frac{T_b - T_{fo}}{T_b - T_{fi}}\right) R'_b} = \frac{H \Delta T_{ln}}{R'_b}$$

Equations (5.73) and (5.74) are then equivalent; the former one is more practical when the heat transfer is known instead of the inlet temperature. Another interesting expression is to relate the heat exchange directly to the inlet temperature. From Eq. (5.57):

$$T_f = T_{fi} + \frac{q'_b H}{2 \dot{m} C_p}$$

Replacing in Eq. (5.72), we have:

$$q'_b = \frac{T_b - T_f}{R'^{\star}_b} = \frac{T_b - T_{fi} - \frac{q'_b H}{2 \dot{m} C_p}}{R'^{\star}_b}$$

$$q'_b \left(1 + \frac{\gamma^*}{2}\right) = \frac{T_b - T_{fi}}{R'^{\star}_b}$$

$$q'_b = \frac{T_b - T_{fi}}{R'^{\star}_b \left(1 + \frac{\gamma^*}{2}\right)} \qquad (5.76)$$

with:

$$\gamma^* = \frac{H}{\dot{m} C_p R'^{\star}_b} = \gamma \frac{R'_b}{R'^{\star}_b} \qquad (5.77)$$

5.2.3 Impact of Internal Resistance

The exponential distribution of temperature is a well-known expression in heat transfer for internal flows. However, for vertical boreholes, there are a downward and an upward leg that can also exchange heat that is not taken into account in the last relations. In the case where pipe-to-pipe heat transfer is taken into account, the downward flow will be governed in the *steady-flux* regime by the following equation:

$$- \dot{m} C_p \frac{dT_{f1(z)}}{dz} = \frac{T_{f1} - T_b}{R'_1} + \frac{T_{f1} - T_{f2}}{R'_{12}}$$

The upward leg will be governed by:

$$\dot{m} C_p \frac{dT_{f2(z)}}{dz} = \frac{T_{f2} - T_b}{R'_1} + \frac{T_{f2} - T_{f1}}{R'_{12}}$$

These two differential equations can be complicated to solve for general expressions of T_b (see Hellstrom [5] and Cimmino [17]), but in the case where the borehole temperature is uniform, the last two equations can easily be solved by Laplace transform. Again the solution was given by Hellstrom [5] and redone by Zeng et al. [18], where they solved the system in dimensionless form:

$$-\frac{d\theta_d(\bar{z})}{d\bar{z}} = \frac{\theta_d}{S_1} + \frac{\theta_d - \theta_u}{S_{12}}$$

$$\frac{d\theta_u(\bar{z})}{d\bar{z}} = \frac{\theta_u - \theta_d}{S_{12}} + \frac{\theta_u}{S_1} \tag{5.78}$$

where:

$$S_1 = \frac{\dot{m}C_p R'_1}{H}, S_{12} = \frac{\dot{m}C_p R'_{12}}{H}$$

The outlet temperature expression is given by:

$$\theta_o = \frac{\cosh(\eta) - \xi \sinh(\eta)}{\cosh(\eta) + \xi \sinh(\eta)} \tag{5.79}$$

where:

$$\eta = \sqrt{\frac{1}{S_1^2} + \frac{2}{S_1 S_{12}}} \tag{5.80}$$

which are better expressed in terms of the borehole resistances:

$$\eta = \frac{H}{\dot{m}C_p \sqrt{R'_b R'_a}} = \frac{H}{2\dot{m}C_p R'_b \xi} = \frac{\gamma}{2\xi} \tag{5.81}$$

$$\xi = \sqrt{\frac{R'_a}{4R'_b}} \tag{5.82}$$

The temperature profile, which will be referred as the Zeng's profile, is given by:

$$\theta(\bar{z})_d = \frac{\cosh(\eta(1-\bar{z})) + \xi \sinh(\eta(1-\bar{z}))}{\cosh(\eta) + \xi \sinh(\eta)} \tag{5.83a}$$

$$\theta(\bar{z})_u = \frac{\cosh(\eta(1-\bar{z})) - \xi \sinh(\eta(1-\bar{z}))}{\cosh(\eta) + \xi \sinh(\eta)} \tag{5.83b}$$

It is easy to verify that if there is no mutual interference $R'_{12} \to \infty$, (5.79) becomes (5.63), and the Zeng's profile becomes the exponential profile. Recall that:

$$R'_a = \frac{4R'_b R'_{12}}{4R'_b + R'_{12}}$$

So, if the short-circuit is small $R'_{12} \to \infty$:

$$R'_a \approx 4R'_b \Rightarrow \xi = 1$$

$$\theta_o = \exp(-2\eta) = \exp(-\gamma)$$

Effective resistance is easily found from the analysis from the last section. Replacing (5.79) in (5.69) instead of (5.63) gives:

$$\frac{1 + \theta_o}{1 - \theta_o} = \frac{\cosh(\eta)}{\xi \sinh(\eta)}$$

$$R'^*_b = R'_b \eta \coth(\eta) \tag{5.84}$$

The quality of the heat exchange can also be measured by the exchanger *efficiency* [19]:

$$\epsilon = \frac{T_{fo} - T_{fi}}{T_b - T_{fi}} = \frac{T_{fo} - T_b}{T_b - T_{fi}} + \frac{T_b - T_{fi}}{T_b - T_{fi}} = 1 - \theta_o \tag{5.85}$$

From which we can relate the effective resistance:

$$R'^*_b = R'_b \frac{\gamma}{2} \left[\frac{1 + \theta_o}{1 - \theta_o}\right] = R'_b \gamma \frac{2 - \epsilon}{2\epsilon} \tag{5.86}$$

Example 5.5 A 150-meter-long borehole has a radius of 7.5 cm. Inside the borehole is a single U-tube with a 3.35 cm diameter pipe. The shank spacing is 5.5 cm. The borehole is filled with a grout, with a conductivity of 1.5 W/m-K. The ground conductivity is 2.1 W/m-K and its initial temperature 10°C. The pipe resistance is $R'_p = 0.076$ K-m/W . Heat is extracted at a rate of 50 W/m. If the ground volumetric heat capacity is 2.43 MW/m^{3}K, find the fluid outlet after 100 hours. The liquid is water ($\rho = 1000$ kg/m^3, $C_p = 4200$ J/kg-K), and its flow rate is 0.19 l/s. Use the line-source method to evaluate the borehole and internal resistances; calculate the outlet temperature assuming:

(a) linear approximation
(b) exponential assumption
(c) Zeng's profile

Solution After 100 hours, the Fourier number is:

$$\alpha = \frac{2.1}{2.43e6} = 8.63\text{e-}7 \text{m}^2/\text{s} = 0.0031 \text{m}^2/\text{hr}$$

$$Fo = 0.0031 \cdot 100/0.075^2 = 55.2$$

In that range of Fourier number, all models previously discussed will give similar results. Using the ILS model, we would get:

$$T_b = T_o - \frac{50}{2.1} G_{ils}(55.2) = 0.856 \,^\circ\text{C}$$

Using the line-source method to evaluate the borehole resistance gives:

$$\sigma = \frac{1.5 - 2.1}{1.5 + 2.1} = -0.167$$

(continued)

Example 5.5 (continued)

$$R'_g = \frac{1}{4\pi 1.5}\left[\log\left(\frac{7.5^2}{3.35\cdot 5.5}\right) + 0.167\log\left(1 - \left(\frac{5.5}{7.5}\right)^4\right)\right]$$

$$= 0.0562\ \text{K-m/W}$$

$$R'_{b,ls} = 0.0562 + 0.038 = 0.094\ \text{K-m/W}$$

$$R'_{a,ls} = \frac{1}{\pi 1.5}\left[\log\left(\frac{2\cdot 5.5}{3.35}\right) + 0.167\log\left(\frac{1 + \left(\frac{5.5}{7.5}\right)^2}{1 - \left(\frac{5.5}{7.5}\right)^2}\right)\right] + 0.152$$

$$= 0.509\ \text{K-m/W}$$

(a) Linear Approximation

The mean fluid temperature can now be evaluated. For the linear approximation, (5.4) gives:

$$T_f = 0.856 - 50\cdot 0.094 = -3.855\ ^\circ\text{C}$$

and from (5.56):

$$T_{fo,l} = -3.855 + \frac{50\cdot 150}{2\cdot 798} = 0.844\ ^\circ\text{C}$$

(b) Exponential Approximation

$$\gamma = \frac{H}{\dot{m}C_p R'_b} = 1.99$$

The effective resistance is given by (5.69):

$$R'^*_b = 0.094\cdot 0.997\cdot\coth 0.997 = 0.1236\ \text{K-m/W}$$

From which we have:

$$T_f = 0.856 - 50\cdot 0.1236 = -5.322\ ^\circ\text{C}$$

$$T_{fo,e} = -5.322 + \frac{50\cdot 150}{2\cdot 798} = -0.623\ ^\circ\text{C}$$

(c) Zeng's Profile

A similar calculation can be using:

$$\xi = \sqrt{\frac{R'_a}{4R'_b}} = 1.16$$

$$\eta = \frac{\gamma}{2\xi} = 0.858$$

$$R'^*_b = 0.094\cdot 0.858\cdot\coth 0.858 = 0.116\ \text{K-m/W}$$

From which we have:

$$T_f = 0.856 - 50\cdot 0.1236 = -4.96\ ^\circ\text{C}$$

$$T_{fo,z} = -4.96 + \frac{50\cdot 150}{2\cdot 798} = -0.259\ ^\circ\text{C}$$

The same results can be found from the alternative formalism (5.65), which may be interesting in some applications.

For the linear approximation, we have:

$$\theta_o = \frac{1 - \gamma/2}{1 + \gamma/2} = 0.0015$$

$$T_{fo,l} = 0.856 - 50\cdot 0.094\left(\frac{1.99\cdot 0.0015}{1 - 0.0015}\right) = 0.844\ ^\circ\text{C}$$

For the exponential approximation, we have:

$$\theta_o = \exp(-\gamma) = 0.136$$

$$T_{fo,e} = 0.856 - 50\cdot 0.094\left(\frac{1.99\cdot 0.136}{1 - 0.136}\right) = -0.623\ ^\circ\text{C}$$

For the Zeng's profile:

$$\theta_o = \frac{\cosh(\eta) - \xi\sinh(\eta)}{\cosh(\eta) + \xi\sinh(\eta)} = 0.1063$$

$$T_{fo,e} = 0.856 - 50\cdot 0.094\left(\frac{1.99\cdot 0.1063}{1 - 0.1063}\right) = -0.259\ ^\circ\text{C}$$

The three corresponding full temperature profiles are shown on the following figure and compare to a numerical simulation done on COMSOL ™. It is seen that the outlet temperature found numerically is:

$$T_{Comsol} = -0.068\ ^\circ\text{C}$$

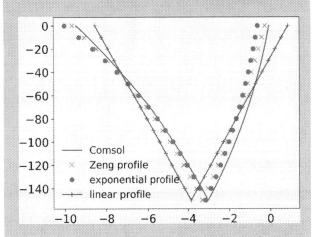

(continued)

(continued)

Example 5.5 (continued)

Comments: *It can be observed that the Zeng's profile follow very well the exact numerical solution even though it is based on the assumption of uniform borehole temperature. It is quite different than the linear profile, but in this example, the flow rate is lower than what is often used in practice, so that the often-used linear approximation works often quite well. For comparison, the same problem is solved on the following figure where the flow rate is three times higher:*

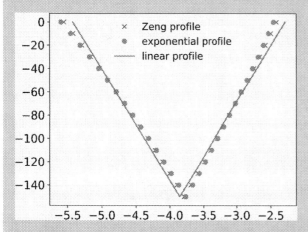

It is seen that the three solutions are similar. However, care should be used taking it especially for low flow rate and/or long boreholes. Exit temperatures higher than the ground temperatures ($\theta_o < 0$) can sometimes be observed. Finally, the Zeng's profile and the exponential profile have similar behavior, which implies that, at least in this case, the short-circuit effect is small.

∎

Example 5.6 In order to increase the effect of short-circuiting, a case where the pipes are almost touching is now analyzed. All parameters of the previous example are the same except that the shank spacing is now 1.5 cm, the pipe is smaller, $d_{po} = 2.6$ cm, and the flow rate is reduced to 0.1 l/s. The pipe resistance is also reduced, $R'_p = 0.05$ K-m/W

Solution T_b remains the same. The borehole resistances, using the line source, will give:

(continued)

$$R'_g = \frac{1}{4\pi 1.5}\left[\log\left(\frac{7.5^2}{2.6\cdot 1.5}\right) + 0.167\log\left(1 - \left(\frac{1.5}{7.5}\right)^4\right)\right]$$

$$= 0.142 \text{ K-m/W}$$

$$R'_{b,ls} = 0.142 + 0.025 = 0.167 \text{ K-m/W}$$

$$R'_{a,ls} = \frac{1}{\pi 1.5}\left[\log\left(\frac{2\cdot 5.5}{3.35}\right) + 0.167\log\left(\frac{1 + \left(\frac{5.5}{7.5}\right)^2}{1 - \left(\frac{5.5}{7.5}\right)^2}\right)\right] + 0.1$$

$$= 0.277 \text{ K-m/W}$$

$$\gamma = \frac{H}{\dot{m}C_p R'_b} = 2.145$$

$$\xi = \sqrt{\frac{R'_a}{4R'_b}} = 0.642$$

$$\eta = \frac{\gamma}{2\xi} = 1.67$$

– linear approximation

$$\theta_{o,l} = \frac{1 - \gamma/2}{1 + \gamma/2} = -0.0347$$

$$T_{fo,l} = 0.856 + 50\cdot 0.167\left(\frac{-2.145\cdot 0.0347}{1 + 0.0347}\right) = 1.46°C$$

For the exponential approximation, we have:

$$\theta_{o,e} = \exp(-\gamma) = 0.117$$

$$T_{fo,e} = 0.856 + 50\cdot 0.167\left(\frac{-2.145\cdot 0.117}{1 + 0.117}\right) = -1.51°C$$

For the Zeng's profile:

$$\theta_{o,z} = \frac{\cosh(\eta) - \xi\sinh(\eta)}{\cosh(\eta) + \xi\sinh(\eta)} = 0.2515$$

$$T_{fo,z} = 0.856 - 50\cdot 0.1063\left(\frac{1.99\cdot 0.2515}{1 - 0.2515}\right) = -5.14°C$$

The outlet temperature found numerically is:

$$T_{fo,comsol} = -5.67$$

Again, the whole profile is shown on the figure:

(continued)

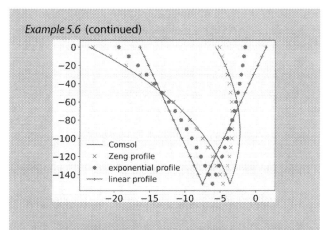

Example 5.6 (continued)

Comments: *With the given parameters, the short-circuit effects are amplified, and only the Zeng's profile give reasonable results. The linear approximation gives an non-physical outlet temperature higher than the borehole temperature. Again, in most applications, the conditions are not as severe.*

∎

5.2.3.1 Double U-Tubes

Effective resistances can be calculated in a similar way in the case of a double U-tube configuration. The analysis was also done by Zeng et al. [18] where they analyzed the parallel and series configurations. Only the parallel case will be given here. In such a case, care should be taken when the mass flow rate is given since the total borehole flow rate is twice the flow rate per pipe. Looking at Eqs. (5.64)–(5.69), it is easy to verify that the general expression (5.69) is still valid as long as the total mass flow rate is used. In Zeng's paper, they used the flow rate per pipe so the number 2 in (5.69) is replaced by 4. Here, the total flow rate per borehole will be used. Taking advantage of the symmetry of the problem, the heat balance in the four pipes can be written as:

$$\pm \dot{m}_1 C_p \frac{dT_{f1(z)}}{dz}$$

$$= \frac{T_{f1} - T_b}{R'_1} + \frac{T_{f1} - T_{f2}}{R'_{12}} + \frac{T_{f1} - T_{f3}}{R'_{13}} + \frac{T_{f1} - T_{f4}}{R'_{12}}$$

$$\pm \dot{m}_1 C_p \frac{dT_{f2(z)}}{dz}$$

$$= \frac{T_{f2} - T_b}{R'_1} + \frac{T_{f2} - T_{f1}}{R'_{12}} + \frac{T_{f2} - T_{f3}}{R'_{12}} + \frac{T_{f2} - T_{f4}}{R'_{13}}$$

$$\pm \dot{m}_1 C_p \frac{dT_{f3(z)}}{dz}$$

$$= \frac{T_{f3} - T_b}{R'_1} + \frac{T_{f3} - T_{f1}}{R'_{13}} + \frac{T_{f3} - T_{f2}}{R'_{12}} + \frac{T_{f3} - T_{f4}}{R'_{12}}$$

$$\pm \dot{m}_1 C_p \frac{dT_{f4(z)}}{dz}$$

$$= \frac{T_{f4} - T_b}{R'_1} + \frac{T_{f4} - T_{f1}}{R'_{12}} + \frac{T_{f4} - T_{f2}}{R'_{13}} + \frac{T_{f4} - T_{f3}}{R'_{12}}$$

where a minus sign is used in the case of a downward pipe and a plus sign for the upward case. In the previous expressions, $\dot{m}_1$ is the flow rate per pipe and half the total borehole flowrate. Only two cases are different:

Case 12-34

In this configuration, the fluid goes down in pipes 1 and 3 and comes up in pipes 2 and 4. $T_1 = T_3$, $T_2 = T_4$. The new system is then:

$$-\frac{\dot{m}}{2} C_p \frac{dT_{f1(z)}}{dz} = \frac{T_{f1} - T_b}{R'_1} + 2\frac{T_{f1} - T_{f2}}{R'_{12}}$$

$$\frac{\dot{m}}{2} C_p \frac{dT_{f2(z)}}{dz} = \frac{T_{f2} - T_b}{R'_1} + 2\frac{T_{f2} - T_{f1}}{R'_{12}}$$

It can be observed that the profile given in the last section is still valid provided that:

$$S_1 = \frac{\dot{m} C_p R'_1}{2H}, \quad S_{12} = \frac{\dot{m} C_p R'_{12}}{4H} \qquad (5.87)$$

Replacing these expressions in (5.80):

$$\eta = \frac{H}{\dot{m} C_p} \sqrt{\frac{4(4R'_1 + R'_{12})}{R'^2_1 R'_{12}}}$$

From (5.48) and (5.52):

$$\eta = \frac{H}{\dot{m} C_p \sqrt{R'_b R'_a}}$$

which shows that the usual expression (5.84) is still valid as long as the correct borehole and internal resistances are used and that the total mass flow rate per borehole is used.

Case 13-24

In this configuration $T_1 = T_2$, $T_3 = T_4$

$$-\frac{\dot{m}}{2} C_p \frac{dT_{f1(z)}}{dz} = \frac{T_{f1} - T_b}{R'_1} + \frac{T_{f1} - T_{f3}}{R'_{12}} + \frac{T_{f1} - T_{f3}}{R'_{13}}$$

$$\frac{\dot{m}}{2}C_p\frac{dT_{f3(z)}}{dz} = \frac{T_{f3} - T_b}{R'_1} + + \frac{T_{f3} - T_{f1}}{R'_{12}} + \frac{T_{f3} - T_{f1}}{R'_{13}}$$

Again, the profile given in the last section is still valid provided that:

$$S_1 = \frac{\dot{m}C_p R'_1}{2H}, \; S_{12} = \frac{\dot{m}C_p R'_{12} R'_{13}}{2(R'_{12} + R'_{13})H} \tag{5.88}$$

Replacing these expressions in (5.80), using (5.55) gives:

$$\eta = \frac{H}{\dot{m}C_p}\sqrt{\frac{4(R'_{12}R'_{13} + 2R'_1(R'_{12} + R'_{13}))}{R'^2_1 R'_{12} R'_{13}}}$$

$$= \frac{H}{\dot{m}C_p\sqrt{R'_b R'_a}}$$

The expression of the effective resistances is the same for the case 12-34 and 13-24 except that the numerical value of R'_a will be different (Sect. 5.1.2).

Example 5.7 In Example 4.2, the borehole temperature evolution due to three thermal pulses was calculated with three different models. We would like to evaluate the outlet temperature after 5 hours. Apart from the borehole data already given, the following data are given:

$$x_c = r_b/3(\text{ case B}), \; k_g = 1.7 \text{ W/m-K}, \; k_p = 0.4 \text{ W/m-K}$$

$$\dot{V} = 0.8 \text{ l/s}$$

The pipe is SDR-11 (1.25″) and water is used. (Evaluate the properties at 30°C). Use the line-source method for R'_b, and compare the outlet temperature when:

(a) linear approximation
(b) exponential approximation
(c) Zeng's profile

Use only the ICS model.

Solution With the ICS model, the borehole temperature found in Example 4.2 was:

$$T_b = 9.74 \text{ °C}$$

The calculation of the borehole resistance is easy to evaluate. From Appendix B:

$$r_{pi} = 0.017 \text{ m}, r_{po} = 0.021 \text{ m}$$

(continued)

---

besides the soil conductivity, all other parameters are similar to Example 5.1, and the line-source method will give:

$$R'_b = 0.119 \text{ K-m/W} \quad , \quad R'_a = 0.323 \text{ K-m/W}$$

From which, we found after 5 hours:

– (a) Linear Approximation

$$T_f = T_b - q'R'_b = 9.74 + 2 \cdot 0.119 = 9.977 \text{ °C}$$

$$T_{fo} = T_f + \frac{q'H}{2\dot{m}C_p} = 9.977 - \frac{200}{2 \cdot 3329} = 9.946 \text{ °C}$$

– (b) Exponential Approximation

$$\gamma = \frac{H}{\dot{m}C_p R'_b} = \frac{100}{3329 \cdot 0.119} = 0.25$$

$$R'_e = 0.119 \cdot 0.126 \cdot \text{cotanh}(0.126) = 0.12 \text{ K-m/W}$$

$$T_f = T_b - q'R'_e = 9.74 + 2 \cdot 0.12 = 9.978 \text{ °C}$$

$$T_{fo} = T_f + \frac{q'H}{2\dot{m}C_p} = 9.978 - \frac{200}{2 \cdot 3329} = 9.948 \text{ °C}$$

– (c) Zeng's Profile

$$\xi = \sqrt{\frac{R'_a}{4R'_b}} = 0.82$$

$$\eta = \frac{\gamma}{2\xi} = 0.153$$

$$R'^* = 0.119 \cdot 0.153 \cdot \text{cotanh}(0.153) = 0.12 \text{ K-m/W}$$

$$T_f = T_b - q'R'_e = 9.74 + 2 \cdot 0.12 = 9.979 \text{ °C}$$

$$T_{fo} = T_f + \frac{q'H}{2\dot{m}C_p} = 9.979 - \frac{200}{2 \cdot 3329} = 9.949 \text{ °C}$$

It is seen that all values are almost identical, which explains why linear approximation is very often used. The large number of significant digits is just to illustrate that the answers are different. However, if, for example, the flow rate is diminished by a factor of 10, the results would be very different. Without going into all the details, the result would be:

$$T_{fo,l} = 9.69 \text{ °C}$$

$$T_{fo,e} = 9.80 \text{ °C}$$

$$T_{fo,l} = 9.84 \text{ °C}$$

(continued)

Example 5.7 (continued)
Not only that the differences are larger, but the linear approximation makes no sense since the outlet temperature is lower than the borehole temperature. The difference would be even larger after the second pulse (after 3 hours), where we would have:

$$T_b = 8.57\,°C$$

$$T_{fo,l} = 8.96\,°C$$

$$T_{fo,e} = 7.99\,°C$$

$$T_{fo,l} = 7.62\,°C$$

Again the linear approximation is not physical. The other two values make sense, but the interference effects are now well observed. Of course, such a small flow rate is normally not used, but if the borehole height is very large, care should be taken.

∎

5.3 Coaxial Boreholes

In the last section, the thermal behaviors of classical U-tubes and double U-tubes were analyzed. Another borehole configuration that can be used is the coaxial configuration, where the heat transfer fluid travels through concentric tubes. The fluid can enter the borehole in the center portion of the concentric tube and exit through the annulus portion or the contrary. Several studies were presented in the literature to evaluate the performance of coaxial configuration in comparison with U-tubes configurations, but in this section, only the tools needed to analyze this type of boreholes will be discussed. The reason why this configuration was not treated with the other ones is because it is closely linked to the temperature variation through the borehole. First, let's assume that the fluid enters inside the borehole in the annulus portion of the concentric tubes and exits through the inner portion.

$$T_{f,annulus} = T_{fd}, \qquad T_{f,inner} = T_{fu}$$

Again, as in the case of the U-tube, several assumptions can be made. A simple, yet not trivial, assumption is that the fluid temperature varies linearly. The main difference is that the inner tube is not in contact with the ground so that assumptions about the thermal behavior of the inner tube have to be made. A simple one is that the boundary between the inner portion and the annular portion is well isolated so that no thermal

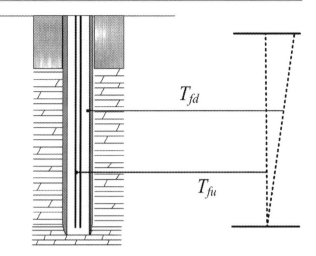

Fig. 5.6 Linear profile

short-circuit occurs. The linear approximation will then give the following temperature profile (Fig. 5.6):

As it is seen on the figure, even though simplistic assumptions are made, the fluid temperature profile is quite different than in the other configurations, and the mean temperature of the fluid at a given section varies with the depth. Since only the annulus temperature exchanges heat with the ground, one has:

$$\bar{T}_{fd} = T_b - q'_b R'_1$$

From the linear approximation, we have:

$$T_{fu} - \bar{T}_{fd} = \frac{T_{fu} - T_{fd,in}}{2} = T_{fu} - \left(T_b - q'_b R'_1\right)$$

which gives:

$$\frac{T_{fo} + T_{fi}}{2} = T_b - q'_b R'_1$$

As usual:

$$\frac{T_{fo} - T_{fi}}{2} = q_b/(2\dot{m}C_p)$$

Summing both expressions:

$$T_{fo} = T_b - q'_b R'_1 \left(1 - \frac{\gamma}{2}\right) \qquad (5.89)$$

$$\gamma = \frac{H}{\dot{m}C_p R'_1}$$

It is easily shown that the same expression would be obtained in the case where the flow enters the inner part of the concentric tube. Comparing with (5.59), it is seen that the linear approximation for the coaxial configuration leads to a similar expression than the linear approximation of the U-tube system where the borehole resistance is given by $R'_b = R'_1$. Looking at Fig. 5.7, this resistance can be expressed by:

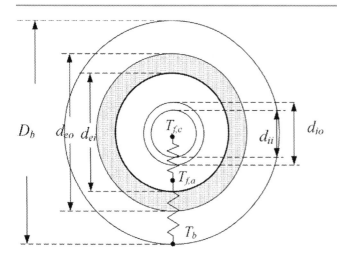

Fig. 5.7 Thermal resistances—Coaxial

$$R'_1 = R'_{conv} + R'_{cond,outerpipe} + R'_{cond,grout} + R'_{contact} \tag{5.90}$$

$$R'_1 = \frac{1}{2\pi r_{eo}\hbar_{eo}} + \frac{\log\left(\frac{r_{eo}}{r_{ei}}\right)}{2\pi k_{pe}} + \frac{\log\left(\frac{r_b}{r_{eo}}\right)}{2\pi k_g} + R'_{contact} \tag{5.91}$$

Coaxial boreholes are not always grouted [20]. In that case, the contact with the ground may be not perfect, and a thermal contact resistance can be present. In the case of perfect contact with no grout, the borehole resistance can be quite small, which brings some authors to claim that the thermal performance of coaxial boreholes is better [20] although others observed the contrary [21]. Comparative analysis will be postponed to later chapters; for now, only the different tools are presented. Like the U-tube configuration, a probably better assumption is to assume a variable heat flux through the annulus portion with a uniform temperature, which can be, as in the case of the U-tube, the borehole temperature T_b. In such a case, assuming again no short-circuiting, (5.66) is still valid only for the annulus portion, and the following result will be found:

$$T_{fo} = T_b - q'_b R'_1 \left(\frac{\gamma \exp(-\gamma)}{1 - \exp(-\gamma)} \right) \tag{5.92}$$

which is again the same result as the U-tube case where the borehole resistance is R'_1.

5.3.1 Short-Circuit Effect

Even though thermal short-circuit can reduce the performance of U-tube configurations, it is observed that in many cases, the effect is small at least for moderate-to-high flow rates. In the case of coaxial tubes, this assumption that

was made in the previous section is less valid due to the much better thermal path between the upward and downward portions of the tubes. The short-circuit resistance is:

$$R'_{12} = \frac{1}{2\pi r_{ii}\hbar_{ii}} + \frac{\log\left(\frac{r_{io}}{r_{ii}}\right)}{2\pi k_{pi}} + \frac{1}{2\pi r_{io}\hbar_{io}} \tag{5.93}$$

In coaxial applications, the pipe material can be different. The inner pipe (k_{pi}) is chosen to be a poor conductor, and the outer pipe (k_{po}) has, if possible, a much higher conductivity and/or a thinner thickness.

5.3.1.1 Annulus, In; Inner Tube, Out

To analyze its impact, the energy balance between both portions has to be done. Assuming first that the inward portion is the annulus, the following equations have to be solved:

$$-\dot{m}C_p \frac{dT_{fd}(z)}{dz} = \frac{T_{fd} - T_b(z)}{R'_1} + \frac{T_{fd} - T_{fu}}{R'_{12}} \tag{5.94a}$$

The upward leg will be governed by:

$$\dot{m}C_p \frac{dT_{fu}(z)}{dz} = \frac{T_{fu} - T_{fd}}{R'_{12}} \tag{5.94b}$$

This system can be solved [22] if we assume a uniform borehole temperature. With the following boundary conditions:

$$T_{f1}(0) = T_{f,i}, \qquad T_{f1}(H) = T_{f2}(H)$$

The result is again given in terms of the reduced temperature:

$$\theta(\bar{z}) = \frac{T_f(\bar{z}) - T_b}{T_{fi} - T_b}$$

$$\theta(\bar{z})_d = \exp(-\gamma\bar{z}/2) \left[\frac{\cosh(\eta(1 - \bar{z})) + \xi \sinh(\eta(1 - \bar{z}))}{\cosh(\eta) + \xi \sinh(\eta)} \right] \tag{5.95a}$$

$$\theta(\bar{z})_u = \exp(-\gamma\bar{z}/2) \left[\frac{\cosh(\eta(1 - \bar{z})) - \xi \sinh(\eta(1 - \bar{z}))}{\cosh(\eta) + \xi \sinh(\eta)} \right] \tag{5.95b}$$

where the variables have similar roles than in the U-tube configuration:

$$\xi = \sqrt{\frac{R'_a}{4R'_1}}$$

$$\eta = \frac{\gamma}{2\xi}$$

The resistance that plays the same role than the internal resistance in the U-tube configuration is:

$$R'_a = \frac{4R'_1 R'_{12}}{4R'_1 + R'_{12}} \tag{5.96}$$

It is seen that the exit temperature has the same expression than before:

$$\theta_o = \frac{\cosh(\eta) - \xi \sinh(\eta)}{\cosh(\eta) + \xi \sinh(\eta)} \quad (5.97)$$

From (5.69), one finds the usual result:

$$R'^*_b = R'_b \eta \coth(\eta) \quad (5.98)$$

with:

$$\eta = \frac{H}{\dot{m}C_p \sqrt{R'_1 R'_a}} \quad (5.99)$$

which is the same result as before owing that $R'_b = R'_1$.

5.3.1.2 Annulus, Out; Inner Tube, In

A similar analysis assuming the flow entering the inner portion of the concentric will lead to similar results:

$$-\dot{m}C_p \frac{dT_{fd}(z)}{dz} = \frac{T_{fd} - T_{fu}}{R'_{12}} \quad (5.100a)$$

The upward leg will be governed by:

$$\dot{m}C_p \frac{dT_{fu}(z)}{dz} = \frac{T_{fu} - T_b}{R'_1} + \frac{T_{fu} - T_{fd}}{R'_{12}} \quad (5.100b)$$

The solution of the problem is almost the same:

$$\theta(\bar{z})_d = \exp(\gamma \bar{z}/2) \left[\frac{\cosh(\eta(1 - \bar{z})) + \xi \sinh(\eta(1 - \bar{z}))}{\cosh(\eta) + \xi \sinh(\eta)} \right] \quad (5.101a)$$

$$\theta(\bar{z})_u = \exp(\gamma \bar{z}/2) \left[\frac{\cosh(\eta(1 - \bar{z})) - \xi \sinh(\eta(1 - \bar{z}))}{\cosh(\eta) + \xi \sinh(\eta)} \right] \quad (5.101b)$$

So is the exit temperature as well as the effective resistance:

$$\theta_o = \frac{\cosh(\eta) - \xi \sinh(\eta)}{\cosh(\eta) + \xi \sinh(\eta)} \quad (5.102)$$

The theoretical result just found seems to indicate that there is no thermal performance difference between both configurations. It was experimentally observed that the thermal performance can change from one configuration to another [23, 24]. This can be explained if the far-field temperature varies or in the unsteady cases which is not the case here. In [23], for example, it was suggested that the annulus-in configuration is better in the early stage of the simulation, and the difference between both configurations tends to zero with time. In intermittent applications, this unsteady effects can become important, and some authors suggest to use the annulus-in configuration. Beier et al. [25, 26] solve a similar system except that they replace the borehole temperature by the far-field temperature.

$$-\dot{m}C_p \frac{dT_{fd}(z)}{dz} = \frac{T_{fd} - T_o(z)}{R'_t(t)} + \frac{T_{fd} - T_{fu}}{R'_{12}} \quad (5.103a)$$

The upward leg will be governed by:

$$\dot{m}C_p \frac{dT_{fu}(z)}{dz} = \frac{T_{fu} - T_{fd}}{R'_{12}} \quad (5.103b)$$

with:

$$R'_t(t) = R'_1 + R'_s(t)$$

$R'_s(t)$ is the unsteady soil resistance, which can be calculated with any thermal response model:

$$R'_s = \frac{G}{k_s}$$

In their solution, the far-field temperature can vary, but in the special case where it is uniform, the solution is:

$$\Theta_{down}(\bar{z}) = \frac{T_{fd}(\bar{z}) - T_o}{T_{fi} - T_o} = C_1 \exp(a_1 \bar{z}) + C_2 \exp(a_2 \bar{z}) \quad (5.104a)$$

$$\Theta_{up}(\bar{z}) = \frac{T_{fu}(\bar{z}) - T_o}{T_{fi} - T_o} = C_3 \exp(a_3 \bar{z}) + C_4 \exp(a_4 \bar{z}) \quad (5.104b)$$

with:

$$C_1 = \frac{(\delta_2 - 1) \exp(a_2)}{(\delta_2 - 1) \exp(a_2) - (\delta_1 - 1) \exp(a_1)}$$

$$C_2 = 1 - C_1 \quad , \quad C_3 = \delta_1 C_1 \quad , \quad C_4 = \delta_2 C_2$$

$$\delta_1 = \frac{N_{s1} + N_{12} + a_1}{N_{12}}$$

$$\delta_2 = \frac{N_{s1} + N_{12} + a_2}{N_{12}}$$

$$a_1 = \frac{-(N_{s1} - N_{s2}) + \sqrt{(N_{s1} - N_{s2})^2 - 4\Delta}}{2}$$

$$a_2 = \frac{-(N_{s1} - N_{s2}) - \sqrt{(N_{s1} - N_{s2})^2 - 4\Delta}}{2}$$

$$\Delta = (N_{s1} + N_{12})(N_{s2} + N_{12}) - N_{12}^2$$

and finally:

$$N_{12} = \frac{H}{\dot{m}C_p R'_{12}} \quad (5.105)$$

$$N_{s1} = \begin{cases} \dfrac{H}{\dot{m}C_p(R'_1 + R'_s)} & , \text{ Annulus-In} \\ \\ 0 & , \text{ Center-In} \end{cases} \quad (5.106a)$$

$$N_{s2} = \begin{cases} \dfrac{H}{\dot{m}C_p(R'_1 + R'_s)} & , \text{ Center-In} \\ \\ 0 & , \text{ Annulus-In} \end{cases} \quad (5.106b)$$

It was observed in [22] that solutions (5.104a) and (5.104b) are equivalent than Eqs. (5.95a) and (5.95b) if R'_s is removed from Beier's expression or if it is included in the former expressions. The inclusion of the soil resistance in the analytical solution was shown to give better results [22], but it is less trivial to infer the borehole resistance although it is possible to find the asymptotic value of the difference between the mean fluid and mean borehole temperatures.

5.3.2 Uniform Heat Flux

It can be seen that the assumption of a uniform borehole temperature is a better approximation in the case of U-tube than for the coaxial configuration [22]. This can easily be accepted if the ideal temperature variation for each case (Figs. 5.4 and 5.6) is compared. A better approximation is to assume a uniform heat transfer along the borehole.

$$q'(z) = \frac{q}{H} = C \quad , \quad \text{constant}$$

In such a case, the solution was given by Hellstrom [5] and reproduced in [22]. In the latter case, the solution was given in terms of the dimensionless temperature defined as:[3]

$$\Phi = \frac{\bar{T}_b - T_f}{T_{ref}} \quad , \quad T_{ref} = \frac{q}{\dot{m}C_p} \tag{5.107}$$

Annulus-in

The dimensionless form of (5.94) (annulus-in) becomes:

$$-\frac{d\Phi_d(\bar{z})}{d\bar{z}} = \frac{\Phi_d(\bar{z}) - \Phi_b(\bar{z})}{\bar{R}'_1} + \frac{\Phi_d(\bar{z}) - \Phi_{fu}(\bar{z})}{\bar{R}'_{12}} = \bar{q}'_d(\bar{z}) \tag{5.108a}$$

$$\frac{d\Phi_u(\bar{z})}{d\bar{z}} = \frac{\Phi_u(\bar{z}) - \Phi_d(\bar{z})}{\bar{R}'_{12}} = \bar{q}'_u(\bar{z}) \tag{5.108b}$$

where:

$$\bar{q}'(\bar{z}) = \frac{q(z)}{H \cdot C} \quad , \quad \bar{R}' = \frac{\dot{m}C_p R'}{H} \tag{5.109}$$

From the assumption made:

$$\bar{q}'_d(\bar{z}) + \bar{q}'_u(\bar{z}) = 1 \tag{5.110}$$

and summing (5.108) gives:

$$\frac{d(\Phi_u - \Phi_d)(\bar{z})}{d\bar{z}} = 1$$

$$\Phi_u(\bar{z}) - \Phi_d(\bar{z}) = \bar{z} + c_1$$

Since $\Phi_u(1) = \Phi_d(1)$:

---
[3]The sign convention was opposite in [22].

$$\Phi_u(\bar{z}) - \Phi_d(\bar{z}) = \bar{z} - 1$$

From (5.108b):

$$\bar{q}'_u(\bar{z}) = \frac{\bar{z} - 1}{\bar{R}'_{12}}$$

and from (5.110):

$$\bar{q}'_d(\bar{z}) = 1 - \frac{\bar{z} - 1}{\bar{R}'_{12}}$$

Integrating Eqs. (5.108) will give:

$$\Phi_u(\bar{z}) = \frac{(1 - \bar{z})^2}{2\bar{R}'_{12}} + c_2$$

$$\Phi_d(\bar{z}) = -\bar{z} + \frac{(1 - \bar{z})^2}{2\bar{R}'_{12}} + c_3$$

Since $\Phi_u(1) = \Phi_d(1) = \Phi(1)$:

$$\Phi_u(\bar{z}) = \frac{(1 - \bar{z})^2}{2\bar{R}'_{12}} + \Phi(1)$$

$$\Phi_d(\bar{z}) = (1 - \bar{z}) + \frac{(1 - \bar{z})^2}{2\bar{R}'_{12}} + \Phi(1)$$

The temperature $\Phi(1)$ is found from the mean borehole temperature:

$$T_b(z) - T_{f,d} = R'_1 q'$$

$$\Rightarrow \Phi_b(\bar{z}) = \Phi_d - \bar{R}'_1$$

Taking the mean value:

$$\bar{\Phi}_b = \int_0^1 \Phi_b(\bar{z})d\bar{z} = \left(-\frac{(1 - \bar{z})^2}{2} - \frac{(1 - \bar{z})^3}{6\bar{R}'_{12}} + \Phi(1)\bar{z} \right)\Big|_0^1$$

$$- \bar{R}'_1 = \frac{\bar{T}_b - \bar{T}_b}{T_{ref}} = 0$$

$$\Rightarrow \Phi(1) = \bar{R}'_1 - \frac{1}{2} - \frac{1}{6\bar{R}'_{12}}$$

$$\Phi_d(\bar{z}) = \bar{R}'_1 + \frac{1}{2} - \frac{1}{6\bar{R}'_{12}} - \bar{z} + \frac{(1 - \bar{z})^2}{2\bar{R}'_{12}} \tag{5.111a}$$

$$\Phi_u(\bar{z}) = \bar{R}'_1 - \frac{1}{2} - \frac{1}{6\bar{R}'_{12}} + \frac{(1 - \bar{z})^2}{2\bar{R}'_{12}} \tag{5.111b}$$

The effective resistance can be obtained as:

$$R'^*_b = \frac{\bar{T}_b - T_f}{q'} = T_{ref} \frac{\Phi_d(0) + \Phi_u(0)}{2q'}$$

$$R'^*_b = \frac{H}{\dot{m}C_p} \left(\bar{R}'_1 + \frac{1}{3\bar{R}'_{12}} \right)$$

$$\Rightarrow R'^{*}_{b} = R'_{1}\left(1 + \frac{1}{3\bar{R}'_{1}\bar{R}'_{12}}\right) \qquad (5.112)$$

Center-in A similar development will lead to:

$$\Phi_{d}(\bar{z}) = \bar{R}'_{1} + \frac{1}{2} - \frac{1}{6\bar{R}'_{12}} + \frac{(1-\bar{z})^{2}}{2\bar{R}'_{12}} \qquad (5.113a)$$

$$\Phi_{u}(\bar{z}) = \bar{R}'_{1} - \frac{1}{2} - \frac{1}{6\bar{R}'_{12}} + \bar{z} + \frac{(1-\bar{z})^{2}}{2\bar{R}'_{12}} \qquad (5.113b)$$

and the effective resistance is the same as in the previous case.

Example 5.8 Compute the effective resistance for the coaxial ground heat exchanger ("Annulus-in configuration") having the following characteristics:

$$H = 150 \text{ m}, \quad \dot{m} = 2 \text{ kg/s},$$

$$\hbar_{ii} = \hbar_{io} = \hbar_{eo} = 2000 \text{ W/m}^{2}\text{-s}$$

The outer pipe is a 4″-SDR13.5 HDPE ($k = 0.4$ W/m-K) tube and the inner pipe a 2″-SDR11 HDPE tube having the same conductivity. The outer pipe is in direct perfect contact with the ground, i.e.:

$$r_{b} = r_{eo}, \quad R'_{contact} = 0$$

The fluid is water ($C_{p} = 4800$, J/kg-K). Compare the effective resistance assuming uniform temperature and uniform heat flux.

Solution The borehole dimensions are given from SDR Tables (Appendix B):

$$r_{b} = r_{eo} = 5.715 \text{ cm}, r_{ei} = 4.868 \text{ cm},$$

$$r_{io} = 3.016 \text{ cm}, r_{ii} = 2.467 \text{ cm}$$

$$R'_{12} = \frac{1}{2000\pi 0.049} + \frac{\log(3.016/2.467)}{2\pi 0.4} + \frac{1}{2000\pi 0.060}$$

$$= 0.086 \text{ K-m/W}$$

$$R'_{1} = \frac{1}{2000\pi 0.097} + \frac{\log(5.715/4.868)}{2\pi 0.4} = 0.065 \text{ K-m/W}$$

$$\frac{H}{\dot{m}C_{p}} = 0.018 \text{ K-m/W} \Rightarrow \bar{R}'_{12} = \frac{0.086}{0.018} = 4.777, \quad \bar{R}'_{1} = 3.647$$

$$\gamma = \frac{1}{\bar{R}'_{1}} = 0.274, \quad \xi = \sqrt{\frac{0.064}{4 \cdot 0.065}} = 0.497$$

$$\eta = \frac{0.274}{2 \cdot 0.497} = 0.276$$

Uniform temperature Eq. (5.98):

$$R'^{*}_{b1} = 0.065\frac{0.276}{\tanh(0.276)} = 0.067 \text{ K-m/W}$$

Uniform heat flux Eq. (5.112):

$$R'^{*}_{b2} = 0.065\left(1 + \frac{1}{3 \cdot 3.647 \cdot 4.777}\right) = 0.067 \text{ K-m/W}$$

It is seen that both results are almost similar and not very different than the initial borehole resistance(R'_{1}). The dimensionless temperature variation for both cases is shown in the following figure:

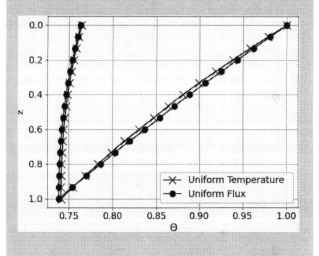

Example 5.9 Redo the last example if the height is 250 m, the flow rate is 1 kg/s, and the outer pipe is made of high conductivity plastic ($k = 1.5$ W/m-K). All other parameters are the same.

Solution The inner resistance does not change. The borehole resistance is now:

$$R'_{1} = \frac{1}{2000\pi 0.097} + \frac{\log(5.715/4.868)}{2\pi 1.5} = 0.019 \text{ K-m/W}$$

Actually, the heat transfer coefficient should change with the flow rate, but as long as the flow is turbulent (which is the case), it should not change too much the results. All other results become:

$$\frac{H}{\dot{m}C_{p}} = 0.06 \text{ K-m/W} \Rightarrow \bar{R}'_{12} = \frac{0.086}{0.06} = 1.433, \quad \bar{R}'_{1} = 0.312$$

(continued)

(continued)

Example 5.9 (continued)

$$\gamma = 3.21 \quad , \quad \xi = 0.731 \quad , \quad \eta = 2.193$$

$$R'^*_{b1} = 0.019 \frac{2.193}{\tanh(2.193)} = 0.042 \text{ K-m/W}$$

$$R'^*_{b2} = 0.019 \left(1 + \frac{1}{3 \cdot 1.433 \cdot 0.312} \right) = 0.032 \text{ K-m/W}$$

where it is seen that not only that the effective resistance is much higher than the borehole resistance but both approaches give different results. This can also be emphasized by the temperature variation:

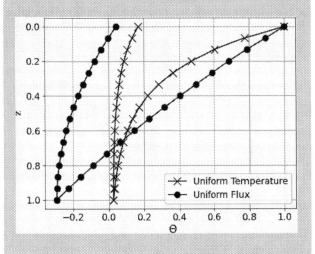

Example 5.10 As a final example, we can compare the last results with the one proposed by Beier et al. In such a case, the real temperatures should be given at a specified time. This will be done with $Fo = 100$, $T_o = 0\,°C$, $k_{soil} = 2.5\,\text{W/m-K}$, and $q' = -50\,\text{W/m}$.[a]

Solution The soil resistance can be calculated with any thermal response factor. Using the ILS, for example, we will find:

$$R_s = \frac{G_{ILS}(100)}{2.5} = 0.172 \text{ K-m/W}$$

To evaluate the absolute temperature from the Beier's profile, the inlet temperature has to be evaluated. This can be easily obtained from:

$$\Theta_{up}(0) = \frac{T_{fo} - T_o}{T_{fi} - T_o} = \frac{T_{fo} - T_{fi} + T_{fi} - T_o}{T_{fi} - T_o} = \frac{q/(\dot{m}C_p)}{T_{fi} - T_o} + 1$$

---
[a]Remember that a negative heat exchange means heat rejection.

$$T_{fi,Beier} = T_o + \frac{q}{\dot{m}C_p(\Theta_{up}(0) - 1)} = T_o + \frac{q}{\dot{m}C_p(C_3 + C_4 - 1)}$$

It is found that:

$$\Theta_{up}(0) = 0.747 \quad \Rightarrow \quad T_{fi} = 11.81\,°C$$

$$T_{fo,Beier} = 0 + 0.747(11.81 - 0) = 8.81\,°C$$

The borehole mean temperature needed for the other models is:

$$T_b = T_o + 50 \cdot 0.172 = 8.62\,°C$$

and the inlet temperature assuming uniform temperature follows from a similar way:

$$T_{fi,uniformT} = T_b + \frac{q}{\dot{m}C_p(\theta_{up}(0) - 1)} = 12.2\,°C$$

$$\theta_{up}(0) = 0.167$$

$$T_{fo,uniformT} = T_b + 0.167(12.2 - 8.62) = 9.22\,°C$$

$$T_{ref} = \frac{-12500}{4180} = -3K$$

$$\Phi_{up}(0) = 0.044$$

$$T_{fo,uniformq} = 8.62 - 0.044 \cdot (-3) = 8.75\,°C$$

The final temperature profile is thus given by:

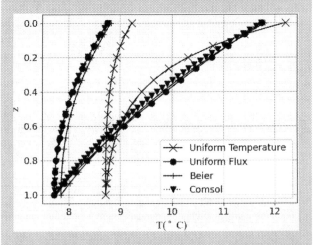

It is seen that in this much more severe case, i.e., deep borehole, low flow rate, and small borehole resistance, the hypothesis of uniform temperature is not good but assuming uniform heat transfer is better.

Problems

Pr. 5.1 The following table gives the monthly loads associated with a typical building.

	Monthly building heating loads (kW)	Monthly building cooling loads (kW)
Jan	9	0
Feb	7	0
Mar	5	0
Apr	4	0
May	2	0
June	0	0
July	0	2
Aug	0	3.5
Sept	0	0.5
Oct	2	0
Nov	5	0
Dec	6	0

(a) After the coldest month of the year (January) of the second year, evaluate the mean fluid temperature at the end of 4 hours extra pulse (see Fig. Pr_5.1), where the heat pump works at full load. Use the ICS model and the real number of hours for each month.

$$T_o = 9\,°\text{C} \quad,\quad COP_h = 3.4 \quad,\quad COP_c = 4.8 \quad,$$

$$H = 100 \text{ m} \quad,\quad q_{full,load} = 11 \text{ kW}$$

$$k_s = 3.1 \text{ W/m-K} \quad,\quad \alpha_s = 0.09 \text{ m}^2/\text{d} \quad,$$

$$R'_b = 0.12 \text{ K-m/W} \quad,\quad r_b = 8 \text{ cm}$$

(b) Redo the same calculation if the 12 first pulses are aggregated in one single yearly pulse (Fig. Pr_5.2).

Answer: (a) -13.67 °C, (b) -13.48 °C (small difference)

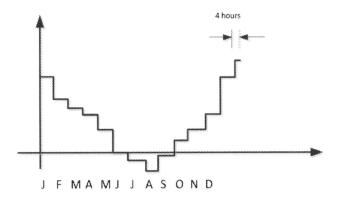

Fig. Pr_5.1 Twelve average monthly pulses

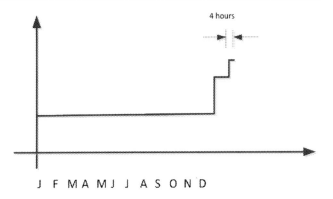

Fig. Pr_5.2 Aggregated annual pulse

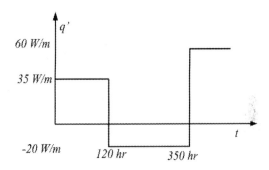

Fig. Pr_5.3 Three pulses heat exchange

Pr. 5.2 The heat per unit length rejected to the ground by a heat pump is given in Fig. Pr_5.3:

$$T_o = 10\,°\text{C} \quad,\quad k_s = 2.5 \text{ W/m-K} \quad,\quad \alpha_s = 0.1 \text{ m}^2/\text{d}$$
$$k_g = 1.2 \text{ W/m-K} \quad,\quad x_c = 2.0 \text{ cm} \quad,\quad H = 100 \text{ m}$$

The pipe is SDR-11, 1 inch nominal ($k_p = 0.4$ W/m-K). The fluid properties are:

$$\rho = 1000 \text{ kg/m}^3, \ C_p = 4.19 \text{ kJ/kg-K}, \ \nu = 8.54e - 7 \text{ m}^2/\text{s} ,$$
$$Pr = 5.86, \ k_f = 0.61 \text{ W/m-K},$$

(a) Evaluate the outlet temperature after 250 hours and 400 hours with $\dot{m} = 2000$ kg/hr using:
  (i) The line-source approach for the borehole resistance and a, assuming a linear temperature profile
  (ii) The Sharquawi relation for the borehole resistance and a, assuming a linear temperature profile
  (iii) The line-source approach for the borehole resistance and a, assuming an exponential temperature profile
  (iv) The Sharquawi relation for the borehole resistance and a, assuming an exponential temperature profile
  (v) The line-source approach for the borehole resistances and taking account the interference between pipes

Answer: (a)

$T_b(250) = 6.89°C$

(i) $T_{fo} = 4.93°C$,(ii) $T_{fo} = 5.13°C$, (iii) $T_{fo} = 4.84°C$,(iv) $T_{fo} = 5.04°C$,(v) $T_{fo} = 4.80°C$

$T_b(400) = 19.03 \ °C$

(i) $T_{fo} = 24.91°C$, (ii) $T_{fo} = 24.3°C$, (iii) $T_{fo} = 25.17°C$,iv) $T_{fo} = 24.58°C$,v) $T_{fo} = 25.30°C$

(b) Redo the same with $\dot{m} = 1000$ kg/hr.

Answer: (b)

(i) $T_{fo} = 5.76 \ °C$,(ii) $T_{fo} = 5.97 \ °C$, (iii) $T_{fo} = 5.42 \ °C$,(iv) $T_{fo} = 5.60 \ °C$,(v) $T_{fo} = 5.24 \ °C$

(i) $T_{fo} = 22.39 \ °C$,(ii) $T_{fo} = 21.79 \ °C$, (iii) $T_{fo} = 23.41 \ °C$,(iv) $T_{fo} = 22.88 \ °C$,(v) $T_{fo} = 23.96 \ °C$

Pr. 5.3 A coaxial vertical ground heat exchanger has the same dimensions as the one in Problem 2.1. The total length of the borehole is 150 m. If 10 kW of heat in injected in the ground, evaluate the outlet temperature after 48 hours using the ICS as the ground model and the uniform heat flux inside the borehole.

$$k_{soil} = 2.75 \text{ W/m-K} \quad, \quad \alpha_{soil} = 0.1 \text{ m}^2/\text{day} \quad, \quad T_o = 8 \,°C$$

Answer $T_{fo} = 23.3 \,°C$

References

1. Bergman, T.L., Incropera, F.P.: Fundamentals of Heat and Mass Transfer, 7th edn. Wiley, New York (2011)
2. Paul, N.D.: The Effect of Grout Thermal Conductivity on Vertical Geothermal Heat Exchanger Design and Performance. Master's thesis. South Dakota State University, New York (1996)
3. Remund, Ch.P.: Borehole thermal resistance: laboratory and field studies. ASHRAE Trans. **105**, 439 (1999)
4. Sharqawy, M.H., Mokheimer, E.M., Badr, H.M.: Effective pipeto-borehole thermal resistance for vertical ground heat exchangers. Geothermics **38**(2), 271–277 (2009). https://doi.org/10.1016/j.geothermics.2009.02.001
5. Hellstrom, G.: Ground Heat Storage: Thermal Analysis of Duct Storage Systems. Part I Theory. PhD thesis. University of Lund, Department of Mathematical Physics, Sweden (1991)
6. Claesson, J., Javed, S.: An analytical method to calculate borehole fluid temperatures for time-scales from minutes to decades. ASHRAE Trans. **117**, 279–288 (2011)
7. Lamarche, L., Kajl, S., Beauchamp, B.: A review of methods to evaluate borehole thermal resistances in geothermal heat-pump systems. Geothermics **39**(2), 187–200 (2010)
8. Saqib, J., Spitler, J.: Accuracy of borehole thermal resistance calculation methods for grouted single U-tube ground heat exchangers. Appl. Energy **187**(Supplement C), 790–806 (2017)
9. Rottmayer, S.P.: Simulation of ground coupled vertical U-tube heat exchangers. Master's thesis. University of Wisconsin, Madison (1997)
10. Beier, R.A.: Vertical temperature profile in ground heat exchanger during in-situ test. Renew. Energy **36**(5), 1578–1587 (2011)
11. Beier, R.A. et al.: Vertical temperature profiles and borehole resistance in a U-tube borehole heat exchanger. Geothermics **44**(Supplement C), 23–32 (2012)
12. Claesson, J., Hellström, G.: Multipole method to calculate borehole thermal resistances in a borehole heat exchanger. HVAC&R Res. **17**(6), 895–911 (2011)
13. Carslaw, H.S., Jaeger, J.C.: Conduction of Heat in Solids, 2nd edn. (1959)
14. Pahud, D., Fromentin, A., Hadorn, J.C.: The duct ground heat storage model (DST) for TRNSYS used for the simulation of heat exchanger piles. In DGC-LASEN, Lausanne (1996)
15. Hellström, G., et al.: Experiences with the borehole heat exchanger software EED. Proc. Megastock **97**, 247–252 (1997)
16. Kavanaugh, S.P.: Simulation and experimental verification of vertical ground-coupled heat pump systems. PhD thesis. Oklahoma, Oklahoma State, USA (1985)
17. Cimmino, M.: The effects of borehole thermal resistances and fluid flow rate on the g-functions of geothermal bore fields. Int. J. Heat Mass Transf. **91**, 1119–1127 (2015)
18. Zeng, H.Y., Diao, N.R., Fang, Z.H.: Heat transfer analysis of boreholes in vertical ground heat exchangers. Int. J. Heat Mass Transf. **46**(23), 4467–4481 (2003)
19. Diao, N.R., Zeng, H.Y., Fang, Z.H.: Improvement in modeling of heat transfer in vertical ground heat exchangers. HVAC&R Research **10**(4), 459–470 (2004)
20. Acuña, J.: Distributed thermal response tests: New insights on U-pipe and Coaxial heat exchangers in groundwater-filled boreholes. PhD thesis
21. Wood, C.J., Liu, H., Riffat, S.B.: Comparative performance of 'Utube' and 'coaxial' loop designs for use with a ground source heat pump. Appl. Therm. Eng. **37**, 190–195 (2012). https://doi.org/10.1016/j.applthermaleng.2011.11.015
22. Lamarche, L.: Analytic models and effective resistances for coaxial ground heat exchangers. Geothermics **97**, 102224 (2021). https://doi.org/10.1016/j.geothermics.2021.102224
23. Zanchini, E., Lazzari, S., Priarone, A.: Effects of ow direction and thermal short-circuiting on the performance of small coaxial ground heat exchangers. Renew. Energy **35**(6), 1255–1265 (2010)
24. Holmberg, H., et al.: Thermal evaluation of coaxial deep borehole heat exchangers. Renew. Energy **97**, 65–76 (2016). https://doi.org/10.1016/j.renene.2016.05.048
25. Beier, R.A., et al.: Borehole resistance and vertical temperature profiles in coaxial borehole heat exchangers. Appl. Energy **102**, 665–675 (2013). Special Issue on Advances in sustainable biofuel production and use—XIX International Symposium on Alcohol Fuels—ISAF. https://doi.org/10.1016/j.apenergy.2012.08.007
26. Beier, R.A., et al. Transient heat transfer in a coaxial borehole heat exchanger. Geothermics **51**, 470–482 (2014). https://doi.org/10.1016/j.geothermics.2014.02.006

Design of Secondary Loop Ground-Source Systems

6.1 Design Steps and Criteria

Like in most engineering processes, a GSHP designer will have to follow several steps. A possible description may include:

1. The evaluation of the building heating and cooling loads.
2. The choice of the control strategy and system components.
3. The estimation or in situ measurement of soil characteristics.
4. The choice of the heat transfer fluid.
5. The calculation of the field length and size.
6. The design of the piping network.
7. The evaluation of the investment costs.
8. Iteration to evaluate different configurations.

This chapter will look primarily at step 5: the sizing of the geothermal field. This step is very important because it is the part that has the most impact on the total cost. Obviously, several steps are coupled and different iterations are often done based on the results obtained. For instance, to evaluate the field size, we need the type of pipe used (diameter, thickness, etc.). But this choice will also have an impact on the head losses, and it is possible that this iniltial choice will be reevaluated after step 6. Obviously, the use of computer tools available today makes these different iterations easier.

Assume that the initial step is already done and the building loads are known for a typical meteorological year and that the hourly loads are similar than the one given in Fig. 2.5 in Chap. 2. A quick glance shows that the maximum heating power is of the order of 9 kW and the maximum cooling would be 5 kW. This information could immediately help us make an initial choice of equipment. A common practice in design is to oversize the equipment to ensure that it meets the loads in all conditions. In geothermal applications, oversizing is expensive; it is often rather suggested to undersize, sometimes at 70% of the maximum load. Again, as emphasized

in Chap. 2, the demand curve is a useful tool to have a better idea of how many hours the system is assumed to work at full load. Knowing this number of hours will give an idea on the best capacity limit that should be chosen. If not 100 % of the load is covered, a backup system is necessary. This backup system will again depend on the strategy used to meet the demand. Several choices are possible: one can choose an auxiliary unit that will offset what is lacking in terms of capacity (*parallel mode*); otherwise, the heat pump can stop and the auxiliary system can handle the entire load (*alternative mode*) (Fig. 6.1).

Although the first approach seems an obvious choice, because the extra equipment will be of smaller size, it is possible to see the second, as in the case of air-source heat pumps. Knowing the necessary powers, choosing a heat pump will be done. There exist a significant number of manufacturers and the choice will therefore depend on performance considerations, prices, and sometimes on good or bad experiences that the designer may have previously encountered. A performance chart from a water-air GSHP is shown in Appendix A. As discussed in Chap. 3, a water-air heat pump performance depends mainly on:

- The entering water temperature (EWT)
- The dry air temperature
- The water flow rate
- The air flow rate

The most important of these parameters is the entering water temperature (EWT). The other must remain in acceptable bounds but their impact has less influence. The COP versus the EWT in heating and cooling mode is plotted in Fig. 6.2: Thus, by increasing the temperature in cooling mode or by lowering it in heating mode, the system performance increases. In geothermal applications, it is obviously limited to the ground temperature. It is impossible to have a temperature higher than the ground temperature in heating mode. Moreover, if we want to reach almost the ground temperature,

© The Author(s), under exclusive license to Springer Nature Switzerland AG 2023
L. Lamarche, *Fundamentals of Geothermal Heat Pump Systems*,
https://doi.org/10.1007/978-3-031-32176-4_6

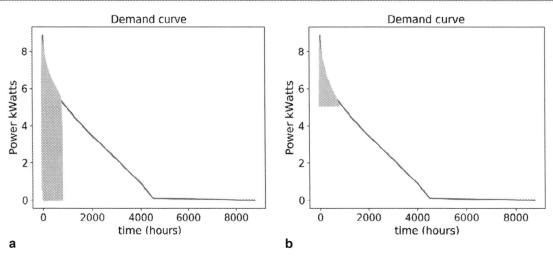

Fig. 6.1 (**a**) Alternative mode. (**b**) Parallel mode

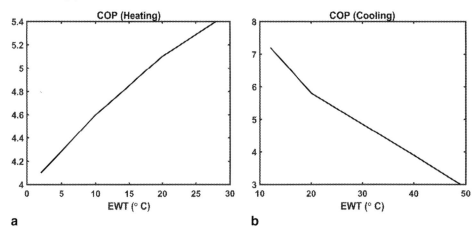

Fig. 6.2 (**a**) Heating COP temperature dependance. (**b**) Cooling COP temperature dependance

the borehole length and the total cost will be very high. So, a compromise between cost and performance has to be made. This compromise is not always easy to do. Kavanaugh and Raferty [1] suggested to take the following values:

$$EWT_{min} \approx T_{ground} - [5 - 10]°C \qquad \text{heating} \quad (6.1a)$$

$$EWT_{max} \approx T_{ground} + [11 - 16]°C \qquad \text{cooling} \quad (6.1b)$$

Regardless of the values chosen, the temperatures and the impact of the total length are thus very important and are often chosen as the main criterion for sizing geothermal fields. Thus, the conventional way to size a geothermal field is to choose the suitable one:

$$EWT_{heating} > EWT_{min} \qquad (6.2a)$$

$$EWT_{cooling} < EWT_{max} \qquad (6.2b)$$

From this criterion, the goal of a sizing method is to find the relation between these temperatures and the problem inputs,

that is, the building loads, the ground properties, etc.... This is done using the tools given in the last chapters.

6.2 The Concept of Borehole Sizing

In the last chapters, different models that can be used to evaluate the borehole temperature knowing the heat exchange in the ground were described. Whatever model used, it can be expressed as (4.48):

$$\Delta T_b(Fo)$$

$$= -\frac{1}{k} \left(q'(0)G(Fo) + (q'(Fo_1) - q'(0))G(Fo - Fo_1) + \ldots \right)$$

$$T_b(Fo) - T_o$$

$$= -\frac{1}{L} \underbrace{\left[\frac{1}{k} \left(q(0)G(Fo) + (q(Fo_1) - q(0))G(Fo - Fo_1) + \ldots \right) \right]}_{TLS}$$

$$(6.3)$$

where TLS will be used to express the *temporal load superposition*. L represents the total length of the bore field, i.e., $L = N_b H$. As it was stated earlier, G can represent any step response that we already discussed in the previous chapters. The goal now is to link this expression to the design criterion which is the EWT or the ground exit fluid temperature.

From Eq. (5.4), assuming the linear profile:

$$T_f = \frac{T_{f,in} + T_{f,out}}{2} = T_b - q_b' R_b' = T_o - \frac{1}{L}\left(TLS + q_b R_b'\right)$$

Isolating the total length gives:

$$L = \frac{TLS + q_b R_b'}{T_o - \frac{T_{in} + T_{out}}{2}} \tag{6.4}$$

This relation is the basic equation for all sizing methods that were proposed.

Example 6.1 Let's look again at Examples 4.2 and 5.7. Evaluate the borehole length such that the heat pump EWT would be 5 °C at the end of the second pulse (before the heat injection). Use only the linear approximation and give the answer when the response factor is:

(a) The ICS
(b) The ILS

Solution From Example 5.7, the borehole resistance was:
$$R_b' = 0.12 \text{ K-m/W}$$

In order to use (6.4), we need the mean fluid temperature. This value will change with the loads. It is evaluated in the worst case, that is, here, at the end of the second pulse:

$$q = 1800 \text{ W} = \dot{m} C_p (T_{f,out} - T_{f,in})$$

$$T_{f,out} = EWT = 5°C \Rightarrow T_{f,in} = 4.46°C \Rightarrow T_f = 4.73°C$$

For the ICS, we have:

$$Fo_1 = \frac{\alpha t}{r_b^2} = 0.7407, \; Fo = 2.22$$

$$G_{ICS}(2.22) = 0.168, \; G_{ICS}(2.22 - 0.7407) = 0.147$$

$$TLS = \frac{1}{2}\left(1000 \cdot 0.168 + 800 \cdot 0.147\right) = 143 \text{ m-K}$$

In (6.4) $q_b(t)$ is the heat load at a given time. Here it is the heat value at the final time. The borehole resistance contribution is:

$$q_b R_b' = 1800 \cdot 0.12 = 215 \text{ m} \cdot K$$

The final result is then:

$$L_a = \frac{143}{5.27} + \frac{215}{5.27} = 27.13 + 40.79 = 67.9 \text{ m}$$

Although the final value is important, it is often instructive to see the impact of all contributions on the total sizing. This will help the designer to identify the parameters that can improve a given design. In this example, the load history and the borehole resistance effects are in the same order of magnitude, the latter being more important. It is easy to verify that replacing 100 m value by 67.9 m in Example 5.7 will give $T_{fo} = 5\,°C$.

To find the total length if the ILS is used, the only difference will be the effect of the TLS:

$$G_{ILS}(2.22) = 0.137, \; G_{ILS}(2.22 - 0.7407) = 0.108$$

$$TLS = \frac{1}{2}\left(1000 \cdot 0.137 + 800 \cdot 0.108\right) = 112\text{m-K}$$

$$L_a = \frac{112}{5.27} + \frac{215}{5.27} = 21.2 + 40.79 = 62 \text{ m}$$

Comments: *We observe a substantial difference between the two approaches, in the order of 10 %. One must however remark that the time involved here (3 hours) is very small compared to time range normally used in geothermal applications (several years) where the difference will often be relatively smaller. Nevertheless, the impact of the short-time behavior of thermal response is now an important aspect of current research. For now, let's just emphasize that the ICS will give a more conservative design than the ILS. As observed in the previous example, the interference effect in this example is very small and the linear approximation is adequate. The impact of the interference in the sizing method will be discussed later on.*

∎

(continued)

6.3   ASHRAE Sizing Method

Although (6.4) is the basis of most of the sizing algorithms that were proposed, the sizing software available will often give different answers to the same problem. Several causes explain this discrepancy:

- They might use different approaches to quantify the superposition of thermal loads in the ground.
- They might use different step responses (ILS, FLS, ICS, numerical response, etc.)
- They might evaluate the mutual interferences between boreholes differently.
- They might use different relations for the borehole resistance R'_b.

In this section, a method known as the ASHRAE method, given in their Handbook [2], for several years was first proposed by Kavanaugh and Rafferty [1]. The method is based on (6.4) where:

- The TLS is based on three pulses.
- The relation of Paul and Remund is used for the borehole resistance.
- The ICS solution is used to evaluate the ground response.
- The concept of temperature penalty is used to take care of interference between boreholes.

The idea of using three pulses is to have a simple formula that can be calculated by hand and still keep the time dependence in the ground. For instance, if we are looking at the temperature at the end of a 10-year cycle, using hourly loads, we can evaluate the fluid temperature obtained at the end of 87,600 pulses (Fig. 6.3). This would take a lot of time. Instead, the mean heat exchange during a year is evaluated and used as a *yearly pulse* (q_a) that will last for several years

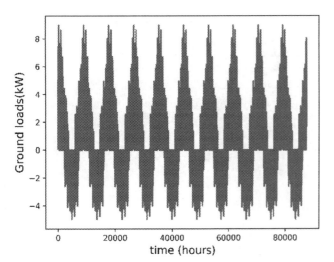

Fig. 6.3 Ten years hourly loads

(taken as 10 originally). At the end of this pulse, a monthly pulse associated with the worst month of the year is added, and finally an hourly pulse associated with the peak load is added at the end (Fig. 6.4):

The yearly pulse was initially chosen as 10 years but that can be easily changed to 20 or 30 if necessary. Doing so, the superposition of temporal loads become:

$$TLS = \frac{1}{k_s}\left(q_a G(Fo_f) + (q_m - q_a)G(\underbrace{Fo_f - Fo_y}_{Fo_1}) \right.$$
$$\left. + (q_h - q_m)G(\underbrace{Fo_f - Fo_m}_{Fo_2}) \right)$$

It was proposed to rearrange the terms as:

$$TLS = q_a \frac{G(Fo_f) - G(Fo_1)}{k_s}$$
$$+ q_m \frac{G(Fo_1) - G(Fo_2)}{k_s} + q_h \frac{G(Fo_f)}{k_s} \qquad (6.5)$$
$$= q_a R'_{g,a} + q_m R'_{g,m} + q_h R'_{g,h}$$

We also have:

$$Fo_f = \frac{\alpha_s t_f}{r_b^2}$$
$$Fo_1 = \frac{\alpha_s (t_f - t_y)}{r_b^2} \qquad (6.6)$$
$$Fo_2 = \frac{\alpha_s (t_f - t_m)}{r_b^2}$$

If the time scale is hour, we have:

$$t_f = N \cdot 365 \cdot 24 + 730 \cdot 24 + b_h \qquad (6.7)$$
$$t_f - t_y = 730 \cdot 24 + b_h \qquad (6.8)$$
$$t_f - t_m = b_h \qquad (6.9)$$

N is the number of years that is kept (10, 20, 30,…) . b_h is an hourly block which is often 4 or 6 hours. Using this formalism, Eq. (6.4) can now be written as:

$$L = \frac{q_a R'_{g,a} + q_m R'_{g,m} + q_h (R'_{g,h} + R'_b)}{T_o - T_p - T_f} \qquad (6.10)$$

In this equation, T_p is a new term that was not yet discussed and is called the temperature penalty which is due to the interference between boreholes. This concept will be discussed in the next section. For now, we only need to remember that this value is zero for a single borehole. Equation (6.10) is

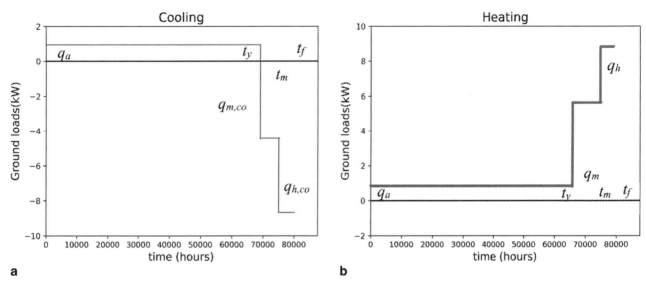

Fig. 6.4 (a) Three pulses equivalent, cooling mode. (b) , Three pulses equivalent, heating mode

valid for cooling mode and heating mode. In the ASHRAE Handbook, they specify two equations for each mode, but they are essentially the same. We must just remember the sign convention: positive in heating mode and negative in cooling mode. We usually compute one length in each mode and keep the longest one:

$$L = \max(L_{cooling}, L_{heating}) \qquad (6.11)$$

It is also important to remember that in each mode, the monthly and hourly loads will change but the yearly pulse will remain the same.

Example 6.2 The monthly heat exchanged by a GSHP is given in the following Table. It is assumed that these loads will remain the same for all subsequent years.

Month	Q(MWh)	Q(MWh)
Jan	2.17	0.0
Feb	2.07	0.0
Mar	1.75	0.0
Apr	1.39	0.0
May	0.9	−1.2
Jun	0.0	−1.6
Jul	0.0	−2.0
Aug	0.0	−1.8
Sep	0.85	−0.8
Oct	1.22	−0.4
Nov	1.64	0.0
Dec	2.02	0.0

(continued)

The peak load is chosen at the heat pump capacity which is 12 kW in heating mode and cooling mode with a COP = 3 (heating) and 5 (cooling). The hourly block is chosen to be 4 hours and a 10-year life time is used. We have the following data:

Ground

$$k_s = 2.5 \text{ W/m-K}, \alpha_s = 0.09 \text{ m}^2/\text{day} , T_o = 10°C$$

Borehole (single U-tube)

$$k_g = 1.7 \text{ W/m-K}, r_b = 7.5 \text{ cm}, d_{po} = 33 \text{ mm}, d_{pi} = 27 \text{ mm}$$

$$x_c = r_b/3 (\text{CASE B}), k_p = 0.4 \text{ W/m-K}$$

Fluid

$$\dot{V} = 0.6 \text{ l/s}, \rho = 1030 \text{ kg/m}^3, Cp = 3.80 \text{ kJ/Kg-K}$$

$$\mu = 0.004 \text{ N-s/m}^2, k_f = 0.44 \text{ W/m-K}$$

The minimum temperature at the heat pump entrance is 0°C and 25°C in cooling mode. Assume that all months have the same number of hours (730). Find the minimum length assuming a single borehole, using

(a) Borehole resistance by Paul
(b) Borehole resistance by Sharqawi
(c) Borehole resistance by the line source

(continued)

Example 6.2 (continued)

Solution First, the three pulses in heating and cooling are evaluated. Remember that these are ground loads, not building loads. Since the monthly loads are on the ground side, the monthly loads are simply:

$$q_m(heating) = 2.17 \cdot 1e6/730 = 2973 \text{ W}$$

$$q_m(cooling) = -2.00 \cdot 1e6/730 = -2740 \text{ W}$$

If the real number of hours was chosen (744 in January, 672 in February, etc.), the monthly peak would be in February since even if the energy consumption is less, the average power would be higher, but the difference is small, and we will continue with these numbers. Be also careful that the cooling loads can be given in absolute values, so the minus sign has to be added. If hourly loads were given, it would have been possible to estimate the hourly pulse, but here it is unknown. In that situation, it is possible to use the capacity of the heat pump. Since it is always given on the building side, the ground side is given by:

$$q_h(heating) = 12000 \cdot \frac{COP - 1}{COP} = 8000 \text{ W}$$

$$q_h(cooling) = -12000 \cdot \frac{COP + 1}{COP} = -14400 \text{ W}$$

Finally, the yearly load is found knowing the total energy delivered to the ground:

$$q_a = \frac{\sum Q_h + \sum Q_c}{8760} = \frac{6210000}{8760} = 709 \text{ W}$$

Again, the addition sign is used if the cooling loads are signed. The Fourier numbers are given by:

$$Fo_f = \frac{\alpha_s t_f}{r_b^2} = \frac{0.00375(8760 + 730 + 4)}{0.075^2} = 58889$$

$$Fo_1 = \frac{\alpha_s t_1}{r_b^2} = \frac{0.00375(734)}{0.075^2} = 489$$

$$Fo_2 = \frac{\alpha_s t_2}{r_b^2} = \frac{0.00375(4)}{0.075^2} = 2.67$$

$$G(Fo_f) = 0.938, \ G(Fo_1) = 0.558, \ G(Fo_2) = 0.179$$

$$R'_{g.a} = \frac{0.938 - 0.558}{2.5} = 0.1522 \text{ m-K/W}$$

$$R'_{g.m} = \frac{0.558 - 0.179}{2.5} = 0.1516 \text{ m-K/W}$$

$$R'_{g.h} = \frac{0.178}{2.5} = 0.071 \text{ m-K/W}$$

The borehole resistance is now calculated. The convective resistance is found from the Reynolds number:

$$Re = \frac{4\dot{m}}{\pi D_i \mu_i} = \frac{4 \cdot 0.618}{\pi 0.027 \cdot 0.004} = 7286$$

Since it is turbulent, the Gnielinski correlation is used (2.43b):

$$Pr = \frac{\nu_f}{\alpha_f}, \ \nu_f = \frac{\mu_f}{\rho_f} = 3.88\text{e-}6 \text{ m}^2/s$$

$$\alpha_f = \frac{k_f}{\rho_f Cp_f} = 1.12\text{e-}7 \text{ m}^2/s, \ Pr = 34.5$$

$$f = 0.0086, \qquad Nu_D = 103.86$$

$$\hbar_f = \frac{Nu_d k_f}{D_i} 1692 \text{ W/m}^2 \text{ K}$$

$$R'_{conv} = 0.007 \text{ K-m/W}$$

$$R'_{cond} = \frac{\log(33/27)}{2\pi 0.04} = 0.08 \text{ K-m/W} \Rightarrow R'_p = 0.087 \text{ K-m/W}$$

The grout resistance is first calculated from Paul and Remund relation (5.8):

$$R'_g = 0.084 \text{ w-m/K} \Rightarrow R'_b = 0.084 + \frac{0.087}{2} = 0.128 \text{ w-m/K}$$

Since $T_p = 0$, the only unknown left is the mean fluid temperature. In heating mode, we have:

$$T_{fo} = 0, \ T_{fi} = T_{fo} - \frac{q_b}{\dot{m}C_p} = 0 - \frac{8000}{2348.4} = -3.4°C$$

$$\Rightarrow T_f = -1.7°C$$

In cooling mode, we have:

$$T_{fo} = 25°C, \ T_{fi} = T_{fo} - \frac{q_b}{\dot{m}C_p} = 25 + \frac{14400}{2348.4}$$

$$= 31.1°C \Rightarrow T_f = 28.0°C$$

From these results, we can now evaluate the required length:

$$L_{heating}$$

$$= \frac{709 \cdot 0.1522 + 2973 \cdot 0.1516 + 8000(0.071 + 0.128)}{10 - (-1.7)}$$

$$= 184 \text{ m}$$

(continued)

(continued)

Example 6.2 (continued)

Although the application is heating dominant, the length in cooling mode is also computed:

$$L_{cooling} =$$

$$\frac{709 \cdot 0.1522 - 2740 \cdot 0.1516 - 14400(0.071 + 0.128)}{10 - 28}$$

$$= 176 \text{ m}$$

$$L = \max(184, 176) = 184 \text{ m}$$

Although the total length is required, it is often instructive to separate each contribution:

$$L_{heating} = L_y + L_{m,h} + L_{h,h} + L_{R_b,h}$$

$$L_{heating} = 9 + 39 + 49 + 87 = 184 \text{ m}$$

$$L_{cooling} = -6 + 23 + 57 + 102 = 176 \text{ m}$$

$$L_a = 184 \text{ m}$$

So the most important contribution to the total length is due to the peak load, which contributes through the hourly ground resistance and the borehole resistance, the latter one given the highest contribution. When we want to improve a typical design, it is important to know which parameter is the most important.

If we use the Sharqawi relation or the line-source relation for the borehole resistance, everything will be the same except for the value of R'_b. The final result will be:

(b)

$$R'_b = 0.131 \text{ W-m/K}$$

$$\Rightarrow L_b = 186 \text{ m}$$

(c)

$$R'_b = 0.133 \text{ W-m/K}$$

$$\Rightarrow L_c = 188 \text{ m}$$

The order of magnitude of the borehole length does not change significantly in the three cases.

Comments: *Since no instructions were given in the problem statement, the ICS model was used as proposed in the original ASHRAE method, but it is, of course, possible to use the ILS or FLS.*

■

Special Case

Example 6.2 (Modified) Redo the same problem if the COP in cooling mode is 3. Assume that the monthly loads remain unchanged.

Solution Since the monthly loads are ground loads, they remain unchanged, it is easily seen that the heating length will remain the same. For the cooling length, only the peak hourly load changes:

$$q_{h,cool} = -12000\frac{4}{3} = -16000 \text{ W}$$

$$T_{fo} = 25 + \frac{16000}{2348.4} = 31.8 \text{ °C}$$

$$T_{f,c} = 28.4 \text{ °C}$$

$$L_{cooling}$$

$$= \frac{709 \cdot 0.1522 - 2740 \cdot 0.1516 - 16000(0.071 + 0.128)}{10 - 28.4}$$

$$= 190 \text{ m}$$

$$L_{cooling} = -6 + 23 + 62 + 111 = 190 \text{ m}$$

$$L = \max(184, 190) = 190 \text{ m}$$

We observe now that, even though the problem is heating dominant ($q_a > 0$), the maximum length is cooling mode due to the high peak load. In this case, it is easy to see that the direct application of Eq. (6.10) will have a negative term that reduces the total length. This means that cooling the ground for 10 or 20 years is beneficial for the cooling mode which gives the predominant length. But one has to observe that this effect will be obtained only after many years, so in the first years of operation, this favorable impact will not be observed and a temperature higher than the prescribed one (25 in this example) will be reached in the first year of operation. To prevent this, it is recommended [3] to put the yearly load to zero in the non-dominant mode, which will finally be given here:

$$L_{cooling} = 0 + 23 + 62 + 111 = 196 \text{ m}$$

Comments: *From this recommendation, it would seem normal to redo the previous example, but the final result will not change since the larger length will still be in heating mode.*

■

Remark In the ASHRAE method, it is suggested to put the yearly load to zero in the non-dominant mode, or equivalently, $L_y = \max(L_y, 0)$

Before looking in more details with the concept of temperature penalty, it may be noted that the original formula given in the ASHRAE Handbook is a little bit different although equivalent than the one given. The original relation looks more like:

$$L = \frac{q_a R'_{g,a} + (q_{load} - W_c)(R'_b + \text{PLF}_m R'_{g,m} + R'_{g,h} F_{sc})}{T_o - T_p - T_f}$$

(6.12)

Besides some minor differences in the subscripts, it is basically the same thing as before. The first term is exactly the same. The factor in front of the second term is the peak building load minus the work of the compressor. This is basically the heat exchange in the ground; remember that the loads are positive in heating and negative in cooling. So this factor is essentially the peak hourly pulse as previously defined. Finally, PLF_m is defined as the monthly part load factor and from its definition:

$$\text{PLF}_m = \frac{\text{Total Energy given by the heat pump in the month}}{\text{Total Energy given by the heat pump if always ON}}$$

(6.13)

the product of the peak load times this factor gives our monthly pulse. Finally, the last variable not yet defined is given by the factor F_{sc} and is described as the *short-circuit factor* which is due to the interference between upward and downward legs of the U-tube. However, the value proposed is quite small (≈ 1.04) and does not change very much the total length calculated. For now, we will leave this factor to unity and come back later to the effect of interferences later on. The modified expression that we will only use here has first been proposed by Bernier [4].

6.4   Temperature Penalty

In the ASHRAE method, the ground step response used is the ICS solution which assumes that there is just one borehole. Actually, the solution is still valid if there are several boreholes as long as there is no interference, that is, if one borehole does not see the effect of the others. That would be the case if the boreholes are far apart or if the loads are balanced, i.e., the yearly load is almost zero. To take into account the effect of field of boreholes, we can replace the ICS solution by a step response of the bore field. This is the approach taken by the "g-functions" proposed by the Swedish researchers. In the ASHRAE method, the solution chosen is to keep the ICS solution associated with a single borehole and to evaluate the temperature variation around a

single borehole due to all others. This temperature variation is called the *temperature penalty*. This correction can be seen as a variation of the unperturbed ground temperature. It has always the same sign than the yearly pulse. The idea is to find the long-range induced temperature from the neighborhood boreholes.

6.4.1   The Kavanaugh and Rafferty Solution

Kavanaugh and Rafferty [5] proposed the following procedure:

Based on Fig. 6.5, they assume that all the heat that would normally diffuse in the ground will be trapped inside a rectangle region that would result in an increase (or decrease) of the temperature. In order to find this temperature variation, they compute:

$$T_p^* = \frac{Q'_{out} L}{m C_p} = \frac{Q'_{out}}{\rho b^2 C_p}$$

(6.14)

where Q'_{out} is the total amount of heat per unit length (J/m) that would be trapped inside the volume and b is the distance between boreholes. To evaluate this number, they suggest the following procedure.

The neighboring region including the borehole is now represented by a cylinder of diameter b. To evaluate the heat that would diffuse out of this cylinder and that is now trapped inside the cylinder due to the adjacent boreholes, they evaluate the "total energy" in the adjacent region out of the inner cylinder if the borehole would be alone. This value was associated with the sum of annular regions having a mean temperature increase of T_i:

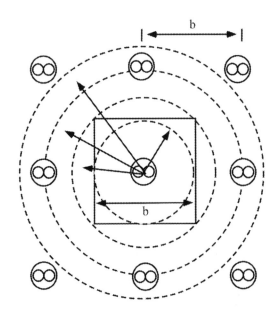

Fig. 6.5 Temperature penalty evaluation

$$Q'_{out} \approx \sum_{i=1}^{N} \rho A_i C_p T_i \qquad (6.15)$$

where

$$A_i = \frac{\pi(r_{oi}^2 - r_{ii}^2)}{4} \qquad (6.16)$$

is the area of ring i outside the inner region and

$$T_i = \frac{q'_a I(X_i)}{2\pi k_s} \qquad (6.17)$$

is the average induced temperature in the ring i, calculated with:

$$X_i = \frac{r_{ci}}{2\sqrt{\alpha t_f}}, \quad r_c i = \frac{r_{oi} + r_{ii}}{2} \qquad (6.18)$$

which leads to:

$$T_p^* = q'_a \frac{\sum_{i=1}^{N} I(X_i)(r_{oi}^2 - r_{ii}^2)}{2\pi k_s b^2} \qquad (6.19)$$

There is a sign difference between the last expression and the one given in the last chapter. The reason is that the penalty temperature is explicitly subtracted from the ground temperature in (6.10). This arbitrary choice implies that this temperature represents the temperature reduction in the ground and not the temperature variation.

In their book, they suggest to use three rings in Eq. (6.19) each one having a 5 feet thickness (≈ 1.5 m).

Example 6.3 In a typical borehole arrangements, the yearly pulse per unit length is expected to be equal to:

$$q'_a = 8 \text{ W/m}$$

The soil conductivity is 2.0 W/m-K, and the diffusivity is $\alpha_s = 0.1$ m²/day. The distance between boreholes is 6 meters. A 10-year period is assumed for the calculation.

Solution

$$X_1 = \frac{3 + 1.5/2}{2\sqrt{2 \cdot 0.1 \cdot 3650}} = 0.0981 \Rightarrow I(X_1) = 2.0375$$

$$X_2 = \frac{3 + 1.5 + 1.5/2}{2 * \sqrt{2 \cdot 0.1 \cdot 3650}} = 0.1374 \Rightarrow I(X_2) = 1.7057$$

$$X_3 = \frac{3 + 3 + 1.5/2}{2 * \sqrt{2 \cdot 0.1 \cdot 3650}} = 0.1767 \Rightarrow I(X_3) = 1.4604$$

$$T_p^* = 1.2735 + 1.4924 + 1.643 = 4.41 K$$

(continued)

Remember that this is a temperature reduction, which means that the ground temperature will "decrease" by approximately 4.4 K.

∎

Since this concept assume that all the energy is trapped inside a given volume, it is valid for the inner boreholes. In order to correct for the boreholes at the periphery of the field, the authors propose the following correction:[1]

$$T_p = T_p^* \frac{N_4 + 0.75N_3 + 0.5N_2 + 0.25N_1}{C_{fHoriz}N_t} \qquad (6.20)$$

where $N_t = N_1 + N_2 + N3 + N_4$, and:

- N_4: number of boreholes surrounded by four other boreholes (inner boreholes)
- N_3: number of boreholes surrounded by three other boreholes (boreholes at the periphery)
- N_2: number of boreholes surrounded by two other boreholes (corner boreholes)
- N_1: number of boreholes surrounded by one borehole (outer borehole inline)

The factor C_{fHoriz} was introduced in [1]. It is associated with the depth of the borehole and defined as:

$$C_{fHoriz} = \frac{2H_{bore}(W_{field} + L_{field}) + (W_{field} \cdot L_{field})}{2H_{bore}(W_{field} + L_{field})} \qquad (6.21)$$

This correction factor is always greater than one and is close to one when the depth is large.

Example 6.4 The uncorrected temperature penalty T_p^* for a given application is given by the last example, $T_p^* = 4.41$ K. What is the corrected temperature penalty for a:

(a) 3×3 square borefield
(b) 9×1 inline borefield

Solution Since no depth is given, the factor C_{fHoriz} will be put arbitrarily equal to 1. In the first case, we have:

$$N_4 = 1, N_3 = 4, N_2 = 4, N_1 = 0$$

(continued)

---

[1]The weighting factors were changed from [5] to [1]. The values were previously 1.0, 0.5, 0.25, and 0.1.

Example 6.4 (continued)

$$T_p = 4.41\frac{1 + 0.75 \cdot 4 + 0.5 \cdot 4}{9} = 2.94 \text{ K}$$

In the second case, we have:

$$N_4 = 0, N_3 = 0, N_2 = 7, N_1 = 2$$

$$T_p = 4.41\frac{0.5 \cdot 7 + 0.25 \cdot 2}{9} = 1.96 \text{ K}$$

∎

The concept of temperature penalty given in the book of Kavanaugh and Rafferty is based on several arbitrary assumptions. The number of rings to evaluate the total energy trapped is quite arbitrary. The corrected weights are proposed too. If the number of rings is increased, the value of the temperature penalty will increase. Actually, the total energy that is trapped inside a given cylinder is quite easy to evaluate. From (4.12):

$$q'(r_i, t) = q'_o e^{-\frac{r_i^2}{4\alpha_s t}}$$

To evaluate the amount of energy that left the cylinder, we integrate from 0 to t_f:

$$Q'_{out} = q'_o \int_o^{t_f} e^{-\frac{r_i^2}{4\alpha_s \tau}} d\tau$$

with the change of variable $u = \frac{r_i^2}{4\alpha_s \tau}$:

$$Q'_{out} = q'_o t_f E_2\left(\frac{r_i^2}{4\alpha_s t_f}\right) = q'_o t_f \left(\exp(-u - u E_1(u))\right)$$

(6.22)

E_1 and E_2 being the generalized exponential integral of order one and two. Using this value in (6.14), it can be verified that it will give the same answer as if an infinite number of rings are used. In the last example, this would give:

$$T_p^* = 39°C$$

The final temperature, thus obtained, becomes very large, and the borehole length calculation will be very large. It might be surprising that with such a discrepancy, the ASHRAE method has been used for a long time with some success. The main reason is that assuming that the whole energy is trapped for all the inner boreholes is too excessive.

6.4.2 Temperature Penalty Using the ILS

A better approach has been proposed by Bose et al. [6], although not in the context of temperature penalty. They compute the temperature variation due to all adjacent holes on one borehole using the ILS response. To do that, some assumptions have to be made; a simple one is that all heat flux are the same in each borehole:

$$T_{p,i} = \sum_{\substack{j=1 \\ j \neq i}}^{N} T_{ij} = \frac{q'}{2\pi k_s} \sum_{\substack{j=1 \\ j \neq i}}^{N} I\left(\frac{d_{ij}}{2\sqrt{\alpha_s t_f}}\right)$$

(6.23)

$$T_p = \text{mean}(T_{p,i})$$

(6.24)

with:

$$d_{ij} = \sqrt{(x_i - x_j)^2 + (y_i - y_j)^2}$$

If, for example, a 3×3 bore field is used, a possible evaluation of the temperature penalty for boreholes 1, 3, 7, and 9 would be:

$$T_{p,1} = T_{p,3} = T_{p,7} = T_{p,9} = \frac{q'}{2\pi k_s}\left[2I\left(\frac{b}{2\sqrt{\alpha_s t_f}}\right) + 2I\left(\frac{2b}{2\sqrt{\alpha_s t_f}}\right)\right.$$

$$\left. + 2I\left(\frac{\sqrt{5}b}{2\sqrt{\alpha_s t_f}}\right) + I\left(\frac{\sqrt{2}b}{2\sqrt{\alpha_s t_f}}\right) + I\left(\frac{2\sqrt{2}b}{2\sqrt{\alpha_s t_f}}\right)\right]$$

A similar expression can be found for:

$$T_{p,2} = T_{p,4} = T_{p,6} = T_{p,8}$$

and a final one for the inner borehole $T_{p,5}$. The final penalty temperature can then be evaluated taking the mean of the nine values.

Example 6.5 Redo the last example with the new concept using the ILS to evaluate the temperature penalty.

Solution For the 3×3 field:

$$T_{p,1} = T_{p,3} = T_{p,7} = T_{p,9} = 5.40 \text{ K}$$

$$T_{p,2} = T_{p,4} = T_{p,6} = T_{p,8} = 6.21 \text{ K}$$

$$T_{p,5} = 7.17°C \Rightarrow T_p = 5.96 \text{ K}$$

For the inline field:

$$T_{p,1} = T_{p,9} = 2.48 \text{ K}$$

$$T_{p,2} = T_{p,8} = 3.45 \text{ K}$$

$$T_{p,3} = T_{p,7} = 3.98 \text{ K}$$

$$T_{p,4} = T_{p,6} = 4.26 \text{ K}$$

$$T_{p,5} = 4.35 \text{ K} \Rightarrow T_p = 3.64 \text{ K}$$

∎

Although this value is less than the value obtained using the whole "trapped energy," this value is still much higher than the one recommended by the ASHRAE method and would lead to larger design. The main problem comes from the fact that for the long-term effect, it is important to take into account the axial effects.

6.4.3 Temperature Penalty Using the FLS

A similar approach than the previous one but one that can take into consideration the axial effects is to use the FLS solution instead of the ILS:

$$T_{p,i} = \sum_{\substack{j=1 \\ j \neq i}}^{N} T_{ij} = \frac{q'}{2\pi k_s} \sum_{\substack{j=1 \\ j \neq i}}^{N_b} {}_1g(\bar{t}, \bar{d}_{ij}, \bar{z}_o) \quad (6.25)$$

$$T_p = \text{mean}(T_{p,i}) \quad (6.26)$$

where again (see Chap. 5):

$$\bar{z}_o = \frac{z_o}{H}, \bar{d}_{ij} = \frac{d_{ij}}{H}, \bar{t} = \frac{9\alpha_s t}{H^2}$$

The solution depends on two new parameters z_o and H. However, if H is very large, the solution should converge to Eq. (6.24).

Example 6.6 Redo the last example with the FLS model. Use $H = 100.0$ m and $z_o = 0$ and 4 m.

Solution For $H = 100$ m and $z_o = 0$, the following results will be found:
For the 3×3 field:

$$T_{p,1} = T_{p,3} = T_{p,7} = T_{p,9} = 4.47 \text{ K}$$

$$T_{p,2} = T_{p,4} = T_{p,6} = T_{p,8} = 5.16 \text{ K}$$

$$T_{p,5} = 7.17°C \Rightarrow T_p = 4.94K \text{ K}$$

For the inline field:

$$T_{p,1} = T_{p,9} = 2.00 \text{ K}$$

$$T_{p,2} = T_{p,8} = 2.83 \text{ K}$$

$$T_{p,3} = T_{p,7} = 3.27 \text{ K}$$

$$T_{p,4} = T_{p,6} = 3.49 \text{ K}$$

$$T_{p,5} = 3.56 \text{ K} \Rightarrow T_p = 2.97 \text{ K}$$

(continued)

(b) For $\bar{z}_o = 0.04$, which is the default value used in Eskilson's g-function, the final result is:
For the 3×3:

$$T_p = 5.11 \text{ K}$$

For the 9×1:

$$T_p = 3.08 \text{ K}$$

∎

The approach of calculating the temperature penalty with the FLS solution is very similar to the method proposed by Bernier et al. [7] where they defined the temperature penalty as:

$$T_p = \frac{1}{2\pi k_s} \left(g_n(\bar{t}, \bar{r}_b, \bar{z}_o) - {}_1g(\bar{t}, \bar{r}_b, \bar{z}_o) \right) \quad (6.27)$$

where $g_n(\bar{t}, \bar{r}_b, \bar{z}_o)$ was defined as the global *g-function* associated with the particular borefield configuration. From our analytical expression (4.53) that was given to calculate the *g-function*, it can be seen that this is exactly the same definition as (6.25) as long as the *g-function* is computed by (4.53). In their work, they use the original tabulated values of Eskilson *g-function*, so this would lead to a small variation.

6.4.4 Approximate Solution for the Temperature Penalty: The Philippe-Bernier Approach

In another paper, Philippe et al. [8] provided a simple correlation to evaluate the penalty based on several parameters. The formula is given by:

$$T_p = \frac{q'}{2\pi k_s} F(\bar{t}, \bar{b}, N_b, A) = \frac{q'}{2\pi k_s} \left(7.8189 + \sum_{i=1}^{36} b_i c_i \right)$$

$$(6.28)$$

A is the aspect ratio of the borefield: the longest direction divided by the other one. The expressions of coefficients b_i, c_i are given in Table 6.1.

6.4.5 Approximate Solution for the Temperature Penalty: The Fossa-Rolando Approach

The previous correlation was mostly proposed to be used in a spreadsheet environment where the evaluation of the FLS solution would be difficult. A similar argument has led Fossa

Table 6.1 Regression coefficients from [8]

i	b_i	c_i
1	$-6.4270e+01$	(b/H)
2	$+1.5387e+02$	$(b/H)^2$
3	$-8.4809e+01$	$(b/H)^3$
4	$+3.4610e+00$	$\log(\bar{t})$
5	$-9.4753e-01$	$(\log(\bar{t}))^2$
6	$-6.0416e-02$	$(\log(\bar{t}))^3$
7	$+1.5631e+00$	Nb
8	$-8.9416e-03$	Nb^2
9	$+1.9061e-05$	Nb^3
10	$-2.2890e+00$	A
11	$+1.0187e-01$	A^2
12	$+6.5690e-03$	A^3
13	$-4.0918e+01$	$(b/H)\log(\bar{t})$
14	$+1.5557e+01$	$(b/H)(\log(\bar{t}))^2$
15	$-1.9107e+01$	$(b/H)Nb$
16	$+1.0529e-01$	$(b/H)Nb^2$
17	$+2.5501e+01$	$(b/H)A$
18	$-2.1177e+00$	$(b/H)A^2$
19	$+7.7529e+01$	$(b/H)^2\log(\bar{t})$
20	$-5.0454e+01$	$(b/H)^2(\log(\bar{t}))^2$
21	$+7.6352e+01$	$(b/H)^2Nb$
22	$-5.3719e-01$	$(b/H)^2Nb^2$
23	$-1.3200e+02$	$(b/H)^2A$
24	$+1.2878e+01$	$(b/H)^2A^2$
25	$+1.2697e-01$	$\log(\bar{t})Nb$
26	$-4.0284e-04$	$\log(\bar{t})Nb^2$
27	$-7.2065e-02$	$\log(\bar{t})A$
28	$+9.5184e-04$	$\log(\bar{t})A^2$
29	$-2.4167e-02$	$(\log(\bar{t}))^2Nb$
30	$+9.6811e-05$	$(\log(\bar{t}))^2Nb^2$
31	$+2.8317e-02$	$(\log(\bar{t}))^2A$
32	$-1.0905e-03$	$(\log(\bar{t}))^2A^2$
33	$+1.2207e-01$	$Nb\,A$
34	$-7.1050e-03$	$Nb\,A^2$
35	$-1.1129e-03$	$Nb^2\,A$
36	$-4.5566e-04$	$Nb^2\,A^2$

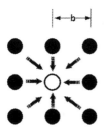

Fig. 6.6 Excess temperature due to neighboring boreholes

and Rolando [9, 10] to propose their own approximate solution. First, they observed that the temperature disturbance was mostly due to the neighboring boreholes, so they first evaluate a first disturbance calculating the effect of the eight nearby sources. They then called their method "Tp eight." Since they wanted to avoid the evaluation of the finite line source, they used the infinite line source (Fig. 6.6):
This is simply given by:

$$\theta_8 = 4\frac{q'_a}{k_s}\left[G_{ils}(Fo_b^*) + G_{ils}(Fo_{\sqrt{2}b}^*)\right] \quad (6.29)$$

with:

$$Fo_b^* = \frac{\alpha_s t_f}{b^2}\left(\frac{H_{ref}}{H}\right), \quad Fo_{\sqrt{2}b}^* = \frac{\alpha_s t_f}{2b^2}\left(\frac{H_{ref}}{H}\right)$$

or:

$$\theta_8 = 2\frac{q'_a}{\pi k_s}\left[I(X_b^*) + I(X_{\sqrt{2}b}^*)\right] \quad (6.30)$$

with:

$$X_b^* = \frac{b}{2\sqrt{\alpha_s t_f}}\sqrt{\frac{H}{H_{ref}}}, \quad X_{\sqrt{2}b}^* = \frac{b}{\sqrt{2\alpha_s t_f}}\sqrt{\frac{H}{H_{ref}}}$$

where:

$$Fo^* = Fo\left(\frac{H_{ref}}{H}\right), \quad H_{ref} = 100 \text{ m} \quad (6.31)$$

is a corrected Fourier number associated with the fact that the correlation was obtained with all boreholes having 100 m.

Recognizing that this was an approximation not taking into account the other boreholes neither their finite length, they proposed some correction factors that they obtained by minimizing their approximation results with the penalty obtained with the FLS approach. They put their correlation in the form:

$$Tp_8 = \theta_8\frac{aN_4 + bN_3 + cN_2 + dN_1}{N_t} \quad (6.32)$$

where N_1, N_2, N_3, N_4 have the same meaning than in the Kavanaugh approach. The values of a, b, c, d depend whether the field was dense (square or rectangular) ("R" configuration) or non-dense: inline, U-configuration, L-configuration, etc. ("non-R" configuration). The values were (Table 6.2):

6.4.6 Approximate Solution for the Temperature Penalty: The Capozza Approach

Another similar approach was proposed by Capozza et al. [11], that is, to evaluate the first approximation of the temperature penalty using the ILS approach (Sect. 6.4.2) and then

Table 6.2 Regression coefficients from [9]

	R configuration	Non-R configuration
a	$1.95005 + \frac{0.105215}{B/H} - 55.6543\left(\frac{B}{H}\right)^2$	0
b	0.280	0.950
c	0.450	$-0.28174\log\left(\frac{B}{H}\right) - 0.23546$
d	0	0.05

derive a correction factor to take into account the axial effect. Again, this correction factor was found from minimizing the error with the original FLS approach:

$$T_{p,Capozza} = \frac{T_{p,ILS}}{Tpratio} \quad (6.33)$$

$$Tpratio = 1$$

$$+ \left\{ b(N_t) + (1 + b(N_t)) \left(\frac{f(N_t) - 1}{0.05} \right) (B/H - 0.05) \right\}$$

$$e^{-(\log(\bar{t}_f) - 3)/\beta}$$

where again $\bar{t}_f = 9\alpha t_f/H^2$, $\beta = 6/\log(0.07) = -2.256$ and b and f are interpolated from Table 6.3:

Example 6.7 Using the approximate formulas proposed by Philippe et al., Fossa-Rolando, and Capozza et al., recalculate the penalty temperature of the last problem.

Solution For the 3×3 arrangement, Eq. (6.28) with A = 1, gives:

$$T_p = 4.52 \text{ K}$$

For Eq. (6.32) we find:

$$\theta_8 = 7.17 \text{ K}$$

$$a = 3.5, b = 0.28, c = 0.45, d = 0$$

$$T_{p8} = \frac{3.5 + 0.28 \cdot 3 + 0.45 \cdot 3}{9} = 5.12 \text{ K}$$

For Eq. (6.33) we find:

$$T_{p,ILS} = 5.96 \text{ K}$$

$$t_{ss} = \frac{100^2}{9 \cdot 0.1} = 11111 \text{days}$$

$$\bar{t}_f = \frac{3650}{11111} = 0.3285$$

$$Tpratio = 1 + \left\{ 1.29 + 2.29 \left(\frac{1.246 - 1}{0.05} \right) (0.06 - 0.05) \right\}$$

$$e^{-(-1.11-3)/-2.256} = 1.23$$

$$T_{Cap} = \frac{5.96}{1.23} = 4.85 \text{ K}$$

(continued)

For the inline arrangement, similar calculations are done, and the Table 6.4 gives the different T_p values described in the last examples.

Without giving specific conclusions with a single example, one may note that the original ASHRAE approach will always underestimate the temperature penalty and the ILS approach will give a maximum value. One of the arguments brought up by the original authors is that water movement is often present which will reduce the temperature imbalance. No further judgment will be made, but the decision comes to whether optimistic or pessimistic approach is considered.

Example 6.8 Redo Example (6.2) where all loads and capacities are now three times the previous ones. Take into account the penalty temperature using the original ASHRAE approach and the FLS approach. For the FLS use $z_o = 4$ m. Give the answer using Paul's resistance only. Assume that the flow rate per borehole does not change.

(continued)

Table 6.3 Regression coefficients from [11]

N_t	b	f
1	0	1
4	1.03	1.24
8	1.26	1.245
16	1.52	1.25
25	1.59	1.30
100	2.66	1.31
225	2.32	1.41

Table 6.4 T_p obtained by different approaches

Case	T_p (3×3)	T_p 9×1
ASHRAE (3 Rings)	2.94	1.96
ASHRAE (∞ Rings)	26	17
Eq. (6.24)	5.96	3.64
Eq. (6.25)($\bar{z}_o = 0$)	4.94	2.97
Eq. (6.25)($\bar{z}_o = 0.04$)	5.11	3.08
Eq. (6.28)	4.52	2.82
Eq. (6.32)	5.12	3.19
Eq. (6.33)	4.95	2.96

(continued)

Example 6.8 (continued)
Solution The new loads are:

Month	Q(MWh)	Q(MWh)
Jan	6.51	0.0
Feb	6.21	0.0
Mar	5.25	0.0
Apr	4.17	0.0
May	2.7	−3.6
Jun	0.0	−4.8
Jul	0.0	−6.0
Aug	0.0	−5.4
Sep	2.55	−2.4
Oct	3.66	−1.2
Nov	4.92	0.0
Dec	6.06	0.0

Since the capacities are also multiplied by 3 and that the COPs do not change, the hourly pulse will be:

$$q_{h,ch} = 24000 \text{ W}, \qquad q_{h,cl} = -47250 \text{ W}$$

From the previous example, the total length is expected to be in the order of 600 m. Since this is too long for a single borehole, a natural choice can be either six boreholes in the order of 100 meters or four of 150 meters. Let's take for this example the first choice. The initial configuration may depend on some landscape constraints, but let's start with a 3×2 arrangement with all boreholes separated by 6 meters. The temperature penalty that will influence the answer is in general an iterative procedure since we need the length of the field to evaluate it. Any reasonable choice would do. Let's try with $T_p = 0$:

First iteration:

$$T_p = 0, \qquad N_b = 6, \qquad \dot{V} = 3.6 \text{ l/s}$$

Since the flow rate in each borehole is the same as before, the Reynolds number will be the same and all resistances will be unchanged:

$$R'_b = 0.128 \text{ m-K/W}$$

However, since the maximum load is three times the previous one and the total flow rate is six times the previous one, the mean fluid temperature will change:

$$T_{fo} = 0, T_{fi} = T_{fo} - \frac{q_b}{\dot{m}C_p} = 0 - \frac{24000}{14090} = -1.7°C$$

$$\Rightarrow T_f = -0.85°C$$

In cooling mode, we have:

$$T_{fo} = 0, T_{fi} = T_{fo} - \frac{q_b}{\dot{m}C_p} = 25 + \frac{43200}{14090} = 28.1°C$$

$$\Rightarrow T_f = 26.53°C$$

If the same flow rate per kW was assumed, the same fluid temperature would have been calculated, but the flow rate per borehole and the convective resistance would change. This resistance is often small and its change will not influence the final result unless the flow becomes laminar. In that case, the convective resistance will increase and care should always be taken to keep the flow turbulent to have a better heat transfer. The three pulses are now:

$$q_a = 2127 \text{ W}$$

$$q_m(heating) = 8918 \text{ w}, \qquad q_m(cooling) = -8219 \text{ w}$$

$$q_h(heating) = 24000 \text{ W}, \qquad q_h(cooling) = -43200 \text{ W}$$

All other values being the same, the result is:

$L_{heating}$

$$= \frac{2127 \cdot 0.1522 + 8918 \cdot 0.1516 + 24000(0.071 + 0.128)}{10 - (-0.85)}$$

$$= 596 \text{ m}$$

$$L_{cooling} = \frac{0 - 8219 \cdot 0.1516 - 43200(0.071 + 0.128)}{10 - 26.53} = 596 \text{ m}$$

$$\Rightarrow H = 100 \text{ m per borehole}$$

From this first evaluation, the penalty temperature is evaluated:

$$q'_a = 3.56 \text{ W/m}$$

Using the proposed ASHRAE method, we find:

$$T_p^* = 1.52, C_{fHoriz} = 1.02 \Rightarrow T_p = 1.52 \frac{2 \cdot 0.75 + 4 \cdot 0.5}{1.02 \cdot 6} = 0.87°C$$

Again, for the same reason as discussed previously for the yearly pulse, the temperature penalty for the non-dominant mode will be beneficial but only after many years. To ensure that the first year is not underestimated, we put it equal to zero in the non-dominant mode:

Second iteration:
Using the value found, a second iteration is done:

(continued)   (continued)

Example 6.8 (continued)

$L_{heating}$

$$= \frac{2127 \cdot 0.1522 + 8918 \cdot 0.1516 + 24000(0.071 + 0.128)}{10 - 0.87 - (-0.85)}$$

$$= 647 \text{ m}$$

$$L_{cooling} = \frac{0 - 8219 \cdot 0.1516 - 43200(0.071 + 0.128)}{10 - 26.53} = 596 \text{ m}$$

$$L = \max(647, 596) = 647 \text{ m}$$

$$L = 32 + 135 + 172 + 307 = 647 \text{ m}$$

$$\Rightarrow H = 108 \text{ m per borehole}$$

After a few iterations, the solution will converge to:

$$L = 643 \text{ m} \Rightarrow H = 107 \text{ m per borehole}$$

If the penalty temperature based on the FLS was used, in order to evaluate (6.25), the height of a single borehole and the value of z_o are needed. Of course, here the one just calculated can be taken, but if we start from scratch, a typical value could have been taken, for example, 100 m, since this is the range we expect and this dependance is weak. In that case, (6.25) gives:

$$T_p = 1.23$$

$L_{heating}$

$$= \frac{2127 \cdot 0.1522 + 8918 \cdot 0.1516 + 24000(0.071 + 0.128)}{10 - 1.23 - (-0.85)}$$

$$= 671 \text{ m}$$

$$L_{cooling} = \frac{-8219 \cdot 0.1516 - 43200(0.071 + 0.128)}{10 - 26.53} = 596 \text{ m}$$

$$L = \max(671, 581) = 671 \text{ m}$$

$$L = 33 + 141 + 178 + 319 = 671 \text{ m}$$

$$\Rightarrow H = 111 \text{ m per borehole}$$

which would change a little bit after few iterations:

$$L = 664 \text{ m} \Rightarrow H = 111 \text{ m per borehole}$$

Comments: *The solution given in this example is given in the script Example6_8.py where all the design steps are written. A more concise solution using the class borefield in the module "design_md.py" is given in Example6_8cl.py.*

■

Remark In the ASHRAE method, it is suggested to put the temperature penalty to zero in the non-dominant mode.

6.5 Alternatives to ASHRAE Method

The concept of temperature penalty has been accepted for a long time, but some debates exist on its definition. Its use is by no means mandatory in the GSHP design procedure. Equation (6.10) can be rewritten by multiplying both sides by the denominator:

$$L(T_o - T_p - T_f) = q_a R'_{g,a} + q_m R'_{g,m} + q_h(R'_{g,h} + R'_b)$$

Using (6.25) as the definition of the temperature penalty:

$$L(T_o - T_f) - \frac{q_a}{2\pi k_s} \frac{1}{N_b} \sum_{i=1}^{N_b} \left(\sum_{\substack{j=1 \\ j \neq i}}^{N_b} 1g(\bar{t}_f, \bar{d}_{ij}, \bar{z}_o) \right)$$

$$= q_a \frac{G(Fo_f)}{k_s} + (q_m - q_a)\frac{G(Fo_1)}{k_s} + (q_h - q_m)\frac{G(Fo_1)}{k_s}q_h R'_b$$

Remember that the temperature penalty concept has been introduced since it proposed to use the thermal response of a single borehole in the *superposition of temporal loads*. Instead of using the ICS as is often used, the FLS solution for a single borehole can be used. It was seen that for the medium time range, the solution is identical, and for the long time period, it is indeed a better solution. The new solution becomes:

$$L(T_o - T_f)$$

$$= \frac{q_a}{2\pi k_s} \left(\underbrace{1g(\bar{t}_f, \bar{r}, \bar{z}_o) + \frac{1}{N_b} \sum_{i=1}^{N_b} \left(\sum_{\substack{j=1 \\ j \neq i}}^{N_b} 1g(\bar{t}_f, \bar{d}_{ij}, \bar{z}_o) \right)}_{g(\bar{t}_f, \bar{r}, \bar{z}_o)} \right)$$

$$+ (q_m - q_a)\frac{G(Fo_1)}{k_s} + (q_h - q_m)\frac{G(Fo_1)}{k_s}q_h R'_b$$

It can be seen that the original ASHRAE method is almost unchanged if the thermal response of the bore field ($g - function$) is used instead of the ICS without the concept of temperature penalty. In order to have a coherent procedure, it is even better if the same thermal response for all pulses is kept:

$$L(T_o - T_f) = \frac{1}{2\pi k_s}\left(q_a g(\bar{t}_f, \bar{r}, \bar{z}_o)+\right.$$

$$(q_m - q_a)g((\bar{t}_f - \bar{t}_y), \bar{r}, \bar{z}_o)$$

$$+(q_h - q_m)g((\bar{t}_f - \bar{t}_m), \bar{r}, \bar{z}_o) + q_h R'_b)$$

$$(6.34)$$

and finally:

$$L = \frac{q_a R'_{g,a} + q_m R'_{g,m} + q_h(R'_{g,h} + R'_b)}{T_o - T_f} \quad (6.35)$$

with:

$$R'_{g,a} = \frac{g(\bar{t}_f, \bar{r}, \bar{z}_o) - g((\bar{t}_f - \bar{t}_y), \bar{r}, \bar{z}_o)}{2\pi k_s},$$

$$R'_m = \frac{g((\bar{t}_f - \bar{t}_y), \bar{r}, \bar{z}_o) - g((\bar{t}_f - \bar{t}_m), \bar{r}, \bar{z}_o)}{2\pi k_s}, \quad (6.36)$$

$$R'_h = \frac{g((\bar{t}_f - \bar{t}_m), \bar{r}, \bar{z}_o)}{2\pi k_s}$$

Since the monthly pulse is in the medium time range, the thermal response calculation should be almost the same as the previous one. The main difference that could take place is in the hourly pulse since, in that case, the FLS solution will give a similar value than the ILS but different than the ICS used originally. However neither solutions are strictly valid for this short-time behavior, and it may be argued that the ILS might be even better than the ICS [12]. The best practice would be to use special *g-function* for short-time behavior (see Chap. 13).

Although the concept of *g-function* based on the FLS-type I superposition principle was used to derive the last equation, it is of course still valid if it is replaced by the original tabulated values of Eskilson [13] or the one generated by Cimmino and Bernier [14, 15]. Modification of the original ASHRAE method following similar concepts that are described in this section was given by Ahmadfard and Bernier [16].

Example 6.9 Redo the last example using the concept of *g-function* instead of the penalty temperature concept.

Solution The *g-function* based on the FLS-I will be used for that example. Using (6.35), it is observed that no iterations are needed to evaluate the penalty temperature. However, the response factor depends now on the borehole height which is unknown. However, this dependence is weak, and if a good first guess is

(continued)

---

known, it should not change a lot. Let's start with the value calculated in the previous example for the initial guess of $H \approx 110$ m. The new dimensionless time is now:

$$\bar{t}_f = \frac{9\alpha_s t_f}{H^2} = \frac{9 \cdot 0.00375(87600 + 730 + 4)}{110^2} = 0.246$$

$$\bar{t}_1 = \frac{9\alpha_s t_1}{H^2} = \frac{9 \cdot 0.00375 \cdot 734}{110^2} = 2.05\text{e-}3$$

$$\bar{t}_2 = \frac{9\alpha_s t_2}{H^2} = \frac{9 \cdot 0.00375 \cdot 4}{110^2} = 1.116\text{e-}5$$

$$g(\bar{t}_f) = 11.2, g(\bar{t}_1) = 3.49, g(\bar{t}_2) = 0.94$$

$$R'_{g,a} = \frac{11.2 - 3.49}{2\pi 2.5} = 0.49 \text{ m-K/W}$$

$$R'_m = \frac{3.49 - 0.94}{2\pi 2.5} = 0.163 \text{ m-K/W}$$

$$R'_h = \frac{0.94}{2\pi 2.5} = 0.06 \text{ m-K/W}$$

$L_{heating}$

$$= \frac{2127 \cdot 0.49 + 892 \cdot 0.163 + 24000(0.06 + 0.128)}{10 - (-0.85)}$$

$$= 645 \text{m}$$

As suggested previously, the yearly pulse will be neglected for the non-dominant mode:

$L_{cooling}$

$$= \frac{-8920 \cdot 0.163 - 43200(0.06 + 0.128)}{10 - 26.53}$$

$$= 571 \text{ m}$$

$$L = \max(645, 571) = 645 \text{ m}$$

$$L = 96 + 134 + 132 + 283 = 645 \text{ m}$$

$$\Rightarrow H = 107 \text{ m per borehole}$$

Since this value is smaller than the one assumed (110 m), a second iteration can be done, but the final result would be almost the same. The answer also depends on $\bar{z}_o = z_o/H$, which was not specified but this dependence is weak.

Comments: *Several observations can be drawn from the last two examples. First, it is observed that*

(continued)

Example 6.9 (continued)

the total length has the same order of magnitude which is of course reassuring. To have a better understanding, we must look in deeper details in each of the contributions. The annual contribution seems much higher now. This is normal since the effect of the long-term mutual interference was before in the denominator, so its effect was scattered in the four terms, whereas now it is concentrated in the first term of the sum. Second, in the last example, the hourly pulse had a higher weight than the monthly pulse (170 vs. 132 m), whereas now the hourly pulse and the monthly pulse are in the same order of magnitude (134 vs. 132 m). Again, the reason for that is that now, the hourly pulse that is modeled now by the FLS has a different short-time behavior than the ICS which was previously used. Remember that the ICS is more conservative, explaining the larger length obtained.

∎

6.6 Swedish Approach

A very popular approach to size geothermal bore field is based on the works developed in Sweden in the 1980s by the researchers of Lund University. The basic principle is the same as was previously discussed. The main differences between its approach and the ASHRAE method are that:

- The TLS is based on a series of monthly pulses starting at a predefined initial month plus a peak hourly load.
- The borehole resistance is based on the multipole method.
- The g-function associated with the particular field was used as the thermal response factor.
- The concept of penalty temperature was then not necessary.

No new formalism is then needed to describe mathematically the Swedish approach. Equation (6.4) is used as the basic relation remembering the remarks just described. Let's rewrite this equation:

$$L = \frac{\sum_{n=1}^{N_p}(q_n - q_{n-1})\dfrac{g\left(\bar{t}_f - \bar{t}_{n-1}, \bar{r}, \bar{z}_o\right)}{2\pi k_s} + q_b(t)R'_b}{T_o - \dfrac{T_{in} + T_{out}}{2}},$$

$$q_0 = \bar{t}_0 = 0 \tag{6.37}$$

where the temporal load superposition has been explicitly written using the *g-function* concept. The number of pulses is the total number of months used in the simulation plus one hourly pulse.

Remark Since, in the non-dominant mode, the temperature range will be more and more favorable over the years, following the ASHRAE approach, it is suggested to stop the simulation the first year for that mode. Actually, in the way the method is described, this is not mandatory, since the length is chosen such that all temperatures stay between the min-max ranges prescribed, whether the temperature is in the first or the last year, but that is the simplest way to mimic this behavior with the TLS approach.

For instance, if the number of simulation years is 10, the peak monthly load in heating is in February and July in cooling mode and the initial month of the simulation is January (first month of the year). If the heating mode is dominant, then the number of monthly pulses in heating mode will be:

$$N_m = 10 \cdot 12 + 2 = 122$$

Adding the hourly pulse, the total amount of pulses would be:

$$N_p = 122 + 1 = 123$$

In cooling mode it would be:

$$N_m = 7, \qquad N_p = 7 + 1 = 8$$

If the cooling mode is dominant, we have:

$$N_{m,he} = 2 + 1 = 3$$

and:

$$N_{m,co} = 127 + 1 = 128$$

Calculating it this way, the simulation is continued until the worst month after the total years of simulation. Using monthly pulses instead of an aggregated yearly pulse, it is possible to see the impact of starting the simulation at a different month of the year. In the last example, if the GSHP system is started in summer, in June for example, the result would be in heating dominant mode:

$$N_m = 7 + 9 \cdot 12 + 2 = 117, \qquad N_p = 117 + 1 = 118$$

In cooling mode it would be:

$$N_m = 2, \qquad N_p = 2 + 1 = 3$$

The general formula that can then be used would, both in heating and cooling mode, be for the dominant mode:

$$N_p = 12N_{years} + m_p - m_{ini} + 2 \qquad (6.38)$$

where m_{ini} is the initial month (1 for January, 2 for February, etc.) and m_p is the month where the peak load occurs, and for the non-dominant mode:

$$N_p = 12N_1 + m_p - m_{ini} + 2 \qquad (6.39)$$

where $N_1 = 0$ if $M_{ini} \le m_p$, and 1 otherwise. This ensures that the simulation continues until the worst month even if the system starts after the worst month of the first year.

If the time base is hour, the time intervals are the number of hours during each month. It is possible to use the real hours (744, 672, etc.) but to simplify it is possible to use equal month of 730 hours. In that special case, the final result would be:

$$\bar{t}_{N_m} = \frac{9\alpha_s 730\,N_m}{H^2}, \quad \bar{t}_f = \bar{t}_{N_m} + b_h \qquad (6.40)$$

where again, b_h is the hourly block.

Example 6.10 Redo the previous example using the Swedish approach. Take again a 10-year simulation period and a 4-hour hourly block, and do the calculations in the following cases:

(a) Using Paul's resistance and starting the simulation in January.
(b) Using Paul's resistance and starting the simulation in June.
(c) Using multipole method (J=10) for the resistance and starting the simulation in January.
(d) Using multipole method (J=10) for the resistance and starting the simulation in June.

Solution

(a) The only difference is in the calculation of the TLS. Without writing the 123 terms, the first one is given. As before, to calculate the g-function, the height of the borehole is needed but it is unknown. From our previous discussion, let's start with an initial guess of 110 m:

$$\bar{r} = \frac{r_b}{H} = 0.00068, \quad \bar{z}_o = \frac{z_o}{H} = 0.036,$$

$$\bar{b} = \frac{B}{H} = 0.0545$$

The problem is heating dominant, so:

$$N_{p,he} = 120 + 1 - 1 + 2 = 122, \quad N_{m,he} = 121$$

(continued)

$$N_{p,co} = 7 - 1 + 2 = 8, \quad N_{m,he} = 7$$

From (6.40):

$$\bar{t}_f = \frac{9\alpha_s 88334}{110^2} = 0.246, \qquad \bar{t}_0 = 0$$

$$g(0.246, 0.00068, 0.036) = 11.19, \quad q_1 = 8917.8\ \text{W}$$

$$s_1 = \frac{(8917.8 - 0)11.19}{2\pi 2.5} = 6350.\ \text{m-K}$$

Doing the same for all the monthly pulses will give

$$s_m = \sum_{n=1}^{N_m}(q_n - q_{n-1})\frac{g(\bar{t}_f - \bar{t}_{n-1}, \bar{r}, \bar{z}_o)}{2\pi k_s} = 3051\ \text{m-K}$$

Adding now the peak load hourly pulse with again a 4-hour block:

$$g(1.11e\text{-}5, 0.00068, 0.036) = 0.94, \quad q_h - q_{N_p}$$

$$= 24000 - 8917.8 = 15082\ \text{W}$$

$$\frac{(15082)0.94}{2\pi 2.5} = 902\ \text{m-K}$$

Adding the effect of the borehole resistance, in the first case, the value already computed before ($R'_b = 0.128$ m-K/W) is kept:

$$24000 \cdot 0.128 = 3065\ \text{m-K}$$

The final length will be given by:

$$L_{heating_a} = \frac{3051 + 902 + 3065}{10.85} = 647\ \text{m}$$

A similar calculation would give:

$$L_{cooling_a} = 553\ \text{m}$$

so

$$L_a = 647\ \text{m} \qquad \text{and} \qquad H = \frac{647}{6} = 108\ \text{m}$$

Again, it is so near the initial value taken (110 m) that a second iteration is not necessary. It might be quite surprising to get a result so similar than in the previous example. Before discussing on this result, we will finish the other cases.

(b) Starting in June will be very similar. The only difference will be that the initial monthly pulse will be:

(continued)

Example 6.10 (continued)

$$\bar{t}_f = \frac{9\alpha_s 84684}{110^2} = 0.236$$

$$L_{heating_b} = \frac{3006 + 902 + 3065}{10.85} = 642 \text{ m}$$

We can observe a small difference. The reason is that the application is heating dominant and starting the system in cooling mode pre-heat the soil which gives a very small increase in performance in that case.

(c) The Swedish method as described is the basis of several design software like EED and GLHEPro, both of which use the multipole method to evaluate the borehole resistance. This resistance can be calculated using the function Rb_multipole given in the library.

$$R'_b = 0.133 \text{ m-K/W}$$

All the rest will be the same, reevaluating the borehole resistance contribution leads to:

$$24000 \cdot 0.133 = 3072 \text{ m-K}$$

$$L_{heating_c} = \frac{3051 + 902 + 3193}{10.85} = 659 \text{ m}$$

$$L_{heating_d} = \frac{3006 + 902 + 3193}{10.85} = 654 \text{ m}$$

To summarize, different methods (other possibilities exist) were used to size the same problem. The results are summarized in Table 6.5.

Several conclusions can be drawn from these results.

1. Comparing cases 3 and 4, it is surprising to see that the aggregation of the yearly loads does not influence very much the final result, at least in this example.

Table 6.5 Results obtained with different approaches

Case	Method used	Tp used	R'_b(m-K/W)	Start month	L (m)
1	(6.10)	(6.20)	Paul-Remund (=0.128)	–	643
2	(6.10)	(6.25)	Paul-Remund (=0.128)	–	664
3	(6.35)	–	Paul-Remund (=0.128)	–	644
4	(6.37)	–	Paul-Remund (=0.128)	1	647
5	(6.37)	–	Paul-Remund (=0.128)	6	642
6	(6.37)	–	Multipole (=0.133)	1	659
7	(6.37)	–	Multipole (=0.133)	6	654

2. Using monthly loads, it is possible to see a small advantage to pre-heat the ground in heating-dominant applications or to pre-cool the ground in cooling-dominant applications. This was not possible with the three-pulse method. The gain is however small at least in this example.
3. The difference between case 2 and cases 3 and 4 is due to the short-time behavior of the ICS solution which gives conservative calculations. A better knowledge of short-time behavior is thus significant in borehole design.
4. The impact of borehole resistance is very important in the final design. A good knowledge of the grout properties and a good calculation procedure are thus important. In situ measurement of this resistance is even better. Differences larger than the one observed in this example (cases 4-5 vs 6-7) can occur.
5. The original ASHRAE method, case 1, seems to underestimate the effect of borehole interference (temperature penalty) and overestimate the short-time ground resistance (due to ICS solution), and both effects seem to cancel out at least in this example.

6.7 Pipe Interference: Effective Resistance

In the previous sections, a systematic approach was described to size a geothermal bore field, given the loads that have to be exchanged with the ground. A nontrivial assumption based on the previous development comes from Eq. (5.4) which was shown in the previous chapter to be not always valid. Fortunately, with the concept of effective resistance, the whole formalism can be kept without changing anything by just replacing the borehole resistance with the effective resistance. For instance, using (5.70) in (6.4) assumes an exponential variation of the inlet temperature, using (5.84) instead assumes the Zeng's profile, etc.

Example 6.11 Redo the last example (c) and (d) taking into account the pipe-to-pipe interference.

Solution Since only the line source or the multipole method can be used to evaluate the internal resistance, only cases (c) and (d) will be described. The whole calculation is the same except that we replace the borehole resistance value. For the first iteration, it was assumed that $H = 110$ m:

$$\eta = \frac{H}{\dot{m}C_p\sqrt{R'_b R'_a}} = \frac{110}{0.618 \cdot 3800\sqrt{0.133 \cdot 0.373}} = 0.20$$

$$R'^*_b = 0.133 \cdot 0.2/\tanh(0.2) = 0.135 \text{ K-m/W}$$

(continued)

In the original sizing approach proposed by ASHRAE, the loss of performance was taken into account through the correction factor (F_{sc}) in Eq. (6.12). This multiplication factor produces an increase of the total length due to the loss of performance. Logically, this factor is larger for low flow rates than for high flow rates. However, the numerical values given seem quite arbitrary without any scientific justification, and what is the most surprising is that it is associated with the short-time ground resistance and not the borehole resistance which does not make too much sense. This resistance is influenced by the soil properties, whereas the short-circuit loss of performance should logically be associated with the internal borehole properties. The use of effective resistance is, thus, much more appropriate.

6.8   A Little Bit of History: The Hart and Couvillon Approach

From the mid-twentieth century, research has been conducted to study the behavior of ground heat exchangers. Early work from Ingersoll et al. [17, 18] as well as Carslaw and Jaeger [19] are often referenced as major contributions to the field. Infinite line source (ILS) and cylindrical line source (ICS) were mostly the analytical models proposed as well as mathematical approximations of the associated complex expressions. From the start, it was implicitly assumed that, although both solutions are similar after some time, the ICS was preferred for shorter time period [20] although neither ones are correct for very short times (see Chap. 13). The main reason for that was that the early analysis was based on the evaluation of the pipe temperature ($T_p(t)$) and not the borehole temperature ($T_b(t)$) in which case the conclusion might be different [12]. An interesting extensive study was also presented by Hart and Couvillon [21], and those who had a look of their work may wonder what can be the link between their work and the theory outlined in this chapter since they introduced very specific concepts. So, it can be interesting to have a glimpse on their work.

In their book, Hart and Couvillon studied mostly the thermal behavior of a single pipe as a heat emitter. They did not take into account the concept of borehole resistance, at least not directly. So their main concern was first to evaluate the thermal response associated with the heat transfer associated with a single pipe. As for the other studies done at that time, they first relied on the infinite line source. They proposed to use the well-known series expansion of the exponential integral but expressed it in terms of a so-called infinite radius r_∞ defined as:

$$\Delta T = -\frac{q_o'}{k_s} G_{h-c} \tag{6.41}$$

$$G_{h-c} = \frac{1}{2\pi} \left[\log\left(\frac{r_\infty}{r}\right) - 0.9818 + \frac{4r^2}{r_\infty^2} \right.$$
$$\left. - \frac{1}{4 \cdot 2!} \left(\frac{4r^2}{r_\infty^2}\right)^2 + \dots + \frac{(-1)^{n+1}}{2n \cdot n!} \left(\frac{2r^2}{r_\infty^2}\right)^n \right] \tag{6.42}$$

where:

$$r_\infty = 4\sqrt{\alpha t} \tag{6.43}$$

and r can be any radius but more often the outside pipe radius that they called r_o. The first two terms are equivalent to Eq. (4.14) since:

$$G \approx \frac{1}{4\pi} \left(\log\left(\frac{4\alpha t}{r^2}\right) - \gamma \right)$$

$$\approx \frac{1}{4\pi} \left(\log\left(\frac{r_\infty}{2r}\right)^2 - \gamma \right)$$

$$\approx \frac{1}{2\pi} \left(\log\left(\frac{r_\infty}{r}\right) - \log(2) - \gamma/2 \right)$$

$$\approx \frac{1}{2\pi} \left(\log\left(\frac{r_\infty}{r}\right) - 0.9818 \right)$$

Hart and Couvillon were interested in evaluating the temperature at the pipe radius assuming that the heat is emitted there, but, for some unknown reasons, they did not use the obvious ICS solution for this specific case. So they used their series expansion of the ILS for large time ($r_\infty > 15r_o$ or equivalently $Fo > 14$)[2] when it is well known that both solutions are almost equal, and for smaller values of time, they used a rather contrived way in which they recognized that for those short values of time, the heat emitted into the ground by the ILS (Eq. (4.15)) is smaller than the constant heat input assumed in the ICS (q_o'). So they looked for a new "larger" infinite radius in which the heat predicted by the ILS would be equal to the heat if emitted at the pipe radius. From

---

[2]Remember that the Fourier number is based on the pipe radius.

this hypothesis, they imposed the following heat balance:

$$tq(t) = \rho C_p \int_{V_o}^{V_\infty} (T_\infty - T(t)) dV = t q_o' L$$

Replacing Eq. (6.42) inside the previous integral and keeping only a significant number of terms, they obtained the following equation for the new value of r_∞:

$$Fo = 0.0649\tilde{r}_\infty + 0.2409 - \frac{1}{2} \log(\tilde{r}_\infty)$$

$$- \left(\frac{1}{2\tilde{r}_\infty^2} - \frac{1}{3\tilde{r}_\infty^4} + \frac{2}{9\tilde{r}_\infty^6} - \frac{2}{15\tilde{r}_\infty^8} \right) \quad (6.44)$$

with:

$$Fo = \frac{\alpha t}{r_o^2}, \quad \tilde{r}_\infty = \frac{r_\infty}{r_o}$$

So basically the solution was given by Eq. (6.42) with r_∞ given by Eq. (6.43) when it is larger than $15r_o$ and, implicitly, by Eq. (6.44) otherwise. So at the end, they obtained something in between the ICS and the ILS solutions as shown in Fig. 6.7:

Besides their very special treatment of the thermal response, it is interesting to see how they modeled a typical U-tube arrangement. As previously said, they did not use the concept of borehole resistance. Instead, they calculated the thermal performance of individual pipes, taken into account the thermal interference between one and the other. For instance, a U-tube is represented by two single pipes having a mutual interference impact. Two boreholes with double U-tubes are represented by eight single pipes, etc. This approach is similar than the one of Li and Lai [22] except that these authors used the composite line source taken into account the difference in conductivity of the grout. Since Hart and Couvillon did not take into account the presence of grout,

we will compare their strategy with a borehole filled with the adjacent ground.

The analysis will be limited to the case of a single borehole with a single U-tube. In the classical approach described previously, the analysis is based on the borehole height. For a single heat pulse, for example, the pipe temperature was given by:[3]

$$T_{p,o} - T_o = -q' \left(\frac{G(Fo)}{k_s} + R'_g \right) = -\frac{q}{L} \left(R'_s + R'_g \right)$$

From our terminology, the total length for several boreholes was given by:

$$L = N_b H$$

So the previous relation for a single borehole is:

$$T_{p,o} - T_o = -\frac{q}{H} \left(R'_s + R'_g \right) \quad (6.45)$$

Hart and Couvillon worked by length of pipe (that they called L_s) and not borehole. For a single borehole, neglecting interference effects, this would be:

$$T_{p,o} - T_o = -\frac{q}{L_s} \frac{G_{h-c}(Fo_o)}{k_s}$$

where Fo_o is the Fourier number based on the outside pipe radius and not the borehole radius. Hart and Couvillon observed that, due to the thermal interference, the total heat produced by a given temperature difference will be smaller than if the two pipes were far apart or, for a given heat pulse, the same temperature difference can be obtained if the heat per unit length is increased, that is, if the length of the borehole is reduced. They then proposed the following correction:

$$T_{p,o} - T_o = -\frac{q}{L_m} \left(\frac{L_m}{L_s} \right) \frac{G_{h-c}(Fo_o)}{k_s} \quad (6.46)$$

with:

$$\frac{L_m}{L_s} > 1$$

as a correction factor. This factor comprises two contributions. The first one is quite obvious; it represents the impact of the temperature variation of one emitter on the other one. This can be interpreted in a manner similar to the penalty temperature. In heating mode, for example, the local temperature around a tube will be lowered by the presence of the other one and can be estimated as:

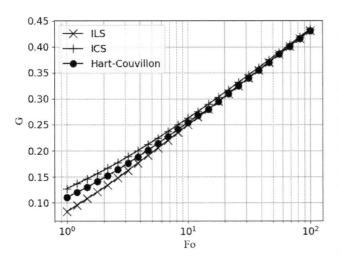

Fig. 6.7 Comparison between ICS and ILS and Hart-Couvillon pipe models

---

[3]Indice p is associated with the "pipe" temperature not to be confused with the "penalty" temperature. Indice p, o will be used.

$$T_{p,o} - (T_o + \delta T) = -\frac{q}{L_m}\frac{G_{h-c}(Fo_o)}{k_s}$$

$$\left(\frac{L_m}{L_s}\right) = \left(\frac{L_m}{L_s}\right)_{he}\left(\frac{L_m}{L_s}\right)_{T\infty} \quad (6.50)$$

with:

$$\delta T = -\frac{q}{L_m}\frac{G_{h-c}(Fo_{sd})}{k_s}$$

where

$$Fo_{sd} = \frac{\alpha t}{(sd)^2}$$

where sd is the distance between the pipes ($2x_c$ in our terminology). Replacing:

$$T_{p,o} - T_o = -\frac{q}{L_m}\frac{G_{h-c}(Fo_o)}{k_s}\underbrace{\left(1 + \frac{G_{h-c}(Fo_{sd})}{G_{h-c}(Fo_o)}\right)}_{(L_m/L_s)_{he}}$$

In their work, they replaced the response factors (G) by their series expansion (Eq. (6.42)) keeping only a significant number of terms. The factor

$$\left(\frac{L_m}{L_s}\right)_{he}$$

was associated with a "correction factor" having similar temperature difference for a given heat exchange. A second much more intricate factor was also suggested, associated with a correction in the far-field temperature T_o. The first one was intended to produce the correct temperature difference but not the correct temperature, if the far-field temperature changes. This seems a little bit strange since, by definition, the far-field undisturbed temperature should remain "undisturbed." This came probably from their definition of a "far-field radius" (r_∞). Without going into too many details, numerical comparisons can be done to verify their assertion. First, we will give, without justification, the definition of their second correction factor, again only for the case of a single U-tube:

$$\left(\frac{L_m}{L_s}\right)_{T_\infty} = \frac{1}{1-F} \quad (6.47)$$

with[4]

$$F = \frac{1}{2\pi}(\theta_m - \cos(\theta_m)\sin(\theta_m))\left(\frac{G_{h-c}(Fo_{sd/2})}{G_{h-c}(Fo_o)}\right) \quad (6.48)$$

and:

$$\theta_m = \cos^{-1}\left(\frac{sd}{2r_\infty}\right) \quad (6.49)$$

and finally:

From the pipe temperature, the fluid mean and outlet temperature were found as usual:

$$T_f - T_{p,o} = q'R'_p$$

Example 6.12 Evaluate the pipe temperature of borehole with a single U-tube having the following characteristics:

- $r_b = 8$ cm.
- $x_c = 4.2$ cm
- $r_{po} = 1.5$ cm
- $k_s = 2.5$ W/m-K
- $\alpha_s = 1e\text{-}6$ m^2/s
- $T_o = 0$ °C

following a heat transfer of -50 W (heat injection) during 190 hours. As previously stated, the grout has the same properties than the ground. Compare two methods:

(a) Using the usual approach with the ICS solution and the line-source method from the borehole resistance
(b) Using the approach of Hart and Couvillon

Solution

(a) First let's evaluate the grout resistance. Using the line-source solution, the value is

$$R'_g = 0.052 \text{ W/m-K}$$

After 190 hours,

$$Fo = \frac{0.0036 \cdot 190}{0.08^2} = 107$$

$$G(107) = 0.439$$

$$T_b = \frac{50}{2.5}0.439 = 8.77 \text{ °C}$$

$$T_{p,o} = 8.77 + 50 \cdot 0.052 = 11.37 \text{ °C}$$

(b) With the Hart-Couvillon approach, we need the heat transfer per unit length of pipe which is simply half the value given (-25 W/m). The Fourier

[4] Again, in their book, the series expansion with a finite number of terms is used.

(continued)

Example 6.12 (continued)
number based on the pipe radius is:

$$Fo_o = \frac{0.0036 \cdot 190}{0.015^2} = 3040$$

The Hart-Couvillon g-function (Eq. (6.42)) would give:

$$G_{H-C}(3040) = 0.7025$$

The first correction factor is given by:

$$\left(\frac{L_m}{L_s}\right)_{he} = \left(1 + \frac{G_{h-c}(Fo_{sd})}{G_{h-c}(Fo_o)}\right)$$

$$Fo_{sd} = \frac{0.0036 \cdot 190}{0.084^2} = 97$$

$$G_{H-C}(97) = 0.43$$

$$\left(\frac{L_m}{L_s}\right)_{he} = 1.61$$

The second correction factor is given by:

$$\frac{sd}{2r_\infty} = \frac{sd}{8\sqrt{\alpha t}} = \frac{1}{8\sqrt{Fo_{sd}}} = 0.0126$$

$$\theta_m = 1.56 \text{ rd}$$

$$Fo_{sd/2} = 4 \cdot 97 = 389$$

$$G_{H-C}(389) = 0.54$$

$$F = \frac{1}{2\pi}(1.56 - \cos(1.56)\sin(1.56))\left(\frac{0.54}{0.7025}\right) = 0.189$$

$$\left(\frac{L_m}{L_s}\right)_{T\infty} = \frac{1}{1 - 0.189} = 1.23$$

$$\left(\frac{L_m}{L_s}\right) = 1.61 \cdot 1.23 = 1.98$$

$$T_{p,o} = \frac{25}{2}1.98 \cdot 0.7025 = 13.95\,°C$$

which is a rather large difference. It is interesting to see that using only the first correction factor, we would have:

$$T_{p,o} = \frac{25}{2}1.61 \cdot 0.7025 = 11.31\,°C$$

The same problem was solved with a finite element software, and the solution was compared with the previous solutions and is given in Fig. 6.8:

It is easily observed that the second correction factors seem to overestimate the temperature increase

(continued)

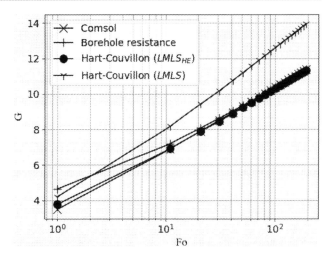

Fig. 6.8 Comparison between ICS and ILS and Hart-Couvillon U-tube models

due to the heat injection. There was unfortunately few analysis or some kind of physical, numerical, or experimental validation on their sizing approach besides their own work [21] or some work done at Caneta Research [23] where the method is described without validation. From this elementary analysis, it seems that the temperature prediction using only the first correction factor would give very good agreement with more classical methods. It can be seen that for short period of time, the values are even better since the heat capacity of the borehole is taken into account, but one has to remember that the grout properties were not taken into account. For a specific treatment of the borehole heat capacity, we can refer to Chap. 13.

■

Problems

Pr. 6.1 A geothermal borehole exchange heat is given by the following pulses (Fig. Pr_6.1):
The following data are given:

$$\alpha_{soil} = 0.1 \text{ m}^2/\text{d}, k_{soil} = 2.00 \text{ W/m-K}, T_o = 7.5\,°C, r_b = 6 \text{ cm}$$
$$R'_b = 0.095 \text{ K-m/W}, (\dot{m}C_p)_{fluid} = 2850 \text{ W/K}$$

(a) Neglecting potential interference between boreholes, find the total length needed to reach 0 °C at the end of the fourth pulse (use the logarithmic approximation of the ILS for the thermal response).

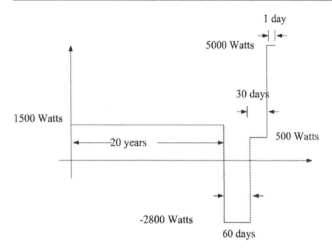

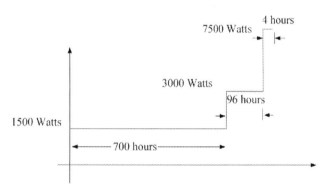

Pr_6.1 Four pulses ground heat exchange

Pr_6.2 Four pulses ground heat extracted

(b) Assuming that two boreholes separated by 6 m. are used, verify the non-interference assumption by evaluating the induced temperature of one borehole to the other using the logarithmic approximation of the ILS. Use an equivalent pulse which represents the aggregated value of the four pulses.

(c) If the internal resistance is $R'_a = 0.1$ K-m/W, evaluate the effective resistance and evaluate if the outlet temperature will differ from 0°C due to the potential pipe-to-pipe interference. Neglect again the borehole-to-borehole interference.

Answers: (a) 180 m, (b) -1.27 °C (not a very good assumption), (c) -0.36°C

Pr. 6.2 Figure Pr_6.2 shows the quantity of heat that is extracted from the ground in a single borehole.
The following data are given:

$\alpha_{soil} = 0.09$ m²/d , $k_{soil} = 2.00$ W/m-K , $T_o = 6$ °C
$\alpha_{grout} = 0.052$ m²/d , $k_{grout} = 2.00$ W/m-K , $d_{po} = 4.3$ cm
$r_b = 0.08$ cm , $x_c = 0.5r_b, (\rho C_p)_f = 4.2$ MJ/m³-K

Estimate the length of the borehole in order to have a mean temperature lower than 15°C after 800 hours.

- Use the ILS solution.
- Use the line-source method for the borehole resistance.
- The pipe is thin and the fluid convection is small ($R'_p \approx$ 0).

Answer: 157 m

Pr. 6.3 The outlet temperature at the end of Problem 5.1 was very low. Evaluate the length needed to obtain −5°C using the non-aggregated pulses. Assume you have only one borehole (no interference).
Answer: 162 m

Pr. 6.4 Using the same load history and using the ASHRAE approach, calculate the total length needed for a 20-year period with a minimum mean temperature of 0°C at the exit of the field. The maximum length of each borehole is 150 m, and use the Bernier-Philippe regression for the temperature penalty if more than one borehole is needed. In that case, use inline boreholes separated by 6 m.
Answer: L = 295 m, 2 boreholes, $T_p = 1$ K

Pr. 6.5 The monthly loads of a building are given in the following table.

	Monthly building heating loads (MWh)	Monthly building cooling loads (MWh)
Jan	21.9	0
Feb	18.25	0
Mar	14.6	0
Apr	10.95	0
May	3.65	7.3
June	0	10.95
July	0	14.6
Aug	0	14.6
Sept	0	7.3
Oct	10.95	0
Nov	14.6	0
Dec	21.9	0

We assume that all months have identical number of hours (730). The hourly pulses on the building side are:

$$q_{h,he} = 100 \text{ kW} , \quad q_{h,co} = 90 \text{ kW}$$

Configuration C ⊛ is used and the following properties are given:

$(\rho C_p)_{soil} = 2.8$ MJ/m^3-K, $k_{soil} = 2.95$ W/m-K, $T_o = 13\,°$C, $r_b = 6$ cm
$COP_{he} = 4$, $COP_{co} = 5$, $(\dot{m}C_p)_{fluid} = 16.8$ kW/K, $h_{conv} = 1000$ W/m^2-k
$d_{po} = 4$ cm, $d_{pi} = 3$ cm, $k_{grout} = 1.5$ W/m-K, $k_{HDPE} = 0.6$ W/m-K

The ASHRAE method is used with the penalty temperature evaluated with the FLS, 6-hour pulse, and the following constraints:

$$T_{fo,min} = 4\,°C, \ T_{fo,max} = 26\,°C, \ N_{years} = 20, \ H < 200 \text{ m}$$

The boreholes are separated by a distance of 6 m. Use the multipole method of order 10 for the borehole resistances, and take into account the internal heat transfer.
Answer: L = 1522 m with a 3 × 3 field

Pr. 6.6 Redo the same thing with the Swedish approach.
Answer: L = 1492 m with a 3 × 3 field

References

1. Kavanaugh, S., Rafferty, K.: Geothermal Heating and Cooling: Design of Ground-Source Heat Pump Systems. ASHRAE, Atlanta (2014)
2. American Society of Heating, Refrigerating and Air-Conditioning Engineers: ASHRAE Handbook, HVAC Applications. The Society, New York (2015)
3. Kavanaugh, J.: Long-term commercial GSHP performance: Part 1: project overview and loop circuit types. ASHRAE J. **54**(6), 48 (2012)
4. Bernier, M., et al.: A multiple load aggregation for annual hourly simulations of GCHP systems. HVAC&R Research **10**(4), 471–487 (2004)
5. Kavanaugh, S., Rafferty, K.: Ground-Source Heat Pumps: Design of Geothermal Systems for Commercial and Institutional Buildings. ASHRAE, Atlanta (1997)
6. Bose, J.E., Parker, J.D., McQuiston, F.C.: Design Data Manual for Closed-loop Ground-Coupled Heat Pump Systems. American Society of Heating, Refrigerating and Air-Conditioning Engineers, Atlanta, Ga. (1985). ISBN: 0910110417
7. Bernier, M., Chahla, A., Pinel, P.: Long-term ground-temperature changes in geo-exchange systems. ASHRAE Transactions **114**(2), 342–351 (2008)
8. Philippe, M., Bernier, M. Marchio, D.: Sizing calculation spreadsheet vertical geothermal Boerfields. ASHRAE Journal **52**(7), 20–28 (2010)
9. Fossa, M., Rolando, D.: Improving the Ashrae method for vertical geothermal borefield design. Energ. Buildings **93**, 315–323 (2015). https://doi.org/10.1016/j.enbuild.2015.02.008
10. Fossa, M.: Correct design of vertical borehole heat exchanger systems through the improvement of the ASHRAE method. Science and Technology for the Built Environment **23**(7), 1080–1089 (2017). https://doi.org/10.1080/23744731.2016.1208537
11. Capozza, A., De Carli, M., Zarrella, A.: Design of borehole heat exchangers for ground-source heat pumps: A literature review, methodology comparison and analysis on the penalty temperature. Energ. Buildings **55**, 369–379 (2012). https://doi.org/10.1016/j.enbuild.2012.08.041
12. Lamarche, L.: Short-term behavior of classical analytic solutions for the design of ground-source heat pumps. Renew. Energy **57**(0), 171–180 (2013)
13. Eskilson, P.: Thermal analysis of heat extraction systems." PhD thesis. Lund University, Sweden (1987)
14. Cimmino, M., Bernier, M., Adams, F.: A contribution towards the determination of g-functions using the finite line source. Appl. Therm. Eng. **51**(1-2), 401–412 (2013)
15. Cimmino, M., Bernier, M.: A semi-analytical method to generate gfunctions for geothermal bore fields. Int. J. Heat Mass Transf. **70**(0), 641–650 (2014)
16. Ahmadfard, M., Bernier, M.: Modifications to ASHRAE's sizing method for vertical ground heat exchangers. Science and Technology for the Built Environment **24**(7), 803–817 (2018)
17. Ingersoll, L.R., Plass, H.J.: Theory of the ground pipe source for the heat pump. ASHVE Transactions **54**, 339–348 (1948)
18. Ingersoll, L.R., et al.: Theory of earth heat exchangers for the heat pump. ASHVE Transactions **56**, 167–188 (1950)
19. Carslaw, H.S., Jaeger, J.C.: Conduction of Heat in Solids. 2nd edn. (1959)
20. Philippe, M., Bernier, M., Marchio, D.: Validity ranges of three analytical solutions to heat transfer in the vicinity of single boreholes. Geothermics **38**(4), 407–413 (2009)
21. Hart, D.P., Couvillion, R.: Earth-coupled heat transfer: offers engineers and other practitioners of applied physics the information to solve heat transfer problems as they apply to earth-coupling. In: National Water Well Association (1986)
22. Li, M., Lai, A.C.K.: New temperature response functions (G functions) for pile and borehole ground heat exchangers based on composite-medium line-source theory. Energy **38**(1), 255–263 (2012)
23. Cane, R.L.D., Forgas, D.A.: Modeling of ground-source heat pump performance. ASHRAE Trans. **97**(1), 909–925 (1991)

7.1 Introduction

Once the approximate total length of the earth loop is known, the piping network configuration has to be set up in order to physically realize the borefield. For a small system with one borehole, it is trivial, but as the number of boreholes increases, several options are possible and might affect the performance and total cost of the installation. As soon as there is more than one borehole, a choice has to be made between a series, parallel, or mix configuration (Table 7.1).

Whatever choice is made, a higher limit in a series configuration is soon reached as the maximum length for practical application is given in Table 7.2:

In practice, even if for residential applications, or high temperature energy storage applications, some series arrangement can be used, all commercial application will have a parallel configuration as stated in article 7.2.7 of the CSA-448-02 norm for ground heat exchangers. This decision among others related to the outdoor hydraulic piping network is of major importance in order to keep a high efficiency system at the end. Poorly designed hydraulic network can ruin the benefit of an expensive GSHP system. The following example illustrates this problem in a simple way:

Example 7.1 A one h.p. inline pump is used in conjunction with a 15 kW heating capacity GSHP having a COP of 3.5. What is the system's COP?

Solution The work of the compressor is:

$$\dot{W}_{comp} = \frac{15}{3.5} = 4.28 \text{ kW}$$

(continued)

The work of the pump is:

$$\dot{W}_{pump} = 0.746 \text{ kW}$$

$$\dot{W}_{tot} = 5.03 \text{ kW}$$

$$COP_{system} = \frac{15}{5.03} = 2.98$$

resulting in an important deterioration of the performance.

Comments *Remember from Chap. 2 that ISO ratings include the energy needed to overcome the pressure losses through the heat pump but not the ground loop.*

∎

The last simple example shows that, if not correctly designed, the performance of the system can deteriorate. Kavanaugh and Rafferty [1] provided a quantitative metric for rating the hydraulic energetic performance of SL-GSHP systems (Table 7.3).

In the last example, the rating performance would be:

$$\frac{746}{15} = 49.7 \text{ W/kW}$$

which would result in a F-grade. In order to have a good knowledge of the hydraulic components that have to be chosen, a good knowledge of piping head losses is mandatory. The piping analysis of a ground heat exchanger uses the same tools as regular piping systems, and several good references can be consulted [2].

© The Author(s), under exclusive license to Springer Nature Switzerland AG 2023
L. Lamarche, *Fundamentals of Geothermal Heat Pump Systems*,
https://doi.org/10.1007/978-3-031-32176-4_7

Table 7.1 Pros and cons of series ans parallel configurations

Series configuration		Parallel configuration	
Advantages	Inconvenience	Advantages	Inconvenience
Uniform flow rate	Larger diameter	Better efficiency	Hard to balance flow rates
Easier to purge	Harder to install	Easy to install	

Table 7.2 Typical lengths per parallel circuit

Diameter	Good	Correct	Bad
19 mm	30–60 m	60–70 m	>70 m
25 mm	45–90 m	90–105 m	>105 m
32 mm	75–150 m	150–180 m	>180 m
38 mm	90–180 m	180–300 m	>300 m

Table 7.3 SL-GSHP system performance ratings (From [1])

Electric pump power/installed capacity (W_e/kW_{th})	Grade
<13	A (Excellent)
13–19	B (Good)
19–25	C (Correct)
25–36	D (Bad)
>36	F (Fail)

7.2 Head Losses: Piping Fundamentals

Simply stated, head losses are related to mechanical energy losses in pipes due to friction. Specific mechanical energy is defined as:

$$e_a[\text{ J/kg}] = gz + \frac{u^2}{2} + \frac{p}{\rho} \qquad (7.1)$$

It comprises three terms: potential, kinetic, and pressure terms. In the expression of the specific energy, g is the gravitational acceleration (9.81 m/s^2), which is a constant at constant altitudes. The same information can then be evaluated if the energy is divided by this constant. The specific energy, often called *head* in this form, is then written as:

$$e_b[\text{ m}] = z + \frac{u^2}{2g} + \frac{p}{\rho g} \qquad (7.2)$$

In this case, the energy unit is length and the symbol h is often used.[1]

In the case of a liquid, the density of the fluid (often water) does not change significantly, and the energy is often associated with pressure by multiplying each term by the density:

$$e_c[\text{ Pa}] = \rho gz + \frac{\rho u^2}{2} + p \qquad (7.3)$$

Whatever unit is used, the loss of mechanical energy must be compensated by pump power.[2] Using the first form of energy, going from one point (1) of a pipe to another one (2), we have:

$$e_{a1} + (gh)_{pump} = e_{a2} + (gh)_{losses}$$

Since most of the time the loop is closed ($e_{a1} = e_{a2}$), the final result for closed loop is:

$$P_{pump}[\text{ W }] = \dot{m}(gh)_{losses} \qquad (7.4)$$

When the energy losses are evaluated as pressure losses, it can be expressed as:

$$P_{pump}[\text{ W }] = \frac{\dot{m}}{\rho}(\Delta P)_{losses} = \dot{V}(\Delta P)_{losses} \qquad (7.5)$$

The losses of specific energy can generally be separated into three distinct parts:

$$(h)_{losses} = (h)_{\text{major losses}} + (h)_{\text{minor losses}} + (h)_{\text{valve losses}} \qquad (7.6)$$

$$(h)_{losses} = \underbrace{f\left(\frac{L}{D}\right)\frac{u^2}{2g}}_{\text{major losses}} + \underbrace{\sum K \frac{u^2}{2g}}_{\text{minor losses}} + \underbrace{\frac{S_g}{\rho g}\sum\left(\frac{\dot{V}}{C_v}\right)^2}_{\text{valve losses}} \qquad (7.7)$$

Minor losses , or singular losses, associated with bends, joints, etc. are sometimes represented by a loss coefficient K but can also be expressed as equivalent lengths (L_e) of pipe:

$$(h)_{\text{minor losses}} = \sum f\left(\frac{L_e}{D}\right)\frac{u^2}{2g} \qquad (7.8)$$

or both.

7.2.1 Evaluation of Major Losses

Major losses, or pipe losses, depend on the friction factor f, total length of piping, internal diameter of the pipe, and mean fluid velocity and are given by the *Darcy-Weisbach* equation [3]:

$$(h)_{\text{major losses}} = f\left(\frac{L}{D}\right)\frac{u^2}{2g} \qquad (7.9)$$

In Eq. (7.9), the friction factor f is the same expression as the one used in the Gnielinski correlation (2.43b). In laminar regime it is given by:

---

[1] Not to be confused with enthalpy.

[2] In a closed loop and/or potential energy in open system.

$$f = \frac{64}{Re_D} \quad \text{Laminar} \quad (7.10)$$

In the turbulent case, it is given by the Colebrook formula:

$$1/\sqrt{f} = -2\log_{10}\left(\frac{\epsilon/D}{3.7} + \frac{2.51}{Re_D\sqrt{f}}\right) \quad \text{Turbulent, rough pipe}$$
$$(7.11)$$

or the *Moody graph*. A simpler formula has been suggested by Jeppson [4]

$$f = \frac{1.325}{\left[\log\left(\frac{\epsilon/D}{3.7} + \frac{5.74}{Re_D^{0.9}}\right)\right]^2} \quad \text{Turbulent, rough pipe} \quad (7.12)$$

For smooth pipe one can use any of these expressions with $\epsilon = 0$. It is also possible to use the Petukhov relation [5]:

$$f = (0.79\log(Re_{D_i}) - 1.64)^{-2} \quad \text{Turbulent, smooth pipe}$$
$$(7.13)$$

or the Blasius equation for turbulent flow

$$f = \frac{0.3164}{Re^{1/4}} \quad \text{Turbulent, smooth pipe} \quad (7.14)$$

The head losses are dependent on the pipe dimensions. Most geothermal applications use high-density polyethylene (HDPE) pipes that are subjected to different ASTM standards. Standard dimension ratio (SDR) and dimensional ratio (DR) are two closely related standards of thickness of the plastic pipes. They are defined as:

$$SDR = DR = \frac{\text{Outside Diameter}}{\text{thickness}} = \frac{O.D.}{t} \quad (7.15)$$

Typical dimensions for DR-11 to DR-17 are given in Appendix B.

Example 7.2 Pipe major losses are often evaluated using log-log graphs giving head loss per given length of pipe. In North America, they were often given in feet of water per 100 feet of pipe, but similar graphs are also used in SI units. Looking at one of these graphs (see [6]), head losses for 20% propylene glycol at 1.9 l/s are 0.7 m/30 m of pipe at 32 degC and 0.9 m/30 m of pipe at 0degC in the case of SDR-11 2" nominal pipes. Verify this result using (7.9).

Solution From Appendix B, we have that for SDR-11:

$$D_i = 0.49\,\text{m} \Rightarrow A_c = \frac{\pi \cdot D_i^2}{4} = 1.91\text{e-}3\,\text{m}^3$$

(continued)

The flow rate is:

$$\dot{V} = 1.9\text{e-}3\text{m}^3/\text{s} \Rightarrow u = \frac{1.9\text{e-}3}{1.91\text{e-}3} \approx 1\,\text{m/s}$$

The propylene-glycol properties at 32 °C give:

$$\nu = 1.40\text{e-}6\,\text{m}^2/\text{s} \Rightarrow Re = \frac{1 \cdot 0.49}{1.4\text{e-}6} = 35000$$

Using the Colebrook relation for smooth pipe:

$$f_1 = 0.0227$$

$$\Rightarrow h_1 = 0.0227\frac{30}{0.49}\frac{1^2}{2 \cdot 9.81} = 0.697\,\text{m}$$

At 0°C, the only parameter that changes is the viscosity:

$$\nu = 3.97\text{e-}6\,\text{m}^2/\text{s} \Rightarrow Re = \frac{1 \cdot 0.49}{3.93\text{e-}6} = 12330$$

$$f_2 = 0.0297$$

$$\Rightarrow h_2 = 0.0297\frac{30}{0.49}\frac{1^2}{2 \cdot 9.81} = 0.909\,\text{m}$$

Comments *HDPE was assumed to be smooth. A value of rugosity of approximately 0.02 mm is sometimes suggested which will have a slight influence on the final result.*

∎

7.2.2 Minor and Valve Losses

Minor losses occur through elbows, tees, fittings, etc. They are often given in terms of a loss coefficient:

$$(h)_{\text{minor losses}} = \sum K\frac{u^2}{2g} \quad (7.16)$$

where the summation is done over all the singular components (in series). Minor losses are sometimes represented as an equivalent length (L_e) of pipe:

$$(h)_{\text{minor losses}} = \sum f\left(\frac{L_e}{D}\right)\frac{u^2}{2g} \quad (7.17)$$

Loss coefficients can be found in several references [7] dealing with pipe sizing. In ASHRAE Handbook Fundamentals [2], for example, many such coefficients are given, mostly related to steel pipes. Kavanaugh and Rafferty [1, 6] report several minor factors associated specifically with HDPE piping used in geothermal applications. They gave their loss coefficients in terms of equivalent lengths.

Valve losses are most often associated with a *valve flow coefficient*, C_v. In the United States, this coefficient is defined as the amount of water at 60 °F ($\approx$16°C) in US gpm (3.78 l/min) flowing through the valve under a pressure drop of 1 psi (6.89 kPa). The pressure loss across a single valve is then defined as:

$$\Delta P = S_g \left(\frac{\dot{V}}{C_v} \right)^2 \tag{7.18}$$

where S_g is the fluid *specific gravity*:

$$S_g = \frac{\rho_{fluid}}{\rho_{water(16°C)}} \tag{7.19}$$

In the SI unit system, a different symbol is sometimes used (K_v) and the unit is m^3/h/(bar)$^{1/2}$. In this book the symbol C_v will be used, and care should be taken to use coherent units. The valve coefficient can be related to a loss coefficient:

$$h_{loss} = K \frac{u^2}{2g} = K \frac{8\dot{V}^2}{\pi^2 D^4 g} = \frac{S_g}{\rho g} \left(\frac{\dot{V}}{C_v} \right)^2$$

From which:

$$K = \frac{S_g \pi^2 D^4}{8\rho C_v^2} \tag{7.20}$$

In the last equation, if the other variables are given in the SI system, the valve coefficient should of course be expressed in the fundamental SI units (m^3/s/(Pa)$^{1/2}$) to have a dimensionless K parameter.

Example 7.3 Evaluate the head losses in a 100 m length smooth pipe having a 4 cm diameter. The flow rate is 1.5 l/s. Along the pipe, one elbow has a loss coefficient of $K = 2.5$ and another one has an equivalent length of $L_e = 2$ m. There is also a ball valve, whose coefficient is 1.08 l/s/(kPa)$^{1/2}$. Do the calculation if:

(a) The fluid is water.
(b) The fluid is a mixture of water and propylene glycol (20% volume fraction).
(c) The fluid is a mixture of water and propylene glycol (40% volume fraction).

Assume that the pipe is smooth and evaluate the properties at 5 °C.

Solution From the water properties, we have:

$$\nu = 1.52\text{e-}6 \text{ m}^2/\text{s}, \quad \rho = 1000 \text{ kg/m}^3$$

The average velocity is:

$$A_c = \frac{\pi \cdot D_i^2}{4} = 1.256\text{e-}3 \text{m}^2,$$

$$u = \frac{1.5\text{e-}3}{1.256\text{e-}3} = 1.19 \text{ m/s}$$

$$Re = \frac{u D_i}{\nu} = 31300$$

From the Colebrook formula, we have:

$$f = 0.0234 \Rightarrow h_{pipe}$$

$$= 0.0234 \left(\frac{100}{0.04} \right) \frac{1.19^2}{2 \cdot 9.81} = 4.248 \text{ m}$$

The minor losses are readily obtained:

$$h_{minor,1} = 2.5 \frac{1.19^2}{2 \cdot 9.81} = 0.181 \text{ m}$$

$$h_{minor,2} = 0.02334 \left(\frac{2.0}{0.04} \right) \frac{1.19^2}{2 \cdot 9.81} = 0.085 \text{ m}$$

The loss from the valve will be given as pressure losses:

$$\Delta P_{valve} = 1.0 \left(\frac{1.5}{1.08} \right)^2 = 1.93 \text{ kPa}$$

The corresponding loss in length is:

$$h_{valve} = \frac{\Delta P}{\rho g} = \frac{1930}{1000 \cdot 9.81} = 0.197 \text{ m}$$

$$h_{total} = 4.71 \text{ m}$$

The power that a pump would have to give to the fluid is:

$$\dot{W}_{fluid} = 1.5 \cdot 9.81 \cdot 4.71 = 69.3 \text{ W}$$

b) Changing the fluid changes only the properties:

$$\nu = 3.26\text{e-}6 \text{ m}^2/\text{s}, \quad \rho = 1024 \text{ kg/m}^3$$

$$Re = 14530, \quad f = 0.0284$$

$$h_{pipe} = 0.0284 \left(\frac{100}{0.04} \right) \frac{1.19^2}{2 \cdot 9.81} = 5.16 \text{ m}$$

For the minor losses, one expression will be affected by the change of friction factor but not the other one:

$$h_{minor,1} = 2.5 \frac{1.19^2}{2 \cdot 9.81} = 0.181 \text{ m}$$

(continued)   (continued)

Example 7.3 (continued)

$$h_{minor,2} = 0.0289 \left(\frac{2.0}{0.04} \right) \frac{1.19^2}{2 \cdot 9.81} = 0.103 \text{ m}$$

The loss from the valve will be given as pressure losses:

$$\Delta P_{valve} = 1.024 \left(\frac{1.5}{1.08} \right)^2 = 1.978 \text{ kPa}$$

$$h_{valve} = \frac{1978}{1024 \cdot 9.81} = 0.197 \text{ m}$$

$$h_{total} = 5.64 \text{ m}$$

$$\dot{W}_{fluid} = 85 \text{ W}$$

From the result, it can be surprising that the change of fluid (or actually, Reynolds number) influences the losses of some piping elements and not the others. Actually, it does (see next section) but for now, it will be left as it is.

(c) With the glycol 40%, the results are:

$$\nu = 9.0\text{e-6 m}^2/\text{s}, \qquad \rho = 1043 \text{ kg/m}^3$$

$$Re = 5306, \qquad f = 0.0379$$

$$h_{pipe} = 6.884 \text{ m}$$

$$h_{minor,1} = 0.181 \text{ m}$$

$$h_{minor,2} = 0.137 \text{ m}$$

$$\Delta P_{valve} = 2.014 \text{ kPa}$$

$$h_{valve} = 0.197 \text{ m}$$

$$h_{total} = 7.4 \text{ m}$$

$$\dot{W}_{fluid} = 113 \text{ W}$$

Comments *Assuming that the minor losses are independent on the Reynolds number is not logical (see next section). The example shows that the impact of a viscous fluid can be very harmful for the energy consumed by the pumps.*

∎

7.2.2.1 Effect of the Reynolds Number on the Loss Coefficient

It is well known that the head loss coefficient is Reynolds number dependent [8, 9]. Hooper [8] suggested a two-parameter 2-K method to take into account the dependance on the Reynolds number as well as the impact of the pipe diameter on the fitting loss coefficient. Darby [9] proposed

a similar three-parameter 3-K method where the Reynolds number dependence is similar but the correction due to the pipe diameter is different. A similar approach is suggested for valve loss coefficients; the ESDU engineering reference [10] suggests a graphical correction factor associated with moderate Reynolds numbers. Without going into these details, a simple correction factor that can be applied to minor losses is to apply the same correction ratio as if the loss was evaluated by an equivalent length:

$$K_{corr} = K_{nom} \frac{f_{fluid}}{f_{water}} \tag{7.21}$$

This correction is expected to be within the uncertainties expected with head loss evaluations. It should be remembered that whatever rule is used to correct or not the loss coefficient, this correction becomes small at large Reynolds number and, in that case, often neglected.

Example 7.3 (modified) Let's redo the last example with the correction factor suggested in the previous section.

(a) For water, nothing changes:

$$f = 0.0234$$

$$h_{total} = 4.71 \text{ m}$$

$$P_{fluid} = 69.3 \text{ W}$$

(b) For glycol, only the minor losses associated with a loss coefficient change:

$$f = 0.0284$$

$$\text{Correction factor} = \frac{0.0284}{0.0234} = 1.21$$

$$h_{minor,1} = 1.21 \cdot 0.181 = 0.22 \text{ m}$$

$$h_{minor,2} = 0.103 \text{ m}$$

$$h_{valve} = 1.21 \cdot 0.197 = 0.239 \text{ m}$$

$$h_{total} = 5.72 \text{ m}$$

$$P_{fluid} = 86.3 \text{ W}$$

(c)

$$f = 0.0379$$

$$\text{Correction factor} = \frac{0.0289}{0.0234} = 1.62$$

$$h_{minor,1} = 1.62 \cdot 0.181 = 0.284 \text{ m}$$

(continued)

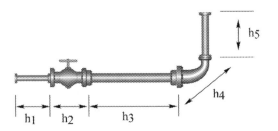

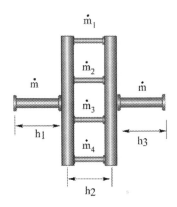

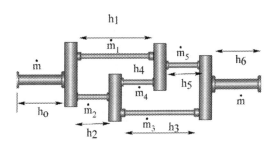

Example 7.3 (continued)

$$h_{minor,2} = 0.137 \text{ m}$$

$$h_{valve} = 1.62 \cdot 0.197 = 0.319 \text{ m}$$

$$h_{total} = 7.634 \text{ m}$$

$$P_{fluid} = 117 \text{ W}$$

Comments *The correction factor thus calculated (21% for 20% glycol solution and 62% for 40% colume) is in the same order of magnitude than the rule of thumb correction factors often seen in technical papers (see, e.g., [11, 12]). It should be noted that in this example, those losses were small in comparison with the major losses and in all cases, the uncertainties associated with head losses are large.* ∎

7.2.3 Series and Parallel Connections

When several piping elements are in series and/or parallel, the total head losses are easy to evaluate. Looking at Fig. 7.1, the total energy that must be given to the fluid is:

$$\dot{W}_{tot} = \sum \dot{m}_i g h_i$$

Since the flow rate is the same, one find:

$$\dot{W}_{tot} = \dot{m}_i g \underbrace{\sum h_i}_{h_{tot}} \qquad (7.22)$$

This is basically what was done in the previous example. In the case of parallel connections (Fig. 7.2):

$$\dot{W}_{tot} = \sum \dot{m}_i g h_i \qquad (7.23)$$

$$= g(\dot{m}h_1 + \underbrace{(\dot{m}_1 + \dot{m}_2 + \dot{m}_3 + \dot{m}_4)}_{\dot{m}} h_2 + \dot{m}h_3)$$

$$\qquad\qquad\qquad (7.24)$$

$$= \dot{m}g\underbrace{(h_1 + h_2 + h_3)}_{h_{tot}} \qquad (7.25)$$

So only the head loss in one of the parallel loops is necessary to evaluate the total losses. The complexity in the analysis is to evaluate the flow rates. An easy case is when one assumes that the flow rate is the same in all loops. In that case the analysis is easy. In the more general case, a more complex approach, called the Hardy-Cross method (see Appendix C) is needed. An example of a more complex

Fig. 7.1 Pipe losses in series

Fig. 7.2 Pipe losses in parallel

Fig. 7.3 Pipe losses in complex arrangement

configuration is shown in Fig. 7.3 where this method would be needed to find the flow rates from which:

$$\dot{W}_{tot} = g(\dot{m}h_o + \dot{m}_1 h_1 + \dot{m}_2 h_2 + \dot{m}_3 h_3 + \dot{m}_4 h_4 + \dot{m}_5 h_5 + \dot{m}h_6)$$

Even in that case, however, it is possible to make some simplifying assumptions. For instance, the symmetry of the configuration can possibly allow us to assume that:

$$\dot{m}_1 = \dot{m}_3 = \dot{m}_4 = \frac{\dot{m}}{3}$$

so that:

$$\dot{m}_2 = \dot{m}_5 = \frac{2\dot{m}}{3}$$

The total energy becomes:

$$\dot{W}_{tot} = g\left(\dot{m}(h_o + h_6) + \frac{\dot{m}}{3}(h_1 + h_3 + h_4) + \frac{2\dot{m}}{3}(h_2 + h_5)\right)$$

Fig. 7.4 Pump performance curve

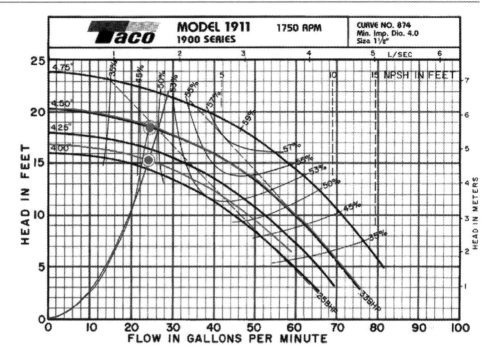

$$\dot{W}_{tot} = \dot{m}g \underbrace{(h_o + h_6 + 1/3(h_1 + h_3 + h_4) + 2/3(h_2 + h_5))}_{h_{tot}}$$

(7.26)

This kind of situations occurs in reverse-return layouts.

7.3 Pump Characteristics

The energy that has to be delivered to the fluid is given by pumps. These pumps have certain efficiency, so that more energy than that calculated in the previous section is needed. The mechanical energy that the pump must deliver on the shaft of the impeller is often called *brake horsepower* and is defined as:

$$\dot{W}_{shaft} = \frac{\dot{W}_{fluid}}{\eta_p}$$

(7.27)

where η_p is the efficiency of the pump. The shaft is normally powered by an electric motor having its own efficiency:

$$\dot{W}_{motor} = \frac{\dot{W}_{shaft}}{\eta_m} = \frac{\dot{W}_{fluid}}{\eta_p \eta_m}$$

(7.28)

Finally, if the motor has a variable speed drive, often called variable frequency drive (VFD), these drives have also losses defined by an efficiency:

$$\dot{W}_{elec} = \frac{\dot{W}_{motor}}{\eta_d} = \frac{\dot{W}_{fluid}}{\eta_d \eta_p \eta_m} = \frac{\dot{W}_{fluid}}{\eta_{tot}}$$

(7.29)

VFD are used to vary the pump flow rate at partial load by varying the frequency associated with AC induction motor.

New brushless electronically commutated motors (ECM) are now available that have an integrated flow rate modulation capability. Once the energy needed to give to the fluid is known, a proper pump must be chosen. Several suppliers offer selection guides to help to choose the pump correctly. Typical characteristics are shown on Fig. 7.4. Since it comes from an American supplier, the IP units are used.

Example 7.4 In the previous example, find the operating point needed to supply the head and flow requirements of the previous example when the fluid is water and if a 1911 pump is used.

Solution Since the figure is in IP units, we convert our results:

$$\dot{V}(gpm) = 15.85 \cdot 1.5 = 23.77 \text{ gpm}$$

$$h_{tot}(ft) = 3.28 \cdot 4.71 = 15.5 \text{ ft}$$

The solution is given by the red dot on the last figure (a 4.1" diameter should be chosen). It is seen that the efficiency is ≈47%. The BHP can be read on the chart or calculated from the efficiency:

$$\dot{W}_{shaft} = \frac{69.3}{0.47} = 147.5 \text{ W} = 0.2 \text{ HP}$$

The electric power will be higher since this efficiency does not take into account the electric efficiency. ∎

7.3.1　Pump Scaling Laws

Classical dimensionless analysis [13] shows that for *geometrically similar* pumps, dimensionless parameters for pumps are given by:

$$C_H = \frac{gh}{\omega^2 D^2} = f_1\left(\frac{\dot{V}}{\omega D^3}, \frac{\epsilon}{D}, \frac{\omega D^2}{\nu}\right)$$
$$C_P = \frac{\dot{W}}{\rho \omega^3 D^5} = f_2\left(\frac{\dot{V}}{\omega D^3}, \frac{\epsilon}{D}, \frac{\omega D^2}{\nu}\right) \tag{7.30}$$

The effect of the relative roughness is often neglected. The last parameter is the Reynolds number, and for high Reynolds numbers, it is also generally neglected. If it is the case, the dimensionless parameter associated with the pump head is given by:

$$\frac{gh}{\omega^2 D^2} = f_1\left(\frac{\dot{V}}{\omega D^3}\right) = f_1(C_q) \tag{7.31}$$

with:

$$C_q = \frac{\dot{V}}{\omega D^3}$$

Although these are the basic dimensionless numbers, any combination of those parameters is also a pertinent number. For instance, the efficiency of the pump is given by:

$$\frac{C_h \cdot C_q}{C_p} = \frac{\rho \dot{V} g h}{\dot{W}} = \eta$$

Then for two similar pumps, we have, if:

$$C_{q1} = C_{q2}$$

$$\Rightarrow C_{h1} = C_{h2}, \quad C_{P1} = C_{P2}$$

from which it is seen that for the same diameter, the pump head will follow the scaling laws:

$$\frac{\dot{V}_1}{\dot{V}_2} = \frac{\omega_1}{\omega_2} \tag{7.32}$$

$$\frac{h_1}{h_2} = \left(\frac{\omega_1}{\omega_2}\right)^2 \tag{7.33}$$

$$\frac{\dot{W}_1}{\dot{W}_2} = \left(\frac{\omega_1}{\omega_2}\right)^3 = \left(\frac{\dot{V}_1}{\dot{V}_2}\right)^3 \tag{7.34}$$

The last relation is often used to stress the importance of flow rate on pumping energy that is related to the third power of the flow rate ratio. This is true however only when the system curve does not change and if the static head is very small. In any cases, the reduction of flow rate has a very high impact on the hydraulic performance.

In this case, it is seen that the head curve can be used without correction if the fluid is different than water. However, the shaft power will be influenced by the density.

Example 7.5 What is the shaft power associated with the previous example for glycol (20% and 40%)?

Solution The flow rate does not change but the new head in the IP system are:

$$h_{tot} = 18.8 \text{ ft} \qquad 20\% \text{ glyclol}$$
$$h_{tot} = 25.0 \text{ ft} \qquad 40\% \text{ glyclol}$$

If the assumption that the pump curve does not change, the same pump with a 4.5" diameter can be used. The new operating point is shown in Fig. 7.4. It is seen that the efficiency does not change so that the predicted shaft power will be:

$$(\text{glycol } 20\%): \dot{W}_{shaft} = \frac{\dot{W}_{fluid}}{0.47} = \frac{86.3}{0.47} = 183 \text{ W}$$
$$= 0.25 \text{ HP}$$

For the 40% glycol solution, the required head is too high for this family of pump. A suggested pump from the selection guide of the supplier is shown from Fig. 7.5 from which it is seen that the efficiency is reduced to 37%:

$$(\text{glycol } 40\%): \dot{W}_{shaft} = \frac{\dot{W}_{fluid}}{0.37} = \frac{117}{0.37} = 316 \text{ W}$$
$$= 0.42 \text{ HP}$$

∎

7.3.2　Pump Correction Factors

In the previous example, the effect of viscosity was taken into account for calculating the increase in head required. It was however neglected for the pump performance curves assuming that they are not affected by the Reynolds number. In practice, it is not the case especially for highly viscous fluid (cold high percentage glycol solutions). In practice, pump providers suggest to increase the flow and head required by a correction factor to prevent the potential losses of performance. Table 7.4 are values given in [22]

From the table nothing would change in the previous example for the 20% glycol solution. For the 40% solution, it would be suggested to choose a pump that would give:

Fig. 7.5 Pump performance curve

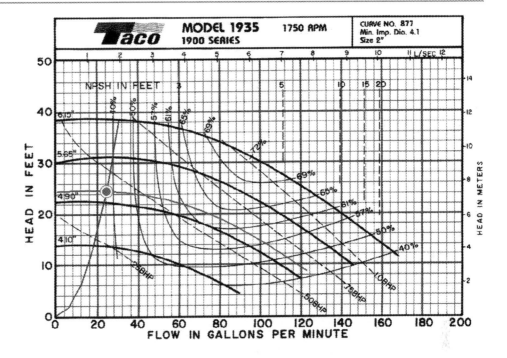

Table 7.4 Pump glycol correction factors (from [22])

4 °C Ethylene glycol (30–50% solutions)		
Flow	Flow correction	Head correction
<100 gpm	1.05	1.02
>100 gpm	1.0	1.02
4 °C Propylene glycol (30–50% solutions)		
<100 gpm	1.1	1.05
>100 gpm	1.0	1.02

$$25.01 \cdot 1.05 = 26.3 \text{ ft}$$

at a flow rate of:

$$23.77 \cdot 1.1 = 26.1 \text{ gpm}$$

Example 7.6 Redo the last example for water, but instead of choosing the pump, assume that you have the pump TACO 1911 4.25 " impeller available. From the data sheet, the head-flow rate curve can be approximated by the following equation:

$$h_{pump}(\text{m}) = 5.4952 - 0.04935\dot{V} - 0.15556\dot{V}^2 - 0.01588\dot{V}^3$$

where $\dot{V}$ is given in l/s. Find the new flow rate and head loss associated with the new configuration.

Solution The new operating point is easily obtained equalizing the system curve to the pump curve using the same set of units. Since the flow rate will change,

(continued)

the Reynolds number will also change and so does the friction factor. In the turbulent regime, the variation is small, and if neglected, the total head losses given in the previous example are given by:

$$h_{total} = 4.71 \text{ m} = \frac{4.71}{1.5^2}\dot{V}^2 = 2.094\dot{V}^2 \quad (\dot{V}((\text{l/s})))$$

So:

$$2.094\dot{V}^2 = 5.4952 - 0.04935\dot{V} - 0.15556\dot{V}^2 - 0.01588\dot{V}^3$$

gives:

$$\dot{V} = 1.544 \text{ l/s} \quad, h = 5.0 \text{ m}$$
$$\dot{V} = 24.47 \text{ gpm}, h = 16.4 \text{ ft}$$

which is indeed the intersection point between the system blue line and the pump curve in the last example. To find this expression, it was assumed that the friction factor does not change. To verify, we can recalculate the Reynolds number:

$$u = \frac{1.544\text{e-}3}{1.256\text{e-}3} = 1.22 \text{ m/s}$$
$$Re = \frac{uD_i}{\nu} = 32212$$
$$f = 0.0232$$

A second iteration would give:

$$2.093\dot{V}^2 = 5.4952 - 0.04935\dot{V} - 0.15556\dot{V}^2 - 0.01588\dot{V}^3$$

(continued)

Example 7.6 (continued)

$$\dot{V} = 1.548 \text{ l/s}, \qquad h = 4.987 \text{ m}$$

which is a very small difference (a greater number of significant digits were kept to illustrate the difference). ∎

7.3.3 Wire-to-Water Efficiencies

Based-mounted pump efficiencies relate the fluid power to the shaft power. The electric efficiency of the electric motor chosen must be taken into account. In the case of small pumps or circulators, they come altogether with the electric motor integrated, and the efficiency given is the total efficiency, often called *wire-to-water efficiency*. The efficiency of these smaller pumps is often smaller than for larger applications. A technical note [15] did a survey over a large number of available circulators and proposed the following correlation:

$$\eta_{tot} = 0.0646 \dot{W}_{fluid}^{0.344} \tag{7.35}$$

A more recent study done by Gagne-Boisvert and Bernier [16] performed a similar analysis, and they proposed three correlations ranked as "low, high, best" efficiencies given by:

$$\eta_{tot} = 0.404 \dot{W}_{fluid}^{0.0886}, \qquad \text{Best effciency} \tag{7.36a}$$

$$\eta_{tot} = 0.321 \dot{W}_{fluid}^{0.115}, \qquad \text{High effciency} \tag{7.36b}$$

$$\eta_{tot} = 0.118 \dot{W}_{fluid}^{0.249}, \qquad \text{Low effciency} \tag{7.36c}$$

The last four correlations are plotted on Fig. 7.6, which shows a large increase in the efficiency of circulators in recent years. It should be emphasized that these results were found using

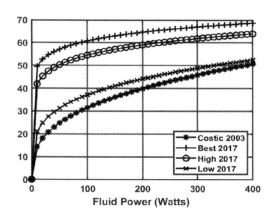

Fig. 7.6 Wire-to-water efficiencies

the best efficiency point given by the manufacturer, a point which is not always obtained and lower efficiencies can be expected.

7.3.4 Impact of Speed Reduction on Pump Curves

As emphasized in the last example, performance curves are often represented by polynomials. The pump laws can be used to estimate the pump performance when the rotational speed changes. Assuming, for example, a third-degree polynomial:

$$h_1 = p_o + p_1 \dot{V}_1 + p_2 \dot{V}_1^2 + p_3 \dot{V}_1^3$$

$$h_2 \frac{h_1}{h_2} = p_o + p_1 \dot{V}_2 \left(\frac{\dot{V}_1}{\dot{V}_2} \right) + p_2 \dot{V}_2^2 \left(\frac{\dot{V}_1}{\dot{V}_2} \right)^2 + p_3 \dot{V}_2^3 \left(\frac{\dot{V}_1}{\dot{V}_2} \right)^3$$

$$h_2 \frac{1}{x^2} = p_o + p_1 \dot{V}_2 \left(\frac{1}{x} \right) + p_2 \dot{V}_2^2 \left(\frac{1}{x} \right)^2 + p_3 \dot{V}_2^3 \left(\frac{1}{x} \right)^3$$

$$h_2 = p_o x^2 + p_1 \dot{V}_2 x + p_2 \dot{V}_2^2 + p_3 \dot{V}_2^3 \left(\frac{1}{x} \right) \tag{7.37}$$

where:

$$x = \frac{\dot{V}_2}{\dot{V}_1} = \frac{\omega_2}{\omega_1}$$

Pump efficiency is also often given as polynomials with respect with flow rates. When the rotational speed is reduced, if the same assumptions are made, the affinity laws stipulate that the efficiency should remain unchanged for the same flow coefficient. The new polynomial can then be evaluated as:

$$\eta = p_o + p_1 \dot{V}_1 + p_2 \dot{V}_1^2 + p_3 \dot{V}_1^3$$

$$\eta = p_o + p_1 (C_{q1} \omega_1 D_1^3) + p_2 (C_{q1} \omega_1 D_1^3)^2 + p_3 (C_{q1} \omega_1 D_1^3)^3$$

$$\eta = p_o + p_1 (C_{q1} \omega_2 D_1^3) \left(\frac{1}{x} \right) + p_2 (C_{q1} \omega_2 D_1^3)^2 \left(\frac{1}{x} \right)^2$$

$$+ p_3 (C_{q1} \omega_1 D_1^3)^3 \left(\frac{1}{x} \right)^3$$

With $C_{q1} = C_{q2}$ and $D_1 = D_2$:

$$\eta = p_o + p_1 \left(\frac{\dot{V}_2}{x} \right) + p_2 \left(\frac{\dot{V}_2}{x} \right)^2 + p_3 \left(\frac{\dot{V}_2}{x} \right)^3 \tag{7.38}$$

Example 7.7 In the last example, what would be the operating point if the rotational speed of the pump is reduced to 80% of its nominal value?

(continued)

Example 7.7 (continued)

Solution Assuming that the system curve does not change, the dimensionless curves should stay constant. At the operating point of the previous problem, one finds that:

$$\omega = \frac{1750 \cdot 2\pi}{60} = 183.26 \text{ rd/s}$$

$$C_q = \frac{\dot{V}}{\omega D^3} = \frac{0.00154}{183.26 \cdot 0.108^3} = 0.0067$$

$$C_h = \frac{gh}{\omega^2 D^2} = \frac{9.81 \cdot 4.99}{183.26^2 \cdot 0.108^2} = 0.125$$

$$\omega_2 = 0.8 \cdot 183.26 = 146.61 \text{ rd/s}$$

At reduced speed, one have:

$$C_{q1} = C_{q2}$$

$$\Rightarrow \dot{V}_2 = 0.0067 \cdot 146.6 \cdot 0.108^3 = 0.00124 \text{m}^3/\text{s} = 1.24 \text{ l/s}$$

$$C_{h1} = C_{h2}$$

$$\Rightarrow h_2 = 0.125 \frac{146.6^2 \cdot 0.108^2}{9.81} = 3.19 \text{ m}$$

Another possibility is to find the new pump equation using the scaling laws. From (7.37), one finds:

$$h_2(\text{m}) = 3.516928 - 0.03948\dot{V} - 0.15556\dot{V}^2 - 0.01985\dot{V}^3$$

with $\dot{V}$ in l/s. Assuming the simple form of the system curve:

$$2.094\dot{V}^2 = 3.516928 - 0.03948\dot{V} - 0.15556\dot{V}^2 - 0.01985\dot{V}^3$$

would give:

$$\dot{V}_2 = 1.24 \text{ l/s}, \qquad h_2 = 3.19 \text{ m}$$

The Reynolds number dependence of the system curve can be modeled more easily with this second approach solving:

$$f(\dot{V})\dot{V}^2 = 3.516928 - 0.03948\dot{V} - 0.15556\dot{V}^2 - 0.01985\dot{V}^3$$

which would give:

$$\dot{V}_2 = 1.21 \text{ l/s}, \qquad h_2 = 3.21 \text{ m}$$

The difference is small enough to be neglected as long as the flow regime is turbulent.   ■

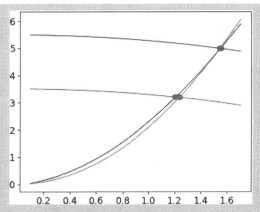

Example 7.8 Nowadays, pumps working at variable speed are common and manufacturers provide sometimes information when the pump is working at reduced speed. An example is shown on Fig. 7.7. If a polynomial fit is done with the curve at 2961 rpm, the following result would be obtained:

$$h_{pump}(\text{ft}) = 86 + 2.72\text{e-}2\dot{V} - 8.29\text{e-}4\dot{V}^2 - 5.53\text{e-}6\dot{V}^3$$

$$\eta_{pump} = 28.06 + 1.02\dot{V} - 6.11\text{e-}3\dot{V}^2 - 5.54\text{e-}6\dot{V}^3$$

where $\dot{V}$ is US gpm. The performance is also given at different speeds, and it is possible to compare with the results given by Eqs. (7.37) and (7.38).

(a) Find the operating point and the efficiency of the pump e-1535 working at 2961 rpm if the system curve is given by:

$$h_{loss}(\text{ft}) = 0.0042\dot{V}^2$$

(b) Using Eqs. (7.37) and (7.38), find the same information if the speed is reduced to 2442 rpm.
(c) Compare the results with the performance given on Fig. 7.7

Solution

(a) The solution is readily obtained by the solution of the equation:

$$0.0042\dot{V}^2 = 86 + 2.72\text{e-}2\dot{V} - 8.29\text{e-}4\dot{V}^2 - 5.53\text{e-}6\dot{V}^3$$

$$\dot{V} = 125 \text{ gpm}, \quad h = 65.6 \text{ ft}, \quad \eta = 71\%$$

(b) The new polynomial for the reduced speed can be obtained from Eqs. (7.37) and (7.38)

$$h_{pump,2}(\text{ft}) = 58.5 + 2.24\text{e-}2\dot{V} - 8.29\text{e-}4\dot{V}^2 - 6.7153\text{e-}6\dot{V}^3$$

(continued)

Performance Curve

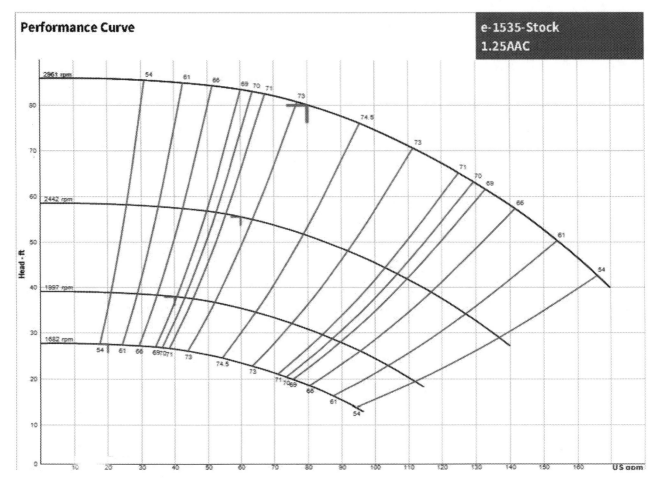

Fig. 7.7 Pump performances at different speeds

Example 7.8 (continued)

$$\eta_{pump,2} = 28.06 + 1.24\dot{V} - 8.93\text{e-}3\dot{V}^2 - 9.87\text{e-}6\dot{V}^3$$

Solving for the head will give:

$$\dot{V} = 103 \text{ gpm}, \quad h = 44.7 \text{ ft}, \quad \eta = 71\%$$

which is on the pump curve for the reduced speed.

Comments   *It is seen that the efficiency is the same as predicted by the scaling laws. This would not be the case if the system curve would have a high static head.*   ∎

7.4   Simple GSHP Analysis

A typical connection for a residential heat pump is shown in Fig. 7.8. Other elements like strainers and flow setters can be present. More than one pump [17] can also be used if the head

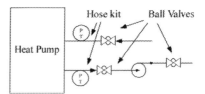

Fig. 7.8 Typical connections

given by a single one is not enough. In a commercial building where more than one heat pump can be present, motorized shut-off valves and check valves can be found. Valves, tees, and other piping elements will produce singular losses as well as the heat pump itself. Typical values of loss coefficients can be found in several references [7], but specific values for HDPE pipes are more rare. Kavanaugh and Rafferty [1, 6] give some values given in Tables 7.5, and 7.6.

In the pipe network, the designer has to choose the configuration and the different pipe diameters. Analysis tools that are presented here help to evaluate the impact of the different choices, but initial one can be done using good practice rules. Some of these values are given in Table 7.7 from [1]. The

Table 7.5 Equivalent lengths (m) from [1]

	Nominal pipe diameter (mm [in])							
	19[¾]	25[1]	32[1¼]	38[1½]	50[2]	76[3]	100[4]	152[6]
Socket U-bend	3.7	2.0	3.4					
Socket 90	1.0	0.8	1.9	2.0	2.1			
Socket tee-branch	1.2	1.6	2.0	3.0	4.0			
Socket tee-straight	0.4	0.4	0.3	0.6	0.9			
Socket reducer		1.9	1.2	1.2	1.3			
Butt U-bend	3.8	6.8	11	13				
Butt 90	2.2	3.0	5.6	3.3	3.7	10	12	16
Butt tee-branch	2.3	2.2	5.2	3.3	4.6	9.4	11	15
Butt tee-straight	1.4	0.8	1.2	1.2	1.2	2.1	2.2	2.3
Socket reducer		1.5	1.7	1.8	1.2.1	3.1	4.1	6.1

Table 7.6 Valves $(m^3/hr/(bar)^{1/2})$ from [1]

	Nominal pipe diameter (mm [in])							
	19[¾]	25[1]	32[1¼]	38[1½]	50[2]	76[3]	100[4]	152[6]
Ball valves	21.6	30.3	40.6	70	91	337	718	1080
Swing check valves	11.2	18	30	39	65	168.6	302.7	856
Hoze kits—3ft	6.9	13.8	29.4	40.7				
Y-strainer	15.6	24.2	37.2	51.9	82.2	134	216	

minimum value suggested is to ensure a minimum linear velocity in the order of 0.6 m/s to facilitate the purging of the air. The maximum flow rate is found from the constraint to limit the linear head loss in the order of 300 Pa/m.

From the energy that must be delivered to transport the fluid through the ground in GSHP systems, the global energy performance can be evaluated, and one can estimate if the system performance is deteriorated by a bad piping layout. In this chapter, we will mostly analyze the impact of the ground loops. However, in real life the energy needed to transport the energy though the building (load side) must also be taken into account. A rigorous analysis would be to:

1. Evaluate the heat pump(s) performance at the operating conditions using either the performance data or the ISO-13256 specifications.
2. Estimate the pump power and energy on the source side (ground).
3. Estimate the pump power and energy on the load side (building) external to the heat pump. This would need an estimation of the external static pressure (ESP) for water-air heat pumps or the head losses for a water-water heat pump.
4. In this case, the energy thus calculated will influence the capacity of system. It will reduce the capacity in cooling and increase it in heating. This correction can be evaluated if necessary.
5. Estimate the loss of performance through different metrics.

Example 7.9 A GSHP is connected to a two-wells field configuration shown in the figure. The bores are 125 m long and there is a 6 m spacing between them. The pipe inside the borehole is HDPE SDR-11 1". The connections (Butt fusion) between the field and the heat pump are made with a HDPE SDR-11 1-1/4" pipe. The total length of the supply piping is 60 m and five 90° elbows are present in the field. The heat pump nominal flow is 0.5 l/s and has a 3 m pressure drop at this flow rate for water. Three 1" globe shut-off valves (Cv = 30.3 $m^3/hr/bar$) are used. A strainer (Cv = 20 $m^3/hr/bar$) is also present. The water-to-air heat pump has a 8 kW heating capacity and a COP of 3.5. It is estimated that the external static pressure is 50 Pa for a flow rate of 700 cfm and that the given COP does not take into account the head losses through the evaporator of the heat pump.

(a) Evaluate the head losses on the ground side assuming a 15% security factor.
(b) Find the pump electric power assuming a pump efficiency of 50% and an electric efficiency of 85%.
(c) Estimate the system COP, neglecting fan power and taking into account the fan power.

Evaluate the properties at 5 °C.

(continued)

Table 7.7 Suggested flow rate range from [6]

Nominal Pipe diameter		Maximum flow rate				Minimum flow rate			
		SDR-11		SDR-13.5		SDR-11		SDR-13.5	
Inch	mm	gpm	l/s	gpm	l/s	gpm	l/s	gpm	l/s
3/4	19	3.9	0.25	4.4	0.28	3.7	0.23	3.9	0.25
1	25	7	0.44	8	0.5	5.7	0.36	6.1	0.4
1 1/4	32	13	0.82	15	0.95	9	0.57	9.8	0.6
1 1/2	40	19	1.2	21	1.3	12	0.76	12.8	0.8
2	50	35	2.2	39	2.5	18	1.12	20	1.3
3	75	100	6.3	110	6.9	40	2.5	43	2.7

Example 7.9 (continued)

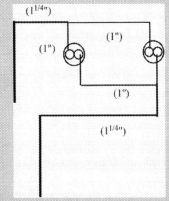

Solution Connecting pipes

The flow rate is the total flow rate and the internal pipe diameter for (Appendix B):

$$\dot{V} = 5\text{e-}4 \text{ m}^3/\text{s}, \quad A_c = \frac{\pi 0.0345^2}{4} = 9.34\text{e-}4 \text{ m}^2$$

$$\Rightarrow u = \frac{\dot{V}}{A_c} = 0.54 \text{ m/s}$$

For water at 5 °C, we have:

$$\nu = 1.51\text{e-}6 \text{ m}^2/\text{s}$$

$$Re_{conn} = \frac{0.54 \cdot 0.0345}{1.51\text{e-}6} = 12155 \qquad \text{turbulent}$$

For smooth pipe, the Colebrook formula gives:

$$f = 0.0298 \Rightarrow h_{pipe} = 0.02988 \frac{60}{0.0345} \frac{0.535^2}{2g} = 0.758 \text{ m}$$

The minor losses for five 90 ° elbows are:

$$h_{elbows} = 0.02988 \frac{5 \cdot 5.6}{0.0345} \frac{0.535^2}{2g} = 0.353 \text{ m}$$

The pressure drop of the three ball valves will give:

$$\Delta P_{valve} = 100 \left(\frac{1.8}{30.3} \right)^2 = 0.353 \text{ kPa}$$

$$\Delta P_{strainer} = 100 \left(\frac{1.8}{20} \right)^2 = 0.810 \text{ kPa}$$

$$h_{valve} = 3 \frac{353}{1000 \cdot 9.81} = 0.108 \text{ m}$$

$$h_{strainer} = \frac{810}{1000 \cdot 9.81} = 0.083 \text{ m}$$

$$h_{tot_{conn}} = 3 + 0.108 + 0.083 + 0.758 + 0.353 = 4.30 \text{ m}$$

Loop Piping

Using a reverse-return configuration, a perfect parallel configuration can be assumed, and the head losses for only one branch can be calculated. From the figure, it can be assumed that the flow will pass through one U-bend, one 90° elbow, and one straight path of a tee-branch :

$$\dot{V} = 2.5\text{e-}4 \text{ m}^3/\text{s}, \quad A_c = \frac{\pi 0.0273^2}{4} = 5.85\text{e-}4 \text{ m}^2$$

$$\Rightarrow u = \frac{\dot{V}}{A_c} = 0.426 \text{ m/s}$$

$$Re_{loop} = \frac{0.427 \cdot 0.0273}{1.51\text{e-}6} = 7672 \qquad \text{turbulent}$$

$$f = 0.034, \quad L = 250 \text{ m}$$

$$\Rightarrow h_{pipe} = 0.034 \frac{250}{0.0273} \frac{0.427^2}{2g} = 2.88 \text{ m}$$

$$h_{minor} = 0.034 \left(\frac{6.8 + 3.0 + 0.8}{0.0273} \right) \frac{0.427^2}{2g} = 0.12 \text{ m}$$

$$h_{loop} = 3.00 \text{ m}$$

We can add the head loss in the header between the borehole:

$$h_{pipe} = 0.034 \frac{6}{0.0273} \frac{0.427^2}{2g} = 0.069 \text{ m}$$

(continued)

(continued)

Example 7.9 (continued)

$$h_{field} = 3.00 + 0.069 = 3.07 \text{ m}$$

$$h_{total} = 4.30 + 3.07 = 7.37 \text{ m}$$

(b) With the safety margin:

$$\dot{W}_{fluid} = 1.15 \cdot 1000 \cdot 0.0005 \cdot 7.37 = 41.5 \text{ W}$$

$$\dot{W}_{shaft} = \frac{41.5}{0.5} = 83 \text{ W}$$

$$\dot{W}_{elec} = \frac{83}{0.85} = 97.7 \text{ W} = 0.13 \text{ H.P.}$$

(c) Neglecting the fan power, the system COP is:

$$\dot{W}_{compressor} = \frac{8}{3.5} = 2.285 \text{ kW}$$

$$COP_{system} = \frac{8}{2.285 + 0.098} = 3.36$$

Taking into account the fan power, we have:

$$\dot{V}_{fan} = 700 \text{cfm} = 0.330 \text{ m}^3/\text{s}$$

$$\dot{W}_{fan} = \frac{0.33 \cdot 50}{0.3} = 55 \text{ W}$$

$$COP_{system} = \frac{8 + 0.055}{2.285 + 0.094 + 0.055} = 3.31$$

The fan energy has been added to the capacity since the system is heating. The external energy ratio is:

$$\frac{93.69}{8} = 11.7 \text{ W/kW}$$

without the fan:

$$\frac{149}{8.055} = 18.5 \text{ W/kW}$$

Comments *It is seen that in this simple example, the external energy is small with respect to the compressor energy but that the fan energy is important with respect to the pump energy. However, as previously stated, it will not be often considered since it is not part of the ground loop analysis. Having an initial design, one can easily analyze the effect of different choices in piping option. It is seen that the flow is turbulent everywhere. Although it has been suggested that laminar flow can be a good design practice [18], it is most often suggested to*

keep the flow turbulent in the borehole to reduce the thermal resistance. However, in the header, laminar flow would be a better choice, but here, the head losses are small and the higher cost associated with a larger pipe would not be worthwhile. A change in the diameters of the piping would show a very small influence of the header piping, at least in this example. A ¾" pipe would increase substantially the total head losses. The use of a glycol solution would increase, as expected, the total head losses but still be ok if the percentage is not too big. A high percentage glycol solution will soon lead to a laminar regime in the borehole which would have an impact on the total length. In this example, the heat pump pressure loss is important; in the ISO-13256 norm, it is taking into account in the calculation of the COP. If it is the case, the analysis would be different. ∎

Example 7.10 Redo the last example, assuming a system twice as big; two similar heat pumps in parallel will be used, so that the head losses in the heat pumps are the same as before, the total flow rate is 1 l/s, and there are four boreholes in parallel.

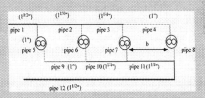

Solution The solution seems quite simple as it follows the same path as the previous example. The head losses for the heat pump connections are the same as before:

$$h_{hp} = 3 + 0.108 + 0.083 = 3.19 \text{ m}$$

Since the total flow rate is larger, a bigger diameter pipe is needed to connect the heat pump to the field. So a 1½" pipe will be chosen.

$$\dot{V} = 1\text{e-3 m}^3/\text{s}, \quad A_c = \frac{\pi 0.0395^2}{4} = 1.22\text{e-3 m}^2$$

$$\Rightarrow u = \frac{\dot{V}}{A_c} = 0.817 \text{ m/s}$$

(continued)

(continued)

Example 7.10 (continued)

$$Re_{conn} = \frac{0.817 \cdot 0.0395}{1.51\text{e-}6} = 21246 \qquad \text{turbulent}$$

$$f = 0.0257 \Rightarrow h_{pipe} = 0.0257 \frac{60}{0.0395} \frac{0.817^2}{2g} = 1.33 \text{ m}$$

$$h_{elbows} = 0.0257 \frac{5 \cdot 3.3}{0.0395} \frac{0.817^2}{2g} = 0.366 \text{ m}$$

$$h_{tot_{conn}} = 3.19 + 1.33 + 0.366 = 4.88 \text{ m}$$

Loop Piping

Since the configuration is reverse-return, it can be assumed that the flow rate will be the same for all loops and equal to the one in the previous example. A simple assumption would be to use the same head loss in individual loops:

$$h_{loop} = 3.01 \text{ m}$$

For the field interconnections, the situation is now a little bit different since the flow rate is different in each part of the header. Neglecting the minor losses except for the loop, knowing that the head losses should be the same in all paths, and choosing the last borehole path (2-3-4-8) yield:

$$\dot{V}_{pipe2} = \frac{3}{4}\text{e-}3 \text{ m}^3/\text{s} , A_c = \frac{\pi 0.0345^2}{4} = 9.34\text{e-}4$$

$$\Rightarrow u = \frac{\dot{V}}{A_c} = 0.803 \text{ m/s}$$

$$Re_{pipe2} = \frac{0.803 \cdot 0.0345}{1.51\text{e-}6} = 18234 \qquad \text{turbulent}$$

$$f = 0.0268 \Rightarrow h_{pipe2} = 0.0268 \frac{6}{0.0345} \frac{0.803^2}{2g} = 0.153 \text{ m}$$

$$\dot{V}_{pipe3} = \frac{1}{2}\text{e-}3 \text{ m}^3/\text{s} , A_c = \frac{\pi 0.0345^2}{4} = 9.34\text{e-}4$$

$$\Rightarrow u = \frac{\dot{V}}{A_c} = 0.535 \text{ m/s}$$

$$Re_{pipe3} = \frac{0.535 \cdot 0.0345}{1.51\text{e-}6} = 12156 \qquad \text{turbulent}$$

$$f = 0.0298 \Rightarrow h_{pipe3} = 0.0298 \frac{6}{0.0345} \frac{0.535^2}{2g} = 0.076 \text{ m}$$

$$Re_{pipe4} = Re_{loop}, f = 0.034 \Rightarrow h_{pipe4} = 0.034 \frac{6}{0.0273} \frac{0.427^2}{2g}$$

$$= 0.069 \text{ m}$$

The head losses in the header becomes:

(continued)

$$h_{head} = 0.153 + 0.076 + 0.069 = 0.298 \text{ m}$$

Adding the loop:

$$h_{field} = 0.298 + 3.01 = 3.30 \text{ m}$$

However, if one follows the path of the second to last borehole (2-3-7-11), the total head losses would be:

$$h_{head} = 0.153 + 0.076 + 0.153 = 0.381 \text{ m}$$

$$h_{field} = 0.381 + 3.00 = 3.38 \text{ m}$$

which does not make sense since the head at both end points should be the same. The reason for that is that the flow path is not equal in all four wells, and if no balancing valves are used, the flow rate will not be equal even though we have a reverse-return setup. To evaluate the flow field, one should use the Hardy Cross method (Appendix C). If the problem is solved using this method, the solution would be:

Pipe	$\dot{V}$ (l/s)	h (m)
2–11	0.748	0.152
3–10	0.5	0.076
4–9	0.252	0.070
5–8	0.252	3.055
6–7	0.248	2.973

The physics will now hold:

$$h_{field} = 0.152 + 0.076 + 0.070 + 3.055$$
$$= 3.353 \text{ m} \quad \text{path 2-3-4-8}$$

$$h_{field} = 0.152 + 0.076 + 2.973 + 0.152$$
$$= 3.353 \text{ m} \quad \text{path 2-3-7-11}$$

Of course, in practice, the simplest approach is sufficient with the order of accuracy needed. Even with the assumption of equal flow, a better approximation would be given by (7.26):

$$h_{field} = h_{loop} + 2\left(\frac{3}{4}h_{pipe2} + \frac{1}{2}h_{pipe3} + \frac{1}{4}h_{pipe4}\right) = 3.34 \text{ m}$$

$$h_{total} = 4.88 + 3.34 = 8.23 \text{ m}$$

(b)

$$\dot{W}_{fluid} = 1.15 \cdot 1000 \cdot 0.001 \cdot 8.23 = 93 \text{ W}$$

(continued)

Example 7.10 (continued)

$$\dot{W}_{shaft} = \frac{93}{0.5} = 186 \text{ W}$$

$$\dot{W}_{elec} = \frac{186}{0.85} = 219 \text{ W}$$

(c) Taking into account only the ground side, the system COP is:

$$COP_{system} = \frac{16}{4.57 + 0.219} = 3.34$$

and the pumping ratio is then:

$$\frac{289}{16} = 13.6 \text{ W/kW} \qquad \blacksquare$$

7.4.1 Direct Return

The advantage of reverse-return configuration is that the flow in each borehole should be almost equal. However, it increases the length of piping and joints. It has been argued [19] that, in the case of a classic hydronic system, direct return can be an interesting option if the variation of flow rates does not discriminate too much the thermal behavior of each parallel path.

Example 7.11 Redo the last example with a direct-return arrangement.

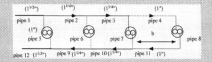

Solution The solution cannot be done using simplifying assumptions. The results are found using the Hardy Cross method.

pipe	$\dot{V}$ (l/s)	h (m)
2–9	0.734	0.147
3–10	0.483	0.071
4–11	0.238	0.064
5	0.266	3.325
6	0.252	3.030
7	0.244	2.888
8	0.238	2.761

The field head losses are then:

(continued)

$$h_{field} = 2(0.147 + 0.071 + 0.064) + 2.761$$

$$= 3.325 \text{ m ,path 2-3-4-8-11-10-9}$$

$$h_{field} = 2(0.147 + 0.071) + 2.888 = 3.325 \text{ m ,path 2-3-7-10-9}$$

$$h_{field} = 2(0.147) + 3.03 = 3.325 \text{ m ,path 2-6-9}$$

$$h_{field} = 3.325 \text{ m ,path 5}$$

which is less than the reverse return, although almost the same. Normally, the connecting pipes should be smaller at least in the order of three borehole distances with probably less than 90-degree turns, but if the same connector losses are kept as in the last example, almost the same head losses are obtained in this configuration. However, the unbalance of flow rate can have an impact on the thermal behavior of the wells that would have to be analyzed. If this impact is not too important, it can be an option but not if the number of boreholes becomes too large.

In practice, the use of balancing vales is not trivial in buried pipes, but if it was possible, a simpler analysis would be possible. A larger valve resistance would be present in the first borehole but not in the last one in order to reach the same flow rate (0.25 l/s) in all boreholes. In that case, it is easy to see that the head losses would be:

$$h_{field} = 2(0.153 + 0.076 + 0.069) + 3.0$$

$$= 3.59 \text{ m, path 2-3-4-8-11-10-9}$$

which is a significant increase. $\blacksquare$

7.4.2 Manifold Option

A manifold is basically a large header, which is often assumed to be at a common pressure similar to a bus strip in electrical circuit but fluid circuits are, as usual, more complex to analyze. So similar tools can be used to evaluate the flow and pressure loss associated with manifolds. In practice, however, there are several differences. In contrary to reducing headers circuits as discussed in the previous sections, all circuits connected to a common manifold are closely connected to a single space, often in the building, or an external vault. The hydraulic circuit between the manifold and the borehole is then longer, but several advantages are associated with the use of manifold like (Fig. 7.9):

- Easy maintenance inside the building.
- Less thermal joints in the ground.

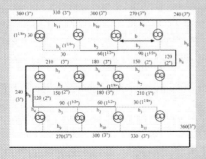

Fig. 7.9 Manifold connections inside a building

- Possibility to add loops.
- Easier to purge.
- Easy to adjust the flow using balancing valves.

Some disadvantages associated with small- to medium-size systems can be associated with a possible increase in energy use for piping and increase in cost due to the purchase of balancing valves. However, for large systems, it is almost always used. Simple manifold analysis can be done assuming that the use of Bernoulli equation can be used as usual. More elaborate theory can be used based on momentum analysis [20, 21], but it will not be presented here.

Example 7.12 As a last example, compare the head losses in the 12-borehole fields associated with the three different configurations shown in the following figures. A total flow rate of 360 l/min of 25% propylene-glycol is given. Each borehole has a 140 m depth and is separated by 6 m. The distance from the building to the field is 45 m. To simplify, neglect all minor losses except a 11 m equivalent for the U-bend at the bottom of each borehole. Evaluate the properties at 5 °C.

(a)

(continued)

Solution As indicated in the previous examples, the simplest approach to evaluate a reverse-return configuration is to take any regular path. For instance, one may evaluate:

$$h_{field} = h_{loop} + \sum_{i=1}^{11} h_i$$

The solution is given in the following table:

	Re	f	D (m)	L (m)	h (m)
Loop	4188	0.041	0.036	291	4.11
1	4188	0.041	0.036	6	0.08
2	7318	0.034	0.041	6	0.15
3	10977	0.031	0.041	6	0.29
4	11709	0.030	0.051	6	0.17
5	14636	0.028	0.051	6	0.25
6	11918	0.030	0.076	6	0.05
7	13904	0.029	0.076	6	0.07
8	15891	0.028	0.076	18	0.27
9	17877	0.027	0.076	6	0.11
10	19863	0.026	0.076	6	0.13
11	21850	0.026	0.076	6	0.15
Total					5.82

Adding the 90 m long 3" pipe would give:

$$Re_{conn} = 23836, \quad f_{conn} = 0.025, \quad h_{conn} = 2.69 \text{ m}$$

$$h_{tot} = 5.82 + 2.69 = 8.51 \text{ m}$$

$$\dot{W} = \frac{360}{60 \cdot 1000} 1029 \cdot 9.81 \cdot 8.51 = 515 \text{ W}$$

(b) In the second configuration, three parallel paths are used, and each one has four boreholes in a reverse-return configuration.

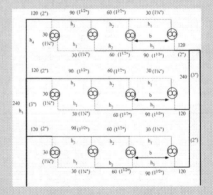

The field head loss is found from one of the three paths.

(continued)

Example 7.12 (continued)

	Re	f	D (m)	L (m)	h (m)
Loop	4188	0.041	0.036	291	4.11
1	4188	0.041	0.036	6	0.08
2	7318	0.034	0.041	6	0.15
3	10977	0.031	0.041	6	0.29
4	11709	0.030	0.051	12	0.33
5	15891	0.028	0.076	12	0.18
Total					5.14

Adding the 90 m long 3" pipe would give:

$$h_{tot} = 5.14 + 2.69 = 7.83 \text{ m}$$

$$\dot{W} = \frac{360}{60 \cdot 1000} 1029 \cdot 9.81 \cdot 7.83 = 474 \text{ W}$$

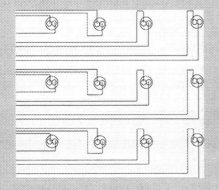

(c) As a final scenario, each borehole is connected with an independent circuit having the same diameter as the borehole pipe. These pipes will be interconnected inside the vault or the building in a common manifold. In such a case, the head losses can differ significantly depending on each distance. In practice balancing valves would be used to even the flow rates, assuming that the balanced total head loss is equal to the farthest borehole. The assumption made here is that this total distance is equal to:

$$L_{tot} = 2 \,[\text{building distance} + 6 \text{ borehole separations}]$$
$$= 162 \text{ m}$$

	Re	f	D (m)	L (m)	h (m)
Loop	4188	0.041	0.036	291	4.11
Conn	4188	0.041	0.036	162	2.29
Total					6.39

$$\dot{W} = \frac{360}{60 \cdot 1000} 1029 \cdot 9.81 \cdot 6.39 = 387 \text{ W}$$

In that last case, the manifold head loss would have to be added to the total losses but was not considered here.

Comments *In this example, the total reverse-return scenario gives the largest losses. Of course, it depends on the pipe diameters chosen and on some arbitrary assumptions, but in all cases, this choice is often suggested to small to medium borefields.*

■

7.5 Pressure Rating

Geothermal application mostly used high-density polyethylene (PE) or cross-linked polyethylene (PEX) piping [14]. Plastic pipes come with different sizing (DR) and pressure rating. The maximum pressure that it can accept depends on its thickness (DR) and quality (design factor). The maximum pressure can be estimated as [23]:

$$P_{max} = \frac{2F \cdot HDB \cdot t}{d_o - t} == \frac{2F \cdot HDB}{SDR - 1} \tag{7.39}$$

where t is the thickness and HDB the *hydrostatic design basis pressure*, which is on the order of 11 Mpa (1600 psi) for polyethylene. F is a safety factor associated with the quality of the compound. In North America, PE3408 (F = 0.5) and PE4710 (F = 0.63) are mostly used.

Example 7.13 Evaluate the maximum pressure for SDR-11 PE3408 and PE4710:

$$P_{max} = \frac{2 \cdot 0.5 \cdot 11}{11 - 1} = 1.10 \text{ MPa} \qquad \text{PE3408}$$

$$P_{max} = \frac{2 \cdot 0.63 \cdot 11}{11 - 1} = 1.38 \text{ MPa} \qquad \text{PE4710}$$

■

In Europe, PE100 polyethylene pipes are often used for geothermal applications. While North American pipes are ASTM D 2837 standard, European piping follows ISO 9080 and ISO 12162 standards where the HDB is replaced by the *minimum required strength* (MRS). Although differences exist, the nomenclature is associated with this value (MRS = 8 MPa for PE80, MRS = 10 MPa for PE100, etc.). In the ISO

(continued)

Fig. Pr_7.1 Three heat pumps in reverse-return connections

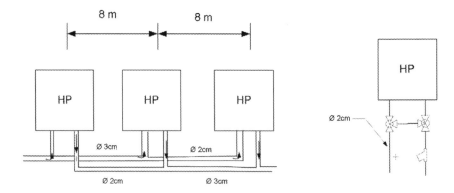

norm instead of the safety design factor (F), the maximum pressure is defined as:

$$P_{max} \frac{2MRS}{C(SDR-1)} \qquad (7.40)$$

where C is safety margin chosen now higher for a conservative approach. For the 1.25 value often suggested, this is equivalent to $F = 0.8$ for the ASTM standard.

Example 7.14 Redoing the last example for PE100, we would find:

$$P_{max} = \frac{2 \cdot 0.8 \cdot 10}{11-1} = 1.60 \text{ MPa} \qquad \text{PE100}$$

a larger value. However, the temperature range and other differences in the ASTM and ISO standard suggest that at the end properties are similar to PE4710 [23]. ∎

Problems

Pr. 7.1 A horizontal ground heat exchanger consists of a 40 m 3"-SDR-11 pipe that supplies fluid to a field with six parallel 1 1/4"-SDR-11 pipes of 200 m length. A 40 m SDR-11 3" pipe returns the flow to the building. The total flow rate is 12 l/s of water with propylene-glycol (40% volume). We can assume that six 90 deg elbows ($L_{eq} = 2.3$ m) are present in the 3" piping section. Evaluate the fluid power needed to circulate this flow. Evaluate the properties at 280 K.
Answer : $\dot{W} = 7$ kW *(flow rate too high in the small tube)*

Pr. 7.2 Three heat pumps are installed as shown in Fig. Pr_7.1. The total flow rate is 1.2 l/s. Each heat pump has a configuration shown on the right, where in normal operation no flow goes through the by-pass. The filter has a loss coefficient equal to 1.5. In normally open position, each valve coefficient is 0.291 (l/s/(kPa)$^{0.5}$. The rest of the pipe

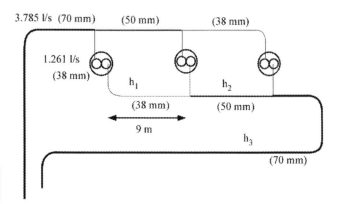

Fig. Pr_7.2 Three boreholes in reverse-return connections

losses and the heat pump coil give a head loss of 4.5 m in nominal conditions when water is used. Find the total power that should be delivered to the fluid. Evaluate the properties at 290 K.
Answer: $h_{tot} = 6.25$ m , $\dot{W} = 71.8$ W. *This value would be obtained if flow is balanced.*

Pr. 7.3 A flow rate of 3.785 l/s of water ($\rho = 1000$ kg/m^3, $\nu = 6e-7$ m^2/s, $C_p = 4180$ J/kg-K) passed through three vertical boreholes having a height of 75 m (Fig. Pr_7.2). Each borehole is separated by 9 m. The header length is 150 m and its inside diameter is 70 mm. The minor losses have an equivalent length of 5 m in the first pipe 1, 4 m in second pipe, and 6 m in third pipe.

(a) Find the total losses associated with the network.
(b) A E-1535 pump is used to transport the water. From Fig. Pr_7.3, what would be the impeller diameter needed? Find the electric power of the pump if the electric efficiency of the motor is 83%.

Answer: (a) 7.92 m, (b) 490 W.

Pr. 7.4 A Bell and Gossett 1510-2G centrifugal pump is chosen to produce a flow rate of approximately 9.2 l/s of a solution of water-propylene glycol (20%) in ten boreholes

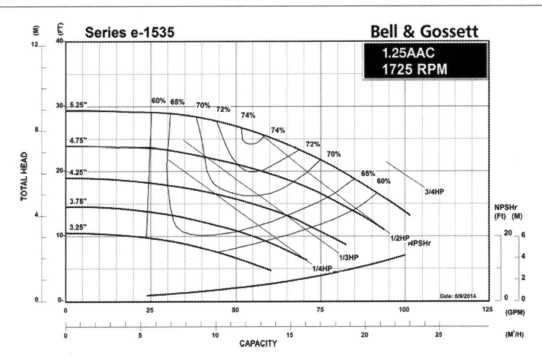

Fig. Pr_7.3 E-1535 pump performances curve

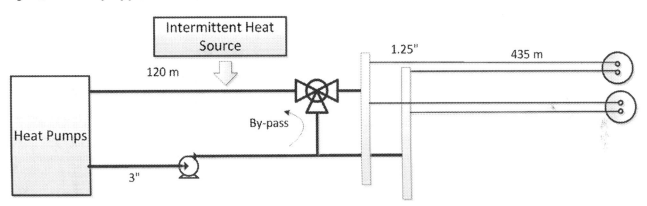

Fig. Pr_7.4 E-1535 pump performances curve

connected to geothermal heat pumps:

$$\rho_f = 1024 \text{ kg/m}^3, \quad v_f = 2.8e{-}6 \text{ m}^2/\text{s}, \quad C_p = 3939 \text{ J/kg-K}$$

A schematic of the system is shown in Fig. Pr_7.4 (SDR-11 piping). In winter, the system is connected to a large grocery store that has to reject heat, and the designer installed a three-way valve for the situation where all the heat on the source side of the heat pumps is provided by the grocery store.

(a) In the normal operation (all the flow goes to the ground), evaluate the flow rate, the head, and the electric power of the pump. The minor losses and the manifold losses are neglected. The pump curve is given by:

$$h_{pump}(\dot V) = 31.8 + 0.031\dot V - 0.0057\dot V^2 - 0.00403\dot V^3$$

with $\dot V$ in l/s and h in m. The pump efficiency (in%) is given by the relation:

$$\eta(\%) = 14.33 + 13.6\dot V - 0.818\dot V^2 - 0.00247\dot V^3$$

and the electric motor efficiency is 80%.
   Answer: $\dot V = 9.26$ l/s, $h = 28.4$ m, $\dot W = 4843$ W
(b) What is the power needed in the by-pass mode?
   Answer: $\dot V = 14.4$ l/s, $h = 19$ m, $\dot W = 10261$ W
(c) If a VFD is installed, what is the rotational speed to keep the same flow rate and what is the new electrical power? (The efficiency of the VFD is 95%.)
   Answer: $\omega = 66.1\% \, \omega_{nom}$, $\dot W = 2855$ W

(d) During a weekend in the off-season, the three-way valve gets stuck in the by-pass position while there is no load and no VFD. Assuming a heat loss coefficient of $U' = 0.3$ W/m-K, on the inside piping, what would be the temperature increase of the water-glycol solution after 2 days?

Answer: $\Delta T = 272$ K (pipe will melt before).

References

1. Kavanaugh, S., Rafferty, K.: Geothermal Heating and Cooling: Design of Ground-Source Heat Pump Systems. ASHRAE Atlanta (2014)
2. American Society of Heating, Refrigerating and Air-Conditioning Engineers. ASHRAE Handbook, Fundamentals. The Society, New York (2017)
3. Cengel, Y.A., Cimbala, J.M.: Fluid Mechanics: Fundamentals and Applications, 3rd edn. McGraw Hill, New York (2014)
4. Jeppson, R.W.: Analysis of ow in pipe networks. R. (1976)
5. Petukhov, B.S.: Heat Transfer and Friction in Turbulent Pipe Flow with Variable Physical Properties. vol. 6, pp. 503–564. Elsevier, Amsterdam (1970). https://doi.org/10.1016/S0065-2717(08)70153-9
6. Kavanaugh, S., Rafferty, K.: Ground-Source Heat Pumps: Design of Geothermal Systems for Commercial and Institutional Buildings. ASHRAE Atlanta (1997)
7. Crane Company. Engineering Division. Flow of Fluids Through Valves, Fittings, and Pipe. Technical Paper No 410. Crane Company, 2018
8. Hooper, W.B.: The 2-K Method Predicts Head Losses in Pipe Fittings. Chem. Eng. **88**(17), 96–100 (1981)
9. Darby, R.: Correlate pressure drops through fittings. Chem. Eng. **106**(7), 101–103 (1999)
10. ESDU (Engineering Sciences Data Unit). ESDU 69022, Pressure losses in valves (2007)
11. Deeppmann, R.L.: Pressure Drop Corrections for Glycol in HVAC Systems (2010). https://www.deppmann.com/home/wp-content/uploads/2010/09/printer_friendly_2010_09_27.pdf. Accessed 18 July 2018
12. American Society of Heating, Refrigerating and Air-Conditioning Engineers. ASHRAE Handbook—HVAC Systems and Equipment. The Society, New York (2016)
13. Munson, B.R. et al.: Fundamentals of Fluid Mechanics, 3rd edn. Wiley, New York (2016)
14. Understanding the Properties of GlycolSolutions Prevents Design Errors in Pumping and Piping Applications. https://metersolution.com/wp-content/uploads/2013/02/Glycol-Pump-Losses.pdf. Accessed 23 Oct 2019
15. ADEME: Rendement des circulateurs selon des donnees de catalogues (2003). http://cregen.free.fr/Notes%20de%20synth%E8ses%20COSTIC/note_technique_65.pdf. Accessed 26 Jan 2019
16. Gagné-Boisvert, L., Bernier, M.: Comparison of the energy use for different heat transfer fluids in geothermal systems. In: Proceedings of the 2017 IGSHPA Conference and Expo, Denver, March, pp. 336–345 (2017)
17. Chiasson, A.D.: Geothermal Heat Pump and Heat Engine Systems: Theory and Practice. Wiley, New York (2016)
18. Gehlin, S., Spitler, J.D.: Effects of ground heat exchanger design flow velocities on system performance of ground source heat pump systems in cold climates. In: ASHRAE Winter Conference (2015)
19. Taylor, S.T., Stein, J.: Balancing variable ow hydronic systems. ASHRAE J. **44**(10), 17–20 (2002)
20. Wang, J.: Pressure drop and ow distribution in parallel-channel configurations of fuel cells: U-type arrangement. Int. J. Hydrogen Energy **33**(21), 6339–6350 (2008)
21. Bajura, R.A., Jones, E.H.: Flow distribution manifolds. J. Fluids Eng. **98**(4), 654–665 (1976)
22. Design and installation of earth energy systems. Standard. Canadian Standard International Organization for Standardization (2016)
23. WLplastics. WL123 HIGH PERFORMANCE PE4710. https://wlplastics.com/wp-content/uploads/Docs/WL123-0117%20High-Performance%20PE.pdf. Accessed 12 Dec 2021

Introduction to Commercial Building Applications

8.1 Introduction

Institutional buildings like schools and commercial buildings are good examples where geothermal heat pumps were often used with success and sometimes with less fortune. These larger buildings share some common elements with residential applications, but some specific issues are associated with them. Some of them will be discussed in this chapter.

8.2 Inside the Building

As discussed briefly in Chap. 2, larger buildings are separated into zones; each of them has their own loads which depend on their geographical location (north, south, etc.) and general purposes (classes, office, etc.). Once the loads have been evaluated, the HVAC systems (here GSHP) have to be installed to cover them. In the literature several options have been proposed, and their choices have impacts on the overall performance of the system. Figure 8.1 shows a typical five-zone building where each individual zone is controlled by its own separate system.

Such configuration is often called unitary system [1, 2] since all systems can be treated as five independent networks. On the figure all zones look the same, but, of course, they can have different number of boreholes and heat pumps. Kavanaugh [2] is in favor of such systems since they are very simple to control. Their individual pumps (often circulators) are in on-off mode, and they often provide less piping inside the building but cannot take into account the load diversity of the building [3]. Another classical configuration is the two-pipe central configuration [1] (Fig. 8.2).

In the figure only one pump is shown, but pumps in parallel can be present. The pump can be used to force the flow inside the building as well as in the bore field, or an hydraulic separator can be used to hydraulically separate each

piping system [4]. A variable frequency drive (VFD) is often used to modify the pump speed at part-load conditions. This configuration can be reverse-returned as shown on the figure or not. In each heat pump connection, a motorized valve should be added to shut off the flow when the heat pump is not working.

Another configuration that has been put forward by Mescher [5] is called the one-pipe system. In this case, as the name implies, a single pipe is used for the supply and return. A direct concern about the configuration is that the temperature at the inlet of the furthest heat pump will be different than the first one which can degrade its performance (Fig. 8.3).

Mescher [5] argues that a proper design will overcome the problem and that the advantages which include a lower investment (less piping, no VFD, etc.), a lower maintenance, simplicity, and overall performance (ENERGY STAR ratings) counterbalance the disadvantages. In their preliminary studies, they indicate that the pumping energy is systematically less than the central system and that the heat pump performance is better at low part load and may be less than the central system at full load.

Another configuration is called sub-central system which shares a common two-pipe supply system, but each heat pump is associated with its own pump or in-line circulator that are controlled with an on-off switch if the heat pump is working or not. These pumps can provide for the total head of the loop (distributive system), or the system can also have a central pump that serves the building and ground loop (distributive w/ primary) [6] (Fig. 8.4).

8.2.1 Primary-Secondary Configuration

The two-pipe central or sub-central configurations can also be separated in a primary-secondary configuration [4, 6] illustrated, for the central configuration in Fig. 8.5.

L. Lamarche, *Fundamentals of Geothermal Heat Pump Systems*, https://doi.org/10.1007/978-3-031-32176-4_8

Fig. 8.1 Unitary GSHP system

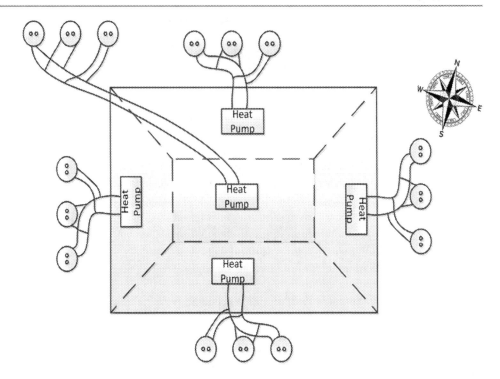

Fig. 8.2 Two-pipe central system

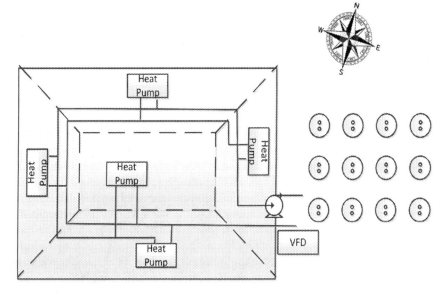

This configuration has the advantage to decouple the flow rate requirements needed for the building and the ground which can be an advantage at part load when the initial piping design was done at full load but leads to an increase in initial and maintenance costs [6].

8.3 Central Configuration Analysis

From the different tools that were discussed in the previous chapters, analysis of unitary system is trivial since we only have to analyze simple systems separately. The case of centralized systems, whether with one- or two-pipe distribution,

is quite different. One of the advantages of the centralized system is the possibility to share different load patterns in a more efficient way.

Figure 8.6 illustrates the concept in a simplistic way. Imagine that at a certain time, a zone demands 4 kW of heat and another one 2 kW of cooling. Assuming that each heat pump has a COP of 3, one has:

$$q_{pumped} = \frac{COP_{he} - 1}{COP_{he}} q_{heating} = 0.666 \cdot 4 = 2.66\,kW$$

$$q_{rejected} = \frac{COP_{co} + 1}{COP_{co}} q_{cooling} = 1.333 \cdot 2 = 2.66\,kW$$

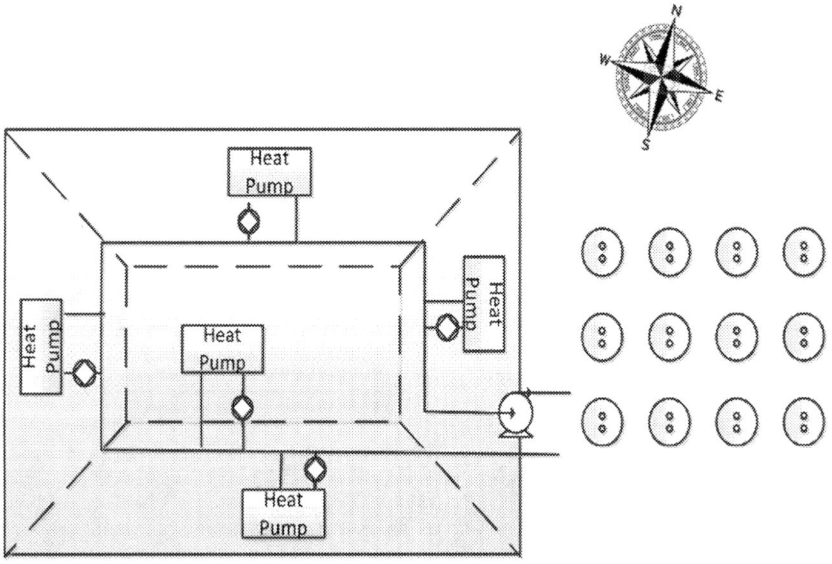

Fig. 8.3 One-pipe central system

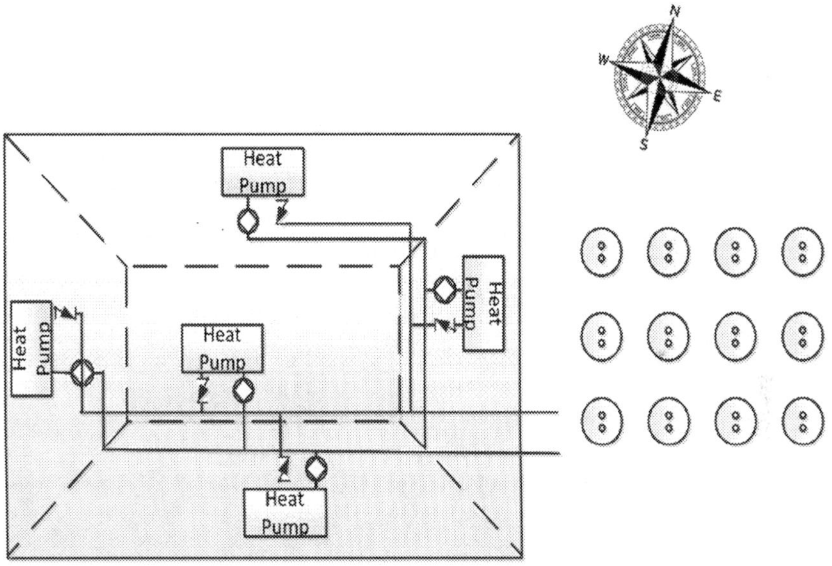

Fig. 8.4 Sub-central system

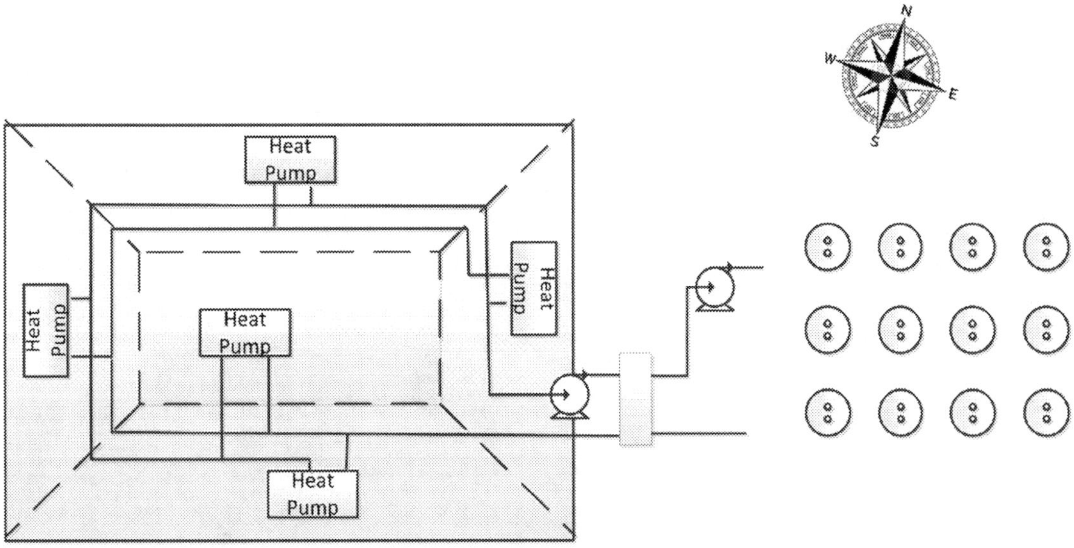

Fig. 8.5 Primary-secondary configuration

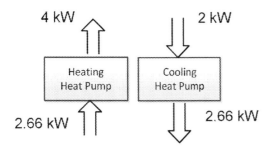

Fig. 8.6 Simple 2-HP system

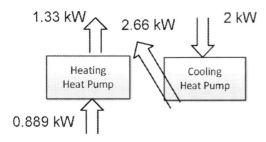

Fig. 8.7 Another simple 2-HP system

so that the total heat exchanged in the ground is zero in the case of a centralized system where as a unitary one would have to exchange that heat in each individual ground field. Of course, in practice, it is not so simple. First, the simultaneity of the loads is important. Imagine that in the last example, this was an hourly load calculated by a computer software and that the heating heat pump has 8 kW capacity and the second one 4 kW of total cooling capacity. That means that each heat pump is working half the time during the one-hour period. If both work during the same half hour, then the previous analysis still remains correct. However, if the first heat pump is working during the first half hour and the second one during the second half hour, then no benefit will, in theory, be observed, and any other combination would be between those two extreme cases. Without going into these very tiny details, it is easy to observe that, in general, a benefit in the total borehole length sizing is possible and can be evaluated with some basic assumptions. Another possibility is illustrated in Fig. 8.7.

Example 8.1 A central reverse-return two-pipe system is shown on the following figure. Each heat pump is designed to receive 0.8 l/s each. A preliminary analysis evaluates the total head losses in the bore field and the building approximately to 12.2 m. At full load, the ground piping head is 5.5 m and each heat pump approximately 5.3 m, the rest being associated with the building interconnections.

(continued)

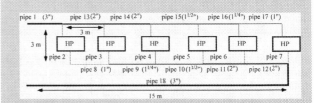

The pump chosen to fulfill the hydraulic requirement is the TACO 1207 that has the following characteristics:

$$h_{pump}(m) = 13.92 - 0.096\dot{V} - 0.017\dot{V}^2 - 0.0079\dot{V}^3, \quad \dot{V}\,(l/s)$$

It's wire-to-water efficiency is given by the following curve:

$$\eta(\%) = 20.5 + 14.5\dot{V} - 0.92\dot{V}^2 - 0.08\dot{V}^3, \quad \dot{V}\,(l/s)$$

which gives a best efficiency of 60% at the design point. The goal of the example is to show the impact of part-load conditions on flow distribution and pump power, so some assumptions will be made to simplify the analysis.

- The heat pump connections (heat pump, ball valves, FCV, etc....) will not be modeled completely but will be expressed by the expression:

$$h_{HP}(m) = 5.3\left(\frac{\dot{V}}{V_{nominal}}\right)^2$$

- The ground field (pipe 1) will not be modeled explicitly and is represented by an equivalent length of 325 m.
- The building interconnections will be modeled as normal pipes.

The hydraulic solution will be done using the Hardy Cross method (see Appendix C). In order to specify a head loss of 5.3 m at nominal condition for the heat pump connections, all connections will be modeled by a singular head loss with a loss coefficient given by:

$$h = \frac{K8\dot{V}^2}{\pi^2 D^4 g} = 5.3\left(\frac{\dot{V}}{V_{nominal}}\right)^2$$

with an arbitrary pipe diameter of 2.73 cm (SDR-11 1") ; the last expression gives a singular loss coefficient of approximately $K = 56$, for all heat pumps.

Evaluate the pumping power with six heat pumps in operation (full load) and five, four, three....heat pumps

(continued)

Example 8.1 (continued)

in operation (part load) with and without VFD. The fluid is water ($\rho = 1000\,\text{kg/m}^3$, $\nu = 8.6\text{e-}7\,\text{m}^2/\text{s}$).

Solution Without VFD

The solution is done with the Hardy Cross model with the six loops indicated on the figure. The final result with six heat pumps is given by:

Pipe	$\dot{V}$ (l/s)	h (m)
1	4.80	5.50
2–7	0.80	5.28
3–6	0.80	5.27
4–5	0.80	5.36
8–17	0.80	0.22
9–16	1.60	0.25
10–15	2.40	0.27
11–14	3.20	0.16
12–13	4.00	0.23
18	4.80	0.25

path 2-8-9-10-11-12
$h_{build} = 5.28 + 0.22 + 0.25 + 0.27 + 0.16 + 0.23 = 6.42$ m
path 13-3-9-10-11-12
$h_{build} = 0.23 + 5.27 + 0.25 + 0.27 + 0.16 + 0.23 = 6.42$ m
path 13-14-4-10-11-12
$h_{build} = 0.23 + 0.16 + 5.36 + 0.27 + 0.16 + 0.23 = 6.42$ m
path 13-14-15-5-11-12
$h_{build} = 0.23 + 0.16 + 0.27 + 5.36 + 0.16 + 0.23 = 6.42$ m
path 13-14-15-16-6-12
$h_{build} = 0.23 + 0.16 + 0.27 + 0.25 + 5.27 + 0.23 = 6.42$ m
path 13-14-15-16-17-7
$h_{build} = 0.23 + 0.16 + 0.27 + 0.25 + 0.22 + 5.28 = 6.42$ m

where pipe 1 represents the whole ground loops and headers. It is seen that with a proper choice of piping, the flow is almost equally distributed even without balancing devices. The total head is given by:

$$h_{pump} = 5.50 + 6.42 + 0.25 = 12.2 \text{ m}$$

$$\dot{W}_{fluid} = 4.8 \cdot 9.81 \cdot 12.2 = 573 \text{ W}$$

$$\eta = 60.0\%$$

$$\dot{W}_{pump} = \frac{4.8 \cdot 9.81 \cdot 12.2}{0.6} = 954 \text{ W}$$

When one heat pump is stopped, the building loop will be somewhat choked which will tend to increase the total head along the pump curve. Redoing the same calculation with the last heat pump shutoff will give:

Pipe	$\dot{V}$ (l/s)	h (m)
1	4.41	4.71
2	0.86	6.21
3	0.87	6.28
4	0.88	6.45
5	0.89	6.61
6	0.90	6.70
7–17	0.0	0.0
8	0.87	0.26
9	1.73	0.29
10	2.62	0.32
11	3.54	0.19
12	4.41	0.28
13	3.54	0.19
14	2.67	0.11
15	1.79	0.16
16	0.90	0.09
18	4.41	0.22

$$h_{build} = 6.21 + 0.26 + 0.29 + 0.32 + 0.19 + 0.28 = 7.55 \text{ m}$$

$$h_{pump} = 7.55 + 4.71 + 0.22 = 12.48 \text{ m}$$

$$\dot{W}_{fluid} = 4.42 \cdot 9.81 \cdot 12.48 = 539 \text{ W}$$

$$\eta = 60\%$$

$$\dot{W}_{pump} = \frac{4.42 \cdot 9.81 \cdot 12.48}{0.60} = 904 \text{ W}$$

A 3% reduction even though the load is reduced by 17%. If a differential pressure transducer monitors the pressure at heat pump 5, it will measure:

$$\Delta p = \rho g h = 51 \text{ kPA}$$

at full load and

$$\Delta p = \rho g h = 66 \text{ kPA}$$

at part load. If a VFD is installed, the reduced rotational speed will modify the pump curve (see Chap. 7), and if the reduction of speed is adjusted until the differential pressure remains the same, the following results will be obtained:

(continued)

(continued)

Example 8.1 (continued)

Pipe	$\dot{V}$ (l/s)	h (m)
1	4.00	3.97
2	0.78	5.11
3	0.79	5.17
4	0.80	5.32
5	0.81	5.45
6	0.82	5.53
7–17	0.0	0.0
8	0.78	0.22
9	1.57	0.25
10	2.37	0.27
11	3.18	0.16
12	4.00	0.23
13	3.21	0.16
14	2.43	0.10
15	1.63	0.14
16	0.82	0.08
18	4.00	0.18

$$h_{build} = 5.11+0.22+0.25+0.27+0.16+0.23 = 6.23 \text{ m}$$

$$h_{pump} = 6.23 + 3.97 + 0.18 = 10.4 \text{ m}$$

$$\dot{W}_{fluid} = 4.00 \cdot 9.81 \cdot 10.4 = 407 \text{ W}$$

$$\eta = 60\%$$

$$\dot{W}_{pump} = \frac{4.0 \cdot 9.81 \cdot 10.2}{0.6} = 685 \text{ W}$$

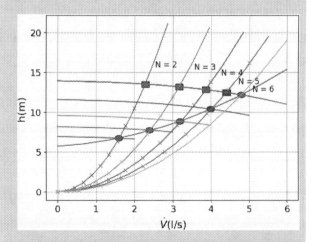

It is observed that the total flow rate is now 83% (5/6) of the initial flow rate. The rotational speed ratio needed to obtain the same differential pressure is found to be:

$$\alpha = \frac{\omega_2}{\omega_1} = 0.91$$

Since the pump scaling laws predict that this ratio is proportional to the flow rate ratio, we might have expected that it would be around 0.83, but one should remember that the system curve has changed when one heat pump is shut off [7,8]. Actually the pump scaling laws work fine when we compare the case with (blue circle) and without VFD (blue rectangle) with the same number of HP.

Flow rate:

$$\frac{4}{4.42} = 0.9 \approx \alpha$$

Head:

$$\frac{10.2}{12.47} = 0.81 \approx \alpha^2$$

Power:

$$\frac{670}{903} = 0.74 \approx \alpha^3$$

since they are on the same system curve. The same analysis can be done with one, two, three, and five heat pumps. The results are shown on the next table and figure. The square is associated with the operating points without VFD.

# Heat pumps	No VFD		VFD		
	h (m)	$\dot{W}$ (W)	h (m)	$\dot{W}$ (W)	η_{w-w}
2	13.5	633	6.7	220	0.48
3	13.2	750	7.7	334	0.55
4	12.8	837	8.8	476	0.58
5	12.5	904	10.4	685	0.59
6	12.2	953	12.2	953	0.60

Joining the three values of reduced head values by a second-degree polynomial and extrapolating to zero flow give a value of 6.4 m, which is close to the mean nominal head loss from the heat pump as predicted by Bernier and Lemire [7].

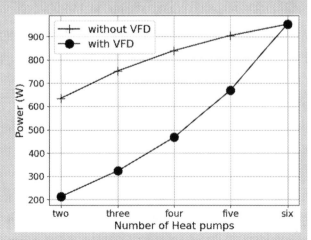

(continued)

In [7], Bernier and Lemire have shown that the reduction in pump power cannot be predicted by the pump laws and is highly dependent on the ratio of the ratio of the head losses in each heat pump to the total head losses. They propose a correlation to predict the electrical power at reduced speed when a VFD is used given by the following relation:

$$
\begin{aligned}
Z(\tilde{V}, X) &= \frac{\dot{W}_{elec}}{\dot{W}_{shaft,nom}} \\
&= 0.22464 - 0.50812\tilde{V} + 0.58266\tilde{V}^2 \\
&\quad + 0.7912\tilde{V}^3 - 0.23834X + 0.06207X^2 \\
\\
&\quad + 0.15208X^3 + 2.05530\tilde{V}X - 0.15778\tilde{V}X^2 \\
&\quad - 1.6652\tilde{V}^2 X - 0.17377\tilde{V}^2 X^2
\end{aligned}
$$
(8.1)

where:
$$
\tilde{V} = \frac{\dot{V}}{\dot{V}_{nom}}
$$

and:
$$
X = \frac{\Delta p_s}{\Delta p_{tot}}
$$

Another approach was used by Sfeir et al. [3, 7] where the electrical power was evaluated by the classical expression:

$$
\dot{W}_{elec} = \frac{\dot{m}_i g h_i}{\eta_{w-w}}
$$
(8.2)

where $\dot{m}_i$ represents the total flow rate associated with a given number of heat pumps (i.e., half of the nominal flow rate when half the number of heat pumps are in operation when a VFD is used) and h_i the total head losses for the same operation case. This total head was evaluated as:

$$
h_i = h_{field,i} + h_{hp,i} = h_{field,nom}\tilde{V}_i^2
$$
$$
+ h_{hp,nom} = h_{total,nom}\left((1 - X)\tilde{V}_i^2 + X\right)
$$
(8.3)

The main difference comes from the fact that in Eq. (8.1), a total efficiency variation was assumed, whereas using (8.2), this efficiency variation has to be evaluated for the given conditions.

Example 8.2 Compare the part-load electrical power calculated in the last example with the one predicted by the Lemire and Bernier relation and Eq. (8.2).

Solution From Example 8.1 the heat pump-to-total loss ratio at full load is given by:

$$
X = \frac{6.42}{12.2} = 0.527
$$

(continued)

With five heat pumps in operation with VFD, the reduced flow rate is simply:

$$
\tilde{V} = \frac{5}{6}
$$

and Eq. (8.1) gives:

$$
\frac{\dot{W}_{elec}}{\dot{W}_{shaft,nom}} = 0.8
$$

For some unknown reasons, the correlation proposed is not the ratio of electrical power at part load but the ratio of electrical power at part load to shaft power at nominal conditions. For this reason, the ratio at full load ($\tilde{V}$) is not equal to 1. In our example, since the wire-to-water efficiency was used, the nominal shaft power is unknown. A reasonable assumption was to take:

$$
\dot{W}_{shaft,nom} = \frac{\dot{W}_{elec}}{Z(1, X)} = \frac{954}{1.12} = 853 \text{ W}
$$

So for the five heat pumps, the electrical power becomes:
$$
\dot{W}_{elec} = 0.8 \cdot 853 = 683 \text{ W}
$$

Using Eq. (8.3) gives:

$$
h_i = 12.2((1 - 0.527)0.833^2 + 0.527) = 10.44 \text{ m}
$$

$$
\dot{W}_{elec} = \frac{4.0 \cdot 9.8 \cdot 10.44}{0.59} = 694 \text{ W}
$$

where the wire-to-water efficiency was taken from the previous example. The final results are shown in the following figure along with the results found in the previous example:

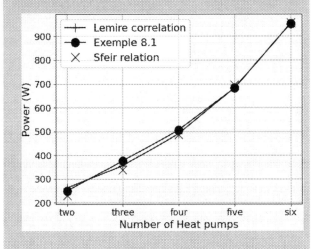

(continued)

As seen in the previous results, using a variable speed pump, the total flow rate is almost equal to the nominal value of flow rate in a heat pump times the number of heat pumps even though no flow balancing units are installed. Although the flow in each heat pump is correct, the total flow rate in the boreholes will get smaller and smaller, and a value too small can harm the heat transfer behavior in the ground, so normally a minimum value will be specified. In [9], a 30% value was arbitrarily chosen. In the previous example, the flow rate would still be correct for two heat pumps but not if just one is used. If a 50% minimum value was chosen, the flow rate for the two heat pumps case would give a total flow rate too small, and the control system will tap it at the minimum value. In such a case, the flow rate in each heat pump will be higher than desired. It is also possible to install a bypass [10] in order to receive the supplementary flow rate. The bypass flow rate can be controlled with a three-way valve or a differential pressure bypass valve (DPBV). A DPBV is a two-way valve which is more often used in hydronic systems to reduce flow noise when some radiator valves are closing. They are normally not necessary when a variable speed pump is used except when a minimum flow rate is prescribed as it would be the case here. In an ideal bypass valve, the flow is zero if the differential pressure is lower than an adjustable pressure setting and allows extra flow rate if it is higher. A typical flow characteristic has the form shown in Fig. 8.8: where the slope would be almost zero for an ideal valve but have a nonzero value in real cases.

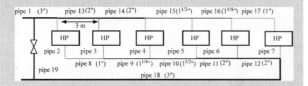
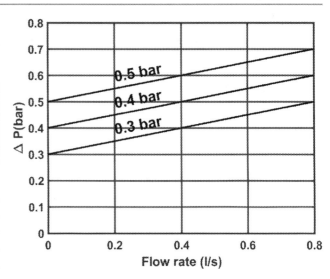

Fig. 8.8 Differential by-pass valve characteristics

same network to that in the previous example is solved; if it is found to be higher, the solution is redone with an extra loop in the network and where the bypass line is modeled as:

$$\Delta P(\text{kPa}) = S.P.(\text{kPa}) + 24\dot{V}(\text{l/s})$$

which is close to the characteristics shown in Fig. 8.8. The set point value is typically higher than the differential pressure observed between heat pumps to take into account the return pipe head loss. For three to five heat pumps, nothing changes when there is VFD present. However when no VFD is installed, the pressure rise due to the closure of one or more heat pumps will open the bypass valve, and some flow will go through the bypass valve to reduce the flow rate. For the five-heat pump case, the following results will be obtained:

Pipe	No VFD, No By-pass $\dot{V}$ (l/s)	No VFD, No By-pass h (m)	No VFD, By-pass $\dot{V}$ (l/s)	No VFD, By-pass h (m)	VFD $\dot{V}$ (l/s)	VFD h (m)
1	4.42	4.67	4.60	5.07	4.00	3.97
2	0.87	6.24	0.84	5.80	0.78	5.11
3	0.87	6.31	0.84	5.86	0.79	5.17
4	0.88	6.49	0.85	6.03	0.80	5.32
5	0.89	6.64	0.86	6.18	0.81	5.45
6	0.90	6.74	0.87	6.27	0.82	5.31
7–17	0.0	0.00	0.00	0.00	0.0	0.0
8	0.87	0.26	0.84	0.24	0.78	0.22
9	1.74	0.29	1.68	0.28	1.57	0.25
10	2.62	0.32	2.53	0.30	2.37	0.27
11	3.52	0.19	3.39	0.17	3.18	0.16

(continued)   (continued)

Example 8.3 (continued)

12	4.42	0.28	4.26	0.26	4.00	0.23
13	3.55	0.19	3.43	0.18	3.21	0.16
14	2.68	0.11	2.58	0.11	2.42	0.10
15	1.80	0.16	1.73	0.15	1.62	0.14
16	0.90	0.09	0.87	0.09	0.82	0.08
18	4.42	0.22	4.26	0.20	4.00	0.18
19	0.00	0.00	0.34	7.26	0.00	0.00
$\dot{W}_{pump}$	905 W		929 W		684 W	

The other cases will behave similarly except for the two-heat pump cases where the pump control will not allow a flow rate less than 50% of the nominal value (by choice), and this will trigger the bypass valve.

Pipe	No VFD, No B-P $\dot{V}$ (l/s)	h (m)	No VFD,B-P $\dot{V}$ (l/s)	h (m)	VFD, No B-P $\dot{V}$ (l/s)	h (m)	VFD, B-P $\dot{V}$ (l/s)	h (m)
1	2.29	1.47	3.96	3.78	2.60	0.61	2.40	1.38
2	1.13	10.7	1.00	8.22	0.79	5.18	0.76	4.74
3	1.15	11.0	1.00	8.53	0.81	5.39	0.77	4.94
4-5-6-7-14-15-16-17	0.00	0.00	0.00	0.00	0.00	0.00	0.00	0.00
8	1.13	0.42	1.00	0.33	0.79	0.22	0.76	0.20
9	2.29	0.49	2.00	0.38	1.60	0.25	1.53	0.23
10	2.29	0.25	2.00	0.00	1.60	0.13	1.53	0.00
11	2.29	0.09	2.00	0.00	1.60	0.05	1.53	0.00
12	2.29	0.09	2.00	0.00	1.60	0.05	1.53	0.00
13	1.15	0.03	1.00	0.02	0.81	0.01	0.77	0.01
18	2.29	0.07	2.00	0.05	1.60	0.04	1.53	0.03
19	0.00	0.00	1.96	8.98	0.00	0.00	0.88	5.21
W_{pump}	635 W		849 W		220 W		278 W	

∎

Another obvious configuration that allows the correct flow rates in the bore field regardless of the number of heat pumps working is the primary-secondary configuration previously discussed.

Example 8.4 In Example 8.1, six heat pumps were in operation at full load, and the impact on the pumping work of shutting one or more heat pumps at partial load was analyzed with and without VFD. Instead of shutting off a heat pump, it is possible to operate them at a lower capacity using variable speed compressor [11–13]. Compare the pumping power difference when five heat pumps (83% heat load) at full capacity and six heat pumps working at 83% of their capacity and 83% of their source flow rate.

Solution The solution is quickly obtained from the previous example script using the VFD to reduce the total flow rate keeping the six heat pumps operating. Since the system curve will now be the same as the original one, the new operating point will follow this curve, and the rotational speed will be reduced at 83% of its original value. The final result will be given by the following numbers:

Pipe	$\dot{V}$ (l/s)	h (m)
1	4.00	3.81
2–7	0.66	3.65
3–6	0.66	3.64
4–5	0.67	3.71
8–17	0.66	0.16
9–16	1.33	0.18
10–15	2.00	0.20
11–14	2.67	0.11
12–13	3.33	0.17
19	4.00	0.18

$$h_{build} = 3.81 + 0.16 + 0.18 + 0.20 + 0.11 + 0.17 = 4.47 \text{ m}$$

$$h_{pump} = 4.47 + 3.81 + 0.18 = 8.45 \text{ m}$$

$$\dot{W}_{fluid} = 4.00 \cdot 9.81 \cdot 8.45 = 331 \text{ W}$$

$$\dot{W}_{pump} = \frac{4.00 \cdot 9.81 \cdot 8.45}{0.6} = 551 \text{ W}$$

where the new operating point (blue square) is compared with the old one (blue circle) on the figure:

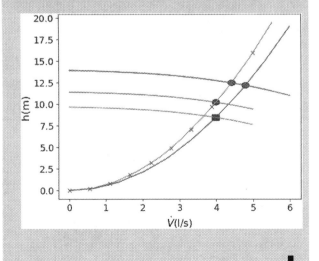

∎

(continued)

8.3.1 Primary-Secondary Configuration

Example 8.5 The previous example can be configured in a primary-secondary configuration. A possible configuration could look like:

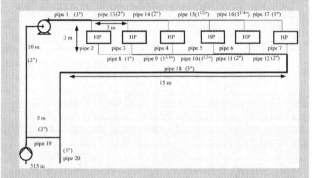

The pipe one length has been removed from the total equivalent length of the bore field. A short common pipe between the two loops ensures a small pressure drop [6]. The pump inside the building loop is chosen to overcome the pressure drop in the building loop. The TACO 2007 pump is chosen:

$$h_{pump}(m) = 7.84 + 0.00041\dot{V} - 0.0235\dot{V}^2 - 0.0042\dot{V}^3, \quad \dot{V}\,(l/s)$$

Its wire-to-water efficiency is given by the following curve:

$$\eta(\%) = 21.45 + 22.7\dot{V} - 2.5\dot{V}^2 + 0.036\dot{V}^3, \quad \dot{V}\,(l/s)$$

The secondary pump should be chosen to overcome the head losses in the ground loop. From the previous examples, they are in the same order of magnitude of the building losses, and the same circulating pump is arbitrarily chosen for the example. Estimate the pumping energy at full load and part load.

Solution The solution is done with the Hardy Cross method with an added loop. The solution at full load is:

Pipe	$\dot{V}$ (l/s)	h (m)
1	4.80	0.17
2–7	0.80	5.28
3–6	0.80	5.27
4–5	0.80	5.36
8–17	0.80	0.22
9–16	1.60	0.25
10–15	2.40	0.27

11–14	3.20	0.16
12–13	4.00	0.23
18	4.80	0.25
19	0.57	0.01
20	5.37	6.51

The pumping energy delivered by the two pumps is:

$$\dot{W}_1 = 4.8 \cdot 9.81 \cdot 6.84 = 322 \text{ W},$$

$$\dot{W}_2 = 5.37 \cdot 9.81 \cdot 6.52 = 343.7 \text{ W}$$

$$\dot{W}_{fluid} = 666 \text{ W}$$

which is higher than in Example 8.1 since there is a higher flow rate in the bore field. If a different secondary pump was chosen to perfectly match the field head losses, the fluid energy would have been exactly the same, but that is quite difficult to obtain due to the inherent inaccuracy of the head losses. One of the advantages of smaller head pumps is that they can possibly have a higher efficiency which is the case here. Both have an efficiency of approximately 78% which gives:

$$\dot{W}_{pumps} = \frac{322}{0.78} + \frac{343.7}{0.78} = 866 \text{ W}$$

which is smaller than the electric power of Example 8.1.

When five heat pumps are in operation, the results without VFD are:

Pipe	$\dot{V}$ (l/s)	h (m)
1	4.18	0.13
2	0.82	5.59
3	0.83	5.65
4	0.84	5.82
5	0.85	5.96
6	0.85	6.04
7–17	0.0	0.0
8	0.87	0.24
9	1.65	0.27
10	2.48	0.29
11	3.33	0.17
12	4.18	0.25
13	3.36	0.17
14	2.54	0.10
15	1.70	0.15
16	0.85	0.08
18	4.18	0.20
19	1.19	0.01
20	5.38	6.51

(continued)

(continued)

Example 8.5 (continued)

$$\dot{W}_{fluid} = 293 + 344 = 637 \text{ W}$$

$$\dot{W}_{pump} = \frac{293}{0.75} + \frac{344}{0.76} = 835 \text{ W}$$

which is again smaller than for the primary configuration, due to the higher efficiency. It is also noted that the increase in flow rate in each heat pumps is smaller than previously since the major head losses in the building loop are now due to the heat pumps themselves. However, if the nominal flow rate is looked for, a VFD drive can be used. Here only the building pump or both can be controlled, and different strategies can be used for each if needed. Since the hydraulic resistance is more influenced by a heat pump shutoff, the speed reduction is smaller. For the five heat pumps, it will be 95% speed instead of 91% for the primary system. On the contrary, since the system curve of the field does not change, a small speed reduction of the field pump will not change much the flow rate. In theory, following the pump scaling law, the speed reduction for this case should be around 5/6 = 0.83 to reduce the flow rate accordingly if it is what is looked for. It should be remembered that reducing the flow rate in the borehole can harm the thermal performance, so a global analysis should be done, but, at least, the configuration allows the possibility to adjust it independently. Changing both pumps speed to 95% speed reduction and the case where the building pump has a 95% reduced speed and the field pump a 83% are given in the following table:

	$\alpha_{pump,1} =$ 0.95, $\alpha_{pump,2} =$ 0.95		$\alpha_{pump,1} =$ 0.83, $\alpha_{pump,2} =$ 0.95	
Pipe	$\dot{V}$ (l/s)	h (m)		
1	4.0	0.12	4.0	0.12
2	0.79	5.10	0.79	5.09
3	0.79	5.16	0.79	5.15
4	0.80	5.31	0.80	5.30
5	0.81	5.44	0.81	5.43
6	0.82	5.52	0.82	5.51
7–17	0.0	0.0	0.0	0.0
8	0.79	0.22	0.79	0.22
9	1.58	0.25	1.58	0.25
10	2.37	0.27	2.37	0.27
11	3.18	0.16	3.18	0.16
12	4.0	0.23	4.0	0.23
13	3.21	0.16	3.21	0.16

(continued)

14	2.42	0.10	2.42	0.10
15	1.62	0.14	1.62	0.14
16	0.82	0.08	0.82	0.08
18	4.0	0.18	4.0	0.18
19	1.12	0.01	0.41	0.01
20	5.12	5.96	4.41	4.56
	$\dot{W} = 728$ W		$\dot{W} = 596$ W	

In Example 8.3, a differential bypass valve was used to ensure a minimum flow rate of 50% in the field. With a primary-secondary setup, it is easier to achieve. Redoing the case with two heat pumps running the building pump at 88% of its nominal speed and the field pump at 50% of its nominal speed (to achieve 50% of nominal flow) results in:

Pipe	$\dot{V}$ (l/s)	h (m)
1	1.6	0.02
2	0.8	5.29
3	0.8	5.40
4-5-6-7-14-15-16-17	0.00	0.00
8	0.8	0.22
9	1.6	0.25
10	1.6	0.13
11	1.6	0.05
12	1.6	0.05
13	0.8	0.01
18	1.6	0.04
19	0.9	0.00
29	2.5	1.68

$$\dot{W}_{fluid} = 93 + 42 = 135 \text{ W}$$

$$\dot{W}_{pump} = \frac{93}{0.55} + \frac{42}{0.67} = 232 \text{ W}$$

Again, it is a smaller value than in Example 8.3.

∎

It was seen in the previous example that the primary-secondary configuration can bring some advantages for control strategies and especially if pumps with higher efficiencies can be selected. However, the higher number of pumps and valves can lead to higher installation and maintenance costs [6].

8.3.2 One-Pipe Configuration

The main argument to use independent pumps in a primary-secondary layout or circulators in a sub-central configuration is to decouple the ground to the heat pump loop. At full load, the pumping energy given to the fluid should in theory be the same if the pumps are chosen correctly and the total electrical power also if the total efficiencies of all pumping elements are equal. However, since circulators are known to have lower efficiencies, a simple analysis would predict a higher electrical cost for the sub-central arrangement at least at full load. This option will not be analyzed further. An interesting system proposed by Mescher [5] is the one-pipe system. The main argument is economical since it requires less piping elements. The economic analysis will be discussed in the next chapters, but for now, we can look at the energy impact. The main disadvantage of this option is that the temperature at the entrance of the last heat pump will be different than the first, so it is difficult to analyze only the pumping energy alone when looking at one-pipe systems. In [5], it is noted that recent efficient heat pumps are less influenced by entrance temperature. However, looking, for example, at Fig. 8.9, it is seen that the cooling capacity from a recent high-performance 5 tons heat pump is moderately affected by the temperature. If one choose the nominal capacity at 25 °C at 18.5 kW, and if a reduction of 10% is acceptable, then, the maximum temperature should be approximately 31 °C, a 6 degree increase. This would influence directly the minimum flow rate acceptable. In cooling mode, for example, it can be observed that the minimum flow rate would be:

$$\dot{m}_{total,min} = \frac{\sum_{i=1}^{N-1} q_{cond,i}}{C_p(T_{N,max} - T_{in})} \quad (8.4)$$

where $q_{cond,i}$ is the heat at the condenser, T_{in} the entrance temperature of the first heat pump, and $T_{N,max}$ the maximum entrance temperature desired for the last heat pump.

Example 8.6 Evaluate the pumping energy at full load and partial load for the six heat pumps shown in the following figure:

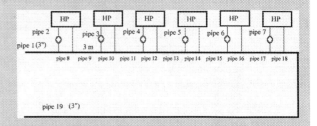

All heat pumps are TS 060 Climate Master heat pumps having a 0.8 l/s nominal flow rate, whose performances are shown in Fig. A.16. The fluid is water, and the entrance temperature from the ground is 25 °C. The ground head losses are the same as in the previous example, whereas the heat pump connections require less valves and connections and are modeled by the following expression:

$$h_{HP} = 4.4 \left(\frac{\dot{V}}{\dot{V}_{nominal}} \right)^2$$

The maximum temperature desired at the last heat pump is 31 °C.

Solution The first step is to evaluate the flow rate. As previously stated, it is not trivial since the capacity and the COP will vary from heat pump to heat pump. As a first guess, it can be assumed that they remain unchanged. At 25 °C and a flow rate of 0.8 l/s ($\approx$ 0.8 kg/s) for water, the cooling capacity and COP are found to be:

(continued)

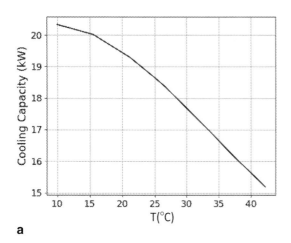

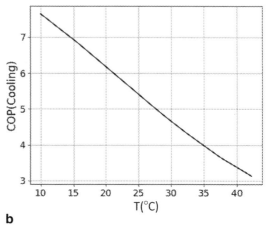

Fig. 8.9 (a) Heat pump cooling capacity vs. entrance temperature from [14]. (b) Cooling COP vs. entrance temperature from [14]

Example 8.6 (continued)

$$q_{evap}(25\,°C) = 18.63 \text{ kW}, \quad COP(25\,°C) = 5.4$$

$$q_{cond}(25\,°C) = 22.1 \text{ kW}$$

$$\dot{m}_{min} = \frac{5 \cdot 22.1}{4.18 \cdot 6} = 4.4 \text{ kg/s}$$

From this value, it is possible to evaluate the temperature at the entrance of all heat pumps. For example, for the second one, we would find:

$$T_{out,hp,1} = 25 + \frac{22.1}{0.0008 \cdot 1000 \cdot 4.18} = 31.6\,°C$$

$$T_{in,2} = \frac{0.8 \cdot 31.6 + (4.4 - 0.8)25}{4.4} = 26.2\,°C$$

From this value a new value of cooling capacity and COP can be found, and the same procedure can be repeated for all heat pumps. At the end a 30.9 °C inlet temperature will be obtained, a value very near the 31 °C that we would get neglecting the capacity variation. It is seen that this value is less (at least in this example) than the total flow rate of the previous example (4.8 l/s) which would diminish the head losses in the ground. However, since it is a minimum value and since this value could also influence the thermal behavior of the ground heat exchanger, the value of 4.8 l/s will be used, which will ameliorate the thermal performance of the heat pumps. The primary pump follows the following law:

$$h_{pump}(m) = 7.80 + 0.21\dot{V} - 0.103\dot{V}^2 - 0.00139\dot{V}^3, \quad \dot{V}\,(l/s)$$

and its efficiency follows the same trend than in Example 8.1. Finally all circulators are 1/8 HP with curve pump given by:

$$h_{circ}(m) = 6.96 - 2.99\dot{V} - 0.2096\dot{V}^2 - 0.00676\dot{V}^3, \quad \dot{V}\,(l/s)$$

The solution given by the Hardy Cross method is:

Pipe	V (l/s)	h (m)
1	4.80	5.50
2–7	0.80	4.45
8-10-12-14-16-18	4.00	0.04
9-11-13-15-17	4.80	0.05
19	4.80	0.3

The primary pump fluid energy is:

$$\dot{W}_{fluid,prim} = 4.8 \cdot 9.81 \cdot 6.28 = 295 \text{ W}$$

(continued)

and for the circulators:

$$\dot{W}_{fluid,circ} = 0.8 \cdot 9.81 \cdot 4.4 = 35 \text{ W}$$

The total energy given to the fluid is then at full load:

$$\dot{W}_{fluid} = 504 \text{ W}$$

The electrical power for the primary pump, with a 60% wire-to-water efficiency, is:

$$\dot{W}_{pump,prim} = 492 \text{ W}$$

which is a little bit less than with the two-pipe arrangement (Example 8.1) due to the lower head losses inside the building. However, since the efficiency of the circulators are smaller, the electrical power will increase:

$$\dot{W}_{circ} = 6 \cdot 93 = 559 \text{ W}$$

$$\dot{W}_{elec,tot} = 1051 \text{ W}$$

This is slightly higher than the two-pipe system (in this example). At part load, the circulator is simply stopped, and a check valve is sometimes used although not as necessary as the sub-central case since the flow will go naturally in the large pipe. The total flow rate can also be adjusted, if needed, either by using a VFD on the primary pump or by using two pumps in parallel (see [5]). A minimum flow rate can also be easily obtained without the need of bypass. In this example, we will assume that a VFD is present in order to keep the total flow rate in the same ratio as the number of heat pumps. The final results are shown:

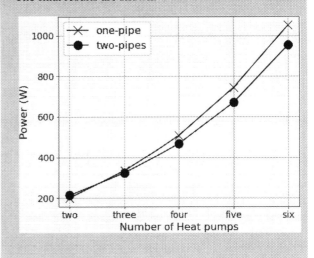

(continued)

Example 8.6 (continued)

As it is seen, at part load, the lower efficiency of the circulator is offset by the lower head losses in the building.

Comments *In [5], the saving in pumping energy is even more significant since the heat pumps have different flow rates and pressure drops. In a one-pipe system, each individual loop will deliver the sufficient head. In a two-pipe system, balancing valves will give the highest head necessary in each loop given, in that case, a clear advantage of the one-pipe system. So, it is important not to generalize conclusions to all applications. It is important also to note that, in this particular example, the compressor of the last heat pump at full load will have a 10% increase in energy consumption compared with the first one. So a global picture has to be analyzed before concluding.*

∎

8.4   Motors and VFD Efficiencies

In the last examples, an arbitrary wire-to-water efficiency relationship was used. Sfeir et al. [3] suggested expressions to evaluate motor and VFD efficiencies, so does Kavanaugh and Lambert [15]. A more recent study [16] gives a slightly modified expression that is also used in EnergyPlus™ simulation program[17]. The expression is given by:

$$\eta_{VFD} = \frac{a_{VFD} \cdot x_{VFD}}{b_{VFD} + x_{VFD}} + c_{VFD} \cdot x_{VFD} \qquad (8.5)$$

where $0 < x_{VFD} = \frac{\omega}{\omega_{max}} < 1.0$ is the part-load fraction. Values of parameters a, b, c can be found in the EnergyPlus documentation (Table 8.1):

The efficiencies obtained are greater than the relation given in Sfeir et al. [3] especially at partial load. Similar expressions are given in EnergyPlus reference guide [17] for electric motor where fitted expressions were found from DOE MotorMaster+ database [18]. The motor efficiency was evaluated from (Tables 8.2 and 8.3):

$$\eta_{motor} = \eta_{motor,max} \frac{\eta_{motor}}{\eta_{motor,max}} \qquad (8.6)$$

$$\eta_{motor_{max}} = \frac{a_{m,max} \cdot Y_{motor}}{b_{m,max} + Y_{motor}} + c_{m,max} \qquad (8.7)$$

where:

$$Y_{motor,max} = \log(\dot{W}_{mot})(\dot{W} \text{ in hp}). \qquad (8.8)$$

Table 8.1 VFD efficiency coefficients (from [17])

VFD rated power (hp)	a_{VFD}	b_{VFD}	c_{VFD}
3	0.978856	0.034247	-0.007862
5	0.977485	0.028413	-0.002733
10	0.978715	0.022227	0.001941
20	0.984973	0.017545	-0.000475
30	0.987405	0.015536	-0.005937
50	0.987910	0.018376	-0.001692
60	0.971904	0.014537	0.011849
75	0.991874	0.017897	-0.001301
100	0.982384	0.012598	0.001405

Table 8.2 Motor maximum efficiency coefficients (from [17])

Case	$a_{m,max}$	$b_{m,max}$	$c_{m,max}$
High-efficiency	0.196205	3.653654	0.839926
Mid-efficiency	0.292280	3.368739	0.762471
Low-efficiency	0.395895	3.065240	0.674321

Table 8.3 Motor part-load efficiency coefficients (from [17])

Motor Rated power (hp)	a_{mot}	b_{mot}	c_{mot}
1	1.092165	0.08206	-0.0072
5	1.223684	0.08467	-0.135186
10	1.146258	0.045766	-0.110367
25	1.137209	0.050236	-0.08915
50	1.088803	0.029753	-0.064058
75	1.07714	0.029005	-0.04935
100	1.035294	0.012948	-0.024708
125	1.030968	0.010696	-0.023514

$$\frac{\eta_{motor}}{\eta_{motor,max}} = \frac{a_{mot} \cdot x_{motor}}{b_{mot} + x_{motor}} + c_{mot} \cdot x_{motor} \qquad (8.9)$$

with:

$$x_{motor} = \frac{\dot{W}_{mot}}{\dot{W}_{mot,max}}$$

Finally in [3], typical pump efficiencies were also proposed:

$$\eta_{pump} = 0.322 + 1.066\tilde{V} - 0.533\tilde{V}^2 \qquad \text{High Efficiency} \qquad (8.10)$$

$$\eta_{pump} = 0.172 + 1.066\tilde{V} - 0.533\tilde{V}^2 \qquad \text{Medium Efficiency} \qquad (8.11)$$

$$\eta_{pump} = 0.022 + 1.066\tilde{V} - 0.533\tilde{V}^2 \qquad \text{Low Efficiency} \qquad (8.12)$$

with:

$$\tilde{V} = \frac{\dot{V}}{\dot{V}_{max,eff}}$$

Example 8.7 In Example 8.1, let's assume that the pump efficiency was 0.8 and that a 1.5 HP high-efficiency motor was used with a VFD controller. Compare the overall efficiency at full load (six heat pumps) and part load (five heat pumps).

(continued)

Example 8.7 (continued)
Solution The motor maximum efficiency is given by (8.8):

$$Y = \log(1.5) = 0.4055 \Rightarrow \eta_{max} = 0.8595$$

Since the nameplate HP is the mechanical shaft power, we have from Ex. 8.1:

$$\dot{W}_{fluid} = 573 \text{ W} \Rightarrow \dot{W}_{shaft} = \frac{573}{0.8} = 716.25 \text{ W}$$

$$x_{mot} = \frac{716.25}{1.5 \cdot 745.7} = 0.64$$

$$\eta_{mot} = 0.839926 \frac{1.1086 \cdot 0.64}{0.082386 + 0.64} - 0.023198 \cdot 0.64 = 0.8314$$

The maximum value ($x_{vfd} = 1$) for the VFD (3 HP) will be used for its efficiency:

$$\eta_{vfd} = \frac{0.978856}{0.034247 + 1} - 0.007862 = 0.938$$

$$\eta_t = 0.8 \cdot 0.8314 \cdot 0.938 \approx 0.62$$

which is very near the value assumed in the example. With five heat pumps running, the VFD will work at 90% of its regular speed (see Example 8.1):

$$\dot{W}_{fluid} = 400 \text{ W} \Rightarrow \dot{W}_{shaft} = \frac{400}{0.8} = 500 \text{ W}$$

$$x_{mot} = \frac{500}{1.5 \cdot 745.7} = 0.447$$

$$\eta_{vfd} = \frac{0.978856 \cdot 0.9}{0.034247 + 0.9} - 0.007862 \cdot 0.9 = 0.936$$

$$\eta_{mot} = 0.839926 \frac{1.1086 \cdot 0.447}{0.082386 + 0.447} - 0.023198 \cdot 0.447 = 0.795$$

$$\eta_t = 0.8 \cdot 0.795 \cdot 0.936 \approx 0.596$$

∎

It was shown previously how to evaluate the pumping energy knowing the instantaneous different flow rates in the circuit. But in practice, the instantaneous loads are not known. Building loads are often known at design conditions, and even though a detailed building simulation is done, detailed information is often given as hourly loads, which does not mean that it is uniform during that time period.

(continued)

Example 8.8 Using (8.1), evaluate for $X = 0.4$, the reduction in pumping energy for the previous examples, when the hourly load is half of the maximum load in the four following cases:

(a) Six heat pumps are used during half an hour.
(b) Three heat pumps are used during the full hour.
(c) Four heat pumps are used during half an hour and two during the other half.
(d) Five heat pumps are used during half an hour and one during the other half.

In all cases, the final energy delivered will be the same.

Solution Again we will modify the proposed relation using instead:

$$Z_{mod}(\dot{V}, X) = \frac{Z(\dot{V}, X)}{Z(1, X)}$$

So the solution is quite straightforward: case (a)

$$0.5 Z_{mod}(1, X) = 0.5$$

case (b)

$$Z_{mod}(0.5, X) = 0.33$$

case (c)

$$0.5 Z_{mod}(0.67, X) + 0.5 Z_{mod}(0.33, X) = 0.35$$

case (d)

$$0.5 Z_{mod}(0.83, X) + 0.5 Z_{mod}(0.17, X) = 0.43$$

the mean value being equal to 0.4.

(b) Redo the same, using Eqs. (8.2, 8.3).
As stated previously, the solution will depend on the global efficiency variation:

$$\frac{\dot{W}_{part-load}}{\dot{W}_{full-load}} = \frac{\dot{m}_{part-load} g h_{part-load}}{\dot{m}_{full-load} g h_{full-load}} \frac{\eta_{full-load}}{\eta_{part-load}}$$

Assuming that it does not change much, we have:

$$\frac{\dot{W}_{part-load}}{\dot{W}_{full-load}} = \frac{PLF h_{part-load}}{h_{full-load}} = PLF((1-X)PLF^2 + X)$$

case a)

(continued)

Example 8.8 (continued)

$$\frac{\dot{W}_{part-load}}{\dot{W}_{full-load}} = 0.5$$

case (b)

$$\frac{\dot{W}_{part-load}}{\dot{W}_{full-load}} = 0.5(0.6 \cdot 0.5^2 + 0.4) = 0.275$$

case (c)

$$\frac{\dot{W}_{part-load}}{\dot{W}_{full-load}} = 0.5\left[0.666(0.6 \cdot 0.666^2 + 0.4)\right] +$$
$$0.5\left[0.333(0.6 \cdot 0.333^2 + 0.4)\right] = 0.3$$

case (d)

$$\frac{\dot{W}_{part-load}}{\dot{W}_{full-load}} = 0.5\left[0.833(0.6 \cdot 0.833^2 + 0.4)\right] +$$
$$0.5\left[0.166(0.6 \cdot 0.166^2 + 0.4)\right] = 0.375$$

and a mean of 0.36. Again taking into account the variation of efficiencies, the mean value would be similar to the one found previously. Even though other scenarios are possible, it is seen that predicting that the pumping energy will be reduced by $2^3 = 8$, as predicted by pump scaling laws, is optimistic. A very conservative approach would be to reduce the pumping energy by a factor of 2. Sfeir et al. [3] assume that the flow rate is constant during the whole period predicting, in that special case, a reduction by a factor of 3, factor which is highly dependent on the ratio of the heat pump to total head losses. ∎

8.5   Practical Evaluation of Pumping Energy

In the previous examples, a very detailed flow analysis was used to predict the hydraulic performance of some complex piping configurations. In practice, several simplifying assumptions are done, but rigorous analysis is helpful to verify them. As discussed briefly in the last chapter, the usual practice is to evaluate the longest path of the circuit estimating the largest head losses, and adjusting valves are used in the other paths to adjust the nominal flow rates. However in reverse-return, it is not straightforward to estimate the "longest path" as shown in the next example.

Example 8.9 Estimate the head losses for the following three-zone building. Two 6 tons heat pumps (16 gpm each) are used to condition zone North-East, one 5 tons (13 gpm) for the South-West zone, and one 5 tons (13 gpm) for the central zone. The total flow rate is then 58 gpm. The initial layout chosen is shown in the following figure:

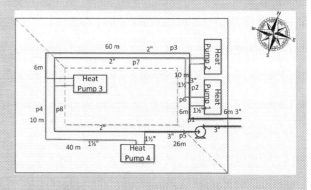

Besides the global connections shown on the figure, each heat pump is connected to the main two pipes by a 4 m, 1½" pipe connected to a 1¼" hoze kit, and Y-strainer, two ball valves, and a flow control valve to adjust for the correct flow rate. The heat pump pressure loss is 4.0 psi for heat pumps 1 and 2 and 3.4 psi for heat pumps 3 and 4. The minor loss calculations are done with the following assumptions:

Pipe	Diameter (in)	Flow rate (gpm)	Length (m)	Fittings
1	3	58	12	90°, tee-straight
2	3	42	10	tee-straight
3	2	26	66	2 × 90°, reducer, tee-straight
4	1.5	13	50	2 × 90°
5	3	58	26	
6	1.5	16	10	90°, tee-straight
7	2	32	66	90°, tee-straight
8	3	45	50	90°, tee-straight

We assume that water is the fluid used (properties at 5 °C).

b) If the ground head loss is estimated at 7.5 m and the wire-to-water efficiency is 0.5, estimate the electric power of the pump and the pumping power to heating capacity ratio.

(continued)

Example 8.9 (continued)

Solution As stated earlier, in practice the global network is not solved. In place, the path with the largest head losses is evaluated. It might be evident in some cases but not the case here, so each path will be examined. First let's evaluate the head losses in the heat pump connections.

Heat pumps 1 and 2:

$$\Delta P = 4.0 \text{ psi} = 27579 \text{ Pa}$$

$$h_{hp} = \frac{27579}{1000 \cdot 9.8} = 2.81 \text{ m}$$

$$\dot{V} = 16 \text{ gpm} = 1.0\text{e-}3 \text{ m}^3/\text{s} = 3.64 \text{ m}^3/\text{hr}$$

For the 4 m, 1 1/2 " SDR-11, $D = 0.039$ m:

$$A = \frac{\pi 0.039^2}{4} = 0.00122 \text{ m}^2$$

$$u = \frac{1.0\text{e-}4}{0.00122} = 0.82 \text{ m}/s$$

$$Re = \frac{0.82 \cdot 0.039}{1.52e - 6} = 21440 \Rightarrow f = 0.026$$

$$h_{pipe} = 0.026 \left(\frac{4}{0.039}\right) \frac{0.82^2}{2g} = 0.09 \text{ m}$$

From the last chapter, the hoze kit valve Cv is 29.4 m³/hr/bar$^{1/2}$:

$$\Delta p_{hoze} = 1e5 \left(\frac{3.64}{29.4}\right)^2 = 1528 \text{ Pa}$$

$$h_{hoze} = 0.155 \text{ m}$$

A similar calculation for the Y-strainer (Cv = 37.2) and ball valves (Cv = 40.6) will give:

$$h_{str} = 0.097 \text{ m}$$

$$h_{ballvalve} = 0.082 \text{ m}$$

Neglecting for now the flow control valve, we have:

$$h_{tot-hp1,2} = 2.81+0.090+0.156+0.0972\cdot0.081 = 3.32 \text{ m}$$

A similar calculation for heat pumps 3 and 4 will give:

$$h_{tot-hp3,4} = 2.39+0.063+0.103+0.064+2\cdot0.054 = 2.73 \text{ m}$$

The main network head losses are finally found as:

Pipe	L (m)	L_{eq} (m)	h (m)
1	12	12.1	0.285
2	10	2.1	0.081
3	66	10.7	1.401
4	50	6.6	0.888
5	26	0	0.308
6	10	4.5	0.327
7	66	4.9	1.866
8	50	12.1	0.470

Inside the building, the path with the largest head losses will be evaluated; in the other paths, flow control valves will increase the head losses to be equal to the largest one for each nominal flow rate. Since pipes 1 and 5 are common to each path, only the other ones will be used for the building:

Path1 (HP1-P6-P7-P8)

$$h_{path,1} = 3.32 + 0.327 + 1.866 + 0.470 = 5.98 \text{ m}$$

Path2 (P2-HP2-P7-P8)

$$h_{path,2} = 0.081 + 3.32 + 1.866 + 0.470 = 5.73 \text{ m}$$

Path3 (P2-P3-HP3-P8)

$$h_{path,3} = 0.081 + 1.401 + 2.73 + 0.470 = 4.67 \text{ m}$$

Path4 (P2-P3-P4-HP4)

$$h_{path,4} = 0.081 + 1.401 + 0.888 + 2.73 = 5.10 \text{ m}$$

$$h_{buiding} = 5.98 \text{ m}$$
$$h_{header} = h_{pipe1} + h_{pipe5} = 0.593 \text{ m}$$
$$h_{total} = 5.98 + 0.593 + 7.5 = 14.1 \text{ m}$$

b) The fluid pumping power is:

$$\dot{W}_{pump} = 3.66 \cdot (14.1) \cdot 9.8 = 505 \text{ W}$$

$$\dot{W}_{elec} = \frac{505}{0.5} = 1010 \text{ W}$$

The total capacity is 22 tons which gives:

$$H.C. = 77.3 \text{ kW}$$

from which:

$$ratio = \frac{1010}{77.3} = 13.1$$

∎

(continued)

Example 8.10 Ground source heat pumps are used to heat and cool the three zones. The building simulations that give the hourly loads associated with the building are given in the file "loads.txt," where the first two columns give the heating and cooling loads (in kW) for the first zone and the third and the fourth the loads for the second zone, etc. The ground temperature is 10 °C, its conductivity 2.75 W/m-K, and its diffusivity 0.1 m²/day. A 15 cm borehole diameter is suggested with normal grout ($k_g = 1$ W/m-K) and normal pipe separation ($x_c = r_b/3$). Estimate the size of the ground heat exchanger and the pumping energy for the building assuming three unitary independent systems.

Take 0 °C as the minimum EWT in heating mode and 25 °C as the maximum temperature in cooling mode. Use propylene glycol 20% as the heat transfer fluid in the ground. Use the ASHRAE method with a 20 years period and the line-source method for the borehole resistance.

Solution From the hourly loads, the maximum heat and cooling demand can be extracted for each zone. It is easily found that for the three zones, we have:

	Zone 1	Zone 2	Zone 3
max(heating) (kW)	29.4	9.6	19.6
max(cooling) (kW)	4.9	22	6.5
max(heating) (MBTU/hr)	100	33	67
max(cooling) (MBTU/hr)	17	76	22

As a first design, we will choose heat pumps to meet 100% of the load. From Appendix, two NV060 heat pumps were chosen for Zone 1 and two NV036 for Zones 2 and 3. The characteristics were evaluated at 0 °C (32 F) for heating and 25 °C (77 F) for cooling. The following results were obtained:

	Heat pump 1	Heat pump 2
gpm	13	8
$\dot{V}$ (l/s)	0.820	0.505
ΔP (psi)	3.40	2.59[a]
ΔP (kPa)	23.4	17.9
CAP(heating) (MBTU/hr)	60.1	37.5

	Heat pump 1	Heat pump 2
CAP(cooling) (MBTU/hr)	61.4	39.8
CAP(heating) (kW)	17.60	10.98
CAP(cooling) (kW)	17.96	11.66
COP (heating)	3.37	3.90
EER (cooling)	18.5	21.3
COP (cooling)	5.42	6.23

[a]Those values were taken for heating mode (0 °C) since it was the worst case.

Those values were given for water. Since propylene glycol is used, the following correction factors will be used as taken from the data sheet:

- Correction for heating capacity: 0.913
- Correction for cooling capacity: 0.969
- Correction for pressure losses: 1.27

The final results are:

	Zone 1	Zone 2	Zone 3
max(heating) (MBTU/hr)	100	33	67
Capacity (heating) (MBTU/hr)	109	68	68
max(cooling) (MBTU/hr)	17	76	22
Capacity(cooling) (MBTU/hr)	118	77	77

From the hourly loads on the building side and the COPs chosen, the ground loads are evaluated, and the five pulses used in the ASHRAE method are calculated.

	Zone 1	Zone 2	Zone 3
q_a (kW)	2.00	0.23	1.08
$q_{m,heating}$ (kW)	6.12	3.10	3.38
$q_{m,cooling}$ (kW)	−1.65	−3.54	−0.95
$q_{h,heating}$ (kW)	18.18	6.14	13.28
$q_{h,cooling}$ (kW)	−5.29	−25.42	−7.23

A preliminary choice for the bore field layout has to be chosen. Let's start with a 2x2 field for Zone 1 and 3x1 for the others with a 1" SDR-11 pipe. For system 1, the following results are obtained:

(continued)

(continued)

Example 8.10 (continued)

$$Fo_f = \frac{\alpha_s t_f}{r_b^2} = \frac{0.0042(175200 + 730 + 4)}{0.075^2} = 175934$$

$$Fo_1 = \frac{\alpha_s t_1}{r_b^2} = \frac{0.0042(734)}{0.075^2} = 545$$

$$Fo_2 = \frac{\alpha_s t_2}{r_b^2} = \frac{0.0042(4)}{0.075^2} = 2.96$$

$$G(Fo_f) = 1.0, G(Fo_1) = 0.567, G)Fo_2) = 0.185$$

$$R'_{g,a} = \frac{1.0 - 0.567}{2.75} = 0.1583 \text{ m-K/W}$$

$$R'_{g,m} = \frac{0.567 - 0.185}{2.75} = 0.1386 \text{ m-K/W}$$

$$R'_{g,h} = \frac{0.185}{2.75} = 0.067 \text{ m-K/W}$$

The borehole resistance is:

$$\dot{m}_{borehole} = \frac{1024 \cdot 2 \cdot 0.82}{4 \cdot 1000} = 0.42 \text{ kg/s}$$

$$Re = \frac{4\dot{m}}{\pi D_i \mu_i} = \frac{4 \cdot 0.42}{\pi 0.027 \cdot 0.0035} = 5586$$

$$Nu_D = 75 \Rightarrow h = 1290 \text{ W/m}^2 \text{ K}$$

$$R'_p = 0.09 \text{ K-m/W}$$

$$R'_g = 0.151 \text{ W/m-K} \Rightarrow R'_b = 0.151 + \frac{0.09}{2} = 0.196 \text{ W-m/K}$$

For heating we have:

$$Tf_i = 0 - \frac{18188}{1.68 \cdot 3937} = -2.74 \,°\text{C}$$

After iteration, the temperature penalty was evaluated as $T_p = 0.847$.

$$L_{system,1}$$

$$= \frac{2000 \cdot 0.1583 + 6120 \cdot 0.1386 + 18188(0.067 + 0.196)}{10 - (-1.37) - 0.847}$$

$$= 565 \text{ m}$$

$$H_{system,1} = \frac{565}{4} = 141 \text{ m}$$

Similar calculations lead to:

$$L_{system,2} = 398, \qquad H_{system,2} = \frac{397}{3} = 132 \text{ m}$$

$$L_{system,3} = 372, \qquad H_{system,3} = \frac{367}{3} = 124 \text{ m}$$

for a total of:

$$L_{|total} = 1334 \text{ m}$$

for the three zones. It would be possible to take into account the interference impact evaluating the effective resistance (Eq. 5.84), but in this specific case, the difference will be small and not taken into account in this example.

The head losses will be evaluated for system 1 only. The heat pump itself gives:

$$h_{h-p} = \frac{1.27 \cdot 23.4 \cdot 1000}{1024 \cdot 9.8} = 2.96 \text{ m}$$

Assuming the connections similar as in the previous example (with a 1.27 factor correction for glycol), we will have:

$$h_{h-p,sing} \approx 0.47 \Rightarrow h_{h-p,tot} = 3.44 \text{ m}$$

For the field, let's assume an arrangement similar to:

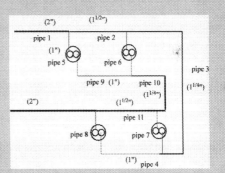

Neglecting minor losses except for the ground loop ($L_{eq} \approx 8.3$ m), the following head losses will be obtained for the path 1-2-3-4-8:

	Re	f	D(m)	L(m)	L_{eq}(m)	h(m)
Loop (8)	5596	0.037	0.027	282	8.3	9.89
1	12394	0.029	0.049	60	0	1.35
2	11620	0.030	0.039	6	0	0.24
3	8866	0.032	0.034	12	0	0.44
4	5596	0.037	0.027	6	0	0.20
Total					12.12	

for a total of:

(continued)

(continued)

Example 8.10 (continued)
$$h_{total} = 3.45 + 11.12 = 15.6 \text{ m}$$

Looking at the Bell https://www.esp-systemwize.com/, the E-90 pump with a 40% wire-to-water efficiency can be chosen.

$$\dot{W}_{system,1} = \frac{1.68 \cdot 9.8 \cdot 15.6}{0.4} = 642 \text{ W}$$

The pumping ratio for system 1 is:

$$\text{ratio}_{system,1} = \frac{642}{2 \cdot 17.25} = 18.4$$

which is correct, but that can be easily improved by changing the field configuration, increasing the number of boreholes, or changing the pipe diameter to $1^1/4$" in the borehole, for example, but we will keep it like this for now. For the two other systems, we have:

$$\eta_{pump,system,2} = 32\%, \quad \text{Bell \& Gossett E-90}$$

$$\eta_{pump,system,3} = 32\%, \quad \text{Bell \& Gossett E-90}$$

$$\dot{W}_{system,2} = \frac{1.03 \cdot 9.8 \cdot 12.57}{0.32} = 398 \text{ W}$$

$$\dot{W}_{system,3} = \frac{1.03 \cdot 9.8 \cdot 12.15}{0.32} = 385 \text{ W}$$

$$\text{ratio}_{system,2} = \frac{398}{2 \cdot 22.7} = 17.6$$

$$\text{ratio}_{system,3} = \frac{385}{2 \cdot 22.7} = 17.0$$

The electric energy consumed by the pumps will depend if they run all the time or not. The main advantage of unitary systems is the ease to stop the circulating pumps when the heat pumps are not working. As in Sfeir et al. [3], a run-time ratio will be defined as:

$$\gamma_{i,j} = \frac{q_{i,j}}{CAP_j} <= 1$$

for all time steps "i" and each system "j." CAP_j is the total capacity of all heat pumps associated with the zone analyzed. The total energy is then:

$$E_{pump,j} = \sum_{i=1}^{Nt} \dot{W}_{pump,j}\gamma_{i,j}\Delta t_i$$

In our example, the total energy is found to be:

(continued)

	Zone 1	Zone 2	Zone 3	Total
E_{pump} (kWh)	693	350	310	1352
$E_{compressor}$ (kWh)	10745	5020	4465	20230
Total				21583

So the pumps are responsible for 6.2% of the total energy, instead of 38%, if they would run continuously.
∎

Example 8.11 Redo the same analysis for a two-loop central configuration.

Solution A configuration similar to Example 8.8 is used with now two heat pumps for each zone:

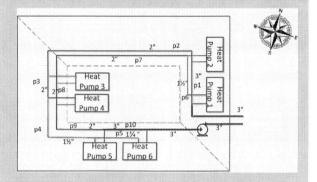

The main difference is now that part of the diversified loads will share heat in the common loop. Once the ground loads are calculated with the given COPs, they will be summed up for each hour, and the final five pulses found will be:

$$
\begin{aligned}
q_a &= 3.33 \text{ kW} \\
q_{m,heating} &= 12.61 \text{ kW} \\
q_{m,cooling} &= -6.14 \text{ kW} \\
q_{h,heating} &= 35.29 \text{ kW} \\
q_{h,cooling} &= -37.94 \text{ kW}
\end{aligned}
$$

The total flow rate is:

$$\dot{m} = 0.00365 \cdot 1024.7 = 3.75 \text{ kg/s}$$

In heating mode, the mean fluid temperature is:

$$Tf_i = 0 - \frac{35299}{3.75 \cdot 3937} = -2.39\,°C$$

(continued)

Example 8.11 (continued)
$$Tf = -1.19\,°C$$

A 4 x 3 field is initially chosen so that the flow rate per borehole is:

$$\dot{m}_{borehole} = \frac{3.75}{12} = 0.31 \text{ kg/s}$$

which will induce a borehole resistance similar than in the other configuration:

$$R'_b = 0.197 \text{ W-m/K}$$

After iteration a 0.94 K penalty temperature is found from which:

$$L = \frac{3328 \cdot 0.1583 + 12605 \cdot 0.1386 + 35298(0.067 + 0.197)}{10 - (-1.18) - 0.94}$$

$$= 1136 \text{ m}$$

$$H = \frac{1136}{12} = 95 \text{ m}$$

A 200 m reduction is observed in the total length of all boreholes which is the main benefit of a common loop system with diversified loads.

Head losses inside the building follow a similar path to that of Example 8.8 with a 1.2 factor correction for the use of glycol:

Heat Pumps	Flow rate (gpm)	L_{eq} (m)	h (m)
1	13	0.42	3.39
2	13	0.42	3.39
3	8	0.17	2.43
4	8	0.17	2.43
5	8	0.17	2.43
6	8	0.17	2.43

And for the inner piping, head losses are:

Pipe	Diameter (in)	Flow rate (gpm)	L (m)	L_{eq} (m)	h (m)
1	3	45	10	2.1	0.113
2	2	32	66	9.5	2.43
3	2	24	6	1.2	0.14
4	1.5	16	10	5.1	0.42
5	1.25	8	6	7.3	0.21
6	1.5	13	6	3.3	0.18
7	2	26	66	4.9	1.59
8	2	34	6	1.2	0.25
9	2	42	4.9	0.77	
10	3	50	6	2.1	0.09

From which, it is easy to find the path with the largest head losses:

Path1 (HP1-P6-P7-P8-P9-P10)

$$h_{path,1} = 3.41 + 0.18 + 1.59 + 0.25 + 0.77 + 0.09 = 6.3 \text{ m} = h_{building}$$

Assuming as in the unitary configuration a 60 m (round-trip) length, a header pipe of 3" with a 90 elbow will give:

$$h_{header} = 1.04 \text{ m}$$

Finally the losses in field will be associated with the configuration chosen. Assuming a case similar to the following figure:

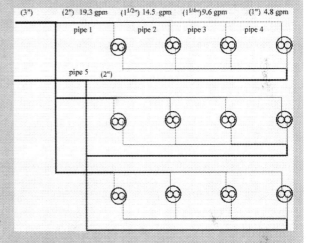

neglecting minor losses except for the borehole itself, we get:

	Re	D(m)	L(m)	L_{eq} (m)	h(m)
Loop	4161	0.027	189	8.3	4.08
1	9216	0.049	6	0	0.08
2	8640	0.039	6	0	0.14
3	6593	0.034	6	0	0.13
4	4161	0.027	6	0	0.12
5	9216	0.049	24	0	0.32
Total					4.89

$$h_{total} = h_{building} + h_{header} + h_{field} = 12.2 \text{ m}$$

$$\dot{W}_{fluid} = 3.75 \cdot 9.81 \cdot 12.2 = 449 \text{ W}$$

Again using the selection tools of Bell and Gosset, the same pump E-90 can be chosen but will work now with a 53% efficiency given an electric power of:

$$\dot{W}_{elec} = \frac{449}{0.53} = 847 \text{ W}$$

(continued)

(continued)

Example 8.11 (continued)

$$\text{ratio} = \frac{847}{80} = 10.6$$

It is seen that the better efficiency will lead to a lower maximum electric power. However, as explained previously, the evaluation of the pumping energy at part load is less trivial than in the independent systems. It is not rare to see the pumps working continuously, in which case the total energy will be:

$$E_{pump} = 7423 \text{ kWh}$$

$$E_{compressors} = 20230 \text{ kWh}$$

where the pumping energy represents now 27% of the total energy even with an apparent efficient design. A major reduction can be expected using a VFD drive[a], but as stated in the previous sections, the exact impact is hard to estimate. A possible solution is to follow a similar path to Example 8.2. From the previous calculation, let's assume that a 1.5 HP motor is used. From Eq. (8.5) we have:

$$\eta_{vfd,full\ charge} = 0.94$$

$$\dot{W}_{elec,full\ charge} = \frac{840}{0.94} = 895 \text{ W}$$

The full charge shaft power will be estimated using the Bernier-Lemire relation (Eq. (8.1)). The fraction of head loss in the heat pump can be estimated as:

$$X = \frac{h_{building}}{h_{total}} = 0.52$$

$$\dot{W}_{shaft,nom} = \frac{895}{Z(1,X)} = 801 \text{ W}$$

and for each hour of the year, evaluate the electric power from:

$$\dot{W}_{elec,i} = \dot{W}_{shaft,nom} Z(\tilde{V}_i, X)$$

with:

$$\tilde{V}_i = \max(\frac{\sum_j^{nzones} \gamma_{i,j} \dot{V}_j}{\dot{V}_{total}}, \tilde{V}_{min})$$

where $\tilde{V}_{min}$ is a minimum flow rate which is often set when using a variable speed system. If a 30% value is taken here, the final result will be:

----

[a]The pump chosen has an ECM motor with variable speed, but it will be treated as a single-speed induction motor for the purpose of the example

$$E_{pump} = 2048 \text{ kWh}$$

$$E_{total} = 22279 \text{ kWh}$$

which will represent now 9% of the total energy, a much better result. Another possibility is to follow a similar path to Sfeir et al. [3], and for each hour evaluate:

$$h_{total,i} = h_{total}(\tilde{V}_i(1 - X)\tilde{V}_i^2 + X)$$

$$\eta_{vfd,i} = \eta_{vfd}(HP, \tilde{V}_i), \qquad \text{Eq. (8.5)}$$

$$\dot{W}_{elec,i} = \frac{\tilde{V}_i \dot{m} g h_{total,i}}{\eta_{w-w} \eta_{vfd,i}}$$

and sum it for each hour of the year, in which case the final results are:

$$E_{pump} = 1469 \text{ kWh}$$

$$E_{total} = 21699 \text{ kWh}$$

which will represent now 7% of the total energy. The difference might be that the variable frequency drive efficiencies given by Eq. (8.5) are probably better than the one used by Lemire and Bernier.

Comments   *The goal of this lengthy example was to outline the different analysis tools, not to give general conclusions on the benefits of different configurations. Arbitrary configurations were chosen, and no "optimization" was performed, but general observations can be made: The advantages of the central loops appear when diversified loads are present which is the case here. The advantage of the unitary system is the ease to control the pump which is shown here in the greater reduction of pumping energy although the specific efficiencies were lower. The final decision is highly case dependent, and economic arguments would be involved.*

∎

Pr. 8.1 Three heat pumps are connected with a two-pipe arrangement as shown in Fig. Pr_8.1:

	Heating Capacity (x 10^3 BTU/hr)	Flow rate (gpm)	Δ p (psi)	COP (heating)
Heat pump 1	110	30	1.6	3
Heat pump 2	90	24	1.3	3
Heat pump 3	75	18	1.0	3

(continued)

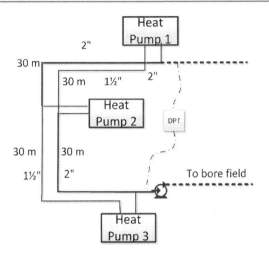

Fig. Pr_8.1 Three heat pumps in a two-pipes reverse-return configuration

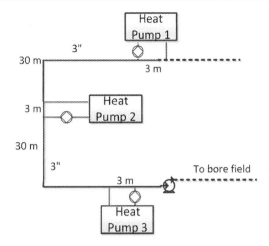

Fig. Pr_8.2 Three heat pumps in a one-pipe configuration

Each heat pump is equipped with an automatic flow-limiting valves (AFLV) in order to maintain the desired flow rate in case of pressure fluctuations. The minimum pressure loss of the AFLV is 3.5 ft of water. The other minor losses are neglected. The bore field pressure losses are estimated assuming a 350 m circuit with a 300 Pa/m head losses. The flow rate is given by a pump whose head curve is given by:

$$h(m) = 20.34 + 0.198\dot{V} - 0.179\dot{V}^2 - 0.0061\dot{V}^3 \quad (\dot{V} \text{ in l/s})$$

$$\eta_{w-w}(\%) = 17.96 + 26.28\dot{V} - 4.07\dot{V}^2 - 0.157\dot{V}^3 \quad (\dot{V} \text{ in l/s})$$

(a) Evaluate the total head losses and pumping electric power at full load. Fluid is water ($\rho = 1000$ kg/m^3, $\mu = 0.0016$ Pa-s, $C_p = 4.2$ kJ/kg-K), and the pipes are smooth.
(b) Estimate the total head losses and pumping electric power when the first heat pump is shut off. The head losses in the field are estimated as:

$$h_{field} = h_{field,nom}\left(\frac{\dot{V}}{\dot{V}_{nom}}\right)^2$$

(c) Estimate the total head losses and pumping electric power when the first heat pump is shut off and if the speed is reduced in order to maintain the pressure differential equal than at full load.

Answer: (a) 17 m, 1111 W, (b) 19.5 m, 818 W, (c) 9.9 m, 383 W

Pr. 8.2 A one-pipe configuration is used in the same building. The layout looks like that in Fig. Pr_8.2. No AFLV are now used, and the other losses from the remaining valves

are again assumed to be included in the heat pump pressure losses.

(a) Estimate the fluid pumping energy and electric pumping energy when all heat pumps are working assuming the same total flow rate (and head losses) in the field than in the previous problem. The pump curve for the field is given by the following equation:

$$h(m) = 15.5 + 0.177\dot{V} - 0.183\dot{V}^2 - 0.0074\dot{V}^3 \quad (\dot{V} \text{ in l/s})$$

$$\eta_{w-w}(\%) = 17.96 + 30.3\dot{V} - 5.45\dot{V}^2 - 0.243\dot{V}^3 \quad (\dot{V} \text{ in l/s})$$

and the heat pump circulators efficiencies are given by Eq. (7.36b).
(b) What is the temperature difference between the inlet temperature at heat pumps 1 and 3?
(c) Redo the same calculation when heat pump 1 is shut off. Use the same estimation to that in the previous problem for the losses in the field.

Answer: (a) 570.8 W, 898.8 W; (b) –4.22 K (c) 549.3 W, 851.7 W

References

1. American Society of Heating, Refrigerating and Air-Conditioning Engineers: ASHRAE Handbook, HVAC Applications. The Society, New York (2015)
2. Kavanaugh, J.: Long-term commercial GSHP performance: part 1: project overview and loop circuit types. ASHRAE J. **54**(6), 48 (2012)
3. Sfeir, A. et al.: A methodology to evaluate pumping energy consumption in GCHP systems. ASHRAE Trans. **111**(1), 714–729 (2005)
4. Chiasson, A.D.: Geothermal Heat Pump and Heat Engine Systems: Theory and Practice. Wiley, New York (2016)

5. Mescher, K.: Simplified GCHP system. ASHRAE J. **51**(10), 24–34 (2009)
6. Mays, C.J.: Ground-coupled heat pump systems: a pumping analysis. Master's Thesis. Kansas State University, 2012
7. Bernier, M., Lemire, N.: Non-dimensional pumping power curves for water loop heat pump systems. ASHRAE Trans. **105**, 1226 (1999)
8. Wood, W.R.: Beware of pitfalls when applying variable-frequency drives. Power (United States) **131**(2) (1987)
9. Gagné-Boisvert, L., Bernier, M.: Integrated model for comparison of one-and two-pipe ground-coupled heat pump network configurations. Sci. Technol. Built Environ. **24**(7), 726–742 (2018)
10. Taylor, S.T., Stein, J.: Balancing variable ow hydronic systems. ASHRAE J. **44**(10), 17–20 (2002)
11. Del Col, D. et al.: Energy efficiency in a ground source heat pump with variable speed drives. Energy Build. **91**, 105–114 (2015)
12. Madani, H., Claesson, J., Lundqvist, P.: Capacity control in ground source heat pump systems: part I: modeling and simulation. Int. J. Refrig. **34**(6), 1338–1347 (2011)
13. Madani, H., Claesson, J., Lundqvist, P.: Capacity control in ground source heat pump systems part II: comparative analysis between on/off controlled and variable capacity systems. Int. J. Refrig. **34**(8), 1934–1942 (2011)
14. Climate Master, Tranquility Series. https://www.climatemaster.com/download/18.274be999165850ccd5b5b73/1535543867815/lc377-climatemaster-commercial-tranquility-20-single-stage-ts-series-water-source-heat-pump-submittal-set.pdf. Accessed 17 July 2021
15. Kavanaugh, S.P., Lambert, S.E.: A bin method energy analysis for ground-coupled heat pumps. ASHRAE Trans. **110**, 535 (2004)
16. Krukowski, A.: Standardizing data for VFD efficiency. ASHRAE J. **55**(6), 16 (2013)
17. EnergyPlus Documentation, Engineering Reference. https://energyplus.net/sites/default/files/pdfsv8.3.0/EngineeringReference.pdf. Accessed 26 August 2019
18. The MotorMaster+ Software Tool. https://www.energy.gov/sites/prod/files/2014/04/f15/motormaster_brochure.pdf. Accessed 26 Aug 2019

9.1 Introduction

In the previous chapters, several methods were presented to analyze and design geoexchange systems. In these expressions, thermal properties of the ground like the conductivity, the diffusivity, and the undisturbed subsurface temperature are of utmost importance. Even though typical soil formation where ground heat exchangers are installed can be highly non-homogeneous, most of the analysis tools used to simulate or design ground heat exchangers use an average effective value of these parameters. The most important property that has to be known with a relatively good accuracy is the thermal conductivity.

9.2 Effective Conductivity

As already stated, this property relates the thermal gradient to the heat flux:

$$q'' = -k\nabla T \qquad (9.1)$$

The thermal diffusivity α [m²/s] is associated to the unsteady behavior of the medium. It is related to the conductivity from its definition:

$$\alpha = \frac{k}{\rho C_p} = \frac{k}{\rho C_p} = \frac{k}{C} \qquad (9.2)$$

where $C = \rho C_p$ [J/m³-K] is the *volumetric heat capacity*. From this definition, it can be seen that from the knowledge of the conductivity, the diffusivity can be assessed from knowledge of the volumetric heat capacity. Since this property has a smaller variability [1], the diffusivity can be known with a certain confidence level if the conductivity is known.

Thermal conductivity cannot be measured explicitly. It is evaluated from indirect measurements of temperature and heat flux. For example, in a one-dimensional steady-state heat transfer in a plane geometry, the temperature profile is linear, and Eq. (9.1) becomes:

$$q'' = k\frac{T_1 - T_2}{L} \qquad (9.3)$$

where T_1 and T_2 are the surface temperature of the sample. Standard laboratory procedures using this principle on guarded hot plate apparatus [2, 3] can be used to evaluate the thermal conductivity of samples of rock that can be collected on site. In an inhomogeneous medium, several samples will have different conductivity, and some kind of averaging has to be done to evaluate the effective conductivity. The choice is however not unique. A simple example is shown in Fig. 9.1.

The figure shows a sample with two different conductivities. In case a, the effective conductivity is:

$$q'' = k_1\frac{T_1 - T_2}{L_1} = k_2\frac{T_2 - T_3}{L_2} = \frac{T_1 - T_3}{L_1/k_1 + L_2/k_2} = k_{eff}\frac{T_1 - T_3}{L_1 + L_2}$$

From which, the effective conductivity is given by:

$$\frac{1}{k_{eff}} = \frac{x_1}{k_1} + \frac{x_2}{k_2} \qquad (9.4)$$

with $x_i = L_i/L$.

In case b, the heat flux will be given by:

$$q = A_1 k_1 \frac{T_1 - T_2}{L_3} + A_2 k_2 \frac{T_1 - T_2}{L_3}$$

$$q'' = \frac{q}{(A_1 + A_2)} = k_{eff}\frac{T_1 - T_2}{L_3}$$

where:

$$k_{eff} = x_1 k_1 + x_2 k_2 \qquad (9.5)$$

In each case, x_i can be replaced by:

$$x_i = V_i/V \qquad (9.6)$$

L. Lamarche, *Fundamentals of Geothermal Heat Pump Systems*,
https://doi.org/10.1007/978-3-031-32176-4_9

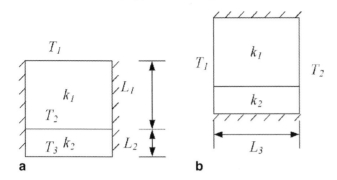

Fig. 9.1 (**a**) Series. (**b**) Parallel

the *volumetric component fraction*. These two values are just two possibilities of extreme cases of an inhomogeneous medium and lead to the maximum and minimum numerical values. They can be generalized as the *harmonic average value*:

$$k_{eff,har} = \left(\sum \frac{x_i}{k_i} \right)^{-1} = k_{eff,minimum} \qquad (9.7)$$

for the first case, and the *arithmetic average value*:

$$k_{eff,ari} = \sum x_i k_i = k_{eff,maximum} \qquad (9.8)$$

for the second one. Many others possibilities for effective conductivity are proposed in the literature [4]. One that is often used for homogeneous mixture [5] is the *geometric average value* defined as:

$$k_{eff,geo} = \prod k_i^{x_i} \qquad (9.9)$$

In the special case where there is two components and that $x_1 = x_2 = 0.5$, the geometric average gives:

$$k_{eff,geo} = \sqrt{k_{eff,ari} k_{eff,har}}$$

9.3   In Situ Measurements

Measuring the effective conductivity from rock samples is not unique as it was pointed out in the previous section, and it depends not only on rock samples but also on how the heat is propagated into the media. It would be interesting to have a better idea on that conductivity, and one method that is often used is to measure the effective conductivity in situ using a *thermal response test* (TRT).

The concept of measuring the thermal conductivity for GSHP applications is relatively new and has been proposed by Mogensen [6] in the 1980s in Sweden. The idea is quite simple. It is proposed to exchange a constant amount of heat with the ground. In Chap. 4, it was seen that in the case of a constant pulse of heat, the mean fluid temperature was given by:

$$T_f = T_o - q'_b \left(\frac{G(Fo)}{k_s} + R'_b \right) \qquad (9.10)$$

where G can be any of the thermal response factor described in Chap. 4. The idea is then to measure the mean fluid temperature and relate it to the conductivity which in that case represents the effective conductivity, at least in the vicinity of the tested borehole.

9.4   Slope Method

One possible approach is to use the infinite line source model in the previous expression. If the Fourier number is large enough, it can be replaced by its simplified expression (Eq. 4.14):

$$T_f = T_o - q'_b \left[\frac{1}{4\pi k_s} \left(\log \left(\frac{4\alpha t}{r_b^2} \right) - \gamma \right) + R'_b \right]$$

From the properties of the logarithms, it becomes:

$$T_f = T_o - q'_b \left[\frac{\log(t)}{4\pi k_s} + \frac{\log \left(\frac{4\alpha}{r_b^2} \right) - \gamma}{4\pi k_s} + R'_b \right]$$

The minus sign in the last expression comes from the sign convention which was chosen as positive for heating applications, that is, for heat extraction. Although not mandatory, most of the time, heat is injected in the ground during a TRT, which represents a negative heat flux value in our convention. For that reason, instead of keeping too many negative signs, we will invert the sign convention and use the heat factor as the injection heat into the ground in this chapter:

$$T_f = T_o + q'_{inj} \left[\frac{\log(t)}{4\pi k_s} + \frac{\log \left(\frac{4\alpha}{r_b^2} \right) - \gamma}{4\pi k_s} + R'_b \right] \qquad (9.11)$$

with $q'_{inj} = -q'_b$. Looking at the previous expression, it is seen that if the mean fluid temperature is measured and plotted with respect to $\log(t)$, a straight line will be obtained:

$$T_f = mx + b \qquad (9.12)$$

where $x = \log(t)$:

$$m = \frac{q'_{inj}}{4\pi k_s} \qquad (9.13)$$

and

$$b = T_o + q'_{inj} \left[\frac{\log \left(\frac{4\alpha}{r_b^2} \right) - \gamma}{4\pi k_s} + R'_b \right] \qquad (9.14)$$

From these two expressions, it is found that:

$$k_s = \frac{q'_{inj}}{4\pi m} \qquad (9.15)$$

and

$$R'_b = \frac{b - T_o}{q'_{inj}} - \frac{\log\left(\frac{4\alpha}{r_b^2}\right) - \gamma}{4\pi k_s} \qquad (9.16)$$

In (9.15), the unit of time is not important since a change in time units will give the same slope and produce only a vertical shift of the curve. However, in Eq. (9.16), it is very important that the diffusivity used is compatible with the time unit to evaluate the intercept b.

As pointed out in Chap. 4, the mean fluid temperature is not always uniform, but most of the time, it is measured at the inlet and outlet of the borehole. The impact of this choice will be analyzed later. So a possible choice for this temperature is:

$$T_f = \frac{T_{fi} + T_{fo}}{2} \qquad (9.17)$$

9.5 Test Procedure

A typical thermal response is simple but needs to respect several constraints in order to reach the minimum accuracy. Most of the time, the heat is produced by an electric heater. The whole system is often located in a small trailer and has the following components:

1. Heating elements
2. Control system to maintain the heat constant
3. A circulation pump
4. Temperature sensors
5. A flow rate sensor
6. An air purge system
7. A data acquisition equipment

The first step is to measure the undisturbed ground temperature. This step can be done in two different ways:

(a) Using a thermal probe to measure the temperature profile along the borehole.
(b) After activating the circulation pump, the mean temperature is measured without heat injection.

With the first approach, it is possible to have a profile of temperature which might influence the GSHP system design particularly if the boreholes are long and/or if the temperature gradient is important. The second approach only gives a mean temperature along the borehole. The measure can also be influenced by the heat generated from fluid friction. However, it is simpler to measure the temperature this way than with the first approach. This measurement has to be done several days after the borehole is drilled since the drilling will induce a non-negligible temperature increase.

Once the undisturbed temperature is measured, the test can start. The ASHRAE Handbook [7] proposes several suggestions for TRT's testing:

1. Temperature sensor accuracy has to better than ± 0.3 K.
2. Flow rate must be adjusted to reach a ΔT of 4–7 K.
3. A waiting period of 3 (k > 1.7 W/m K) to 5 (k < 1.7 W/m K) days has to be followed between the drilling and the test.
4. The TRT should last at least 36–48 hours.
5. The heat injection should be between 50 and 80 W/m.
6. The standard deviation of the fluctuations of inlet power should be within 1.5% of the mean power value.

Example 9.1 The following figure gives the mean fluid temperature versus the logarithm of time (time in hours) for a particular thermal response test. The heat injection rate was 55.4 W/m. The borehole diameter was 15 cm. The undisturbed temperature was 10.4 °C.

(a) Find the thermal conductivity of the soil.
(b) Find the borehole resistance. In that case, the volumetric heat capacity is assumed to be in the order of 2.5 MJ/m³ K.

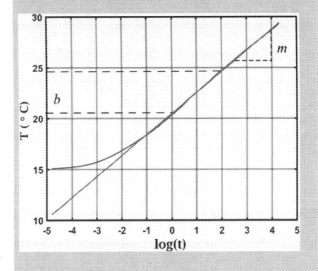

(continued)

Example 9.1 (continued)
Solution

(a) The solution is done graphically. The slope on the curve is given by:

$$m = \frac{24.5 - 20.5}{2} = 2K, \quad b = 20.5\,°C$$

$$k_s = \frac{55.4}{4\pi 2} = 2.2 \text{ W/m-K}$$

(b) The borehole resistance is found from (9.16).

$$\alpha = \frac{2.2}{2.5e6} = 8.8\text{e-}7 \text{m}^2/\text{s}$$

Since the time is in hours, it should be replaced by:

$$\alpha = 8.8\text{e-}7 \times 3600 = 0.00317 \text{m}^2/\text{hr}$$

Finally:

$$R'_b = \frac{20.5 - 10.4}{55.4} - \frac{\log\left(\frac{4 \cdot 0.00317}{0.075^2}\right) - 0.5772157}{4\pi 2.2} = 0.173 \text{ K-m/W}$$

∎

9.5.1 Parameter Estimation from Regression

In practical applications, the analysis will be done from computer using regression analysis. One has to be sure that the correct region is taken for linear regression and not use the early points where the linear approximation is not valid. Different options are possible. Knowing that the linear approximation is valid for large Fourier number, we have:

$$Fo = \frac{\alpha t}{r_b^2} > 5 \tag{9.18}$$

One problem is that the diffusivity is unknown. It is possible however to assume a typical value. In the last example we could have, for example:

$$\alpha \approx 0.004 \text{ m}^2/\text{hr} \qquad \text{Hyp}$$

$$t > \frac{5 \cdot 0.075^2}{0.004} > 7 \text{ hr}, \qquad \log(t) > 1.95$$

which is in the linear region in the figure.

Another possibility is to start with a certain number of points to do the regression and to stop when the conductivity evaluation does not change too much.

Example 9.2 The same problem as in the previous example is analyzed from linear regression; the field measurements are given in the file "TRT_test1.txt."

Solution

(a) In the first approach, the points where the Fourier number is too small are taken out. In the example an assumed value of conductivity of 3.0 W/m² K is used.

$$\alpha_{assumed} \approx \frac{3 \cdot 3600}{2.5e6} = 0.0042 \text{ m}^2/\text{hr}$$

Since the Fourier number should be higher than the critical value of 5, an arbitrary value of 6 is used to take into account the uncertainty on the diffusivity.

$$Fo > 5 \rightarrow Fo_{min} \approx 6 \rightarrow t_{min} = \frac{6 \cdot 0.075}{0.0042} = 7.8 \text{ hr}$$

From the data in the file, the 72 first points are taken out. From the rest of the points, the linear regression gives:

$$m = 1.968, \qquad b = 20.837\,°C$$

From Eqs. (9.15), (9.16), we have:

$$k_s = 2.24 \text{ W/m-K}, \qquad R'_b = 0.180 \text{ K-m/W}$$

(b) In the second approach, several regressions are done. In each case, more points are taken out of the regression. At the end a minimum number of points are kept. An example of a portion of a Python program is given as follows:

```
# second approach
n1 = 10 # Minimum number of points to take out
nf = 50 # Minimum number of points to keep for
          the regression
nte = nt - nf - n1 + 1
#
ksol = zeros(nte)
Rb = zeros(nte)
for i in range(0,nte):
    j1 = n1+i
    x = log(t[j1:nt])
    y = Tc[j1:nt]
    p = polyfit(x,y,1)
    m = p[0]
```

(continued)

Example 9.2 (continued)

    b = p[1]
    ksol[i] = qp/(4.0*pi*m)
    alhr = ksol[i]/CC*3600.0 # m2/hr
    Rb[i] = (b-To)/qp - (log(4*alhr/rb**2)
gam)/(4*pi*ksol[i])
# The mean from the nlast evaluations
nlast = 20
ks = mean(ksol[nte-nlast:nte-1])
Rbn = mean(Rb[nte-nlast:nte-1])

In the following figure the conductivity estimated from the slope is plotted against the number of points removed. It is seen that the final result is quite sensible to the number of points kept for the regression. However, from approximately 100 points removed, the value reaches an almost constant value. So this value can be accepted with a certain confidence. This value corresponds to a Fourier number of approximately 9 which is higher than the one chosen in the first approach. Keeping the average value of the last regressions will give finally:

$$k_s = 2.26 \text{ W/m-K}, \qquad R'_b = 0.181 \text{ K-m/W}$$

(continued)

which is very similar to the first approach.

In the last example, if too many points are removed, the regression does not give accurate results, and situations like the one shown in Fig. 9.2 can be obtained.

Unfortunately, thermal response tests do not always give confident regions, and sometimes, the experimentalist has to

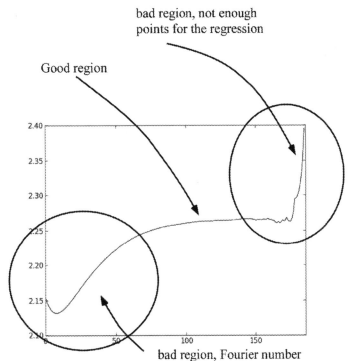

Fig. 9.2 Accuracy of regression with respect to number of points

make some sound decisions about how many points he should keep for the analysis.

One of the advantages of the slope method is to decouple the unknown parameters. For example, if the volumetric heat capacity is unknown, it should, in principle, not influence the thermal conductivity. The same is true for the initial ground temperature. This is not the case however for the borehole resistance evaluation.

The slope thus obtained on a plot showing the fluid temperature T_f with respect to the response factor G will then give the thermal conductivity and the intercept, the thermal resistance. The model used can still be the infinite line source but can be other one like the infinite cylindrical source or any other. Since these models are not valid for a short time, the early points in the TRT's experiment have still to be removed before the analysis is done.

Example 9.3 Redo the previous example with the first approach for:

(a) If the volumetric heat capacity varies of $\pm 10\%$
(b) If the undisturbed ground temperature is known at ± 0.5 K

Solution In theory, the assumed capacity nor the soil temperature should influence the slope so the conductivity is the same. In practice, since the number of points kept in the regression depends on the minimum Fourier number which is slightly dependent on the capacity, a small change can be observed, but this change is of course artificial. In this example, the first 140 points were removed, based on the previous problem. However a variation in the assumed capacity or initial temperature leads to a variation in the inferred borehole resistance. The results are:

	Estimated resistance m-K/W		
	$(\rho C_p) =$ 2.25 MJ/m^{3}K	$(\rho C_p) =$ 2.5 MJ/m^{3}K	$(\rho C_p) =$ 2.75 MJ/m^{3}K
To = 9.9 °C	0.186(3.0%)	0.177 (2.0%)	0.168(7.0%)
To = 10.4 °C	0.190 (5.0%)	0.181 (0%)	0.172 (5.0%)
To = 10.9 °C	0.193 (6.9%)	0.184 (1.9%)	0.175 (3.1%)

∎

9.5.2 Estimation Using Exact Models

In the so called "slope method," the thermal response was modeled using the logarithmic approximation of the infinite line source model that produces a simple method to estimate the parameters by hand. Using a computer for regression analysis, it is not mandatory to use the logarithmic approximation anymore. Equation (9.10) can be rewritten:

$$T_f = \underbrace{\frac{q'_{inj}}{k_s} G(Fo)}_{x} + \underbrace{T_o + q'_{inj} R'_b}_{b} \qquad (9.19)$$

$$= mx + b \qquad (9.20)$$

Example 9.4 Redo the last example, using:

(a) The ILS model
(b) The ICS model
(c) In either case, estimate the impact of an uncertainty of $\pm 10\%$ on the volumetric heat capacity and on the borehole radius.

Solution The goal of the problem is to see the sensitivity of the result to some unknown parameters. However, one of the factors that has the most influence is the number of points rejected before the analysis. From the last examples, the first 140 points are rejected to do the comparison.

Unlike the slope method, the solution needs the knowledge of the thermal capacity, not only for the borehole resistance evaluation but also for the conductivity calculation since the Fourier number has to be evaluated.

Nine different estimations are done for the three values of the volumetric capacity and the three values of radius. For each calculation, an iterative calculation is done since the method asks now for the diffusivity which depends on the unknown conductivity. Starting with an estimated value, the solution is redone until the new conductivity is "similar" to the old one. The results are compared with the values found in Example 9.2 which serve as reference (k = 2.265 W/m-K, $R'_b = 0.181$ K-m/W). The nine results found for the conductivity using the ILS model are:

(a)

	Estimated conductivity W/m-K		
	$(\rho C_p) =$ 2.25 MJ/m^{3}K	$(\rho C_p) =$ 2.5 MJ/m^{3}K	$(\rho C_p) =$ 2.75 MJ/m^{3}K
$r_b = 6.75$ cm	2.250 (0.6%)	2.247 (0.8%)	2.243 (0.9%)
$r_b = 7.5$ cm	2.249 (0.7%)	2.245 (0.9%)	2.241 (1.1%)
$r_b = 8.25$ cm	2.247 (0.8%)	2.243 (1.0%)	2.239 (1.2%)

The differences between the results and the reference values are given in parenthesis. For example, the value found in the middle of the Table shows 0.9% using the nominal values of the heat capacity and the

(continued)

Example 9.4 (continued)

borehole radius. Since the same number of points is used in the regression, the difference comes only on the model used, and, although the value found from the slope method is used as reference, it does not mean that it is more accurate. This difference is similar than the difference due to a $\pm 10\%$ variations of the two other unknown parameters. These parameters will however have a larger impact on the borehole resistance as expected from the last example:

	Estimated resistance m-K/W		
	$(\rho C_p) =$ 2.25 MJ/m³K	$(\rho C_p) =$ 2.5 MJ/m³K	$(\rho C_p) =$ 2.75 MJ/m³K
$r_b = 6.75$ cm	0.169 (6.7%)	0.176 (2.7%)	0.182 (0.9%)
$r_b = 7.5$ cm	0.172 (4.7%)	0.179 (0.7%)	0.186 (2.9%)
$r_b = 8.25$ cm	0.175 (2.9%)	0.183 (1.1%)	0.189 (4.7%)

(b) Using the ICS model, the following results are obtained:

	Estimated conductivity W/m-K		
	$(\rho C_p) =$ 2.25 MJ/m³K	$(\rho C_p) =$ 2.5 MJ/m³K	$(\rho C_p) =$ 2.75 MJ/m³K
$r_b = 6.75$ cm	2.140 (5.5%)	2.118 (6.5%)	2.096 (7.4%)
$r_b = 7.5$ cm	2.129 (6.0%)	2.107 (7.0%)	2.083 (8.0%)
$r_b = 8.25$ cm	2.119 (6.4%)	2.095 (7.5%)	2.070 (8.6%)

	Estimated resistance m-K/W		
	$(\rho C_p) =$ 2.25 MJ/m³K	$(\rho C_p) =$ 2.5 MJ/m³K	$(\rho C_p) = 2.75$ MJ/m³K
$r_b = 6.75$ cm	0.160 (11.5%)	0.166 (8.2%)	0.171 (5.2%)
$r_b = 7.5$ cm	0.163 (9.8%)	0.169 (6.5%)	0.174 (3.6%)
$r_b = 8.25$ cm	0.166 (8.3%)	0.172 (5.0%)	0.177 (2.1%)

Comments *It is observed that the model used to predict the thermal response is, by far, the most sensitive factor for the evaluation of the borehole parameters. Too many significant points are kept on purpose in the Table to show that there are small differences, but in many cases, this difference would have no significance with respect to the expected accuracy. Again, the calculation of the borehole resistance is more sensitive to uncertain parameters. The increased variations observed for the ICS is of course artificial since the reference values were taken from the slope method which is based on the ILS model.*

■

9.6 Estimation of Parameters

The slope method is simple to apply and has the advantage to decouple the ground conductivity from other parameters. It is however associated to simple models. Several authors [8–10] have suggested to use optimization methods to estimate the unknown parameters like the subsurface conductivity and the borehole resistance and also other parameters like the volumetric heat capacity. The principle is simple: a model having the unknown parameters as input is used to evaluate the mean fluid temperature with respect to time, and the following objective function is minimized:

$$f = \sum_{i=1}^{n} (T_{measured} - T_{calculated})^2 \qquad (9.21)$$

or its equivalent form:

$$err_{rms} = \sqrt{\frac{1}{n} \sum_{i=1}^{n} (T_{measured} - T_{calculated})^2} \qquad (9.22)$$

In this approach, different models can be used; some [11] can take into account the short-time behavior of the borehole allowing the user to keep the early points of the TRT. Some models can take into account the ground water effects in the ground [9]. If the ILS or the ICS are used, the measurements associated with small Fourier numbers still have to be removed since the experimental points should be in the range of validity of the model.

Example 9.5 Redo example 9.4, with the parameter estimation approach using only the ILS model.

Solution The approach is almost the same as in the previous example; the only difference would be that, instead of using the *polyfit* function to find the slope of the curve, the *curve_fit* function is used to minimize (9.22). This function is used to minimize least-square problems using different algorithms. By default, the *Levenberg-Marquardt* [12] is used. Part of the solution will look as:

```
cas = 'a'
if cas == 'a':
        G_vect = vectorize(G_function_ils) # ILS
                                  model is used
else:
        G_vect = vectorize(G_function_ics) # ICS
                                  model is used
```

(continued)

Example 9.5 (continued)

```
def T_theo(x, knew,Rbnew):
        return To + qp*(x/knew + Rbnew)
Rb = Rb_nom
ks = ks_nom
for ic in range(0,3):
        CC = CCv[ic]
        for ir in range(0,3):
                rb = rbv[ir]
                ok = False
                compt = 0
                po = [ks,Rb] # initial guess
                while not ok:
                        als = ks/CC
                        alhr = als*3600
                        Fo = alhr*ti/rb**2
                        x = G_vect(Fo)
                        params,resn = curve_fit
                                (T_theo,x,y,po)
                        ksn = params[0]
                        Rbn = params[1]
                        if (abs(ksn-ks)/ks) < 0.001 and
                                (abs(Rbn-Rb)/Rb) < 0.001:
                                ok = True
                        else:
                                ks = ksn
                                Rb = Rbn
                                compt = compt + 1
                                if compt > compt_max:
                                        err = 1
                                        ok = True
                ksv[ic,ir] = ks
                Rbv[ic,ir] = Rb
```

Since the same model as previously is used and the same number points are kept for the analysis, it is expected that very similar results than the ones in the previous example would be found. Again the flexibility of the parameter estimation approach is to be generalized to any other models, either analytical or numerical. The final results are:

(a)

	Estimated conductivity W/m-K		
	$(\rho C_p) =$ 2.25 MJ/m³K	$(\rho C_p) =$ 2.5 MJ/m³K	$(\rho C_p) =$ 2.75 MJ/m³K
$r_b = 6.75$ cm	2.250 (0.6%)	2.247 (0.8%)	2.243 (0.9%)
$r_b = 7.5$ cm	2.249 (0.7%)	2.245 (0.9%)	2.241 (1.1%)
$r_b = 8.25$ cm	2.247 (0.8%)	2.243 (1.0%)	2.239 (1.2%)

(continued)

	Estimated resistance m-K/W		
	$(\rho C_p) =$ 2.25 MJ/m³K	$(\rho C_p) =$ 2.5 MJ/m³K	$(\rho C_p) =$ 2.75 MJ/m³K
	C = 2.25 MJ/m³K	C = 2.25 MJ/m³K	C = 2.25 MJ/m³K
$r_b = 6.75$ cm	0.168 (6.7%)	0.176 (2.7%)	0.182 (0.9%)
$r_b = 7.5$ cm	0.172 (4.7%)	0.179 (0.7%)	0.186 (2.9%)
$r_b = 8.25$ cm	0.175 (2.9%)	0.183 (1.1%)	0.189 (4.7%)

∎

In principle, when an optimization algorithm is used to minimize an objective function, it could be possible to look for more than one parameter (k) or two (k and R_b). Several authors tried to estimate more than two parameters using such an approach. From our previous experience in GSHP analysis, the ground diffusivity would also be a parameter of interest. Several authors explored the possibility of using this approach to estimate it [8, 13, 14]. Others estimated the grout parameters [11, 15]. In practice it becomes more difficult to estimate too many parameters in an inverse-problem approach since such a problem can become ill-posed [16,17]. Without going into too many theoretical details, it is easy to see that if a thermal system is governed by the following relation:

$$\Delta T = f(\alpha) = \Delta T_i \exp(-\alpha t / r_b^2) = \Delta T_i \exp(-k/(\rho C_p)t/r_b^2)$$

it would not be possible to estimate independently ρ, k, and C_p. However, if the thermal response depends on:

$$\Delta T = f(k, \alpha)$$

then it would be possible, in theory, to estimate both k and α and to deduce $C = \rho C_p = k/\alpha$. From our previous analysis, it was observed that the diffusivity influences the transient response, so if the measurement is in a quasi-steady state, it is likely that the estimation will be poor. An important concept associated with the estimation of several parameters using the same objective function is the *sensitivity coefficient* [17] defined[1] by:

$$J_p = \frac{\partial T}{\partial p} \tag{9.23}$$

where p is a parameter that has to be estimated. Sometimes given in dimensionless form:

$$\tilde{J}_p = \frac{p}{T_{ref}} \frac{\partial T}{\partial p} \tag{9.24}$$

---

[1] When, of course, the measured variable is a temperature

Example 9.6 Taking Eq. (9.10) as the model for the mean fluid temperature in a TRT, evaluate the dimensionless sensitivity coefficient for the conductivity and the volumetric heat capacity defined as:

$$\tilde{J}_C = \frac{C}{q'_b/k}\frac{\partial \Delta T_f}{\partial C}, \quad \tilde{J}_k = \frac{k}{q'_b/k}\frac{\partial \Delta T_f}{\partial k}$$

using the ICS and the ILS.

Solution From (9.10):

$$\tilde{J}_C = \frac{C}{q'_b/k}\frac{\partial \Delta T_f}{\partial Fo}\frac{\partial Fo}{\partial C} = \frac{C}{q'_b/k}\left(\frac{-k\,t}{C^2 r_b^2}\frac{\partial \Delta T_f}{\partial Fo}\right)$$

$$= \frac{C}{q'_b/k}\left(\frac{k\,t}{C^2 r_b^2}\frac{q'_b}{k}\frac{\partial G(Fo)}{\partial Fo}\right) = Fo\frac{\partial G(Fo)}{\partial Fo}$$

For the conductivity, it is given by:

$$\tilde{J}_k = \frac{k}{q'_b/k}\frac{\partial \Delta T_f}{\partial k} = \frac{k^2}{q'_b}\left(\frac{q'_b}{k^2}G(Fo) + \frac{\partial \Delta T_f}{\partial Fo}\frac{\partial Fo}{\partial k}\right)$$

$$= \frac{k^2}{q'_b}\left(\frac{q'_b}{k^2}G(Fo) - \frac{q'_b}{k}\frac{t}{Cr_b^2}\frac{\partial G(Fo)}{\partial Fo}\right)$$

$$= G(Fo) - Fo\frac{\partial G(Fo)}{\partial Fo} = G(Fo) - \tilde{J}_C$$

Starting with the ICS, from Eq. (4.9) we have:

$$\tilde{J}_{C,ICS} = \frac{2\,Fo}{\pi^3}\int_o^\infty \frac{e^{-\xi^2 Fo}}{\xi(J_1^2(\xi) + Y_1^2(\xi))}d\xi$$

$$\tilde{J}_{k,ICS} = G_{ICS}(Fo) - \tilde{J}_{C,ICS}$$

For the ILS, from Eq. (4.12) we have:

$$\tilde{J}_{C,ILS} = Fo\frac{\partial}{\partial Fo}\left(\frac{1}{4\pi}\int_{1/4Fo}^\infty \frac{e^{-u}}{u}du\right)$$

Using Leibniz's rule [18], for the case where the limits of integration are variable dependent:

$$\tilde{J}_{C,ILS} = Fo\frac{1}{4\pi}\left(-\frac{e^{-1/4Fo}}{1/4Fo}\right)\left(\frac{-1}{4Fo^2}\right) = \frac{1}{4\pi}e^{-1/4Fo}$$

$$\tilde{J}_{k,ILS} = G_{ILS}(Fo) - \tilde{J}_{C,ILS}$$

(continued)

Results for different values of Fourier numbers are shown in the following figure:

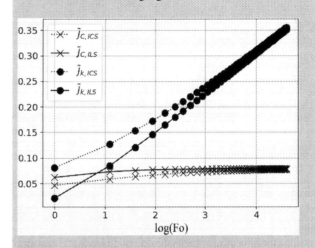

where it is clear that the results are much more sensitive to the conductivity than on the heat capacity. ∎

Although several results are given in the literature for the heat capacity estimation, they are always given with less confidence than for the conductivity. Tikhonov regularization [17, 19] can be used to reduce the problem associated with ill-posed problems, but research has still to be done for good estimation of the heat capacity. Attempts have also been done with oscillatory measurements to estimate it [20–22].

9.7 Multi-injection Rates

In the classical approach of TRT, a control system is used to maintain the heat rate as constant as possible in order to use Eq. (9.10). It is however not necessary to keep this rate constant. Several authors have suggested to use different heat rates during the same TRT [23,24]. In the more general case, the temperature can be found from the temporal superposition (Chap. 4):[2]

$$T_f = T_o + \frac{1}{k_s}\left(q'_o G(Fo) + (q'_1 - q'_o)G(Fo - Fo_1) + \ldots\right) + q'_b R p_b$$

(9.25)

---

[2] Again, since most of the time heat is injected in the ground during TRT, the convention sign is changed in this chapter.

9.7.1   Injection-Recovery Analysis

An interesting special case of the preceding analysis is when there are two pulses where the second one is zero. This period is called the *recovery period*, since the soil recovers its temperature level during that period. Equation (9.25) becomes with $q'_o = q'_{inj}$, $q'_1 = 0$:

$$T_f = T_o + \frac{q'_{inj}}{k_s} G(Fo) + q'_{inj} R'_b \qquad , t < t_1$$

$$T_f = T_o + \frac{q'_{inj}}{k_s} (G(Fo) - G(Fo - Fo_1)) , \; t > t_1$$

$$(9.26)$$

It can be observed that during the recovery period, the temperature response is independent of the borehole resistance. Some authors [25] suggested to use the recovery period to assess the thermal conductivity and the injection period to predict the borehole resistance. Again, in (9.26), any thermal response models can be used, either analytical or numerical, using parameter estimation methods. In the special case where the logarithmic approximation of the ILS model is used, this expression can be simplified as:

$$T_f = T_o + \frac{q'_{inj}}{4\pi k_s} \left(\log \left(\frac{\alpha t}{r_b^2} \right) - \gamma \right) + q'_{inj} R'_b , \; t < t_1$$

$$T_f = T_o + \frac{q'_{inj}}{4\pi k_s} \log \left(\frac{t}{t - t_1} \right) \qquad , t > t_1$$

$$(9.27)$$

In this case, the slope method can be used to evaluate the conductivity in either the injection or the recovery period. In the latter case, the linear equation has the form:

$$\Delta T_f = mx$$

with:

$$m = \frac{q'_{inj}}{4\pi k_s}, \qquad x = \log \left(\frac{t}{t - t_1} \right) \qquad (9.28)$$

Care should be used when graphical solution is used. As it is always the case when the ILS model is used, the short-term period of the recovery period has to be removed from the analysis. This constraint imposes that:

$$\frac{\alpha(t - t_1)}{r_b^2} > 5$$

$$(t - t_1) > \frac{5r_b^2}{\alpha}$$

which implies that:

$$\frac{1}{t - t_1} < \frac{\alpha}{5r_b^2}$$

$$\frac{t_1}{t - t_1} < \frac{\alpha t_1}{5r_b^2}$$

$$\frac{t}{t - t_1} < 1 + \frac{\alpha t_1}{5r_b^2}$$

and finally:

$$x < \log \left(1 + \frac{\alpha t_1}{5r_b^2} \right) \qquad (9.29)$$

On a graphical representation, only the first portion should be used.

Example 9.7 An injection-recovery TRT is performed during 120 hours. In the first 50 hours, heat is injected at a constant rate of 66 W/m. The diameter of the borehole is 126 mm, and the undisturbed temperature is 12 °C. The volumetric heat capacity is assumed to be 1.92 MJ/m^{3}K.

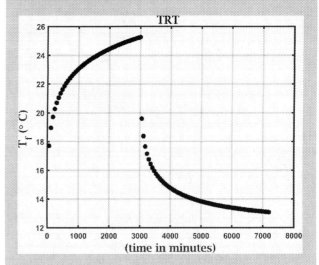

(a) Using the recovery period data, find graphically the equivalent soil conductivity.
(b) Using the injection period, find graphically the borehole resistance.

(continued)

Example 9.7 (continued)
Solution

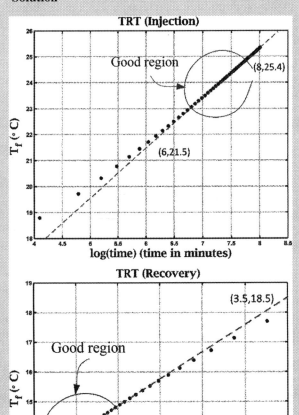

In the recovery period, the Fourier number limit gives (9.29):

$$x < \log\left(1 + \frac{\alpha t_1}{5^2}\right)$$

An order of magnitude of the diffusivity can be found assuming a typical value for the conductivity:

$$\alpha \approx \frac{2.5}{1.92\mathrm{e}6} \times 3600 = 0.0046875 \ \mathrm{m}^2/\mathrm{hr}$$

which gives:

(continued)

$$x < \log\left(1 + \frac{0.0046875 \cdot 50}{50.063^2}\right)$$

$$x < 2.55$$

which is approximately the range used for the graphical line on the figure. The slope is then:

$$m = \frac{18.5 - 13}{3.5 - 0.5} = 1.84$$

$$k_s = \frac{66}{4\pi\,1.84} = 2.86 \ \mathrm{W/m\text{-}K}$$

(b) During the injection period, the Fourier number's limit gives:

$$x > \log\left(\frac{5 \cdot 0.063^2}{\alpha}\right)$$

Now that the conductivity is known, the real diffusivity is:

$$\alpha = \frac{2.86}{1.92\mathrm{e}6} \times 3600 = 0.00537 \ \mathrm{m}^2/\mathrm{hr}$$

$$x > 5.5$$

From the graphical solution, one finds:

$$b = 9.8$$

Since the time is in minutes, it is important to put the diffusivity in $\mathrm{m}^2/\mathrm{min}$.

$$R'_b = \frac{9.8 - 12}{66} - \frac{\log\left(\frac{4\cdot 0.00537/60}{0.063^2}\right) - 0.5772157}{4\pi\,2.862} = 0.049 \ \mathrm{K\text{-}m/W}$$

From the slope in the injection period, it would have been possible to find also the conductivity. The value would have been:

$$m = \frac{25.4 - 21.5}{8.0 - 6.0} = 1.95$$

$$k_s = \frac{66}{4\pi\,1.95} = 2.69 \ \mathrm{W/m\text{-}K}$$

There seems to be a large difference between both approaches. This is due to the great sensibility of the slope method with the number of points that are kept. Doing as in Example 9.2 a moving average keeping less and less points, we would observe the following results:

(continued)

Example 9.7 (continued)

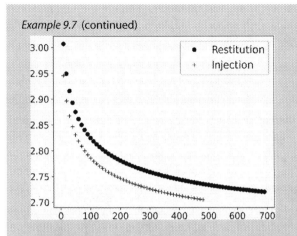

From the figure, it is observed that a more confident value of the conductivity should be around 2.7 W/m-K leading to a borehole resistance of:

$$R'_b = 0.055 \text{ K-m/W}$$

∎

As seen in the previous example, the graphical method is very sensitive to the analyst, and numerical regression methods are suggested if possible.

Example 9.8 Redo the previous example using the parameter estimation method.

Solution The solution follows the same path as in Example 9.5. The only difference is that it is done in two steps where one parameter is estimated separately in each of them. As comparison, the parameter estimation of the two parameters in the injection is also performed as in Example 9.5 at the end. Part of the code is shown here:

```
# estimation of conductivity ( Recovery period)
Fomin = 7 # Arbitrrary choice
t_critr = t1 + Fomin*rb**2/almin
n_critr = sum(t_res < t_critr) # number of points
                                   to remove
y = Tf_res[n_critr:n_res]
ti = t_res[n_critr:n_res]
cas = 'a'
if cas == 'a':
        G_vect = vectorize(G_function_ils) # Infinite
                                   line source model is used
```

(continued)

```
else:
        G_vect = vectorize(G_function) # Infinite
                                   cylindrical source model is used
def T_theo_rec(x, knew):
        return To + qp*x/knew
while not ok:
        als = ks/CC
        almin = als*60
        Fo1 = almin*t1/rb**2
        Fo = almin*ti/rb**2
        x = G_vect(Fo) - G_vect(Fo-Fo1)
        param,resn = curve_fit(T_theo_rec,x,y)
        ksn = param[0]
        if (abs(ksn-ks)/ks) < 0.001:
                ok = True
        else:
                ks = ksn
                compt = compt + 1
                if compt > compt_max:
                        err = 1
                        ok = True
#
# estimation of resistance ( Injection period)
#
t_criti = Fomin*rb**2/almin
n_criti = sum(t_inj < t_criti) # number of points to
                                   remove
y = Tf_inj[n_criti:n_inj]
ti = t_inj[n_criti:n_inj]
Fo = almin*ti/rb**2
x = G_vect(Fo)
ok = False
compt_max = 10
compt = 0
Rb = 0.08 # initial value
def T_theo_inj(x, Rbnew):
        return To + qp*(x/ks + Rbnew)
param,resn = curve_fit(T_theo_inj,x,y)
Rb = param[0]
```

No iteration is needed in the injection period since the diffusivity is now known. The final results are:

$$k_s = 2.72 \text{ W/m-K}$$

$$R'_b = 0.057 \text{ K-m/W}$$

Using only the injection period to estimate both parameters as in Example 9.5 would result in:

(continued)

Example 9.8 (continued)
$$k_s = 2.79 \text{ W/m-K}$$

$$R'_b = 0.059 \text{ K-m/W}$$

which are very similar. The comparison between the measured and simulated points is shown on the figure:

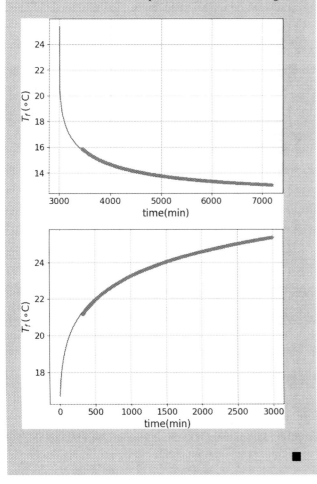

In most of the previous methods, it was observed that the knowledge of the undisturbed ground temperature does not influence the evaluation of the conductivity but has a large influence on the evaluation of the borehole resistance. For the slope method, it is obvious looking Eqs. (9.15, 9.16), but even with a minimization approach, it would also be observed in the injection period even though the uncertain ground temperature appears explicitly in the objective function. However, it is very important in the recovery period:

Example 9.9 Redo the last example assuming that the ground temperature is 11.5 °C.

(continued)

Solution Nothing changes in the solution approach except the value of T_o. Evaluating the conductivity in rejection period and the borehole in the injection period will lead to:

$$k_s = 2.19 \text{ W/m-K}$$

$$R'_b = 0.42 \text{ W/m-K}$$

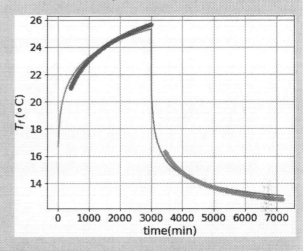

However, even with this wrong value of ground temperature, using only the injection period will give:

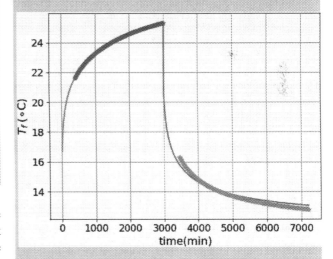

$$k_s = 2.78 \text{ W/m-K}$$

$$R'_b = 0.066 \text{ W/m-K}$$

The value of the conductivity, as in the previous example, is almost the same as previously, but the borehole resistance is overestimated. Although surprising, the result is logical. Since the value of the ground temperature is smaller than the real one, the whole temperature profile would tend to be too low, but the optimization

(continued)

Example 9.9 (continued)
algorithm will increase the borehole resistance which creates a constant offset on the profile so that the good conductivity can be obtained. In the restitution period, the borehole resistance does not appear in the solution, so the algorithm varies the conductivity to reduce the RMS error.

One possibility is to estimate the ground temperature by the optimization algorithm.

```
# estimation of conductivity ( Recovery period)
def T_theo_rec(x, Tonew,knew):
        return Tonew + qp*x/knew
while not ok:
        als = ks/CC
        almin = als*60
        Fo1 = almin*t1/rb**2
        Fo = almin*ti/rb**2
        x = G_vect(Fo) - G_vect(Fo-Fo1)
        param,resn = curve_fit(T_theo_rec,x,y)
        Ton = param[0]
        ksn = param[1]
...
```

The results thus obtained would be:

$$k_s = 2.81 \text{ W/m-K}$$

$$R'_b = 0.059 \text{ K-m/W}$$

a better estimate.   ∎

9.7.2 TRT with Heating Cables

The injection-recovery scenario is interesting since the borehole resistance does not influence the temperature profile in the second interval. However the temperature variations are small, and the accuracy of the temperature measurements is important. It is not obvious to see a real advantage in classical TRT. However, it is important for TRT with heating cables. The subject will not be covered in details, but the principle is simple. In the thermal response test, the heat has to be exchanged (injected or extracted) with the ground, and the fluid temperature response is measured. In the classical approach, the heat is injected (or extracted) outside the borehole where the mean temperature is also measured although a distributed temperature profile is possible [26,27]. In a TRT with heating cable, as the name suggests, heat

is released from a cable that is inserted into the borehole. Since the water does not move, one or several temperature sensors are inserted near the cable to measure the temperature response. Doing so, it is not the mean temperature but a local temperature that is measured, and the exact position of the sensor and the heating cable is important at least in the injection period. In the recovery interval, it was shown by Raymond et al. [25] that the temperature homogenizes rapidly and the exact location of the sensor is not important. It is then very important to exploit this property when using heating cables. However, it makes it difficult to evaluate the borehole resistance. This particularity is an inconvenience associated with the method. The use of heating cables has however several advantages: a lower cost, simplicity, and the possibility to assess more easily than with the classical TRT the conductivity profile in a non-homogeneous soil [28, 29]. For more information, the literature can be consulted [5, 30, 31].

9.8   Mean Fluid Temperature-Effective Resistance

The principle leading to the estimation of the soil conductivity and the borehole resistance is quite straightforward from the theory of the unsteady thermal response due to a heat pulse as explained in the previous sections. However, in all the expressions used so far, the mean fluid temperature is assumed to be given by Eq. (9.17). As pointed out in Chap. 4, this is exactly true only if the heat flux and the borehole temperature are uniform along the borehole. Even though this approximation is not exactly valid, using (9.17) is still widely used. Marcotte and Pasquier [32] were among the first to predict that the estimated resistance thus found will be overestimated and suggested to use a different expression for the mean fluid temperature. They called it the *p-linear average*. It is defined as:

$$\Delta T_p = \frac{p \left(|\Delta T_{in}|^{p+1} - |\Delta T_{out}|^{p+1} \right)}{(1 + p) \left(|\Delta T_{in}|^p - |\Delta T_{out}|^p \right)} \qquad (9.30)$$

where $\Delta T(z) = T_f(z) - T_o$, the difference between the fluid temperature and the undisturbed ground temperature. This average expression assumes that the fluid temperature profile follows the expression:

$$\Delta T(x) = \left[|\Delta T_{in}|^p + \frac{x}{2H} \left(|\Delta T_{out}|^p - |\Delta T_{in}|^p \right) \right]^{1/p} \qquad (9.31)$$

It is easy to verify that $p = 1$ corresponds to the linear approximation used so far. The authors suggested to use $p \rightarrow$

−1, estimating that it corresponds to a more representative temperature profile for a typical borehole setup. Beir [33] verified that this approximation gives indeed a smaller difference between the measured borehole resistance and the real resistance, but this difference depends on the borehole parameters (height, flow rate, etc.). Zhang et al. [14] showed also that this choice of $p \to -1$ is not always the best one, and they proposed a complex method to estimate an unsteady value of this parameter. Lamarche et al. [34] pointed out that this overestimation of the borehole resistance can be associated with the *effective resistance*, a value that can be more important than the local resistance. Without going into too many details, it is easy to verify the impact of the p-average on the estimation of parameters in a TRT. Any method suggested in the previous sections can be used with the replacement of Eq. (9.17) with:

$$\bar{T}_{f,p} = T_o + \Delta T_p \tag{9.32}$$

with ΔT_p given by (9.30).

Example 9.10 The following figure shows the mean temperature response associated with a real TRT [35] in Quebec City.

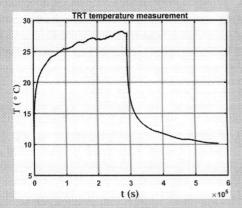

The figure shows the arithmetic mean, but the inlet and outlet temperatures are available (TRT_test3.txt) to evaluate the p-linear temperature if needed. The borehole radius is 5.7 cm, and the heat injection was 66.42 W/m. A temperature profile was done [35], and the mean value was measured as 8.22 °C. The test lasted 1532 hours with 81 hours of heat injection and the rest the recovery period. Using the parameter estimation approach, evaluate the conductivity using the recovery interval and the borehole resistance with the injection period using:

(continued)

(a) The arithmetic mean
(b) The p-linear mean (with $p \to -1$)

Solution The same procedure as done in the last example will be used. The only difference would be the evaluation of the mean temperature. First, a moving average was performed to see how the conductivity was changing when the number of points kept varies using the slope method and the parameter estimation technique.

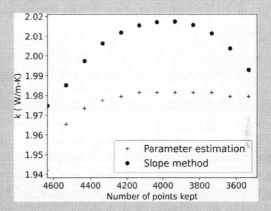

Contrary to Example 9.7, where the points were simulated, here are real measured points showing much more variation, and the moving average shows much more sensibility of the slope method than the estimation technique. So from now only the parameter estimation will be used in this example. With the arithmetic mean temperature, the results obtained are:

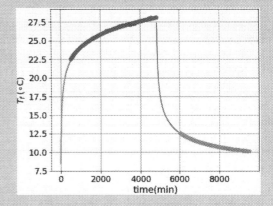

$$k_s = 1.98 \text{ W/m-K}$$
$$R'_b = 0.092 \text{ K-m/W}$$

(continued)

Example 9.10 (continued)
Using the p-linear

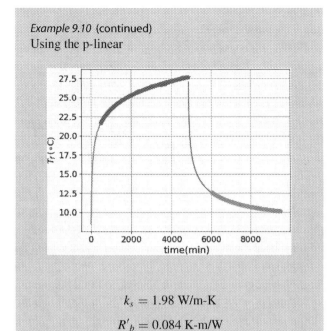

$$k_s = 1.98 \text{ W/m-K}$$

$$R'_b = 0.084 \text{ K-m/W}$$

Although the conductivity does not change much (third decimal), the borehole resistance estimated decreases by approximately 10%. Again, it is often suggested that the higher value found with the arithmetic mean might be more representative to the effective resistance R'^*_b, a more important metric needed in a sounded design, although the difference is often small. ∎

9.9   Error Analysis

As seen in the previous examples, the final value of the conductivity (and borehole resistance) is very sensitive to the knowledge of several parameters (ground temperature, borehole radius, etc....) and to the method used. It was shown that discrepancy in the order of 10% can easily be observed depending on several factors. Thus, the accuracy with which an experimentalist is able to express his results is of primary importance. The classical approach of error propagation [36] evaluation has been applied to TRT by Witte [37] in which he found an error estimate of the order 5% for the conductivity and 10–15% for the borehole resistance for its particular example. Without going into this important issue here, it should be observed that this result was obtained without taking into account the accuracy associated to the model used (slope method). Although discussed in his paper, it was not quantified in the final result. Our previous examples show that this may have an important impact. Again, the reader can refer to the literature for more information about this important subject [37–39].

Problems

Pr. 9.1 A TRT test has been done in a 126 mm diameter borehole. The volumetric capacity is estimated as 1.92 MJ/m³-K. The initial temperature was 12 °C. During the first 50 hours, a constant heat rate of 66 W/m was imposed, and a restitution time of 70 hours was measured afterward. The injection-restitution measurements are shown in the following figures:

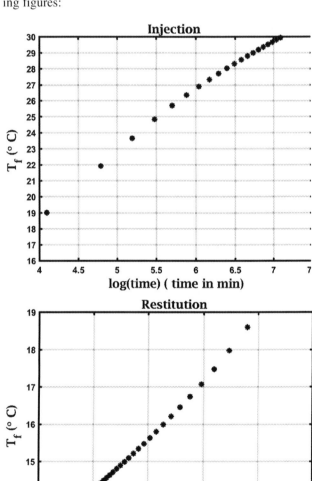

the restitution period, find the conductivity. From the injection period, find the borehole resistance.
Answer: k ≈ 2.6 W/m-K, $R'_b ≈ 0.12$ m-K/W

Pr. 9.2 A thermal response test was started on an experimental setup where 30 W/m of heat was injected in a borehole having a diameter of 140 mm. The injection last 24 hours and was stopped due to a bad breaker. The test was restarted after an 8-hour break, while the power was increased to 60 W/m during another 68 hours. Using only the results during the last part of the measurements, show that the conductivity can be estimated by:

$$k_{soil} = \frac{30}{4\pi m}$$

where m is the slope of the straight line:

$$T_f(t) = mx + b$$

with (t in hours):

$$x \approx \log\left(\frac{t(t-32)^2}{(t-8)}\right)$$

and that the borehole resistance can be found from:

$$R'_b \approx \frac{1}{2}\left(\frac{b-T_o}{30} - \frac{\log\left(\frac{4\alpha}{r_b^2}\right) - 2\gamma}{4\pi k_{soil}}\right)$$

(b) The results of the TRT are given in the file "test_trt.txt." The undisturbed temperature is 8 °C and the volumetric capacity 2.5 MJ/m³-K. Estimate the soil conductivity and borehole resistance.
Answer: $k \approx 3.42$ W/m-K, $R'_b \approx 0.23$ m-K/W

Pr. 9.3 In the last chapter, parameter estimation was used to estimate two parameters, the soil conductivity and the borehole resistance. In theory, the thermal capacity (or the diffusivity) can also be estimated, but when the number of parameters is increased, convergence issues and local minimums can be obtained. Instead, it is suggested, if possible, to monitor the borehole temperature at the same time as the fluid mean temperature. If this is possible, show that the unknown diffusivity can be obtained by:

$$\alpha \approx \frac{r_b^2}{4}\exp\left(\frac{4\pi k_{soil}(b_2 - T_o)}{q'} + \gamma\right)$$

where b_2 is the intercept of the curve

$$T_b(t) = m\log(t) + b_2$$

for $t > t_{min}$ where again the measurements in the early time of the test should be removed.

The file "test_trt2.txt" has three columns; the first is the time in min, the second the mean fluid temperature, and the third the mean borehole temperature measured during a TRT. The heat rate was 55 W/m, the radius was 6.5 cm, and the undisturbed ground temperature was 11 °C. Estimate the conductivity and the borehole resistance from the fluid temperature and the volumetric capacity from the borehole temperature.
$k_{soil} \approx 2.5$ W/m-K, $R'_b \approx 0.14$ m-K/W, $C_{soil} \approx 2.14$ MJ/m³-K

References

1. Clauser, C.: Thermal storage and transport properties of rocks, I: heat capacity and latent heat. In: Gupta, H.K. (ed.) Encyclopedia of Solid Earth Geophysics, pp. 1431–1448. Springer Netherlands (2011). ISBN: 978-90-481-8702-7
2. ASTM Standard C177-13: Standard Test Method for Steady-State Heat Flux Measurements and Thermal Transmission Properties by Means of the Guarded-Hot-Plate Apparatus' (2013)
3. ISO Standard 8302-91: Thermal insulation—Determination of steady-state thermal resistance and related properties—Guarded hot plate apparatus. Technical report. 1991
4. Clauser, C.: Thermal storage and transport properties of rocks, II: thermal conductivity and diffusivity. In: Gupta, H.K. (ed.) Encyclopedia of Solid Earth Geophysics, pp. 1431–1448. Springer, Dordrecht (2011). ISBN: 978-90-481-8702-7
5. Raymond, J., Lamarche, L., Malo, M.: Field demonstration of a first thermal response test with a low power source. Appl. Energy **147**, 30–39 (2015). https://doi.org/10.1016/j.apenergy.2015.01.117
6. Mogensen, P.: Fluid to duct wall heat transfer in duct system heat storages. Document-Swedish Council Build. Res. **16**, 652–657 (1983)
7. American Society of Heating, Refrigerating and Air-Conditioning Engineers: ASHRAE Handbook, HVAC Applications. The Society, New York (2015)
8. Wagner, R., Clauser, C.: Evaluating thermal response tests using parameter estimation for thermal conductivity and thermal capacity. J. Geophys. Eng. **2**(4), 349 (2005)
9. Chiasson, A., OĆonnell, A.: New analytical solution for sizing vertical borehole ground heat exchangers in environments with significant groundwater flow: parameter estimation from thermal response test data. HVACR Res. **17**(6), 1000–1011 (2011)
10. Li, M., Lai, A.C.K.: Parameter estimation of in-situ thermal response tests for borehole ground heat exchangers. Int. J. Heat Mass Transf. **55**(9), 2615–2624 (2012)
11. Li, M., Zhang, L., Liu, G.: Estimation of thermal properties of soil and backfilling material from thermal response tests (TRTs) for exploiting shallow geothermal energy: sensitivity, identifiability, and uncertainty. Renewable Energy **132**, 1263–1270 (2019). https://doi.org/10.1016/j.renene.2018.09.022
12. Pujol, J.: The solution of nonlinear inverse problems and the Levenberg-Marquardt method. Geophysics **72**(4), W1–W16 (2007)
13. Yu, M.Z. et al.: A simplified model for measuring thermal properties of deep ground soil. Exp. Heat Transf. **17**(2), 119–130 (2004). https://doi.org/10.1080/08916150490271845
14. Zhang, L., Zhang, Q., Huang, G., Du, Y.: A p (t)-linear average method to estimate the thermal parameters of the borehole heat exchangers for in situ thermal response test. Appl. Energy **131**, 211–221 (2014)
15. Bozzoli, F. et al.: Estimation of soil and grout thermal properties through a TSPEP (two-step parameter estimation procedure) applied to TRT (thermal response test) data. Energy **36**(2), 839–846 (2011). https://doi.org/10.1016/j.energy.2010.12.031
16. Beck, J.V., Arnold, K.J.: Parameter Estimation in Engineering and Science. Wiley, New York (1977)

17. Ozisik, H.R.B., Orlande, M.N.: Inverse Heat Transfer: Fundamentals and Applications. Taylor & Francis, London (2000). https://doi.org/10.1201/9780203749784

18. Whittaker, E.T., Watson, G.N.: A Course of Modern Analysis, p. 124 (1996)

19. Bozzoli, F. et al.: Estimation of the local heat transfer coefficient in coiled tubes: comparison between Tikhonov regularization method and Gaussian filtering technique. Int. J. Numer. Methods Heat Fluid Flow **27**(3), 575–586

20. Oberdorfer, P.: Heat Transport Phenomena in Shallow Geothermal Boreholes: Development of a Numerical Model and a Novel Extension for the Thermal Response Test Method by Applying Oscillating Excitations. Ph.D. Thesis. Göttingen, Georg-August Universität, Dissertation, 2014

21. Lei, F. et al.: Periodic heat flux composite model for borehole heat exchanger and its application. Appl. Energy **151**, 132–142 (2015). https://doi.org/10.1016/j.apenergy.2015.04.035

22. Giordano, N., Lamarche, L., Raymond, J.: Evaluation of subsurface heat capacity through oscillatory thermal response tests. Energies **14**(18) (2021). https://doi.org/10.3390/en14185791

23. Gustafsson, A.-M., Westerlund, L.: Multi-injection rate thermal response test in groundwater filled borehole heat exchanger. Renewable Energy **35**(5), 1061–1070 (2010). https://doi.org/10.1016/j.renene.2009.09.012

24. Heiko, T.L., Saquib, J., Vistnes, G.: Multi-injection rate thermal response test with forced convection in a groundwater-filled borehole in hard rock. Renewable Energy **48**(Supplement C), 263–268 (2012)

25. Raymond, J., Therrien, R., Gosselin, L.: Borehole temperature evolution during thermal response tests. Geothermics **40**(1), 69–78 (2011)

26. Acuna, J., Mogensen, P., Palm, B.: Distributed thermal response tests on a multi-pipe coaxial borehole heat exchanger. HVAC & R Res. **17**(6), 1012–1029 (2011). https://doi.org/10.1080/10789669.2011.625304

27. Fujii, H. et al.: An improved thermal response test for U-tube ground heat exchanger based on optical fiber thermometers. Geothermics **38**(4), 399–406 (2009)

28. Morchio, S. et al.: Reduced scale experimental modelling of distributed thermal response tests for the estimation of the ground thermal conductivity. Energies **14**(21) (2021). https://doi.org/10.3390/en14216955

29. Raymond, J., Lamarche, L.: Development and numerical validation of a novel thermal response test with a low power source. Geothermics **51**, 434–444 (2014)

30. Raymond, J. et al.: Ten years of thermal response tests with heating cables. In: World Geothermal Congress (2021)

31. Koubikana Pambou, C., Raymond, J., Lamarche, L.: Improving thermal response tests with wireline temperature logs to evaluate ground thermal conductivity profiles and groundwater fluxes. Heat Mass Transf. **55**, 1829–1843 (2019). https://doi.org/10.1007/s00231-018-2532-y

32. Marcotte, D., Pasquier, P.: On the estimation of thermal resistance in borehole thermal conductivity test. Renewable Energy **33**(11), 2407–2415 (2008)

33. Beier, R.A.: Vertical temperature profile in ground heat exchanger during in-situ test. Renewable Energy **36**(5), 1578–1587 (2011)

34. Lamarche, L., Kajl, S., Beauchamp, B.: A review of methods to evaluate borehole thermal resistances in geothermal heat-pump systems. Geothermics **39**(2), 187–200 (2010)

35. Ballard, J.M., Koubikana Pambou, C.H., Raymond, J.: Développement des tests de réponse thermique automatisés et vérification de la performance d'un forage géothermique d'un diamètre de 4.5 po. Technical report. INRS, 2016, 55 pp.

36. Taylor, J.: Introduction to error analysis, the study of uncertainties in physical measurements (1997)

37. Witte, H.J.L.: Error analysis of thermal response tests. Appl. Energy **109**, 302–311 (2013). https://doi.org/10.1016/j.apenergy.2012.11.060

38. Bujok, P. et al.: Assessment of the influence of shortening the duration of TRT (thermal response test) on the precision of measured values. Energy **64**, 120–129 (2014). https://doi.org/10.1016/j.energy.2013.11.079

39. Pasquier, P., Zarrella, A., Marcotte, D.: A multi-objective optimization strategy to reduce correlation and uncertainty for thermal response test analysis. Geothermics **79**, 176–187 (2019). https://doi.org/10.1016/j.geothermics.2019.02.003

10.1 Introduction

In the previous chapters, heat pumps using a secondary loop vertical boreholes were analyzed. Although these systems are the most popular in Europe, in America, horizontal systems continue to be largely used [1, 2]. Despite this observation, there is a general agreement that much less scientific literature exists on horizontal systems than vertical ones [3]. Design approaches remain often associated with rules of thumb and approximate charts [3]. In fact, the main criteria associated with designing horizontal heat exchangers are basically the same as the ones discussed in Chap. 6: The loop must be sized in order that the exit temperature of the field, i.e., the heat pump entering fluid temperature, must be within "acceptable limits," whatever these limits might be. The idea is then to find approximate methods to evaluate the total loop length in order that this temperature is correct for a given field arrangement and given loads. Typical such arrangements are shown in Fig. 10.1:

Even though some common physical aspects relate both technologies, there are some specific factors associated with horizontal loops that need special treatment in the design of such systems. A non-exhaustive list can include:

(a) The ground surface has much more influence on the thermal response than for vertical systems.
(b) The seasonal temperature variations, which are often neglected in vertical systems, are very important in horizontal ones.
(c) Horizontal piping laid in trenches in the subsurface overburden where water content can vary with time and influence the thermal performance of the heat exchange.
(d) Pipes are often buried with indigenous earth instead of grouted backfill.
(e) Phase change due to freezing and thawing may occur.
(f) Several configurations are possible, inline, slinky, etc., which cause pipe-to-pipe interference.

For these reasons, although horizontal loops are easier to install, they are more complicated to analyze. Yet, several studies were done in the past. A well-known report by Claesson and Dunand on horizontal systems [4] is often used as a reference even though several aspects addressed in the previous list are not treated. Mei [5] develops a numerical model for horizontal systems which he compared with experimental measurements. It was extended by Piechowski [6] to include mass transfer due to moisture content. These early works were mostly numerical simulations not necessarily giving design tools for simple sizing procedures. The work of Bose and Parker [7] use the concept of images of the infinite line source model to generate temperature response factors that they called unsteady soil resistances associated with several trench configurations. Using these resistances, classical method like the ASHRAE's three pulses approach can be used to size the heat exchanger as long as the seasonal temperature fluctuations are taken into account in some way or another.

10.2 Soil Resistances of Horizontal Pipes

A simple ground heat exchanger is shown in Fig. 10.2. It consists of a simple back and forth pipe laid out in a trench (Fig. 10.3).

Due to the small depth, the seasonal temperature variations must be taken into account. Since the heat equation is linear (neglecting phase changes), the problem is often split into two parts as shown in Fig. 10.4.

It must be remarked that the superposition is valid if the heat flux is imposed at the pipe boundary but not the temperature. The final solution would then be given by:

$$T(x, y, z, t) = {}^1T(t, x, y, z) + {}^2T(t, z) \qquad (10.1)$$

Several solutions can be used to solve problem 2. It depends on how the boundary conditions are imposed at the air-soil

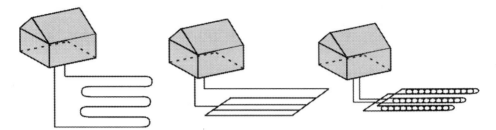

Fig. 10.1 Different horizontal layouts

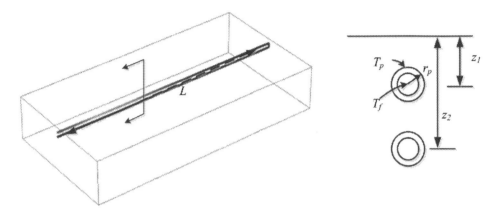

Fig. 10.2 Simple horizontal ground exchanger

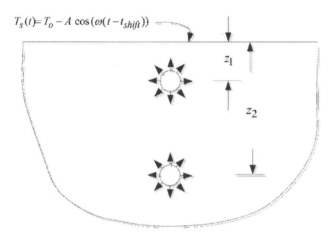

Fig. 10.3 Sectional plane of horizontal piping

interface and whether the soil is an homogeneous solid or if its moisture content is taken into account. The simplest solution is of course to assume a homogeneous ground and assuming a known prescribed temperature profile at its surface as shown in Fig. 10.4. In that case, the classical solution given by Kusuda and Achenback [8] can be used:

$$
{}^{2}T(t, z) = T_o - A \exp\left(-z\sqrt{\frac{\pi}{365\alpha}}\right)
$$

$$
\times \cos\left(\omega\left(t - t_{shift} - \frac{z}{2}\sqrt{\frac{365}{\pi\alpha}}\right)\right) \quad (10.2)
$$

T_o is the mean ground surface temperature which would also be the ground undisturbed temperature if the heat flux gradient is neglected. A is the maximum temperature variation, t_{shift} the coldest day of the year. In this relation, the time is in days and the diffusivity in m2/day. Finally, the frequency should also be given in day^{-1}:

$$
\omega = \frac{2\pi}{365}
$$

10.2.1 Method of Images

The solution of the first problem can be solved with the method of images. In order to simplify the notation, the superscript 1 will be dropped, but all solutions given in this paragraph are related to the homogeneous problem. This method consists of introducing virtual sources or sinks associated with real sources in the ground in order to fulfill the boundary condition at the soil interface. The method is illustrated with a simple source located at a depth of z_o in Fig. 10.5:

In the case of a Dirichlet homogeneous boundary condition, the solution would be to add an image with an opposite sign at a mirror distance from the ground. The final solution would then be to add the effect of both images. The mathematical expression of the thermal response depends if

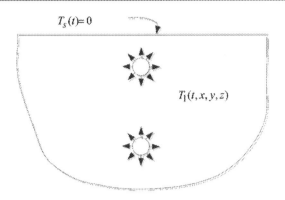

Fig. 10.4 Decoupling in two problems

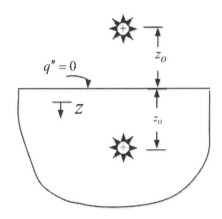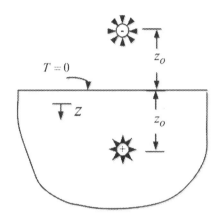

Fig. 10.5 Image of an horizontal heat source

one looks at steady-state or unsteady analysis. In the latter context, Bose and Parker used the ILS model. Their solution for a heat pulse input for a single pipe was then:

$$T_p(t) = -\frac{q'_p}{k_s} \frac{1}{2\pi} \underbrace{(I(X) - I(X_{im}))}_{G_{hori,1}} \qquad (10.3)$$

where

$$X = \frac{r_p}{2\sqrt{\alpha t}}, \qquad X_{im} = \frac{z_o}{\sqrt{\alpha t}} \qquad (10.4)$$

This expression assumes that the difference between the distance of the image and the point on the periphery of the pipe does not change too much as z_o is usually much larger than r_p. $G_{hori,1}$ can be associated with the thermal response factor for a single pipe. Bose and Parker used instead the concept of thermal resistances:

$$T_p(t) = -q'_p \frac{G_{hori,1}}{k_s} = -q'_p R'_{hori,1}(Fo) \qquad (10.5)$$

$$Fo = \frac{\alpha t}{r_p^2}$$

The use of the infinite line source model assumes that the problem does not vary along the pipe, i.e., that the heat transfer per unit length is uniform along the axis of the pipe. The generalization can easily be done for any configurations. For a two-pipe arrangement as shown in Fig. 10.2:

$$T_{p,1}(t) = -\frac{q'_p}{2\pi k_s} (I(X_1) + I(X_2)$$
$$-(I(X_{im,1}) + I(X_{im,2}))) = -q'_p R'_1 \qquad (10.6)$$

where

$$X_1 = \frac{r_p}{2\sqrt{\alpha t}}, \quad X_2 = \frac{|z_2 - z_1|}{2\sqrt{\alpha t}},$$
$$X_{im,1} = \frac{z_1}{\sqrt{\alpha t}}, \quad X_{im,2} = \frac{|z_2 + z_1|}{\sqrt{2\alpha t}} \qquad (10.7)$$

In this special case, the pipe temperature will be different at the second pipe:

$$T_{p,2}(t) = -\frac{q'_p}{2\pi k_s} (I(X_3) + I(X_4)$$
$$-(I(X_{im,3}) + I(X_{im,4}))) = -q'_p R'_2 \qquad (10.8)$$

where

$$X_3 = \frac{r_p}{2\sqrt{\alpha t}}, \quad X_4 = \frac{|z_2 - z_1|}{2\sqrt{\alpha t}},$$

$$X_{im,3} = \frac{z_2}{\sqrt{\alpha t}}, \quad X_{im,4} = \frac{|z_2 + z_1|}{\sqrt{2\alpha t}}$$

(10.9)

A simple assumption is to take the mean value:

$$T_p(t) = \frac{T_{p,1} + T_{p,2}}{2} = -q'_p R'_{hori,2}$$

$$R'_{hori,2} = \frac{R'_1 + R'_2}{2}$$

(10.10)

An example of a figure used generated by the method of images is shown in Fig. 10.6.

for a two-pipe arrangement for $k_s = 0.5$(W/m-k) and $\alpha = 0.001$(m^2/hr). Several other different configurations were given in the book of Bose and Parker. The fluid temperature is found from the pipe temperature from the pipe resistance:

$$T_f = T_p - q' R'_p$$

(10.11)

where the pipe resistance comprises as usual a conduction and a convection resistance:

$$R'_p = \frac{\log\left(\frac{r_{po}}{r_{pi}}\right)}{2\pi k_p} + \frac{1}{2\pi r_{pi} h_{conv}}$$

(10.12)

10.2.2 Inlet-Outlet Temperature

In a vertical system, the borehole as a whole was modeled and inlet-outlet temperature was clearly defined. In the case

of horizontal piping, at least with the image method, it it less obvious. Taking, for example, a two-pipe arrangement as seen previously, is there a difference if the flow is in series or in parallel? The method of images, as described previously, assume that both legs have the same heat exchange. Knowing that q' is the heat per unit length of pipe and $\dot{m}$, the total flow rate, the temperature difference in the series case is simply given by:

$$q = \dot{m} C_p (T_{fo} - T_{fi})$$

and the same relation as in the vertical case is found:

$$T_{fi} = T_f + \frac{q}{2\dot{m}C_p} == T_f + \frac{q'L}{2\dot{m}C_p}$$

(10.13)

as before except that L is the length of pipe. For the parallel case, we have:

$$q_1 = \frac{\dot{m}}{2} C_p (T_{fo} - T_{fi})$$

$$q_2 = \frac{\dot{m}}{2} C_p (T_{fo} - T_{fi})$$

$$q = q_1 + q_2 = \dot{m} C_p (T_{fo} - T_{fi})$$

where now $\dot{m}$ is the total flow rate, not the flow rate in one pipe. So basically, the method of images does not distinguish between configurations although in practice, the pipe diameter would probably be different in each configuration. Finally, one can ask whether it is possible to have two pipes with a parallel configuration, since the flow has to leave the heat pump and come back essentially at the same place. It can be an approximation of a trench making a long turn where the interference between one way and the other would be negligible.

10.3 Horizontal Heat Exchanger Sizing

Ground heat exchanger sizing can be done with similar lines as for the vertical heat exchanger. For example, the three pulses ASHRAE method can be used. Neglecting for now the effect of seasonal temperature variation, Eq. (6.10) can be used:

$$L = \frac{q_a R'_{s,a} + q_m R'_{s,m} + q_h (R'_{s,h} + R'_p)}{T_o - T_f}$$

(10.14)

with:

$$R'_{s,a} = R'_{hori}(Fo_f) - R'_{hori}(Fo_1)$$

$$R'_{s,m} = R'_{hori}(Fo_1) - R'_{hori}(Fo_2)$$

(10.15)

$$R'_{s,h} = R'_{hori}(Fo_2)$$

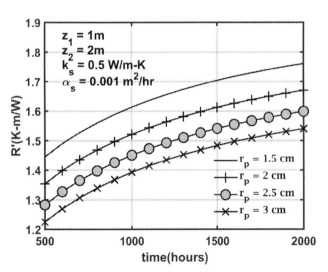

Fig. 10.6 Ground resistance of a two-pipes configuration

The definition of the different Fourier numbers has the same meaning as before:

$$Fo_f = \frac{\alpha_s t_f}{r_{po}^2}, \quad t_f = N \cdot 365 \cdot 24 + 730 \cdot 24 + b_h$$

$$Fo_1 = \frac{\alpha_s (t_f - t_y)}{r_{po}^2}, \quad t_f - t_y = 730 \cdot 24 + b_h \qquad (10.16)$$

$$Fo_2 = \frac{\alpha_s (t_f - t_m)}{r_{po}^2}, \quad t_f - t_m = b_h$$

It is important to remark that, in this analysis, the heat released is the heat per unit length of pipe and not trench. So the length found is the length of the piping and not the borehole as in the vertical case. So a trench with two pipes will have a total length half of the one given by the previous calculation.

Example 10.1 The building loads are represented by the following three ground heating loads:

$$q_a = 1 \, \text{kW}, \quad q_m = 3 \, \text{kW}, \quad q_h = 10 \, \text{kW}$$

Neglecting the seasonal temperature variation, compare the total pipe length needed for a two-pipe horizontal arrangement ($z_1 = 0.75 \, \text{m}$, $z_2 = 1.5 \, \text{m}$) and compare the length with a vertical borehole with the same soil properties. The fluid is water, the soil conductivity is $k_s = 2 \, \text{W/m-K}$, and its volumetric capacity $\rho C_p = 2 \, \text{MJ/m}^2\text{-K}$. In the vertical case, the borehole is 16 cm in diameter with a grout conductivity of 1.5 W/m-K and a B-configuration with a 10-year simulation and a minimum outlet temperature of $4\,^\circ\text{C}$. In both cases, a $1^1/4''$ SDR-11 pipe is used.

Solution The thermal diffusivity is given by:

$$\alpha = \frac{2}{2e6} = 1e\text{-}6 \, \text{m}^2/\text{s} = 0.0036 \, \text{m}^2/\text{hr}$$

$$Fo_f = \frac{\alpha t_f}{r_p^2} = 178873 \rightarrow X_f = 0.00118$$

Equation (10.15) leads to:

$$R'_{hori}(Fo_f) = 0.399 \, \text{K-m/W}$$

$$Fo_1 = \frac{\alpha(t_f - t_a)}{r_p^2} = 1486 \rightarrow X_1 = 0.0129$$

$$R'_{hori}(Fo_1) = 0.366 \, \text{K-m/W}$$

(continued)

$$Fo_2 = \frac{\alpha(t_f - t_m)}{r_p^2} = 8.1 \rightarrow X_2 = 0.1756$$

$$R'_{hori}(Fo_2) = 0.117 \, \text{K-m/W}$$

which gives:

$$R'_{s,a} = 0.399 - 0.366 = 0.032 \, \text{K-m/W}$$
$$R'_{s,m} = 0.366 - 0.117 = 0.249 \, \text{K-m/W}$$
$$R'_{s,h} = 0.117 \, \text{K-m/W}$$

$$T_{fi} = 4 - \frac{10}{0.9 \cdot 4.2} = 1.35 \Rightarrow T_f = 2.68\,^\circ\text{C}$$

The pipe resistance is given by:

$$Re = \frac{4 \cdot 0.9}{\pi \cdot 0.069 \cdot 0.00085} \approx 19000$$

From (2.43b), $h_{conv} = 1190 \, \text{W/m}^2 - \text{K}$:

$$R'_p = \frac{\log\left(\frac{0.042}{0.035}\right)}{2\pi 0.4} + \frac{1}{2\pi 0.034 \cdot 1190} = 0.084 \, \text{K-m/W}$$

$$L_{hori} = \frac{1000 \cdot 0.0325}{9.32} + \frac{3000 \cdot 0.249}{9.32} + \frac{10000 \cdot 0.117}{9.32}$$
$$+ \frac{10000 \cdot 0.084}{9.32}$$

$$L_{hori} = 298 \, \text{m}$$

(b) In the vertical case, if the line source model is used for the borehole resistance, we would get:

$$R'_b = 0.55 + \frac{0.084}{2} = 0.097 \, \text{K-m/W}$$

$$R'_{s,a} = 0.189 \, \text{K-m/W}, \quad R'_{s,m} = 0.166 \, \text{K-m/W},$$
$$R'_{s,h} = 0.051 \, \text{K-m/W}$$

If the penalty temperature is neglected and the ICS model is used, the ASHRAE method would give:

$$H_{vert} = 275 \, \text{m}$$

One should remember that this is the borehole height which represents a total pipe length of 551 m, which is much larger than the horizontal trench.

(continued)

Example 10.1 (continued)
Comments   *This result might seem surprising as it is well known that horizontal piping is known to be larger than vertical configuration. The reasons are simple: first, the negative impact of seasonal variation has not been taken into account, and second, the same ground properties were used; the value of conductivity given is quite low for a vertical borehole in a rock formation and quite high for a dry overburden seen often in horizontal systems. Both impacts will be analyzed later on, but the main conclusion of this example is to see that, if everything else is the same, the impact of the soil-air interface has a positive impact on the heat transfer, especially for the long-term impact which was exceptionally high in this example (1 kW).*

◼

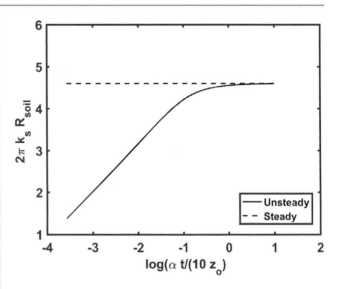

Fig. 10.7 Unsteady and steady-state ground resistance for one-pipe configuration

10.3.1   Time Scale for Horizontal Systems

Before looking at other factors, it is worthwhile to observe that the time scale for the steady regime is much less than for the vertical case. For this latter case, Eskilson [9] suggested that it is given by:

$$t_{c,v} = \frac{H^2}{9\alpha} \qquad (10.17)$$

For a horizontal single pipe, this time will be proportional to the depth:

$$t_{c,h} \propto \frac{z_o^2}{\alpha}$$

Looking at Fig. 10.7, it can be seen that for a single pipe,

$$t_{c,h} \approx \frac{10 z_o^2}{\alpha} \qquad (10.18)$$

For typical soil characteristics, Eq. (10.17) will be in the order of 20–30 years, whereas Eq. (10.18) will be in the order of a few months. The asymptotic value reaches the well-known steady-state resistance associated with this geometry [10]:

$$R'_{ss} = \frac{\log(2 z_o / r_{po})}{2\pi k s} \qquad (10.19)$$

This observation explains the small value of the year's soil resistance in the last example. Indeed, it is not really that it is small but that it does not change too much with time which shows that the impact of multiyear simulation is less important in horizontal configurations. This conclusion may however be different if the air-soil interface is well insulated.

10.4   Impact of the Seasonal Temperature Variation

The exact impact of temperature variation along the pipe is not easy since the temperature level vary with time and space. Simulation tools are available to take into account these variations but simple approaches for design purposes are not trivial. A simple possibility is to replace the undisturbed temperature T_o in the ASHRAE's method by the temperature value given by a predicted model like Kusuda's model (Eq. 10.2). Due to a phase lag between the temperatures at the surface and at a given depth, the minimum value can be at a different time. The idea is then to replace T_o in Eq. (10.14)

$$T_o \rightarrow T_o - A \exp\left(-z_o \sqrt{\frac{\pi}{365\alpha}}\right)$$
$$\times \cos\left(\omega\left(t_f - t_{shift} - \frac{z_o}{2}\sqrt{\frac{365}{\pi\alpha}}\right)\right) \qquad (10.20)$$

Two choices are possible, i.e., either compute the temperature at the day of design or take the worst case:

$$T_o \rightarrow T_o - A \exp\left(-z_o \sqrt{\frac{\pi}{365\alpha}}\right) \qquad (10.21)$$

the latest being more conservative.

Example 10.2 Redo the last example assuming that the day of design is January 15, and the other weather data are:

(continued)

Example 10.2 (continued)
$$A = 10\,^\circ C, t_{shift} = 11$$

Solution Only the denominator will change from the preceding example. In (10.2), the pipe depth has to be given. Here the two pipes are at different depths. Complex analysis taking into account whether the fluid enters the lower or higher pipe can be done [11] but simple assumptions can also be done. One can either compute the temperature variation for each depth and take the mean or estimate the temperature at the mean depth. This is the choice made here: If the temperature at the design day is calculated, the temperature will be:

$$\alpha = 0.0036 \text{ m}^2/\text{hr} = 0.086 \text{ m}^2/\text{d}$$

$$T_{on} = 12 - 10\exp\left(-1.125\sqrt{\frac{\pi}{365 \cdot 0.086}}\right.$$

$$\cos\left(\omega\left(15 - 11 - \frac{1.125}{2}\sqrt{\frac{365}{\pi \cdot 0.086}}\right)\right) = 5.27\,^\circ C$$

$$L_{hori} = \frac{1000 \cdot 0.0325}{2.59} + \frac{3000 \cdot 0.249}{2.59} + \frac{10000 \cdot 0.117}{2.59}$$
$$+ \frac{10000 \cdot 0.084}{2.59}$$

$$L_{hori} = 1072 \text{ m}$$

In the case where the minimum temperature is taken:

$$T_{on} = 12 - 10\exp\left(-1.125\sqrt{\frac{\pi}{365 \cdot 0.086}}\right) = 4.99\,^\circ C$$

$$L_{hori} = \frac{1000 \cdot 0.0325}{2.31} + \frac{3000 \cdot 0.249}{2.31} + \frac{10000 \cdot 0.117}{2.31}$$
$$+ \frac{10000 \cdot 0.084}{2.31}$$

$$L_{hori} = 1205 \text{ m}$$

Comments *The harmful impact of the temperature decrease is important. In this example, the coldest temperature is not much higher than the prescribed outlet temperature targeted. A lowered value with an antifreeze solution would be a more sounded choice. Water was kept here to compare with the previous example.* ∎

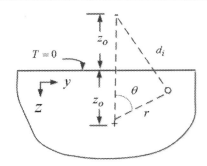

Fig. 10.8 Steady-state analysis for one-pipe configuration

10.5 Steady-State Analysis

As stated before, the time scale when the steady state is reached is much smaller in the case of horizontal systems, and it is worthwhile to pay more attention to the steady-state behavior. The following procedure follows the same step as described in Claesson and Dunand [4]. The method of images discussed earlier is still valid in the steady-state regime if one uses the solution associated with the emission of heat at one source point. This solution is well known and is given by [12]:[1]

$$T = -\frac{q}{4\pi r k_s} \tag{10.22}$$

where $r^2 = (x-\xi)^2+(y-\eta)^2+(z-\zeta)^2$ is the distance from the point source (or sink, remember that positive q means heat extraction). In the case of a line source, the 2D solution is:

$$T = \frac{q'}{2\pi k_s}\log(r) \tag{10.23}$$

where $r^2 = (y - \eta)^2 + (z - \zeta)^2$ is the radial distance from the source in the 2D plane. Referring to Fig. 10.8, we have:

$$T = -\frac{q'}{2\pi k_s}\log\left(\frac{d_i}{r}\right) = -\frac{q}{4\pi k_s}\log\left(\frac{y^2 + (z + z_o)^2}{y^2 + (z - z_o)^2}\right) \tag{10.24}$$

Generalization for multiple pipes can be done the same way. For example, the two-pipe arrangement shown in Fig. 10.9 is found by:

$$T = -\frac{1}{4\pi k_s}\left[q'_1\log\left(\frac{(y - y_1)^2 + (z + z_1)^2}{(y - y_1)^2 + (z - z_1)^2}\right)\right.$$
$$\left. -q'_2\log\left(\frac{(y - y_2)^2 + (z + z_2)^2}{(y - y_2)^2 + (z - z_2)^2}\right)\right] \tag{10.25}$$

---

[1]The solution found in this section is again associated with the temperature field ^1T, associated with the homogeneous boundary condition where the superscript is omitted for clarity.

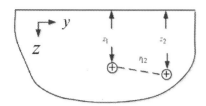

Fig. 10.9 Steady-state analysis for two-pipes configuration

A temperature of particular interest is the temperature at the pipe periphery. For pipe 1, this is:

$$T_{p1} = -\frac{1}{2\pi k_s}\left[q_1' \log\left(\frac{2z_1}{r_p}\right)\right.$$

$$\left. + q_2' \log\left(\frac{\sqrt{(y_1 - y_2)^2 + (z_1 + z_2)^2}}{r_{12}}\right)\right] \quad (10.26)$$

For pipe 2, this is:

$$T_{p2} = -\frac{1}{2\pi k_s}\left[q_1' \log\left(\frac{\sqrt{(y_1 - y_2)^2 + (z_1 + z_2)^2}}{r_{12}}\right)\right.$$

$$\left. + q_2' \log\left(\frac{2z_2}{r_p}\right)\right] \quad (10.27)$$

If both heat transfer rates are the same, these two equations are basically the steady-state equivalent of (10.6) and (10.8). In the more general case, they can be put in the following:

$$T_{p1} = -q_1' R_1' - q_2' R_{12}'$$
$$T_{p2} = -q_1' R_{12}' - q_2' R_2' \quad (10.28)$$

with

$$R_i' = \frac{1}{2\pi k_s} \log\left(\frac{2z_i}{r_p}\right) \quad (10.29)$$

$$R_{ij}' = \frac{1}{2\pi k_s} \log\left(\frac{\sqrt{(y_i - y_j)^2 + (z_i + z_j)^2}}{r_{ij}}\right) \quad (10.30)$$

The mean fluid temperature is found from the usual pipe resistance:

$$T_{f1} = -q_1' R_{t1}' - q_2' R_{12}'$$
$$T_{f2} = -q_1' R_{12}' - q_2' R_{t2}' \quad (10.31)$$

with

$$R_t' = R' + R_p' \quad (10.32)$$

10.5.1 Fluid Temperature Along the Pipe

In the case of a single pipe, the fluid distribution is found from:

$$\dot{m}C_p \frac{dT_f}{dx} = q' = \frac{-T_f}{R_t'}$$

$$R_t' = R_i' + R_p'$$

where the usual exponential solution is:

$$T_f(x) = T_{fi} \exp\left(-\frac{x}{\dot{m}C_p R_t'}\right) \quad (10.33)$$

As in Chap. 5, we have:

$$T_{fo} = T_{fi} \exp\left(-\frac{L}{\dot{m}C_p R_t'}\right) = T_{fi} \exp(-\gamma)$$

with

$$\gamma = \frac{L}{\dot{m}C_p R_t'} = \frac{\mu}{R_t'} \quad (10.34)$$

from Eq. (5.66), it is found that:

$$T_{fo} = -q' R_t'\left(\frac{\gamma \exp(-\gamma)}{1 - \exp(-\gamma)}\right) \quad (10.35)$$

$$T_{fi} = -q' R_t'\left(\frac{\gamma}{1 - \exp(-\gamma)}\right) \quad (10.36)$$

and

$$T_f = \frac{T_{fo} + T_{fi}}{2} = -q_b' R_p'^* \quad (10.37)$$

with

$$R_t'^* = R_t' \frac{\gamma}{2} \coth\left(\frac{\gamma}{2}\right) \quad (10.38)$$

For small values of γ, the linear approximation is still valid:

$$T_{fi} = T_f - q'\frac{\mu}{2}$$

$$T_{fo} = T_f + q'\frac{\mu}{2} \quad (10.39)$$

10.5.2 Fluid Temperature for Several Pipes

Generalization of the temperature profile for several pipes can be easily done if the linear approximation is used. The two-pipe case will be analyzed. For more pipes, the work of Claesson and Dunand [4] can be consulted. Since the analysis is done for each individual pipe, L will be associated with the length of a single pipe (trench).

10.5.2.1 Parallel Configuration
First consider two pipes in parallel. In this case, the inlet fluid temperature is assumed to be the same. From (10.31), we have:

$$T_{fi1} = T_{f1} - \frac{\mu}{2}q_1' = -q_1'\left(R_{t1}' + \frac{\mu}{2}\right) - q_2' R_{12}'$$

$$T_{fi2} = T_{f2} - \frac{\mu}{2}q'_2 = -q'_1 R'_{12} - q'_2 \left(R'_{12} + \frac{\mu}{2}\right)$$

$$(10.40)$$

Putting the two equals gives:

$$q'_1 = \frac{R'_{12} + \frac{\mu}{2} - R'_{12}}{R'_{11} + \frac{\mu}{2} - R'_{12}}q'_2 \qquad (10.41)$$

Claesson and Dunand [4] found their soil resistances with respect to the heat transfer rate per unit length of trench (not pipe), $q'_1 + q'_2$. For comparison with the previous section, the (mean) heat transfer per unit length of pipe will be used here:

$$q' = \frac{q'_1 + q'_2}{2}$$

$$= \frac{R'_{12} + R'_{11} + \mu - 2R'_{12}}{2(R'_{11} + \frac{\mu}{2} - R'_{12})}q'_2 \qquad (10.42)$$

$$= \frac{R'_{12} + R'_{11} + \mu - 2R'_{12}}{2(R'_{12} + \frac{\mu}{2} - R'_{12})}q'_1$$

From which we can define:

$$q'_1 = \frac{2(R'_{12} + \frac{\mu}{2} - R'_{12})}{R'_{12} + R'_{11} + \mu - 2R'_{12}}q' = \zeta_{p1}q'$$

$$q'_2 = \frac{2(R'_{11} + \frac{\mu}{2} - R'_{12})}{R'_{12} + R'_{11} + \mu - 2R'_{12}}q' = \zeta_{p2}q' \qquad (10.43)$$

Putting in (10.40):

$$T_{fi} = -(\zeta_{p1}(R'_{11} + \frac{\mu}{2}) + \zeta_{p2}R'_{12})q'$$

$$T_{fi} = -2\left[\frac{(R'_{12} + \frac{\mu}{2})(R'_{11} + \frac{\mu}{2}) - R'^2_{12}}{R'_{12} + R'_{11} + \mu - 2R'_{12}}\right]q' = -R'_{in}q'$$

$$(10.44)$$

The mean fluid temperature can be found from Eq. (10.39):

$$T_{f1} = T_{fi} + \frac{\mu}{2}q'_1 = T_{fi} + \frac{\mu}{2}\zeta_{p1}q'$$

$$T_{f2} = T_{fi} + \frac{\mu}{2}q'_2$$
$$= T_{fi} + \frac{\mu}{2}\zeta_{p2}q'$$

$$T_f = \frac{T_{f1} + T_{f2}}{2} = -q'\left(R'_{in} - \frac{\mu}{4}(\zeta_{p1} + \zeta_{p2})\right)$$

$$= -q'\left(R'_{in} - \frac{\mu}{4}\right)$$

The final result is then

$$T_f = -2\left[\frac{R'_{12}R'_{11} - R'^2_{12} + \frac{\mu}{2}R'_{12} + \frac{\mu}{4}(R'_{11} + R'_{12})}{R'_{12} + R'_{11} + \mu - 2R'_{12}}\right]$$

$$\times q' = -R'_{eq,p}q' \qquad (10.45)$$

10.5.2.2 Series Configuration

In the series case, the inlet temperature of one pipe is the exit temperature of the first one. Assuming that it enters in pipe one:

$$T_{fi1} = T_{f1} - \frac{\mu}{2}q'_1 = -q'_1\left(R'_{11} + \frac{\mu}{2}\right) - q'_2 R'_{12}$$

$$T_{fi2} = T_{fo1} = T_{f1} + \frac{\mu}{2}q'_1 = -q'_1\left(R'_{11} - \frac{\mu}{2}\right) - q'_2 R'_{12}$$

and also that:

$$T_{fi2} = T_{f2} - \frac{\mu}{2}q'_2 = -q'_1 R'_{12} - q'_2\left(R'_{12} + \frac{\mu}{2}\right)$$

Putting the last two equals leads to:

$$q'_1 = \frac{R'_{12} + \frac{\mu}{2} - R'_{12}}{R'_{11} - \frac{\mu}{2} - R'_{12}}q'_2 \qquad (10.46)$$

From which it is possible to find:

$$q'_1 = 2\frac{R'_{12} + \frac{\mu}{2} - R'_{12}}{R'_{12} + R'_{11} - 2R'_{12}}q' = \zeta_{s1}q'$$

$$q'_2 = 2\frac{R'_{11} - \frac{\mu}{2} - R'_{12}}{R'_{12} + R'_{11} - 2R'_{12}}q' = \zeta_{s2}q' \qquad (10.47)$$

The inlet temperature of the first pipe is then:

$$T_{fi1} = -q'\left(\zeta_{s1}\left(R'_{11} + \frac{\mu}{2}\right) + \zeta_{s2}R'_{12}\right)$$

$$T_{fi1} = -q'2\frac{(R'_{11} + \frac{\mu}{2})(R'_{12} + \frac{\mu}{2}) - R'_{12}(R'_{12} + \mu)}{R'_{12} + R'_{11} - 2R'_{12}}$$

$$(10.48)$$

The outlet temperature of the second pipe is:

$$T_{fo2} = -q'\left(\zeta_{s1}R'_{12} + \zeta_{s2}\left(R'_{12} - \frac{\mu}{2}\right)\right)$$

$$T_{fo2} = -q'2\frac{(R'_{11} - \frac{\mu}{2})(R'_{12} - \frac{\mu}{2}) - R'_{12}(R'_{12} - \mu)}{R'_{12} + R'_{11} - 2R'_{12}}$$

$$(10.49)$$

Summing the two:

$$T_f = \frac{T_{fi1} + T_{fo2}}{2} = -q'R'_{eq,s}$$

$$R'_{eq,s} = 2\frac{R'_{11}R'_{12} - R'^2_{12} + \frac{\mu^2}{4}}{R'_{12} + R'_{11} - 2R'_{12}} \qquad (10.50)$$

In the series configuration, the total flow rate is the same as the flow rate in each pipe, so:

$$T_{fi} = T_f - \mu q'$$
$$T_{fo} = T_f + \mu q'$$

However, in the series configuration, care should be taken that:

$$T_f = \frac{T_{fi1} + T_{fo2}}{2} \neq \frac{T_{f1} + T_{f2}}{2}$$

That is, the global mean fluid temperature is not the mean of the mean temperatures. Indeed, we have:

$$\frac{T_{f1} + T_{f2}}{2} = \frac{T_{fi1} + T_{fo1} + T_{fi2} + T_{fo2}}{4}$$
$$\frac{2T_{fi1} + \mu q'_1 + 2T_{fo2} - \mu q'_2}{4}$$
$$T_f + \frac{\mu(q'_1 - q'_2)}{4}$$

Example 10.3 Compare the inlet and outlet temperatures of a two-pipe arrangement in a similar arrangement as the one shown in Fig. 10.2 for the parallel and series arrangement. In the latter case, compare the case where the inlet is at the bottom and at the top of the circuit. Use the following values:

$$R'_p = 0.04 \text{ K-m/W}, \ r_p = 0.02 \text{ m}, \ L_{trench} = 200 \text{ m},$$
$$k_s = 1.5 \text{ W/m-K}$$

$$z_1 = 2 \text{ m}, \ z_2 = 3 \text{ m}, \ C_p = 4000 \text{ J/Kg-K}, \ \rho = 1000 \text{ kg/m}^3$$

$$u = 0.2 \text{ m/s}, \ q = 4 \text{ kW}, \ T_o = 0 \,°\text{C}$$

Solution Direct comparison is not obvious since the flow rates and the pipe diameters would probably be different in series and parallel configurations. Claesson and Dunand [4] compared both configurations assuming the same inlet temperature and evaluating the total heat rate in both situations. They found that the parallel configuration is better. In heat pump applications, we are merely concerned with the temperature leaving the ground heat exchanger (entering the heat pump) for a given heat transfer.

Series configuration

$$R'_{t1} = \frac{1}{2\pi 1.5} \log\left(\frac{4}{0.02}\right) + 0.04 = 0.602 \text{ K-m/W}$$

(continued)

$$R'_{t2} = \frac{1}{2\pi 1.5} \log\left(\frac{6}{0.02}\right) + 0.04 = 0.645 \text{ K-m/W}$$

$$R'_{12} = \frac{1}{2\pi 1.5} \log\left(\frac{5}{1}\right) = 0.171 \text{ K-m/W}$$

$$\dot{m} = \rho u A_c = 1000 \cdot 0.2 \cdot \pi 0.02^2 = 0.251 \text{ kg/s}$$

$$\mu = \frac{200}{0.251 \cdot 4000} = 0.199 \text{ K-m/W}$$

$$q' = \frac{4000}{2 \cdot 200} = 10 \text{ W/m}$$

From (10.50):

$$R'_{eq,s} = 0.815 \text{ K-m/W}$$

$$T_f = -10 \cdot 0.815 = -8.15 \,°\text{C}$$

$$T_{fi} = -8.15 - 10 \cdot 0.199 = -10.14 \,°\text{C}$$

$$T_{fo} = -8.15 + 10 \cdot 0.199 = -6.16 \,°\text{C}$$

From (10.47), we found also that:

$$\zeta_{s1} = 1.267, \ \zeta_{s2} = 0.732$$

From which

$$q_1 = 2534 \text{ W}, \ q_2 = 1466 \text{ W}$$

Assuming that the fluid enters the lowest pipe first:

$$R'_{t1} = 0.645 \text{ K-m/W}, R'_{t2} = 0.601 \text{ K-m/W}$$

and

$$T_{fi} - 10.14 \,°\text{C}, \ T_{fo} = -6.16 \,°\text{C}$$
$$q_1 = 2344 \text{ W}, \ q_2 = 1655 \text{ W}$$

Parallel configuration

$$R'_{eq,p} = 0.794 \text{ K-m/W}$$

$$T_f = -10 \cdot 0.794 = -7.94 \,°\text{C}$$

$$T_{fi} = -7.94 - 10 \cdot 0.199/2 = -8.93 \,°\text{C}$$

$$T_{fo} = -7.94 + 10 \cdot 0.199/2 = -6.94 \,°\text{C}$$

$$q_1 = 2077 \text{ W}, \ q_2 = 1922 \text{ W}$$

result which seems to indicate that, even though the global resistance is a little bit smaller, the entering heat pump temperature is lower than in the series configuration which would mean poorer performances. This follows from the fact that the flow rate in each portion

(continued)

Example 10.3 (continued)
of the pipe was kept constant which means that the total flow rate in the parallel configuration is twice as big as in the series configuration. Keeping the total flow rate the same would give:

$$\mu = \frac{400}{0.251 \cdot 4000} = 0.398 \text{ K-m/W}$$

$$R'_{eq,p} = 0.794 \text{ K-m/W}$$

$$T_f = -20 \cdot 0.3027 = -7.94 \,^\circ\text{C}$$

$$T_{fi} = -7.94 - 10 \cdot 0.398/2 = -9.93 \,^\circ\text{C}$$

$$T_{fo} = -7.94 + 10 \cdot 0.398/2 = -5.94 \,^\circ\text{C}$$

which is now a better performance. As pointed out earlier, the smaller flow rate in each pipe would probably result in a smaller pipe diameter and a different pipe resistance, but these parameters were kept constant for comparison. Finally, it would be interesting to see what the result would be using the asymptotic value of the unsteady analysis proposed in [7]:

$$R_{horiz,2pipe}(Fo \Rightarrow \infty) = 0.754 \text{ K-m/W}$$

$$R_{img} = 0.754 + 0.04 = 0.794 \text{ K-m/W}$$

$$T_f = 0 - 10 \cdot 0.794 = -5.68 \,^\circ\text{C}$$

$$T_{fi} = -6.08 - \frac{4000}{2 \cdot \dot{m} C_p} = -9.93 \,^\circ\text{C}$$

$$T_{fo} = -6.08 + \frac{4000}{2 \cdot \dot{m} C_p} = -5.95 \,^\circ\text{C}$$

Basically, it is the same as the parallel configuration having the same total flow rate. The reason is easy to understand since in that case, the heat exchange is almost equal in both legs. ∎

Example 10.4 A better comparison would probably be to change the pipe diameter in order to have the same order of velocity. Let's redo the last example with this new configuration:

$$L = 200 \text{ m}, \quad k_s = 1.5 \text{ W/m-K}, \quad z_1 = 2 \text{ m}, \quad z_2 = 3 \text{ m}$$

$$C_p = 4000 \text{ W/K}, \quad \rho = 1000 \text{ W/K}, \quad q = 4 \text{ kW},$$

$$k_p = 0.4 \text{ W/m-K}$$

(continued)

Series Configuration

$$\dot{m} = 0.5 \text{ kg/s}, \quad SDR - 11(2")$$

Parallel Configuration

$$\dot{m} = 0.25 \text{ kg/s}, \quad SDR - 11(1.25")$$

With this choice, the total flow rate at the entrance of the heat pump is the same and the velocity is almost the same in each configuration.[2] Using water.[3] For the series configuration:

$$r_{pi} = 0.025 \text{ m}, \quad r_{po} = 0.030 \text{ m}$$

$$Re = 15107 \Rightarrow \hbar = 1331 \text{ W/m}^2 - \text{K}$$

$$R' = 0.0798 + 0.0048 = 0.084 \text{ K-m/W}$$

$$R'_{t1} = \frac{1}{2\pi 1.5} \log\left(\frac{4}{0.03}\right) + 0.084 = 0.603 \text{ K-m/W}$$

$$R'_{t2} = \frac{1}{2\pi 1.5} \log\left(\frac{6}{0.03}\right) + 0.084 = 0.646 \text{ K-m/W}$$

$$R'_{12} = \frac{1}{2\pi 1.5} \log\left(\frac{6}{1}\right) = 0.170 \text{ K-m/W}$$

$$\mu = \frac{200}{2000} = 0.1 \text{ K-m/W}$$

$$R'_{eq,s} = 0.80 \text{ K-m/W}$$

$$T_f = -10 \cdot 0.8 = -8 \,^\circ\text{C}$$

$$T_{fi} = -8.0 - 10 \cdot 0.1 = -9.0 \,^\circ\text{C}$$

$$T_{fo} = -8.0 + 10 \cdot 0.1 = -7.0 \,^\circ\text{C}$$

In the parallel configuration:

$$r_{pi} = 0.017 \text{ m}, \quad r_{po} = 0.021 \text{ m}$$

$$Re = 10800 \Rightarrow \hbar = 1409 \text{ W/m}^2 - \text{K}$$

$$R' = 0.0798 + 0.006 = 0.086 \text{ K-m/W}$$

$$R'_{t1} = \frac{1}{2\pi 1.5} \log\left(\frac{4}{0.021}\right) + 0.086 = 0.642 \text{ K-m/W}$$

$$R'_{t2} = \frac{1}{2\pi 1.5} \log\left(\frac{6}{0.021}\right) + 0.086 = 0.686 \text{ K-m/W}$$

$$\mu = \frac{200}{1000} = 0.2 \text{ K-m/W}$$

---

[2]The flow rates are however low for the diameter chosen.
[3]Remember that the temperatures found are relative.

(continued)

Example 10.4 (continued)

$$R'_{eq,p} = 0.8344 \text{ K-m/W}$$

$$T_f = -10 \cdot 0.8344 = -8.344 \,^{\circ}\text{C}$$

$$T_{fi} = -8.344 - 10 \cdot 0.2/2 = -9.34 \,^{\circ}\text{C}$$

$$T_{fo} = -8.344 + 10 \cdot 0.2/2 = -7.34 \,^{\circ}\text{C}$$

The outlet temperature is now lower due to higher resistances associated with the smaller diameter and the series configuration gives a better performance. However, the price of the piping will be higher. ∎

In the work of Claesson and Dunand [4], the total heat rate is given but each pipe can exchange heat at different rates compared to the method of image. However, it was limited to the steady case. Extension to the unsteady regime can be found in Fontaine et al. [13] and Lamarche [11]. Even in the case of steady state, the work of Claesson and Dunand assumes that the temperature varies linearly in each part of the pipes. In very long ducts, this assumption can lead to wrong results. This assumption can easily be removed at least in the steady case for the two-pipe system. This was done for the series arrangement in Lamarche [11]. It follows very similar steps than what was done for the U-tube arrangement, except that in this latter case, the analysis was restricted to the inside of the borehole and now it includes the ground. The temperature variation can easily be obtained by two-energy heat balance on each pipe:

$$\dot{m}C_p \frac{dT_{1(x)}}{dx} = q'_1 = \frac{T_o - T_1}{R_1'^{\Delta}} + \frac{T_2 - T_1}{R_{12}'^{\Delta}}$$

$$-\dot{m}C_p \frac{dT_{2(x)}}{dx} = q'_2 = \frac{T_o - T_2}{R_2'^{\Delta}} + \frac{T_1 - T_2}{R_{12}'^{\Delta}} \quad (10.51)$$

The Δ formalism makes reference to the Delta shape in a U-tube. It is less meaningful here but will still be kept. The system (Eqs. 10.28) can easily be inverted to find:

$$R_1'^{\Delta} = \frac{R'_{t1}R'_{t2} - R_{12}'^2}{R'_{t2} - R'_{12}} \quad (10.52)$$

$$R_2'^{\Delta} = \frac{R'_{t1}R'_{t2} - R_{12}'^2}{R'_{t1} - R'_{12}} \quad (10.53)$$

$$R_{12}'^{\Delta} = \frac{R'_{t1}R'_{t2} - R_{12}'^2}{R'_{12}} \quad (10.54)$$

and solving (10.51)

$$\theta_1 = \exp\left(\frac{\tilde{x}(\xi_1 - \xi_2)}{2}\right)$$

$$\times \left[\frac{\eta \cosh(\eta(1 - \tilde{x})) + \xi_m \sinh(\eta(1 - \tilde{x}))}{\eta \cosh(\eta) + \xi_m \sinh(\eta)}\right] \quad (10.55\text{a})$$

$$\theta_2 = \exp\left(\frac{\tilde{x}(\xi_1 - \xi_2)}{2}\right)$$

$$\times \left[\frac{\eta \cosh(\eta(1 - \tilde{x})) - \xi_m \sinh(\eta(1 - \tilde{x}))}{\eta \cosh(\eta) + \xi_m \sinh(\eta)}\right] \quad (10.55\text{b})$$

with

$$\theta = \frac{T - T_o}{T_{in} - T_o}$$

and

$$\tilde{x} = \frac{x}{L}, \quad \xi_1 = \frac{\mu}{R_1'^{\Delta}}, \quad \xi_2 = \frac{\mu}{R_2'^{\Delta}}, \quad \xi_{12} = \frac{\mu}{R_{12}'^{\Delta}}$$

$$\xi_m = \frac{\xi_1 + \xi_2}{2}, \quad \eta = \sqrt{\xi_m^2 + 2\xi_{12}\xi_m}$$

The exit temperature is then:

$$\theta_o = \frac{T_2(0) - T_o}{T_{in} - T_o} = \frac{\eta \cosh(\eta) - \xi_m \sinh(\eta)}{\eta \cosh(\eta) + \xi_m \sinh(\eta)} \quad (10.56)$$

The *effective resistance* associated with this two-pipe arrangement can easily be obtained:

$$q' = \frac{\dot{m}C_p}{2L}(T_{out} - T_{in}) = \frac{\mu}{2}(T_{in} - T_o)(\theta_o - 1)$$

$$T_{in} - T_o = -q' \frac{2\mu}{1 - \theta_o} \quad (10.57)$$

Similarly:

$$T_{out} - T_o = -q' \frac{2\mu\theta_o}{1 - \theta_o} \quad (10.58)$$

Summing both:[4]

$$R'^* = \frac{T_o - T_f}{q'} = \frac{\mu(1 + \theta_o)}{(1 - \theta_o)} = \frac{2R_1'^{\Delta}R_2'^{\Delta}}{R_1'^{\Delta} + R_2'^{\Delta}}\eta\text{cotanh}\,(\eta) \quad (10.59)$$

---

[4]Unfortunately, there was a mistake in [11] (Eq. 21).

Example 10.5 The solution of Example 10.3 has been solved with a FEM software and the solution is given in file "Exemple10_3.txt" for the series arrangement. Compare the effective resistance and the outlet temperature given by the FEM, the one found from Example 10.3 and the one predicted by the previous section.

Solution The solution found from COMSOL was:

$$T_{comsol,in} = -10.37\,°C, \quad T_{comsol,out} = -6.40\,°C$$

From the last example:

$$T_{Claesson,in} = -10.14\,°C, \quad T_{Claesson,out} = -6.16\,°C$$

$$T_{Imag,in} = -9.93\,°C, \quad T_{Imag,out} = -5.95\,°C$$

To evaluate the inlet from Eqs. (10.55a, 10.55b), we can use Eq. (10.57):

$$R_1'^{\Delta} = \frac{R'_{t1}R'_{t2} - R'^2_{12}}{R'_{t2} - R'_{12}} = 0.757\ \text{K-m/W}$$

$$R_2'^{\Delta} = \frac{R'_{t1}R'_{t2} - R'^2_{12}}{R'_{t1} - R'_{12}} = 0.833\ \text{K-m/W}$$

$$R_{12}'^{\Delta} = \frac{R'_{t1}R'_{t2} - R'^2_{12}}{R'_{12}} = 2.104\ \text{K-m/W}$$

$$\xi_1 = \frac{0.98}{0.757} = 0.262,\ \xi_2 = \frac{0.98}{0.833} = 0.238,\ \xi_{12} = \frac{0.98}{2.104} = 0.094$$

$$\xi_m = 0.251,\ \eta = 0.332$$

$$\theta_o = \frac{\eta\cosh(\eta) - \xi_m\sinh(\eta)}{\eta\cosh(\eta) + \xi_m\sinh(\eta)} = 0.6104$$

$$T_{in} = 0 - 10\frac{2\cdot 0.199}{1 - 0.6104} = -10.21\,°C$$

$$T_{out} = 0 - 10\frac{2\cdot 0.199\cdot 0.6104}{1 - 0.6104} = -6.23\,°C$$

A closer value, however, is not very different. The same conclusion can be observed if the effective resistance is calculated:

$$R'^*_{comsol} = \frac{0 - (T_{comsol,in} + T_{comsol,out})/2}{10}$$

$$= 0.838\ \text{K-m/W}$$

$$R'^*(Eq.\ 10.59) = \frac{2R_1'^{\Delta}R_2'^{\Delta}}{R_1'^{\Delta} + R_2'^{\Delta}}\eta\coth(\eta)$$

(continued)

$$= 0.822\ \text{K-m/W}$$

and from Example 10.3:

$$R'^*(Claesson) = 0.815\ \text{K-m/W}$$

$$R'^*(Imag) = 0.794\ \text{K-m/W}$$

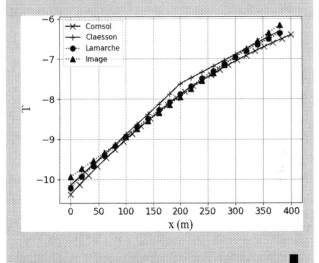

Example 10.6 As a final example, let's redo the last example with more severe conditions with a low velocity and a larger soil conductivity. All the parameters are kept unchanged except for the velocity which is set at $u = 0.06$ m/s and the soil conductivity which is now $k_s = 3$ W/m-K.

Solution The results are:

$$T_{comsol,in} = -15.47\,°C, \quad T_{comsol,out} = -2.21\,°C$$

$$T_{Claesson,in} = -15.25\,°C, \quad T_{Claesson,out} = -1.99\,°C$$

$$T_{Imag,in} = -10.8\,°C, \quad T_{Imag,out} = 2.45\,°C$$

$$T_{in}(Eq.\ 10.57) = -15.53\,°C, \quad T_{out}(Eq.\ 10.58) = -2.27\,°C$$

$$R'^*_{comsol} = 0.884\ \text{K-m/W}$$

$$R'^*(Eq.\ 10.59) = 0.890\ \text{K-m/W}$$

$$R'^*(Claesson) = 0.862\ \text{K-m/W}$$

(continued)

Example 10.6 (continued)

$$R'^*(Imag) = 0.417 \text{ K-m/W}$$

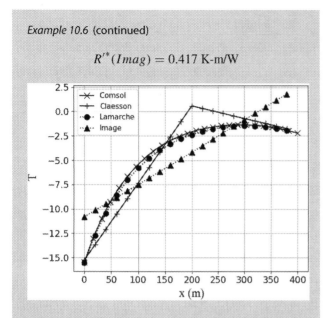

It easily seen that the classical method of image where the heat transfer is assumed identical gives non-physical results ($T_{out} > T_o$) since the heat transfer is negative in the second part of the pipe. The linear approximation of Claesson and Dunand gives relatively good inlet and outlet temperatures, but the temperature profile is crude although not always necessary. Finally, those conditions are rarely met in practice but again care should be taken in these extreme cases. ∎

Problems

Pr. 10.1 A horizontal GSHP is used to heat and cool a building. The total heating and cooling capacity is 8 kW and 10 kW, respectively. The monthly PLF is 0.1 in heating and 0.25 in cooling. The equivalent full-load hours are 400 in heating and 800 in cooling. The ground conductivity is 1.2 W/m-K and its diffusivity is 0.05 m²/day. The heating COP is 3.0 and the cooling COP is 4.0. The outside pipe diameter is 5 cm and the pipe resistance is $R'_p = 0.08$ K-m/W. The heat transfer fluid used is a mixture of 30% of propylene glycol and water:

$$T_o = 11\,°C, A_s = 14K, t_{shift} = 5, \text{ warmest day: } 180$$

Compare the trench length needed for the two configurations shown with a minimum EWT of -2 °C during winter and a maximum of 30 °C during summer. Use the recommended flow rate of 0.054 l/s/kW (Fig. Pr_10.1).

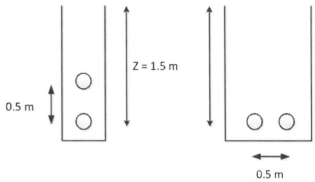

Fig. Pr_10.1 Two two-pipes different layouts

Answer: left: L = 179 m, right: L = 169 m

Pr. 10.2 A heat pump has a cooling capacity of 10 kW and is connected to a horizontal trench as shown in Fig. Pr_10.2. The COP of the heat pump increases with the flow rate. However, the pumping energy also increases.

It is asked to compare the system COP when the flow rate changes. The heat pump COP are given in the following table:

$\dot{V}_{cond}$ (l/s)	COP
0.30	4.0
0.35	4.4
0.40	4.7
0.45	5.0
0.50	5.2

$d_{p,i} = 2.5$ cm, $d_{p,o} = 3.0$ cm, $C_{p,fluid} = 4.18$ kJ/kg-K
$\nu_{fluid} = 8.57e-7$, $\rho_{fluid} = 1000$ kg/m³, $k_{fluid} = 0.61$ W/m-K
$k_{soil} = 1.2$ W/m-K, $\alpha_{soil} = 0.05$ m²/d, $T_o = 12\,°C$
$t_{shift} = 5$, $A_s = 14K$, $z = 1$ m, $x = 0.75$ m

Besides the friction losses inside the pipe, the singular losses associated with the heat pump condenser are modeled by:

$$\Delta P_{cond}(kPa) = 10.1 \left(\frac{\dot{V}}{\dot{V}_{nom}}\right)^2$$

with $\dot{V}_{nom} = 0.4$ l/s. The water-to-wire efficiency of the pump is estimated as 20%. The trench length is calculated in order to have a maximum temperature of 30 °C at the entrance of the heat pump when the heat pump works at full capacity during 120 hours.

(a) What is flow rate giving the best system's COP?
(b) What would it be if the pump's efficiency is 30%?

Answer: a) $\dot{V}$ = 0.4 l/s, COP_{sys} = 4, b)$\dot{V}$ = 0.45 l/s, COP_{sys} = 4.25

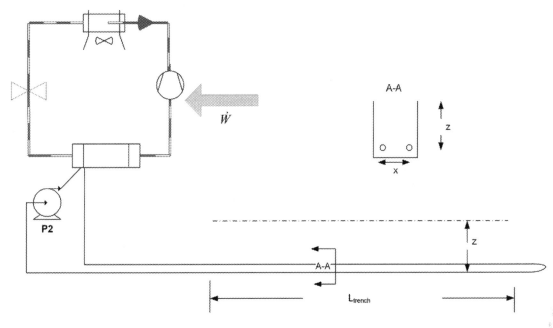

Fig. Pr_10.2 Heat pump coupled with a two-pipes horizontal configuration

Pr. 10.3 Redo the same problem if the length is calculated in the steady state (use a very long pulse time) and compare the results with the one predicted with the soil resistance suggested by Claesson and Dunand and the one given by Eq. (10.59).

Answer:

	V_{opt}(l/s)	COP (Image)	COP (Claesson)	COP (Eq. 10.59)
20%	0.35	3.81	3.73	3.69
30%	0.40	3.99	3.93	3.90

References

1. Lund, J.W., Boyd, T.L.: Direct utilization of geothermal energy 2015 worldwide review. Geothermics **60**, 66–93 (2016)
2. Raymond, J., Lamarche, L., Malo, M.: Field demonstration of a first thermal response test with a low power source. Appl. Energy **147**, 30–39 (2015). https://doi.org/10.1016/j.apenergy.2015.01.117
3. Rees, S.: Advances in Ground-Source Heat Pump Systems. Woodhead Publishing is an imprint of Elsevier, Duxford (2021)
4. Claesson, J., Dunand, A.: Heat Extraction from the Ground by Horizontal Pipes: A Mathematical Analysis. Technicalreport Report No. D1, ISBN 9154038510. Swedish Council for Building Research, 1983, 215 pp.
5. Mei, V.C.: Horizontal ground-coil heat exchanger theoretical and experimental analysis. Report. Report No. ORNL/CON-193, Dec, Oak Ridge National Laboratory, Oak Ridge, 1986. https://www.osti.gov/servlets/purl/7007955
6. Piechowski, M.: Heat and mass transfer model of a ground heat exchanger: theoretical development. Int. J. Energy Res. **23**(7), 571–588 (1999)
7. Bose, J.E,, Parker, J.D., McQuiston, F.C.: Design data manual for closed-loop ground-coupled heat pump systems. American Society of Heating, Refrigerating and Air-Conditioning Engineers, Atlanta (1985). ISBN: 0910110417
8. Kusuda, T., Achenbach, P.R.: Earth temperature and thermal diffusivity at selected stations in the United States. Report. National Bureau of Standards Gaithersburg MD, 1965
9. Eskilson, P.: Thermal analysis of heat extraction systems. Ph.D. Thesis. Lund University, Sweden, 1987
10. Bergman, T.L., Incropera, F.P.: Fundamentals of Heat and Mass Transfer, 7th edn. Wiley, New York (2011)
11. Lamarche, L.: Horizontal ground heat exchangers modelling. Appl. Thermal Eng. **155**, 534–545 (2019)
12. Carslaw, H.S., Jaeger, J.C.: Conduction of Heat in Solids, 2nd edn. (1959)
13. Fontaine, P.O. et al.: Modeling of horizontal geoexchange systems for building heating and permafrost stabilization. Geothermics **40**(3), 211–220 (2011)

11.1 Introduction

The basic principle of ground-source heat pumps is to extract heat in winter and reject excess heat in summer. As previously discussed, most systems used secondary loops where water or a mixture of water and antifreeze is used in closed-loop systems to perform such a task. In some cases, water can be present in the ground and this water can be effectively used as a source or sink for heat exchange. Lakes or ponds can be used as water sources and these systems are named "surface-water" systems. In this chapter, water contained in the ground in aquifers will be examined and these systems are called "groundwater heat pumps" (GWHP). Most of groundwater applications use open-loop configurations but "standing-column " applications can also be considered as GWHP. A schematic of an open-loop ground heat pump is depicted in Fig. 11.1.

Since the heat transfer liquid is the water found in the aquifer, its inlet temperature is then given, and the design of such system differs from the closed loop where the heat pump entering water temperature (EWT) was the main criterion to size the ground loop (Fig. 11.2).

Since here, the EWT is basically known, except when an external heat exchanger is used, the flow rate is the most important design parameter. Design considerations based on this principle will be studied later on. In open-loop system, the major concern is on the water resource itself and whether it is appropriate to be a good source of energy for a GWHP system. Such analysis is normally performed by geologists but basic knowledge on groundwater theory is important for any engineer working in this field, and some notions will be given in the next section.

11.2 Introduction to Groundwater Hydrology

Water present beneath the ground surface is called *underground water* or simply *groundwater*. Most of the water on Earth ($\approx 93\%$) is salted water in the oceans [1]. Available fresh water is estimated as more than 28 million km^3 from which approximately 4 million (14 %) is groundwater (Table 11.1).

The water in the ground is not alone and forms a *porous media* where solid particles are mixed with water and air. The fraction of void volume is called *porosity*. When the void portion is filled with water, the medium is a *saturated zone*, whereas an *unsaturated zone* is when water and air are present in the openings. When the water is free to move in a saturated zone, its top surface will be at atmospheric pressure, and it forms an *unconfined aquifer*. If the water is trapped between two impermeable strata, it forms a *confined aquifer*. Semiconfined or *leaky aquifer* is when the aquifer is trapped at the bottom by an impermeable stratum but has a low permeability region called *aquitard* above. When a well is drilled in an unconfined aquifer, the water will stay at the same height as before which represents the *water table*. On the contrary, since the water in the confined aquifer is pressurized, it will tend to rise (Fig. 11.3).

Water movement in an aquifer is governed by the law of mass conservation also known as continuity equation in fluid mechanics. The classical form of the equation is well known. The integral form of the mass conservation equation can be written as:

$$\frac{d}{dt}\int_V dm + \int_S \rho(\vec{u}\cdot\vec{n})dS = 0 \qquad (11.1)$$

When the fluid is water, it is usually considered as incompressible and the first term is then zero. In porous media, even though the fluid portion is incompressible, the matrix of fluid and solid particles is not and an increase of pressure will induce an increase of water volume in the same matrix volume. This effect is associated with the *storage coefficient*. This variation of water volume is most often expressed in terms of total head variation as:

$$dm = S_s \rho h d\,V \qquad (11.2)$$

where

$$h = \frac{p}{\rho g} + z \qquad (11.3)$$

h is the *total head* and z is the height from an arbitrary chosen datum. $S_s[L^{-1}]$ is called the *specific storage coefficient*. It represents the amount of water volume that would be added (or released) to the porous matrix due to an increase (decrease) in hydraulic head. V is the confined aquifer volume from which the water is added (or drained). The first term of the continuity equation is then not zero but can be expressed as:

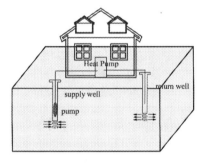

Fig. 11.1 Open-loop configuration

$$\frac{d}{dt} \int_V \rho d\,V = \frac{d}{dt} \int S_s \rho h d\,V = S_s \int \rho \frac{\partial h}{\partial t} d\,V \qquad (11.4)$$

Replacing in Eq. (11.1) leads to:

$$S_s \int_V \rho \frac{\partial h}{\partial t} d\,V + \int_S \rho(\vec{u} \cdot \vec{n}) dS$$

$$= S_s \int_V \rho \frac{\partial h}{\partial t} d\,V + \int_V \nabla \cdot (\rho \vec{u}) d\,V = 0 \qquad (11.5)$$

The specific storage coefficient is related to the *storativity* defined as:

$$S = S_s b \qquad (11.6)$$

where b is the aquifer thickness and the storativity S is a dimensionless parameter.

11.2.1 Darcy's Law

The flow field is usually found from the continuity equation coupled with the momentum equation. In the case of the flow in porous media, the velocity field can be derived directly from Darcy's law which stipulates that:

Table 11.1 Freshwater of the hydrosphere (from [1])

Part of the hydrosphere	Volume (x 10^3 km^3)	Percent of total volume
Glaciers	24,000	85
Groundwater	4000	14
Lakes and reservoirs	155	0.6
Soil moisture	83	0.3
Vapors in atmosphere	14	0.05
River water	1.2	0.004
Total	28,253	100

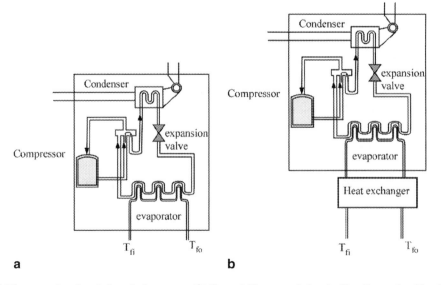

a   **b**

Fig. 11.2 (**a**) Ground-Water entering directly into the heat pump. (**b**) Ground-Water coupled to the Heat Pump via a Heat Exchanger

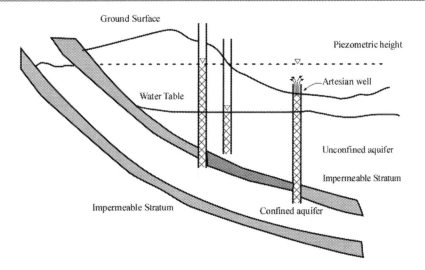

Ground Surface

Piezometric height

Water Table

Artesian well

Unconfined aquifer

Impermeable Stratum

Impermeable Stratum

Confined aquifer

Fig. 11.3 Aquifer

$$\vec{u} = -K\nabla h \qquad (11.7)$$

K [L-t^{-1}] is called the *hydraulic conductivity*. Darcy's law is very similar to Fourier's law in heat transfer where the hydraulic conductivity replaces the thermal conductivity. This law can be used in (11.5), assuming a constant density, to give:

$$\int_V \left(S_s \frac{\partial h}{\partial t} - \nabla \cdot (K\nabla h) \right) dV = 0$$

or

$$S_s \frac{\partial h}{\partial t} = \nabla \cdot (K\nabla h) + Q \qquad (11.8)$$

where the classical *diffusion equation* for the head evaluation is in a confined aquifer. A further difference between the usual concept of mass conservation and aquifer analysis is the possibility to have wells injecting ($+$) or removing ($-$) water that can be interpreted as source terms; Q [t^{-1}] represents this water quantity per unit time and per unit volume of the aquifer. The hydraulic conductivity depends on the soil structure as well as on the fluid used. The *permeability* k [L^2] defined as:

$$k = K\frac{\mu}{\rho g} \qquad (11.9)$$

is a related parameter associated with the soil matrix only. With a constant conductivity, the diffusion equation is the familiar form:

$$\frac{1}{\alpha_h}\frac{\partial h}{\partial t} = \nabla^2 h + \frac{Q}{K} \qquad (11.10)$$

where the *hydraulic diffusivity* is given as:

$$\alpha_h = \frac{K}{S_s} \qquad (11.11)$$

that can also be written as:

$$\alpha_h = \frac{K \cdot b}{S} = \frac{T}{S} \qquad (11.12)$$

where the product $K \cdot b$ is called the *transmissivity* (T) of the aquifer.

11.3 Aquifer Testing

In order to evaluate the resource of an aquifer for a possible use in GWHP applications, several information can be found from well testings that are very similar to TRT tests used in GCHP systems. Comparison of the measured response to prediction models can give information about parameters such as the conductivity and the transmissivity of the aquifer. Tests can be categorized as unsteady and steady tests.

11.3.1 Steady-State Analysis

During a well test, a constant rate of water is pumped out of the aquifer, and the variation of the hydraulic head in the aquifer can be modeled by the diffusion equation (11.8). The theoretical response will be prescribed by its solution of the equation with the associated boundary conditions. Assuming that the undisturbed initial head value is known at infinity, the solution will be given by the infinite line source which does not have a steady-state solution. However, in practice, at a certain distance, the head variation from the initial value is very small. This distance is called the radius of influence r_o. The steady-state solution expression of the diffusion equation in cylindrical coordinates (assuming only a radial dependence) is given by:

$$\frac{1}{r}\frac{d}{dr}\left(r\frac{dh}{dr}\right) = 0$$

$$h(r_o) = h_o \qquad\qquad (11.13)$$

$$\dot{V}(r_w) = \dot{V}_o$$

whose solution is:

$$h(r) = h_o + C\log\left(\frac{r}{r_o}\right)$$

The second constant is found from the prescribed inlet flow rate which is related to the head by Darcy's law:

$$\dot{V}(r_w) = -2\pi r_w b u_w$$

$$= 2\pi r_w b K \frac{dh}{dr}\Big|_{r=r_w}$$

$$= 2\pi C \cdot T$$

The final solution is then:

$$h(r) = h_o + \frac{\dot{V}_o}{2\pi T}\log\left(\frac{r}{r_o}\right)$$

The difference between the head at a given location and the undisturbed value is called the *drawdown* and is the given by:

$$s(r) = h_o - h(r) = \frac{\dot{V}_o}{2\pi T}\log\left(\frac{r_o}{r}\right) \qquad (11.14)$$

Using directly (11.14) is not practical since the concept of radius of influence r_o is arbitrary and difficult to know in practice. The use of this equation is done in relation with *observation wells*. Interpreting steady-state response tests is done when more than one well is drilled, one for the pumping and the others for observation. Assuming a second well (Fig. 11.4), for example, one finds that:

$$s(r_1) = \frac{\dot{V}_o}{2\pi T}\log\left(\frac{r_o}{r_1}\right)$$

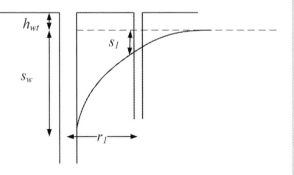

Fig. 11.4 Observation well

from which:

$$s(r) = s_1 + \frac{\dot{V}_o}{2\pi T}\log\left(\frac{r_1}{r}\right) \qquad (11.15)$$

If a second observation well is present:

$$s(r) = s_2 + \frac{\dot{V}_o}{2\pi T}\log\left(\frac{r_2}{r}\right) \qquad (11.16)$$

and the transmissivity of the aquifer can be found:

$$T\frac{\dot{V}_o}{2\pi(s_1 - s_2)}\log\left(\frac{r_2}{r_1}\right) \qquad (11.17)$$

This steady-state analysis is referred as the Thiem method [2].

Example 11.1 A 0.5 m diameter well is drilled into a confined aquifer with a thickness of 20 m. The pumping flow rate is 100 m³/hr. At steady state, the drawdown measured at a distance of 15 m is 1.7 m and 0.6 m at 45 m. Find:

(a) the transmissivity,
(b) the hydraulic conductivity of the aquifer,
(c) the radius of influence,
(d) the drawdown at the well radius.

Solution

(a) Using (11.17), one finds:

$$T = \frac{100}{2\pi(1.7 - 0.6)}\log\left(\frac{45}{15}\right) = 15.9 \text{ m}^2/\text{hr}$$

(b)
$$K = T/b = 0.79 \text{ m/hr}$$

(c)
$$r_o = r_1 \cdot \exp(2\pi T \cdot s_1/\dot{V}_o) = 82 \text{ m}$$

(d)
$$s_w = 1.7 + \frac{100}{2\pi\,15.9}\log\left(\frac{15}{0.25}\right) = 5.8 \text{ m}$$

Comments: *This last value will not be the one that is measured at the pumping well since the well itself produces a local hydraulic resistance. This is the value that can be expected just at the exit of the well*

∎

The drawdown inside the borehole will be different from the one at the outside due to the borehole resistance, a concept very similar to the thermal resistance in classical GCHP systems. The only difference is that, in pumping test, this resistance is nonlinear:

$$s_t = s_w + R_w \dot{V}_o = s_w + \underbrace{C\dot{V}_o}_{R_w} \dot{V}_o = s_w + C\dot{V}_o^2 \quad (11.18)$$

where C is the *well loss coefficient*. From (11.14), this can be written as:

$$s_t = \frac{\dot{V}_o}{2\pi T} \log\left(\frac{r_o}{r_w}\right) + C\dot{V}_o^2 = \beta\dot{V}_o + C\dot{V}_o^2 \quad (11.19)$$

with:

$$\beta = \frac{1}{2\pi T} \log\left(\frac{r_o}{r_w}\right) \quad (11.20)$$

Example 11.2 In the previous example, if a 8 m drawdown is measured in the pumping well, what is the loss coefficient?

Solution

$$s_t = s_w + C\dot{V}_o^2$$

$$8 = 5.8 + C \cdot 100^2$$

$$C = 2.2\text{e-}4 \ \text{hr}^2/\text{m}^5$$

$$C = 0.79 \ \text{min}^2/\text{m}^5$$

∎

A large loss coefficient is harmful for pumping energy and clogging problems; thus, Walton [3] suggested the following table (Table 11.2):

11.3.2 Unsteady Test Analysis

Equation (11.10) has the same form as the heat equation and known solutions of this equation can be used in this context. First, it can be written in terms of the drawdown as:

$$\frac{1}{\alpha_h} \frac{\partial s}{\partial t} - \nabla^2 s = -\frac{Q}{K} \quad (11.21)$$

As previously stated, the hydraulic diffusivity has a direct analogy with the thermal diffusivity. Other analog relations can be seen as (Table 11.3):

In the case of a confined aquifer, a frequent assumption is that the flow is radial:

Table 11.2 Well loss coefficient and well condition

Well loss coefficient (min²/m⁵)	Well condition
< 0.5	Properly designed
0.5 to 1.0	Mild deterioration
1.0 to 4.0	Severe deterioration
> 4.0	Difficult to operate

Table 11.3 Thermal-hydraulic analogy

Thermal	Hydraulic
α	α_h
k	T
ρC_p	S

$$\frac{1}{\alpha_h} \frac{\partial s}{\partial t} - \frac{1}{r} \frac{\partial}{\partial r}\left(r \frac{\partial s}{\partial r}\right) = -\frac{Q}{K} \quad (11.22)$$

Then, for a pumping test where a constant pumping rate is imposed at the start of the test, the solution is similar to the solution of the TRT test where the infinite cylindrical source (ICS) solution can be used if the imposed flow rate is prescribed at the well radius. Assuming a singular flow source at the origin leads to the infinite line source (ILS) solution as in the thermal case. The singular source term at the origin can also be interpreted as an imposed flow rate at a given radius and letting this radius going to zero. So the homogeneous form of the diffusion equation can also be solved case with the following boundary condition:

$$\lim_{r \to 0} r \frac{\partial s}{\partial r} = -\frac{\dot{V}_o}{2\pi T} \quad (11.23)$$

The ILS solution was suggested by Theis [4] based on the thermal analogy and has been used extensively in the literature of well tests. Using the formalism of Chap. 4, we could express the solution as:

$$s(t) = \frac{\dot{V}_o}{4\pi T} E_1(r^2/(4\alpha_h t)) = \frac{\dot{V}_o}{T} \underbrace{\frac{1}{4\pi} E_1(\tilde{r}^2/(4Fo_h))}_{G_{ILS}(\tilde{r}, Fo_h)} \quad (11.24)$$

with:

$$Fo_h = \frac{\alpha_h t}{r_w^2}, \qquad \tilde{r} = \frac{r}{r_w} \quad (11.25)$$

or:

$$s(t) = \frac{\dot{V}_o}{2\pi T} I(X_h) \quad (11.26)$$

with:

$$X_h = \frac{r}{2\sqrt{\alpha_h t}} \quad (11.27)$$

In the formalism used in hydrogeology, the same mathematical expression was given by Theis in the equivalent form:

$$s(t) = \frac{\dot{V}_o}{4\pi T} W(u) \quad (11.28)$$

with:

$$u = \frac{r^2}{4\alpha_h t} \qquad (11.29)$$

The logarithm approximation will give:

$$s(t) \approx \frac{\dot{V}_o}{4\pi T} \left[\log\left(\frac{4\alpha_h t}{r^2}\right) - \gamma + \ldots \right] \qquad (11.30)$$

$$s(t) \approx m \log t + s_o \qquad (11.31)$$

As in the thermal case, this approximation is only valid for sufficiently long time. The limit $u < 0.01$ is often suggested which gives:

$$u = \frac{\tilde{r}^2}{4Fo} < 0.01 \Rightarrow Fo > 25 \text{ for } \tilde{r} = 1$$

which is larger than the range used in thermal analysis; the important thing is that only the linear region must be kept for the slope method. Outside the pumping well, the intercept (s_o, used here so as not confuse with b, the aquifer thickness) is given by:

$$s_o = \frac{\dot{V}_o}{4\pi T} \left(\log\left(\frac{4T}{Sr^2}\right) - \gamma \right) \qquad (11.32)$$

Inside the borehole, the borehole resistance has to be taken into account:

$$s_{t,o} = \frac{\dot{V}_o}{4\pi T} \left(\log\left(\frac{4T}{Sr^2}\right) - \gamma \right) + C\dot{V}_o^2 \qquad (11.33)$$

As in the TRT's test, the slope of the drawdown history versus the logarithm of time will give the *transmissivity*:

$$T = \frac{\dot{V}_o}{4\pi m} \qquad (11.34)$$

In TRT's test, the y-intercept value is used to evaluate the borehole resistance knowing the volumetric heating capacity; the same can be done here for the well loss coefficient:

$$C = \frac{s_{t,o}}{\dot{V}_o^2} - \frac{\log\left(4T/Sr_w^2\right) - \gamma}{4\pi T \dot{V}_o}$$

In pumping tests, hydrogeologists use more often the x-intercept value:

$$C = -\frac{1}{4\pi T \dot{V}_o} \left(\log\left(4T t_o/Sr_w^2\right) - \gamma \right) \qquad (11.35)$$

where $t_o = \exp(x_o)$ is the time value corresponding to the x-intercept on the log plot. As in the TRT's tests, the value of the storage coefficient has to be known or at least estimated. Since in pumping tests, observation wells are often present,

they can be used since in that case there is no flow in the well, and the x-intercept can be used to find the storage coefficient:

$$S = \frac{4T t_o \exp(-\gamma)}{r^2} = \frac{2.246 T t_o}{r^2} \qquad (11.36)$$

Example 11.3 The results of a 23-hour pumping test is given in the file "pumping_test1.txt." The first column is the time in minutes and the following three columns present the drawdown measured in the pumping well (second column) and two observation wells situated at 5.0 m and 25 m. The flow rate is 8 l/s and the pumping well diameter is 300 mm. Find:

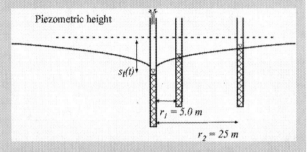

(a) The transmissivity
(b) The storage coefficient
(c) The well loss coefficient

Solution

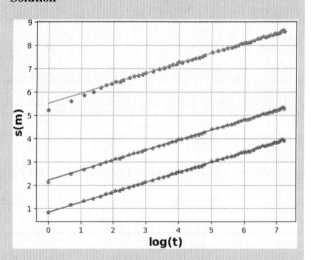

The points in the first 40 minutes were removed for the analysis; with the remaining points, the linear regression gave for the three curves:

$$s_t = 0.431 \log(t) + 5.516$$

$$s_1 = 0.431 \log(t) + 2.217$$

$$s_2 = 0.431 \log(t) + 0.837$$

(continued)

Example 11.3 (continued)

The transmissivity can then be easily found from any of the three curves since the slope is almost the same:

$$T = \frac{0.008}{4\pi 0.419} = 0.00147 \text{ m}^2/\text{s} = 0.089 \text{ m}^2/\text{min}$$

From (11.36), it is possible to find the storage coefficient but only for the observation wells.

$$x_{o,1} = \frac{-2.217}{0.431} = -5.14 \rightarrow t_{o,1} = \exp(-5.14)$$
$$= 0.0058 \text{ min}$$

$$S = \frac{4 \cdot 0.089 \cdot 0.0058 \exp(-\gamma)}{5.0^2} = 4.64\text{e-}5$$

$$\alpha_h = \frac{T}{S} = \frac{0.089}{4.64\text{e-}5} = 1910 \text{ m}^2/\text{min}$$

A similar value would have been found using the second well. Using the pumping well regression, the loss coefficient can be found:

$$x_{o,w} = \frac{-5.516}{0.431} = -12.8 \rightarrow t_{o,w}$$
$$= \exp(-12.8) = 2.76\text{e-}6 \text{ min}$$

$$C = -\frac{1}{4\pi 0.089 \cdot 0.48}\left(\log\left(\frac{4 \cdot 0.089 \cdot 2.76\text{e-}6}{4.64\text{e-}40.015^2}\right) - \gamma\right)$$
$$= 1.20 \text{ min}^2/\text{m}^5$$

which would represent a bad well design from Table 11.2. From these values, it is possible to verify the validity of the ILS solution:

$$u_w = \frac{r^2}{4\alpha_h 40} = \frac{0.15^2}{4 \cdot 1910 \cdot 40} = 7\text{e-}8$$

$$u_1 = \frac{r^2}{4\alpha_h 40} = \frac{5.0^2}{4 \cdot 1910 \cdot 40} = 8.2\text{e}-5$$

$$u_2 = \frac{r^2}{4\alpha_h 40} = \frac{25^2}{4 \cdot 1910 \cdot 40} = 2.0\text{e}-3$$

which are all smaller than 0.01.

It can be interesting to compare these values to steady-state analysis assuming that at the end of the testing period, almost steady-state conditions are observed. Using the value after 23 hours, one finds:

$$s_1(23hr) = 5.295 \text{ m}, \qquad s_2(23hr) = 3.913$$

(continued)

$$T = \frac{0.008}{2\pi(3.913 - 5.295)}\log\left(\frac{5}{25}\right) = 0.00148 \text{ m}^2/\text{s}$$
$$= 0.089 \text{ m}^2/\text{min}$$

which shows that the steady-state assumption is quite valid after that time. The radius of influence can be found from (11.14):

$$r_o = \exp(2\pi T/\dot{V}_o)r_1 = 2382 \text{ m}$$

Of course, the storage coefficient which only affects the transient behavior cannot be evaluated. ∎

11.3.3 Graphical Methods

As pointed out in the last section, unsteady analysis of confined aquifer is very similar to thermal response tests, where, for example, recovery periods are also suggested as alternative methods for transmissivity measurements and the method presented is basically the same as the slope method in TRT's tests. Historically, before the growing use of computers, several equivalent approaches were proposed, based on graphical representation of the drawdown response. Most books on groundwater talks extensively on so-called "match-points" methods [5] based on log-log plots of the time-drawdown which are similar to curve fitting approaches. In the Cooper-Jacob [6] method, the authors suggested to plot the drawdown response on a semi-log paper where the decimal logarithm is used instead of the natural logarithm. Rewriting (11.31) in decimal logarithm, one finds:

$$s(t) \approx \frac{\dot{V}_o}{4\pi T}\log(10)\log_{10}(t) + s_o = \frac{2.303\dot{V}_o}{4\pi T}\underbrace{\log_{10}(t)}_{x} + s_o$$
(11.37)

Equation (11.34) is now given by:

$$T = \frac{2.303\dot{V}_o}{4\pi m}$$
(11.38)

Being even more specific, Cooper and Jacob suggested to measure on the graph the variation of drawdown for a decimal log cycle (i.e., $\Delta x = 1$):

$$m = \frac{\Delta s}{\Delta x} = \Delta s$$

$$T = \frac{2.303\dot{V}_o}{4\pi \Delta s}$$
(11.39)

This is the classical form that is used in the Cooper-Jacob method but of course, regression methods for slope analysis are more rigorous. It is easy to verify that on the decimal semi-log plot, Eq. (11.36) remains unchanged, except that $t_o = 10^{x_o}$.

11.3.4 Leaky Aquifer

A confined aquifer is assumed to be contained by impermeable stratum. In a leaky aquifer, water can be delivered from an unconfined aquifer above a semipermeable stratum called *aquitard*. The added flow is assumed to be proportional to difference between overlying aquifer having an assumed constant water level h_o and the main aquifer head h. From Darcy's law, the velocity entering the main aquifer will be given as:

$$u_{in} = -u_z = K' \frac{h_o - h}{b'}$$

and the source term associated in the diffusion equation is the volume of water per volume of aquifer which is given as:

$$\frac{Q}{K} = \frac{u_{in} \cdot dA}{K b \cdot dA} = \frac{s}{B^2}$$

with

$$B = \left(\frac{K b b'}{K'}\right)^{1/2}$$

where b' is the aquitard thickness and K' its hydraulic permeability. It is given by Hantush and Jacob [7] and Jacob [8]:

$$\frac{1}{\alpha_h} \frac{\partial s}{\partial t} = \frac{1}{r} \frac{\partial}{\partial r}\left(r \frac{\partial s}{\partial r}\right) - \frac{s}{B^2} \tag{11.40}$$

The solution of this equation was given by Hantush and Jacob [7]:

$$s(t) = \frac{\dot{V}_o}{4\pi T} W(u, r/B) \tag{11.41}$$

The function $W(u, r/B)$ called the Hantush *leaky well function* is defined as:

$$W(u, r/B) = \int_u^\infty \frac{1}{y} \exp\left(-y - \frac{r^2}{4B^2 y}\right) dy \tag{11.42}$$

It is easily seen that:

$$\lim_{B\to\infty} W(u, r/B) = W(u) = E_1(u)$$

And the unconfined aquifer solution:

$$\lim_{t\to\infty} W(u, r/B) = W(0, r/B)) = 2K_o(r/B)$$

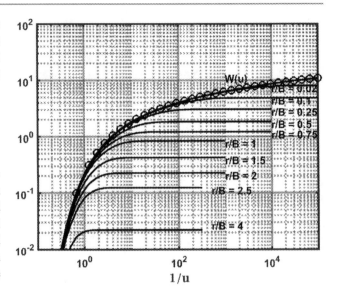

Fig. 11.5 Influence of Leaky Aquifer on Drawdown

with K_o being the modified Bessel function of the second kind of order 0 (Fig. 11.5).

Example 11.4 The measurements of a pumping test in a leaky aquifer are given in file "pumping_test2.txt" where the first column is the time in minutes. The drawdown measured in meters comes from an observation well situated at a 12.2 m distance from the pumping well. The flow rate is 17 m³/min. Using the method of parameter estimation, one finds:

(a) The transmissivity
(b) The storage coefficient
(c) The parameter r/B

Solution Part of the code used is shown:

```
M = np.loadtxt("pumping_test2.txt")
t = M[:,0] # time in minutes
sf = M[:,1] # drawdown in meters
qo = 17
rw = 12.2
# Initial data
Ti = 2.4 # m2/min
Si = 0.004
rbi = 0.03
G_vect = np.vectorize(leaky_function)
Si = 0.003
def s_theo(x,Tnew,Snew,rb):
    al = Tnew/Snew
```

(continued)

Example 11.4 (continued)

```
        u = rw**2/(4*al*t)
        s = qo/(4*pi*Tnew)*G_vect(u,rb)
        return s
#
#
po = [Ti,Si,rbi]
params,resn = curve_fit(s_theo,t,sf,po)
Tn = params[0]
Sn = params[1]
rbn = params[2]
```

The result is:

$$T = 2.3 \text{ m}^2/\text{min}$$

$$S = 0.0038$$

$$r/B = 0.0347$$

The comparison from the measured and correlated values is shown in the following figure:

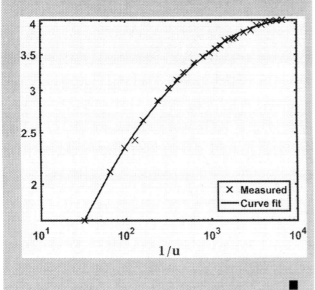

Example 11.5 A 10 kW heat pump has a COP of 4 in heating mode. If the inlet temperature is 5°C, what is the minimum flow rate if the outlet water ($C_p = 4.2$ kJ/kg-K) temperature should be higher than 2°C?

Solution

$$q_{evap} = 10 \text{ kW} \frac{COP - 1}{COP} = 7.5 \text{ kW}$$

$$\dot{m} = \frac{7.5}{4.2(5-2)} = 0.595 \text{ kg/s}$$

∎

Similar constraints can be specified in cooling mode based on environmental issues asking for a maximum temperature not to be exceeded. Choosing a particular flow rate is of particular importance when a heat exchanger is used (Fig. 11.2). In such a case, a compromise is needed between higher heat pump performance and pumping energy.

Example 11.6 The performance of a heat pump used in an open-loop configuration is given in the following table:

EWT (°C)	COP$_h$	EWT (°C)	COP$_h$
−1	3.3	10	8
3	3.9	15	7
6	4.4	20	5.5
7	4.5	23	5
9	4.7	27	4.5
12.5	4.9	30	4.3

The heat pump heating capacity is 12 kW and the total cooling capacity is 16 kW. The heat pump flow rate is 50 l/min in heating mode and 41.5 l/min in cooling mode. The groundwater is assumed to be 9°C all year long and the water temperature should always be higher than 4°C.

(a) If a heat exchanger having a 0.8 efficiency is used, what would be the flow rate suggested for each mode of operation if the targeted COPs are 4.5 in heating mode and 7 in cooling mode ?
(b) Assuming a counter-flow heat exchanger, what is the necessary area of the heat exchanger? The thin inner tube has a diameter of 2.1 cm and the outer isolated tube has a 3 cm diameter. Use the following properties for the water on each side of

11.4   Open-Loop Design Approaches

Once the properties of the aquifer are estimated, if an open-loop heat pump is installed, as discussed in the introduction of this chapter, the major part of the design constraints is associated with the flow rate chosen. In heating mode, an obvious constraint is that the water should not freeze.

(continued)

Example 11.6 (continued)
    the heat exchanger:

$$C_{pf} = 4.2 \text{ kJ/kg-K}, \rho_f = 1000 \text{ kg/m}^3,$$
$$v_f = 1.38\text{e-6 m}^2\text{/s}$$
$$k_f = 0.57 \text{ W/m-K}, Pr_f = 10$$

Solution – Heating mode:

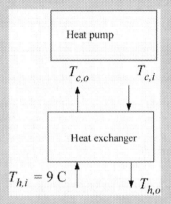

Let's choose an inlet temperature of 7°C which will give a COP of 4.5. This temperature is also the outlet temperature of the heat exchanger on the cold side:

$$q_{evap} = 12\frac{3.5}{4.5} = 9.33 \text{ kW} = q_{he}$$

$$\dot{m}_c = \frac{50}{60} = +0.833 \text{ kg/s},$$

$$C_c = 0.833 \cdot 4.2 = 3.5 \text{ kW/K}$$

$$T_{co} = 7\,°\text{C}, \quad T_{ci} = 7 - \frac{9.33}{3.5} = 4.33\,°\text{C}$$

From the efficiency of the heat exchanger, we have:

$$q_{max} = \frac{q_{he}}{0.8} = 11.67 \text{ kW}$$

From (2.91)

$$q_{max} = C_{min}(T_{hi} - T_{ci}) = C_{min}(9 - 4.33)$$

$$C_{min} = 2.5 \text{ kW/K}$$

Since this value is less than C_c, it can be chosen to be C_h and:

$$\dot{m}_h = \frac{2.5}{4.2} = 0.595 \text{ kg/s}$$

– Cooling mode:

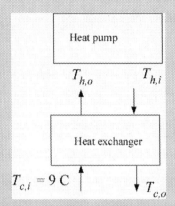

The targeted value of 15°C is possible and a similar analysis as before leads to:

$$q_{cond} = 16\frac{8}{7} = 18.286 \text{ kW} = q_{he}$$

$$\dot{m} = \frac{41.5}{60} = 0.692 \text{ kg/s},$$
$$C_h = 0.692 \cdot 4.2 = 2.905 \text{ kW/K}$$

$$T_{ho} = 15\,°\text{C}, \quad T_{hi} = 15 + \frac{18.286}{2.905} = 21.29\,°\text{C}$$

$$q_{max} = \frac{q_{he}}{0.8} = 22.86 \text{ kW}$$

$$C_{min} = \frac{22.86}{21.29 - 9} = 1.86 \text{ kW/K} = C_c$$

$$\dot{m}_c = \frac{1.86}{4.2} = 0.443 \text{ kg/s}$$

(b)
– Heating mode
    From Table 2.2, we have:

$$C_r = \frac{2.5}{3.5} = 0.714$$

$$NTU = \frac{1}{0.614}\log\left(\frac{0.7}{0.8 \cdot 0.714 - 1}\right) = 2.67$$
$$UA = 2.67 C_{min} = 6.67 \text{ kW/K}$$

The convection coefficient can be estimated with the Reynolds number:
– Heat pump side:

$$Re = \frac{4\dot{m}}{\pi d_i \mu} = 36600$$

From Eq. (2.43b):

$$Nu_{di} = 290 \Rightarrow \hbar_{heatpump} = 7.9 \text{ kW/m}^2\text{-K}$$

(continued)

(continued)

Example 11.6 (continued)

- Ground side:

$$Re = \frac{4\dot{m}}{\pi(d_i + d_o)\mu} = 10768$$

From Eq. (2.65):

$$Nu_{dh} = 82 \Rightarrow \hbar_{groundside} = 5.2 \, \text{kW/m}^2\text{-K}$$

Since the tube is thin, the conduction resistance is neglected and:

$$\frac{1}{U} = \frac{1}{7.9} + \frac{1}{5.2} = \frac{1}{3.1}$$

$$A = \frac{6.67}{3.1} = 2.13 \text{m}^2$$

- Cooling mode

$$C_r = \frac{1.86}{2.905} = 0.64$$

$$NTU = \frac{1}{0.54} \log\left(\frac{0.7}{0.8 \cdot 0.54 - 1}\right) = 2.47$$

$$UA = 2.47 C_{min} = 4.61 \text{kW/K}$$

The global coefficient is found to be:

$$U \approx 2.48 \, \text{kW/m}^2\text{-K}$$

$$A = \frac{6.67}{2.48} = 1.85 \text{m}^2$$

Since the heat exchanger will not change from one mode to another, the maximum area will be chosen. In such a case, the cooling mode performance will be higher than expected. ∎

11.4.1 Pumps in Open-Loop Systems

As for secondary systems, the energy extracted from the ground is transported from a calorimetric fluid that has to be pumped. The main difference when using groundwater is that the pump must not only balance the friction head losses but also the static head since the pipe is not pressurized. This requires pumping with larger head delivered. Two types of pumps are usually used: either a vertical line shaft pump where the motor is at the ground level and separated by a long vertical line shaft or a submersible pump where the hermetically sealed motor is submerged inside the well. The total head that has to be supplied by the pump is then given by:

$$h_{total} = h_{static} + h_{losses} \quad (11.43)$$

where the h_{losses} are evaluated as before and where the heat exchanger can be an important part of it and h_{static} is given by:

$$h_{static} = h_{wt} + s_{t,max} \quad (11.44)$$

where $s_{t,max}$ is the maximum drawdown experienced at the pumping well and h_{wt} is the distance of the undisturbed water table from the ground surface (Fig. 11.4).

Example 11.7 A pumping test has been performed before installing the open-loop heat pump described in the previous example. It has been observed that a confined aquifer having the following properties is present:

$$\beta = 1000 \, \text{s/m}^2, \quad C = 1 \, \text{min}^2/\text{m}^5$$

and that the undisturbed water table is at a distance of 10 m from the surface of the well. Find the COP of the system in heating mode, assuming that all the friction losses are given by: $h_{losses} = 1.4\text{e}7\dot{V}^2$ m on the ground side when the flow rate is in m³/s and $\Delta p(kPa) = 13(\dot{V}/\dot{V}_{nom})^2$ on the heat pump side. The wire-to-water efficiency of the submersible pump is 0.35 and the one of the in-line pump is 0.55.

Solution In heating mode, the flow rate is 0.595 kg/s = 5.95e-4 m³/s:

$$h_{losses} = 1.4e7 \cdot (5.95\text{e-}4)^2 = 4.96 \, \text{m}$$

The maximum drawdown is given by Eq. (11.19) with C changed in s²/m⁵:

$$s_t = 1000\dot{V} + 3600\dot{V}^2 = 0.596 \, \text{m}$$

The total static height is then $h_s = 10 + 0.596 = 10.596$ m and:

$$h_{total} = 15.55 \, \text{m}$$

$$\dot{W}_{sub,pump} = \frac{0.595 \cdot 9.8 \cdot 15.55}{0.35} = 259.3 \, \text{W}$$

$$h_{evap} = \frac{13 \cdot 1000}{g\rho} = 1.326 \, \text{m}$$

$$\dot{W}_{in-linepump} = \frac{0.833g \cdot 1.316}{0.55} = 20 \, \text{W}$$

$$\dot{W}comp = \frac{12000}{4.5} = 2666.67 \, \text{W}$$

(continued)

Example 11.7 (continued)

$$\dot{W}tot = 2.945 \text{ kW}$$

$$COP_{sys} = \frac{12}{2.93} = 4.07$$

Comments: *Here it was assumed that the COP = 4.5 does not take into account the head losses in the evaporator. If it was the case, the results would have been different.*

∎

Problems

Pr. 11.1 The following table (taken from [5]) presents the measurements of a pumping test in a leaky aquifer where the first column is the time in minutes and the second column the drawdown in feet measured at an observation well situated at a distance of 100 feet from the test well.

	Time (min)	s(ft)
0	0.20	1.76
1	0.50	2.75
2	1.00	3.59
3	2.00	4.26
4	5.00	5.28
5	10.00	5.90
6	20.00	6.47
7	50.00	6.92
8	100.00	7.11
9	200.00	7.20
10	500.00	7.21
11	1000.00	7.21

The flow rate was 1000 gpm. Keeping only the number of points where $u < 0.01$, find from parameter estimation the transmissivity, the storage coefficient, and the ratio r/B.
Answer: T = 0.015 m²/s, S = 8.7e-5, r/B = 0.044

Pr. 11.2 In Example 11.7, the heat pump can work at 50 % of its capacity where the flow rate on the heat pump side is also reduced by half. Assuming that the COP relation with inlet temperatures does not change, evaluate the new system COP in heating mode if:

(a) The flow rate on the ground side remains unchanged.
   Answer: $COP_{HP} = 4.64$, $COP_{sys} = 3.76$

(b) The flow rate on the ground side is also reduced by half.
   Answer: $COP_{HP} = 4.60$, $COP_{sys} = 4.19$

Pr. 11.3 An open-loop GSHP is used in cooling mode. The following data are given:

$$T_{water} = 14\,°C, \quad TC = 20 \text{ kW},$$
$$HP \text{ flow rate } = 0.8 \text{ kg/s}, \quad UA = 4000 \text{ W/m}^2\text{-K}$$

$$COP = 11.5 - 0.42T_{in} + 0.006T_{in}^2$$

Aquifer parameters:

$$\beta = 1000 \text{ s/m}^2, \quad C = 2 \text{ min}^2/\text{m}^5, \quad h_{wt} = 10 \text{ m}$$

Head losses on the source side of the heat exchanger:

$$h_{losses}(m) = 15e6\dot{V}^2 \qquad (\dot{V} \text{ in m}^3/\text{s})$$

Pump efficiency 50 %. Neglecting the pump work on the heat pump side, evaluate the total power (pump + compressor) for the following flow rates:

$$\dot{m} = [0.6, 0.7, 0.8, 0.9] \text{ kg/s}$$

Answer (W): 3967, 3936, 3945, 3985 (0.7 kg/s is the best)
(b) What is the source outlet temperatures?
Answer (°C): 23.4, 22.1, 21.0, 20.2

References

1. L'vovich, M.I.: World Water Resources and their Futureand their. American Geophysical Union, Washington, D.C. (1979)
2. Thiem, G.: Hydrologische Methoden: Dissertation zur Erlangung der Wurde eines JM Gebhardt, Leipzig (1906)
3. Walton, W.C.: Selected analytical methods for well and aquifer evaluation. In: Bulletin (Illinois State Water Survey) no. 49 (1962)
4. Theis, C.V.: The relation between the lowering of the piezometric surface and the rate and duration of discharge of a well using groundwater storage. Eos, Transactions American Geophysical Union **16**(2), 519–524 (1935)
5. Water-Supply paper 1545-C. Shortcuts and special problems in aquifer tests: compiled by Ray Bentall. Pp. 117; pls. 4; figs. 31; tbls. 17. U.S. Geol. Surv., Washington, D.C. J. Hydrol. **3**(1), 72 (1965). https://doi.org/10.1016/0022-1694(65)90077-6
6. Cooper, H.H., Jacob, C.E.: A generalized graphical method for evaluating formation constants and summarizing well-field history. Eos, Transactions American Geophysical Union **27**(4), 526–534 (1946)
7. Hantush, M.S., Jacob, C.E.: Non-steady radial ow in an infinite leaky aquifer. Eos, Transactions American Geophysical Union **36**(1), 95–100 (1955)
8. Jacob, C.E.: Radial ow in a leaky artesian aquifer. Eos, Trans-actions American Geophysical Union **27**(2), 198–208 (1946)

Economic Analysis

12.1 Introduction

As often stated, ground-source heat pumps are one of the most efficient systems when properly designed. Another accepted fact is their high initial costs. Arguments to move forward in choosing such a technology have to be not only from environmental sensitivity but also from economic feasibility. Even in the case when the decision to choose such a technology is made, several questions arise as follows: Is it better to have a hybrid system or a full geothermal system? Is it better to have longer loops with better efficiencies than smaller cheaper ones in the long run? Even though exact answers are difficult to give, since they are often based on arbitrary assumptions, basic economic analysis tools are mandatory to give us insight on clever design.

12.2 Life Cycle Cost Analysis

The simplest approach to evaluate the economic viability of a project is the simple *payback period*. It is simply the ratio of the initial cost to the expected yearly savings.

> *Example 12.1* A \$30,000 geothermal is assumed to give a \$3000 energetic cost savings every year, but for a \$200 increase in insurance, what is the payback period?
>
> $$PP = \frac{30000}{3000 - 200} = 10.7 \text{ years}$$
>
> A longer period is often expected. ∎

Simple payback is too simplistic since it does not take into account the variation of costs due to inflation and also the present worth of future savings and expenses. Most economic

analysis is performed using a *life cycle cost* (LCC) approach [1]. In a life cycle cost analysis (LCCA), one tries to evaluate as precisely as possible all future monetary flows. The *present value* of a given cash flow(A) is evaluated using the following formula:

$$PV(A, N, d) = A\frac{1}{(1 + d)^N} \tag{12.1}$$

where N is the number of periods often chosen as years and d is called the *discount rate*. The same argument can be used to evaluate the future value of a present cash flow:

$$FV(A, N, d) = A(1 + d)^N \tag{12.2}$$

This value depends not on the number of terms where the cash flow is done but on the number of terms remaining in the analysis. Let's assume that the LCC is done over a period of N terms; the future value of a payment that will occur in the jth period will then be:

$$
\begin{aligned}
FV(A, K, d) &= PV(A, j, d)(1 + d)^N \\
&= \tfrac{A}{(1+d)^j}(1 + d)^N = A(1 + d)^{N-j} \\
&= A(1 + d)^K
\end{aligned} \tag{12.3}
$$

In the case where a cash flow is known to be repeated every terms, like mortgage payments, for example, the present value of all payments is given by:

$$
\begin{aligned}
PV(\textstyle\sum A, N, d) &= \frac{A}{(1 + d)} + \frac{A}{(1 + d)^2} + \cdots \frac{A}{(1 + d)^N} \\
&= A\underbrace{\left(\frac{1}{d}\left[1 - \left(\frac{1}{1 + d}\right)^N\right]\right)}_{PWF(N,0,d)}
\end{aligned} \tag{12.4}
$$

L. Lamarche, *Fundamentals of Geothermal Heat Pump Systems*,
https://doi.org/10.1007/978-3-031-32176-4_12

where PWF is called the present worth factor of the recurrent payment or also the *present value annuity factor*.[1] The closed form is obtained from the properties of geometric series. It was assumed that the first installment was at the end of the first period. If it was at the beginning, we would get:

$$PV_{beginning}\left(\sum A, N, d\right) = A + PWF(N-1, 0, d)$$
$$(12.5)$$

The same expression can be used to evaluate what are the future payments that must be disbursed (M_i) in order to pay off the present value of a debt or mortgage (M). It is given by:

$$M_i = \frac{M}{\frac{1}{r}\left[1 - \left(\frac{1}{1+r}\right)^N\right]} = \frac{M_i}{PWF(N, 0, r)} \qquad (12.6)$$

where r is now the interest rate and not the discount rate. N is the number of periods. This is the reason why the inverse is often referred to as the *capital recovery factor*:

$$CRF(N, r) = \frac{1}{PWF(N, o, r)} \qquad (12.7)$$

Even though simple annual analysis will be used in our examples, mortgage payments are rarely made once a year. If monthly installments are done, they can be calculated by:

$$M_i = \frac{M}{\frac{1}{r_{e,m}}\left[1 - \left(\frac{1}{1+r_{e,m}}\right)^{12N}\right]}$$

where $r_{e,m}$ is an equivalent monthly interest. In this introduction, we will restrict our analysis to simple annual analysis, but the concept of *effective interest rate* is of primary importance in economic analysis. The effective interest rate is given [2] by:

$$r_e = \left(1 + \frac{r}{J}\right)^C - 1 \qquad (12.8)$$

where

- J: is the number of times the interest is compounded per year.
- K: is the number of installments per year.
- C: is the number of compounded period per installment (J/K).

---

[1]Equation (12.1) is sometimes called single-payment present worth factor, whereas (12.4) stands for equal-payment present worth factor.

Example 12.2 An amount of $10000 is deposited in an account with a 10% nominal interest rate.

(a) If the interest is compounded at the end of the year, what is the final amount in the account?
(b) If the interest is compounded twice in the year, what is the final amount in the account?
(c) What is the equivalent interest rate (effective rate) that would result in the same amount than the last answer if the interest was applied once at the end of the year?

Solution

(a)
$$M_f = 10000(1 + 0.1) = \$11000$$

(b) After 6 months, one has:

$$M_i = 10000(1 + 0.05)$$

And after a year:

$$M_f = 10000(1 + 0.05)^2 = \$11025$$

(c) The effective yearly interest rate is given by (12.8), with $J = 2, K = 1, C = 2$:

$$r_{e,ann} = (1 + 0.1/2)^2 - 1 = 0.1025$$

$$M_f = 10000(1 + 0.1025) = \$11025$$

∎

An equivalent monthly interest compounded semiannually is often used for mortgage loans. A monthly effective rate would then be ($K = 12, J = 2$):

$$r_{e,m} = \left(1 + \frac{r}{2}\right)^{1/6} - 1$$

Instead of evaluating the future periodic payments required to pay a present debt, it is sometimes needed to evaluate the future periodic payments to accumulate a given future value, for example, if we want to have a specific amount of money when our children will reach college or if we want to pay cash for a car after several years. This is directly found when the known amount (M_f) is the final value of a series of equal payments. Finding the future value of Eq. (12.4) where again, the discount rate is replaced by an interest rate:

$$M_f = FV(PV(\sum M_i, N, r)) = M_i PWF(N, 0, r)(1+r)^N$$

$$M_i = \frac{M_f}{PWF(N, 0, r)(1+r)^N} = M_f SFF(N, 0, r)$$
$$(12.9)$$

where SFF is sometimes referred to a *sinking fund factor*.

Often the cash flows will also be affected by inflation. In this case, the present value after N terms will be given by:

$$PV(A, N, i, d) = A \frac{(1+i)^{N-1}}{(1+d)^N} \qquad (12.10)$$

In the last expression, the inflation is assumed to be added at the end of a given period. In the case where it is applied at the beginning, the $N - 1$ is replaced by N. The future value depends on two indices: the number of terms when the payment is done influencing the payment value due to inflation and the number of terms remaining. Again assuming N periods of analysis and the future value of a payment that will occur after j terms, one has:

$$\text{Future value} = \underbrace{A \frac{(1+i)^{j-1}}{(1+d)^j}}_{PV}(1+d)^N$$

$$= A(1+i)^{j-1}(1+d)^{N-j}$$

$$= A(1+i)^{j-1}(1+d)^K$$

If again the same amount is used for all periods, the sum is given by:

$$PV(\sum A, N, i, d) = \frac{A}{(1+d)} + \frac{A(1+i)}{(1+d)^2} + \dots \frac{A(i+i)^{N-1}}{(1+d)^N}$$

$$= A \cdot PWF(N, i, d)$$
$$(12.11)$$

where

$$PWF(N, i, d) = \begin{cases} \dfrac{1}{d-i}\left[1 - \left(\dfrac{1+i}{1+d}\right)^N\right] & , i \neq d \\[4ex] \dfrac{N}{1+d} & , i = d \end{cases}$$
$$(12.12)$$

where again it is assumed that the inflation is calculated at the end of each period. In the case where it is applied at the beginning of the term, the last expression will be multiplied by $(1+i)$. The future value of a recurrent series of cash flow can easily be obtained by combining the last expression with (12.3):

$$FV(\sum A, i, d) = PV(\sum A, i, d)(1+d)^N$$

$$= A \cdot PWF(N, i, d)(1+d)^N \qquad (12.13)$$

$$= A \cdot FWF(N, i, d)$$

where *FWF* is the future worth factor also called *compound amount factor* (CAF). Equal periodic cash flows or cash flows associated with known constant inflation rate are thus easy to evaluate over a long period. Variable cash flows have to be discounted on a periodic basis. An example of this kind of payment is associated with interest amounts.

Example 12.3 A $10000 loan is paid in three years at a 6% interest rate. What is the discounted value ($d = 8\%$) of all the interest payments?

Solution The three annual payments will be given by (12.6):

$$M_i = \frac{10000}{PWF(3, 0, 0.06)} = \$3741.10$$

The total amount of interest can easily be calculated from:

$$3 \cdot 3741.10 - 10000 = \$1223.30$$

The present value of interest can be evaluated terms by terms. In the first year, the interest is $600, so after the first year, the loan balance is:

$$10000 + 600 - 3741.10 = \$6858.90$$

In the second year, the interest is $0.06 \cdot 6858.90 = 411.53$, so after the second year, the loan balance is:

$$6858.90 + 411.53 - 3741.10 = \$3529.33$$

Finally, in the last year, the interest is $0.06 \cdot 3529.33 = 211.76$, so:

$$3529.33 + 211.76 - 3741.10 = 0$$

The total present value of all interest cash flows is:

$$\frac{600}{1 + 0.08} + \frac{411.53}{(1 + 0.08)^2} + \frac{211.76}{(1 + 0.08)^3} = \$1076.48$$

∎

Duffie Beckman [3] give a closed-form expression for the total present value of interest payments:

$$PV(\textstyle\sum int, r, d) = M_{in} \left[\frac{PWF(N_{min}, 0, d)}{PWF(N_p, 0, r)} \right.$$
$$\left. + PWF(N_{min}, r, d)\left(r - \frac{1}{PWF(N_p, 0, r)}\right) \right]$$
$$= M_{in} \cdot PWF_{int}(N_p, N_y, r, d) \tag{12.14}$$

where:

N_p: Number of years of loan
N_y: Number of years of life cycle analysis
N_{min}: $\min(N_p, N_y)$
M_{in}: initial loan
d: discount rate
r: interest rate

Example 12.4 Redo the last example using (12.14).

Solution

$$PWF(3, 0, 0.08) = 2.577, PWF(3, 0, 0.08) = 2.673$$

$$PWF(3, 0.06, 0.08) = 2.7266$$

$$PV(int) = 10000 \left[\frac{2.577}{2.673} + 2.7266\left(0.06 - \frac{1}{2.673}\right) \right]$$

$$= \$1076.48$$

∎

The *life cycle cost* (LCC) [1] of a given project can be written as:

$$LCC = \sum_{k=1}^{N}\sum_{j=1}^{M} PV(C_{k,j}) + I - PV(V_r) \tag{12.15}$$

where N is the number of years of LCCA; M is the number of cash flows; $C_{k,j}$ is a j_{th} cost of the k_{th} year; I is the initial down payment, not discounted since it is done at the beginning of the project; and $V_{r,d}$ is the discounted value of the residual value at the end of the project.

Example 12.5 The future energy costs for the next 5 years of a building are estimated to be:

$$\$5400, \$5550, \$5670, \$5780, \$5880$$

(continued)

The owner has the possibility to buy a new equipment at a cost of $12,000 that will save him some energy costs. The expected costs of the new system are:

$$\$3240, \$3200, \$3150, \$3060, \$2900$$

Compare the LCC of the new project with the old system over the 5-year period. The discount rate is 8% and the residual value is neglected.

Solution Since the cash flows are not constant, the analysis will be done year by year. The calculation is quite simple:

Year	Costs ($)		Discounted costs		Cumulative cash flow		Cumulative LCC	
	Base case	New project	Base case	New project	Base case	New project	Base case	New project
0	0	0	0	0	0	12000	0	12000
1	5400	3240	5000	3000	5400	15200	5000	15000
2	5540	3200	4750	2743	10940	18440	9750	17743
3	5670	3150	4501	2501	16510	21590	14251	20244
4	5780	3060	4248	2249	22390	24650	18499	22493
5	5880	2900	4002	1974	28270	27550	22501	24467

$$Cost = LCC(new) - LCC(base) = 24467 - 22501 = \$1966$$

It is seen that since the LCC of the new system is higher than the base case, the replacement is not profitable for a 5-year period. It is interesting to note that it is profitable looking only at undiscounted costs. ∎

12.3 Net Present Value and Net Future Value

The *net present value* (NPV) is given by the present value of all payments ($-$) and all gains ($+$) associated with the project. It is evaluated in a similar way as the LCC but the cash flows have a different sign: Costs are associated with a minus sign in the NPV and a positive sign in the LCC analysis. It can be calculated by:

$$NPV = \sum_{k=1}^{N}\sum_{j=1}^{M} PV(A_{k,j}) - I + PV(V_r) \tag{12.16}$$

or:

$$NPV = \sum_{k=1}^{N}\sum_{j=1}^{M} A_{k,j}\frac{(1+i_j)^{k-1}}{(1+d)^k} - I + V_{r,d} \tag{12.17}$$

where again N is the number of years of LCCA, M is the number of cash flows, i_j is the inflation rate associated with the cash flow, and d is the discount rate. Most of the time, the cash flow will be uniform ($A_{j,k}$ will not depend on k or will vary with a known inflation rate). In such a case, the NPV will be found from:

$$NPV = \sum_{j=1}^{M} A_j \cdot PWF(N, i_j, d) - I + V_{r,d} \quad (12.18)$$

If L cash flows are associated with interest payments, it can be calculated as:

$$NPV = \sum_{j=1}^{M-L} A_j \cdot PWF(N, i_j, d)$$

$$+ \sum_{j=1}^{L} M_{in.j} \cdot PWF_{int}(N_p, N_y, r_j, d) - I + V_{r,d}$$

$$(12.19)$$

A project where the NPV is positive is assumed to be profitable, that is, its future discounted revenues will exceed its discounted expenses, provided that the assumptions made in the analysis prove to be correct. The NPV approach is appropriate when we wish to compare a new project to a reference one which is often the "doing nothing" case. Evaluating the NPV where savings are associated with revenues is referred to as *discounted cash flow analysis* (DCFA) [4]. The LCC is appropriate when we wish to compare different new projects although both can be used in many situations.

Example 12.6 Redo the last example by evaluating the NPV of the new project.

Solution The solution is trivial. Again since the cash flows are not recurrent, (12.16) will be used. Two approaches can be used to evaluate the cash flows. For example, in the first year, the $3200 expense can be seen as a negative cash flow, but in that case, the $5400 expense of the base case will be treated as a gain. Or a relative approach can be used where the difference will be associated with the gain of the given year. In such a case, the gains for the new project are now:

$$\$2160, \$2340, \$2520, \$2720, \$2980$$

$$NPV = -12000 + \frac{2160}{1.08} + + \frac{2340}{1.08^2} + \frac{2520}{1.08^3} + \frac{2720}{1.08^4} + \frac{2980}{1.08^5} = -\$1966$$

which is negative and the project is not profitable. ∎

Example 12.7 In the last two examples, both approaches gave, of course, the same results. If, for ex-

(continued)

---

ample, in the previous example the existing equipment has reached its end of life and the owner has to buy a new system, he has the choice between the system proposed in the last example and another one cheaper ($10000) but less efficient where the expected future costs are:

$$\$3780, \$3790, \$3780, \$3740, \$3670.$$

Based on the LCCA, which one is better?

Solution Again the NPV can be used where the highest one (positive or negative) compared to the base case will be chosen, but here the LCC calculation is more logical since it does not have to be compared to a base case. The solution is now:

Year	Costs ($)		Discounted costs		Cumulative cash flow		Cumulative LCC	
	New project 1	New project 2	New project 1	New project 2	New project 1	New project 2	New project 1	New project 2
0	0	0	0	0	12000	10000	12000	10000
1	3240	3780	3000	3500	15200	13780	15000	13500
2	3200	3790	2743	3249	18440	17570	17743	16749
3	3150	3780	2501	3001	21590	21350	20244	19750
4	3060	3740	2249	2749	24650	25090	22493	22499
5	2900	3670	1974	2498	27550	28760	24467	24997

Based on LCCA, project 1 is better than project 2.

∎

Of course, the evaluation of LCC and NPV is highly influenced by the discount rate used. This value is important and investors often ask for a value higher than a minimum value, the *minimum acceptable rate of return* (MARR) [2].

Example 12.8

(a) The initial cost of a project is $2200 and it is expected to generate incomes of $800, $1000, $ and 1200 in the next three years. Every year, the new project will cost an extra tax of $50 (nondeductible) which is expected to increase at a 4% inflation rate. What is the VAN for a discount rate of 7%?
(b) Redo the calculation if the income is $1000 each year.

Solution

(a) The calculation is simple and is given by:
gains:

$$\frac{800}{1.07} + \frac{1000}{(1.07)^2} + \frac{1200}{(1.07)^3} = \$2600.66$$

(continued)

Example 12.8 (continued)
costs:

$$\frac{50}{1.07} + \frac{50(1.04)}{(1.07)^2} + \frac{50(1.04)^2}{(1.07)^3} = \$136.29$$

(12.17) gives:

$$NPV_a = -2200 + 2600.66 - 136.29 = \$264.36$$

which is positive.

(b) Only the gains change:

$$\frac{1000}{1.07} + \frac{1000}{(1.07)^2} + \frac{1000}{(1.07)^3} = \$2624.32$$

$$NPV_b = -2200 + 2624.32 - 136.29 = \$288.02$$

A small difference from the previous result, since even though the total amount of money is the same, the time value of money is not. In this second case, since all positive flows are equal, and since the costs are also recurrent except that they increase with a known constant inflation rate, the same answer can be more easily calculated using (12.18):

$$-2200+1000PWF(3,0,0.07)-50PWF(3,0.04,0.07) = \$288.02$$

∎

The *net future value* (NFV) can be defined in a similar way:

$$NFV = \sum_{k=1}^{N} \sum_{j=1}^{M} FV(A_{k,j}) - FV(I) + V_r \qquad (12.20)$$

where again compact form can be used when the cash flows are predictable.

Example 12.9 Evaluate the net future value in the previous example.

Solution Again in (a) the analysis is done terms by terms:

(a) gains:

$$800 \cdot (1.07)^2 + 1000 \cdot (1.07)^1 + 1200 = \$3185.92$$

costs:

$$50 \cdot (1.07)^2 + 50 \cdot (1.04) \cdot (1.07)^1 + 50 \cdot (1.04)^2 = \$166.97$$

(12.18) gives:

$$NFV_a = -2200 \cdot (1.07)^3 + 3185.92 - 166.97 = \$323.86$$

which is positive.

(b) gains:

$$1000 \cdot (1.07)^2 + 1000 \cdot (1.07)^1 + 1000 = \$3214.90$$

$$NFV_b = -2200 \cdot (1.07)^3 + 3214.90 - 166.97 = \$352.84$$

As previously presented, this case can be solved using:

$$-FV(2200, 3, 0.07) + 1000FWF(3, 0, 0.07)$$
$$-50FWF(3, 0.04, 0.07) = \$352.84$$

∎

12.3.1 Equivalent Annual Cost (EAC)

The *equivalent annual cost* (EAC) is simply the equivalent uniform amount of money that will produce the same NPV over the entire lifespan:

$$EAC \cdot PWF(N, 0, d) = NPV \qquad (12.21)$$

or

$$EAC \cdot FWF(N, 0, d) = NFV \qquad (12.22)$$

Since it will have the same sign, a positive EAC will mean that the project is profitable.

Example 12.10 In the previous examples, find the EAC (case a, only).

Solution The solution is straightforward:

$$EAC = \frac{NPV}{PWF(3, 0, 0.07)} = \frac{264.37}{2.62} = \$100.73$$

or:

$$EAC = \frac{NFV}{FWF(3, 0, 0.07)} = \frac{323.86}{3.21} = \$100.73$$

The same calculation for case (b) would give:

$$EAC = \$109.75$$

∎

(continued)

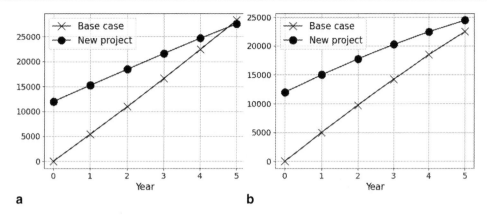

Fig. 12.1 (**a**) Undiscounted cash flow. (**b**) Discounted cash flow

12.3.2 Levelized Cost of Energy (LCOE)

The *levelized cost of energy* (LCOE) is essentially the life cycle cost of the energy produced but the present value of the energy produced:

$$LCOE = \frac{LCC}{PV_{energy}} \qquad (12.23)$$

where it has been referred as *Levelelized Cost of Heat* (LCOH) [5] in the special case of thermal applications. PV_{energy} is the present value of energy, that is, the total energy estimated during the life cycle discounted back to its present value [6]:

$$PV_{energy} = \sum_{k=1}^{N} \frac{E_k}{(1+d)^k} \qquad (12.24)$$

In the case where the energy flow is assumed to remain constant, it is given by:

$$= E_k PWF(N, 0, d) \qquad (12.25)$$

12.3.3 Discounted Payback Period (DPB)

The simple payback period was previously defined as the period of time that a new project will be profitable without taking account the varying time value of money. The *discounted payback period* is the same evaluation with the present value of future costs. The undiscounted and discounted cumulative cash flows of Example 12.5 are represented in Fig. 12.1 showing that the simple payback period is less than 5 years, whereas the DPB is longer than 5 years.

12.4 Internal Rate of Return (IRR) and Modified Internal Rate of Return (MIRR)

The discount rate that will render the net present value null is called the *internal rate of return* (IRR) which can easily be calculated from Eqs. (12.17, 12.18):

$$0 = \sum_{j=1}^{M} A_j \cdot PWF(N, i_j, d) - I + V_{r,d} \qquad (12.26)$$

or:

$$0 = \sum_{j=1}^{M-L} A_j \cdot PWF(N, i_j, d)$$
$$+ \sum_{j=1}^{L} M_{in,j} \cdot PWF_{int}(N_p, N_y, r_j, d) - I + V_{r,d}$$

$$(12.27)$$

Example 12.11 Find the IRR for the last example (case a) only.

Solution Solving:

$$0 = -2200 - 50 PWF(3, 0.04, 0.07) + \frac{800}{1 + irr}$$
$$+ \frac{1000}{(1 + irr)^2} + \frac{1200}{(1 + irr)^3}$$

which will give:

$$irr = 0.13 \quad (13\%)$$

∎

The IRR is often seen as an equivalent rate that the investor would obtain if he invests its initial money at a rate equal to IRR. However, to obtain this result, he would have to reinvest all its earnings at the same rate.

Example 12.12 To illustrate this fact, let's look again at the simple last example. The non-discounted gains were:

$$G = 800 + 1000 + 1200 = \$3000$$

And the costs were:

$$C = 2200 + 50 + 50 \cdot 1.04 + 50 \cdot 1.04^2$$
$$= 2200 + 50 + 52 + 54.08 = \$2356.08$$

If the initial payment was invested at a rate of 13%, the total gain after 3 years would be:

$$2200(1 + 0.13)^3 - 2200 = 3175.04 - 2200 = \$975.06$$

This is higher than what seems to be earning by the project:

$$-2200 + (800 - 50) + (1000 - 52) + (1200 - 54.08) = \$643.92$$

To obtain the same result, one would have to invest the first year's profit at the same rate during two years, the second year's profit during one year, etc.:

$$\text{Second year extra profit} = (800 - 50)(0.13)$$

$$\text{Third year extra profit} = (800-50)(1.13)(0.13)+(1000-52)(0.13)$$

So:

$$643.92 + (800 - 50)(0.13) + (800 - 50)(1.13)(0.13)$$
$$+ (1000 - 52)(0.13)$$

$$= 3175.04 - 2200 = \$975.06$$

or more concisely:

$$-2200 + (800 - 50)(1.13)^2 + (1000 - 52)(1.13)$$
$$+ (1200 - 54.08)$$
$$= \$975.06$$
$$= 2200(1.13)^3 - 2200$$

which can also be written as:

$$2200(1 + irr)^3 = FV(\text{profit year } 1, 2, irr)$$
$$+ FV(\text{profit year } 2, 1, irr)$$
$$+ FV(\text{profit year } 3, 0, irr)$$

(continued)

Another way to write is:

$$(1 + irr)^3 \left(2200 + \frac{\text{cost year 1}}{(1 + irr)} + \frac{\text{cost year 2}}{(1 + irr)^2} + \frac{\text{cost year 3}}{(1 + irr)^3} \right)$$

$$= FV(\text{gain year } 1, 2, irr) + FV(\text{gain year } 2, 1, irr)$$
$$+ FV(\text{gain year } 3, 0, irr)$$

∎

It can be observed from the last example that the last equations can be put in the form:

$$(1 + IRR)^3 (2200) = FV(profits, d = IRR)$$

or

$$(1 + IRR)^N \cdot (PV(costs, d = IRR) = FV(gains, d = IRR)$$
(12.28)

which means that the projected internal rate of return (IRR) will be obtained if it is possible to reinvest at the projected rate of return all future profits. Some economists [7] suggested that this is too severe and propose to find a *modified internal rate of return* (MIRR) where the reinvested rate for the future positive gains in (12.28) is chosen as the *reinvestment rate* (RR), whereas the present value of the possible losses will be discounted at a different weight often chosen as the *financial rate* (FR) which is associated with the rate they can obtain with creditors. Doing so the MIRR can now be calculated from:

$$(1 + MIRR)^N \cdot (PV(\text{negative cash flows}, d = FR) = FV(\text{positive cash flows}, d = RR)$$

$$MIRR = \sqrt[N]{\frac{FV(\text{positive cash flows}, d = RR)}{PV(\text{negative cash flows}, d = FR)}} - 1$$
(12.29)

where the initial investment is a negative flow not discounted since it is done at the beginning.

Example 12.13 In the last example, the IRR found was 13%.

(a) Find the MIRR if the reinvestment rate is 7% and the financial rate is 5%?
(b) Find the MIRR if the reinvestment rate is 0% and the financial rate is 5%?

(continued)

Example 12.13 (continued)
Solution The cash flows are:

$$\text{Year 1: } 800 - 50 = 750$$
$$\text{Year 2: } 1000 - 50(1.04) = 948$$
$$\text{Year 3: } 1200 - 50(1.04)^2 = 1145.92$$

(a)

$$FV = 750(1.07)^2 + 948(1.07) + 1145.92 = 3018.96$$

$$PV = 2200$$

$$MIRR = \sqrt[3]{(3018.96/2200)} - 1 = 0.111 \quad (11.4\%)$$

(b)

$$FV = 750 + 948 + 1145.92 = 3018.96$$

$$PV = 2200$$

$$MIRR = \sqrt[3]{(2843.92/2200)} - 1 = 0.0893 \quad (8.93\%)$$

This represents the rate when no reinvestment is done (we put all the money in our pocket) and equal to what we would get placing the money at the same rate:

$$2200(1+0.0893)^3 - 2200 = 2843.92 - 2200 = 643.92$$

∎

Example 12.14 The energy cost of a residential house is $3500 per year. A GSHP costing $35,000 is considered to replace it. An initial investment of 10% is requested and the rest is financed at a 3.5% interest rate. It is assumed that the system will reduce the energy bill by 60% but will generate a $100 increase in property tax. The inflation rate is 4.5% for the price of energy and 3% for the rest. A circulation pump ($300, present value) is expected to be replaced after 10 years and a residual value of $1000 is expected at the end of 20 years. No tax deduction is possible. The mortgage payment is assumed to be done at the end of each year.

(a) Find the NPV for 20 years with an 8% discounted rate.
(b) Find the IRR.
(c) Find the MIRR with reinvestment rate of 7% and financial rate of 5%.

(continued)

Solution The initial investment is:

$$0.1 \cdot 30000 = \$3000$$

Assuming annual payment, the amount is simply:

$$\frac{27000}{PWF(10, 0, 035)} = \$3165.22$$

The initial energy gain is:

$$0.6 \cdot 3500 = \$2100$$

The NPV can be evaluated directly from:

$$\begin{aligned}
NPV = &-3000 - 3165.22 \, PWF(10, 0, 0.08) \\
&-100 \, PWF(20, 0.03, 0.08) \\
&-300 \frac{(1+0.03)^9}{1.08^{10}} \\
&+2100 \, PWF(20, 0.045, 0.08) + \frac{1000}{1.08^{20}} \\
= &\ \$3523
\end{aligned}$$

To find the IRR, the discount rate is varied until the NPV is zero; the value obtained is:

$$IRR = 11\%$$

Instead, a year-by-year analysis can be performed in which case it is possible to see the evolution of the different cash flows. From the following figure (or Excel file generated), it is easy to see that the conventional payback period would be in the 13th year, whereas the discounted payback period would be in the 16th year.

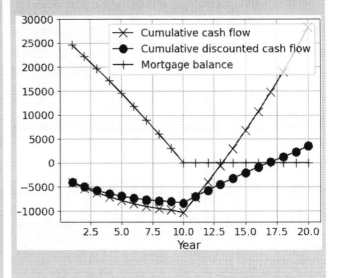

(continued)

Example 12.14 (continued)

Year	Energy gain	Annual payment	Cash flow	Discounted cash flow	Cumulative cash flow	NPV	Mortgage
0					−3000	−3000	27000
1	2100	3165	−1165	−1079	−4165	−4079	24645
2	2195	3165	−1074	−921	−5239	−4999	22219
3	2293	3165	−978	−776	−6217	−5776	19720
4	2396	3165	−878	−645	−7095	−6421	17147
5	2504	3165	−773	−526	−7869	−6948	14496
6	2617	3165	−664	−419	−8533	−7366	11765
7	2735	3165	−550	−321	−9083	−7687	8953
8	2858	3165	−430	−233	−9513	−7920	6057
9	2986	3165	−305	−153	−9818	−8072	3073
10	3121	3165	−566	−262	−10385	−8335	0
11	3261	0	3127	1341	−7258	−6994	0
12	3408	0	3270	1298	−3988	−5695	0
13	3561	0	3419	1257	−570	−4438	0
14	3722	0	3575	1217	3005	−3221	0
15	3889	0	3738	1178	6743	−2043	0
16	4064	0	3908	1141	10651	−902	0
17	4247	0	4087	1104	14738	202	0
18	4438	0	4273	1069	19011	1272	0
19	4638	0	4468	1035	23478	2307	0
20	4847	0	4671	1002	28149	3524	0

(c) Using the year-by-year analysis, it is possible to evaluate the un-discounted cash flows (fourth column) from which the MIRR can be calculated:

$$FV(gains, d = RR)$$
$$= 3126.84(1 + 0.07)^9 + 3269.57(1 + 0.07)^8$$
$$+ 3418.77(1 + 0.07)^7 + 3574.76(1 + 0.07)^6$$
$$3737.83(1 + 0.07)^5 + 3908.30(1 + 0.07)^4$$
$$+ 4086.51(1 + 0.07)^3 + 4272.81(1 + 0.07)^2$$
$$+ 4467.56(1 + 0.07) + 4671.16 = 51935\$$$

$$PV(costs, d = FR)$$

$$= 3000 + \frac{1165.22}{(1 + 0.05)} + \frac{1073.72}{(1 + 0.05)^2} + \frac{978.06}{(1 + 0.05)^3}$$

$$+ \frac{878.05}{(1 + 0.05)^4} + \frac{773.49}{(1 + 0.05)^5} + \frac{664.17}{(1 + 0.05)^6} +$$

$$\frac{549.88}{(1 + 0.05)^7} + \frac{430.40}{(1 + 0.05)^8} + \frac{305.49}{(1 + 0.05)^9} +$$

$$\frac{566.33}{(1 + 0.05)^{10}} = \$8979.25$$

$$MIRR = \sqrt[20]{(51935/8979.25)} - 1 = 0.0917 (9.17\%)$$

∎

12.5   Cost Evaluation for GSHP Systems

If the different economical metrics defined in the previous sections are to be used in the context of ground-source heat pump systems, the knowledge of the installation and operating costs of such systems have to estimated. This is, of course, not easy to do since it depends on many factors including technologies, geographical location, etc. Nevertheless, it is possible to find in the literature several studies giving some information about these costs. It is not the goal here to start an exhaustive survey of such studies. Several sources can be stated. In [8], for example, several prices, mainly associated with installation costs, are presented, some coming from their own surveys, and others coming from RSmeans [9]. Several detailed costs are given as well as distribution of typical prices for commercial buildings either in dollars per surface of building or per capacity installed. Some interesting results are also shown using comparisons between their surveys and some older studies [10]. An interesting conclusion drawn by the authors is that although the increase in the HVAC costs was in the order of 177%, the increase for the ground loop portion was 52%, reducing the percentage of the loop portion of the total cost to 26% instead of 39% in the previous survey. This conclusion may have an impact on some technical choices a designer has to make.

As previously stated, a review of major economic studies will not be described. Some recent studies include [4, 11–17]. In many of these studies, the goal was to estimate the economical feasibility of GSHP systems compared to other technologies. The purpose here would be to evaluate the impact of different designs on the overall economical costs of the system, or, equivalently, to include the economical impact in the final design decision, if possible.

Robert and Gosselin [18] present a design approach based on LCCA where installation costs are evaluated based on assumed typical prices. Again, these prices should be taken with care and not used as estimation costs but will be given solely as possible examples. The main goal is to show the use of the different economical calculations, the possibility to evaluate the sensitivity of design parameters [17] on the total price of the system, and the possible economical impact of government incentives. In [18], installation costs were divided into:

$$C_{inst} = C_{HP} + C_{drill} + C_{excav} + C_{pipe} \qquad (12.30)$$

where
- C_{inst}: Total installation cost
- C_{HP}: Heat pump cost
- C_{drill}: Drilling cost
- C_{pipe}: Pipe cost

Of course other items could have been included (pumps, grout, vault, monitoring sensors, etc.) [14]. A correlation was proposed for the heat pump price and was given as:

$$C_{HP} \ (\$CDN) = 1949.5 \ q_{heating}^{0.665} \qquad (12.31)$$

where $q_{heating}$ is the heating capacity in kW. Again, this price is few years old and should be taken with care. It is possible to apply some arbitrary inflation rate but again, it is not impossible that some items may be cheaper as the technology market penetration increases. The drilling cost was set to $40 (CDN)/m and the excavation price, $65 (CDN)/m^3. A fixed value of $3 (CDN) was chosen as the piping price per meter. In [19], the following correlation was proposed for HDPE piping to take into account the pipe diameter:

$$C_{pipe} = 785.94 \ D(m)^{1.9083} (\$(US)/m)$$
$$\approx 1021.7 \ D(m)^{1.9083} \$(CDN)/m \qquad (12.32)$$

where a 1.3 currency exchange rate was arbitrarily chosen. Some information about the cost of HDPE pipes can also be found in [8] for typical nominal DR 11 diameters. The same information was also available on the Internet [20]. Those prices given in $ US/ft were converted in $CDN/m to compare with (12.32) and a similar regression can be evaluated. The comparison is shown in Fig. 12.2:

$$C_{pipe} \approx 193.57 \ D(m)^{1.116} \$(CDN)/m \quad [8] \qquad (12.33)$$

$$C_{pipe} \approx 248.53 \ D(m)^{1.137} \$(CDN)/m \quad [20] \qquad (12.34)$$

It is seen that now the price seems to be more linear with the diameter. In [8] prices for Butt fusion welds are also indicated, prices which may also influence some design alternatives. A regression similar to the previous ones gives

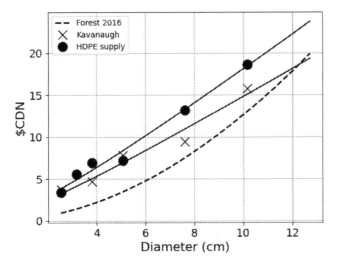

Fig. 12.2 HDPE price versus diameter

the following relation:

$$C_{weld} \approx 272.70 \ D(m)^{0.84} \$(CDN)/weld \quad [8] \qquad (12.35)$$

Similar information can be found for insulation (1" thickness) for several pipe diameters. Finding again a regression expression results in:

$$C_{insulation} \approx 50 \ D(m)^{0.52} \$(CDN)/m \quad [21] \qquad (12.36)$$

The same kind of issues can be raised about the drilling costs. It is sometimes associated with an estimated price per unit length ($/m or $/ft); a sensitivity analysis can be performed by varying this price around a nominal one if necessary, but sometimes, no information is associated with the diameter of the drilling. If is not the case in real life, then it can alter the final choice since, for example, the thermal performance will be associated with a change of diameter, but, if the impact on the price is not estimated, false conclusions can be drawn at the end. The same is true when one has to choose between four 150 m bores and three with a 200 m height. Using only a price per meter will give the same economical impact, whereas in real life, it may not be the same.

It is easier to have a good idea of the price of energy, electricity, or natural gas. However, detailed analysis is not easy since these prices are region dependent, may change between day and night, and include both energy consumption and billing demand with some minimum values applied. Simplified approaches using "typical" mean prices are often applied in order to avoid hour-by-hour complex simulations. Despite all these constraints, a minimal glance at the economical impact of the design parameters are paramount and can be refined when more precise information are available as long as a parameterized analysis is done.

Example 12.15 In Chap. 6, the total length of a ground heat exchanger was estimated using different approaches. The total length varied from 640 m to 663 m depending on the method chosen. In each case, the length was estimated using a high-performance grout having a conductivity of 1.7 W/m-K. This grout is more expensive since special silica sand or graphite particles are included. Using a cheaper grout would reduce the grout price but will increase the excavation costs. Compare the relative installation cost difference between two design alternatives using the previous enhanced grout and a cheaper one having a thermal conductivity of $k = 1.0$ W/m-K. Estimate the drilling cost at $50/m, the enhanced grout at $150/m^3, and the regular grout at $85/m^3. Use Eq. (12.33) for the pipe cost. Use the Swedish method with effective resistance calculated by the multipole method (Example 6.11a)

(continued)

Example 12.15 (continued)

Solution Since both scenarios are designed to reach the same outlet temperature, they should share the same performance, and only the installation cost difference will be estimated without doing a LCC analysis.[a] The analysis is "relative" such that only the prices that are changing are estimated. The heat pump prices, installation inside the building costs, etc. are assumed to stay unchanged in both cases. The enhanced grout solution (Example 6.11) gave a total length of 663 m. Even though a security factor leading to a multiple number of finite length borehole would probably be chosen at the end, this length will be used for the problem. The total volume of grout is:

$$A_{borehole} = \pi 0.075^2 = 0.0177 \text{ m}^2$$

$$A_{pipe} = 2\pi 0.0165^2 = 0.0017 \text{ m}^2$$

$$A_{grout} = 0.0177 - 0.0017 = 0.016 \text{ m}^2$$

$$V_{grout,1} = 10.58 \text{ m}^3$$

$$C_{grout,1} = \$1587$$

$$C_{drilling,1} = 663 \cdot 50 = \$33143$$

$$C_{pipe,1} = 2 \cdot 663 \cdot 5.14 = \$6814$$

$$C_{total,scenario,1} = \$41544$$

The only difference with the second borehole is in the evaluation of the borehole resistance. Using the multipole method, we find:

$$R'_b = 0.193 \text{ K-m/W} \quad , \quad R'_a = 0.489 \text{ K-m/W}$$

$$R'^*_b = 0.195 \text{ K-m/W}$$

From which a longer field will be obtained:

$$L = 798 \text{ m}$$

$$V_{grout,2} = 10.58 \text{ m}^3$$

$$C_{grout,2} = \$1083$$

$$C_{drilling,2} = 798 \cdot 50 = \$39926$$

$$C_{pipe,2} = 2 \cdot 798 \cdot 5.14 = \$8208$$

$$C_{total,scenario,2} = \$49217$$

The grout with a higher quality seems here a better choice. Although the example was done with very

---

[a]Actually, the longer borehole will generate larger pumping power but this aspect will be neglected here.

arbitrary choices, it is interesting to note that varying several parameters, it seems that in almost all the cases, the first choice is almost always better.   ∎

12.6 Putting It All Together

The last example provided a general grasp of the objective of the economical tools presented. Knowing that several prices and parameters (inflation, discount rate, etc.) are not known with accuracy, if it is still possible to see a general qualitative trend of a design choice or a government incentive policy on the life cycle cost or the NPV of the whole system, better decisions can be made. The rest of the chapter will try to illustrate this with some examples.

Example 12.16 In Chap. 8, a comparison of the total field length and pumping energy was done between three unitary systems and one central configuration for the three-zone building. Evaluate the life cycle cost over a 20-year period associated with the unitary configuration. For the economic analysis, use the following data:

- Discount rate: 6%
- Electricity cost: 0.08 $/kWh
- Drilling cost: 40 $/m
- Excavation cost: 65 $/m³
- Grout cost: 85 $/m³
- Glycol cost: 12 $/liter
- Down payment: 20%
- Mortgage rate: 4%
- Mortgage period: 10 years
- Inflation of electricity: 4%

Solution

– Cost Estimation
   First, we will use Eq. (12.31) to estimate the heat pump cost which here will be the same for both configurations. The total capacity is 80 kW, so:

$$C_{pac} = 1949.5 \cdot 80^{0.665} = \$35934$$

– Drilling Cost
   The total length is 1334.65 m. with ten boreholes. As previously discussed, the final design will probably involve boreholes with equal nominal height but we will keep this value for the estimation:

(continued)

(continued)

Example 12.16 (continued)
$$C_{drilling} = 40 \cdot 1334.6 = \$53383$$

– Excavation Cost
Three separated fields are expected to start at a 30 m distance from the building. Eight $(4 + 2 + 2)$ 6 m trenches connect the different boreholes which amounts to 138 m total length. Assuming a 1×2 m^2 front area. This leads to 276 m^3:

$$C_{excavation} = 65 \cdot 276 = \$17940$$

– Piping Cost
Details for system 1 (2x2 field) are:

- Loop $2 \times 565m (1"SDR - 11) \Rightarrow \$4312$

- Head $2 \times 30m (2"SDR - 11) \Rightarrow \$504$

- Reverse-return $2 \times 6m (1.5"SDR - 11) \Rightarrow \$72$

- Reverse-return $2 \times 12m (1.25"SDR - 11) \Rightarrow \$118$

- Reverse-return $2 \times 6m (1"SDR - 11) \Rightarrow \$46$

$$C_{piping,system1} = \$5055$$

A similar analysis will give:

$$C_{piping,system2} = \$3435$$

$$C_{piping,system3} = \$3238$$

$$C_{piping} = \$11728$$

In this example, the heat transfer in the horizontal sections was neglected, so insulation is assumed to be present:

$$C_{iso,system1} = 60 \cdot 10.61 + 12 \cdot 9.14 + 24 \cdot 8.31 + 12 \cdot 7.40$$
$$= \$1035$$

$$C_{iso,system2} = \$688$$

$$C_{iso,system3} = \$688$$

$$C_{iso} = \$2411$$

In the reverse-return configurations, $2 \times$ (nbore-1) welds are required. Using relation (12.35) will give:

$$C_{welding,system1} = \$110$$

$$C_{welding,system2} = \$60$$

$$C_{welding,system3} = \$60$$

$$C_{welding} = \$230$$

Regular grout was used. Assuming again an arbitrary cost of 85$/m^3:

$$C_{grout} = 85 \cdot L_{tot} \cdot \pi(r_b^2 - 2r_{po}^2) = 1805\$$$

Finally, the cost of glycol was also evaluated. Estimating the cost as 12 $/l, the total volume of the piping was:

$$V = 1.873 \text{ m}^3 = 1873 \text{ l}$$

$$C_{glycol} = 1873 \cdot 12 \cdot 0.2 = \$4496$$

Three pumps were needed for the three independent fields. The Bell and Gosset E-90 pump chosen price was found to be around $2000:

$$C_{pumps} = 3 \cdot 2000 = \$6000$$

$$C_{total} = 35934 + 53383 + 17940 + 11728 +$$
$$2411 + 230 + 1805 + 4496 + 6000$$
$$= \$133926$$

– Energy Consumption
As previously stated, the price of energy is highly dependent on the local jurisdiction. Price depends sometimes on the hour of the day. An important part of the total price is often associated with the power demand with some minimum billing demand for every month. To simplify the analysis, a fixed "average price" of electricity is used and is given here as 0.08 $/kWh. The total energy was given in Example 8.9 and was:

$$21583 \text{ kWh} \quad \Rightarrow C_{energy} = 1727\$/year$$

LCC analysis:

$$\text{Down payment} = 0.2 \cdot 133926 = \$26785$$

$$\text{Annual mortage} = \frac{107141}{PWF(10, 0, 0.04)} = \$13209$$

$$LCC = 26785 + 13209 PWF(10, 0, 0.06)$$
$$+ 1727 PWF(20, 0.04, 0.06)$$
$$= \$151358$$

■

(continued)

Example 12.17 A similar evaluation was done with the central configuration. The installed capacity is the same. The total length was reduced to 1135 m.

$$C_{drilling} = 40 \cdot 1135 = 45426\$$$

$$C_{excavation} = 13260\$$$

– Piping Cost

- Loop $2 \times 1135 m (1"SDR - 11) \Rightarrow \$8669$

- Head $2 \times 30 m (3"SDR - 11) \Rightarrow \$799$

- Reverse-return $3 \times 2 \times 6m (2"SDR - 11) \Rightarrow \$302$

- Reverse-return $3 \times 2 \times 6m (1.5"SDR - 11) \Rightarrow \$218$

- Reverse-return $3 \times 2 \times 6m (1.25"SDR - 11) \Rightarrow \$177$

- Reverse-return $3 \times 2 \times 6m (1"SDR - 11) \Rightarrow \$137$

- Reverse-return $3 \times 24 m (2"SDR - 11) \Rightarrow \$604$

$$C_{piping,field} = \$10906$$

In this configuration, extra piping and insulation is necessary inside the building:

$$C_{piping,building} = \$2430$$

$$C_{piping} = \$13337$$

Similar evaluation is done for insulation and welding:

$$\begin{aligned} C_{iso,field} = & \, 60 \cdot 13.10 + 3(12 \cdot 10.61 + 12 \cdot 9.14 \\ & + 12 \cdot 8.31 + 12 \cdot 7.40 + 24 \cdot 10.61) \\ = & \, \$2828 \end{aligned}$$

$$C_{iso,field} = \$2827$$

$$C_{iso} = \$5656$$

$$C_{welding,field} = \$455$$

$$C_{welding,building} = \$224$$

$$C_{welding} = \$679$$

$$V = 2.496 \text{ m}^3 = 2496 \, l$$

$$C_{grout} = \$1537$$

$$C_{glycol} = 2522 \cdot 12 \cdot 0.2 = \$6054$$

$$C_{pumps} = 1 \cdot 2000 = \$2000$$

$$C_{vfd} \approx 2000\$$$

The pump chosen has a ECM motor with variable-speed capability but this arbitrary amount will be kept as a safety factor:

$$\begin{aligned} C_{total} = & \, 35934 + 45426 + 13260 + 13337 + 5656 + 679 \\ & + 1537 + 6054 + 2000 + 1000 \\ = & \, \$124883 \end{aligned}$$

For the electricity consumption, the last value evaluated in Example 8.9 will be used:

$$21700 \text{ kWh} \quad \Rightarrow C_{energy} = 1736\$/year$$

LCC analysis:

$$\text{Down payment} = 0.2 \cdot 124883 = 24977\$$$

$$\text{Annual mortage} = \frac{99907}{PWF(10, 0, 0.04)} = 12318\$$$

$$\begin{aligned} LCC = & \, 24977 + 12318 PWF(10, 0, 0.06) \\ & + 1736 PWF(20, 0.04, 0.06) \\ = & \, \$143132 \end{aligned}$$

It is seen that this value is less than the previous one but the difference is not stringent. Several costs were not well analyzed (plumbing hours, HVAC controls, hangers). Again, it is more important to observe where the differences come from. The first alternative produces less interior piping, while the second less drilling (for this specific case where diversified loads are present). If the individual costs of these items change, other conclusions can be drawn. It is interesting to note that central configurations do not always have variable speed control. In which case, the investment price will not include the VFD drive but the pumping energy will increase in which case the new life cycle cost will be:

$$LCC = \$149751$$

Price which will be higher if the energy cost is higher than the small value used here. Finally, it can be observed that despite all the arbitrary assumptions, the final cost of both scenarios is in the order of 1600 CDN \$/kW ($\approx$1300 US \$/kW) which is in the midrange of the values observed in [8]. ∎

(continued)

Example 12.18 The tools presented in the last examples, although not conclusive, can be better applied if we try to evaluate the impact of different choices of parameters for the design of one system. For example, in the central configuration, evaluate the impact of:

(a) Changing the grout for high conductivity grout ($k = 1.7/$"m-K, cost 150 \$/m^3) .
(b) Changing the plastic for high conductivity plastic ($k = 0.75/$"m-K, cost double normal pipe) .
(c) Changing the maximum EWT to 30 °C.
(d) Changing the minimum EWT to −2 °C, use 30% propylene glycol then.
(e) Changing the pipe location to $r_b - r_o$, configuration C (assume \$5000 extra cost for work and spacers).
(f) Changing the pipe diameter to 3/4" in the borehole.
(g) Changing the pipe diameter to 1.25" in the borehole.

Solution

(a) In the first modification, similar results to Example 12.15 are expected. The price of grout will increase:

$$C_{grout} = 2242 > \$1537$$

but the decrease in drilling and piping costs will be more beneficial

$$C_{drilling} = 37555 < \$45426 \Rightarrow LCC = \$134575$$

(b) A similar conclusion can be expected if a plastic with higher conductivity is used. However, different results are obtained:

$$C_{piping} = 25774 > \$13337$$

but the decrease in drilling and piping costs will be more beneficial

$$C_{drilling} = 43069 < \$45426 \Rightarrow LCC = \$152233$$

The decrease in length does not compensate for the increase of plastic piping. It must be observed, however that, in practice, probably only the plastic in the boreholes would use the higher cost plastic, while here all pipes used the higher-quality one. At the end, the total price will be lowered but the conclusion will probably remain unchanged.

(continued)

(c) The change of maximum and/or minimum temperatures is a subject of interest. Since the total length is imposed here for the heating mode, it is expected that the lower COP used will lower the total performance. Indeed, the energy expected is now:

$$E = 22206 \text{ kWh} > 21700 \text{ kWh}$$

$$\Rightarrow C_{energy} = 1776\$/year > C_{energy} = \$1736/year$$

However, since the energy price is very low, the difference is small, and more heat is rejected into the ground, the yearly pulse is now lower ($q_a = 3.27$ kW) and the total length also:

$$L = 1133m \Rightarrow C_{drilling} = \$45325$$

$$CCV = \$142054$$

Again, the difference with the base case is very small but the final result is surprising. The result would be very different if the cooling mode would be dominant.

(d) Looking at the performance data of the manufacturer, it is observed that the heat pumps can work in cooling mode at high temperature although the performance will suffer. However, in heating mode, the lowest temperature is: 20 °F(≈ -7 °C). At this temperature, only a high flow rate is recommended, a case that can be easily analyzed but here the temperature will be lowered at −1 °C so that the same flow rate is used. A 30% glycol solution will be used for this case. As expected, the compressor and pumping energy will increase:

$$E = 22426 \text{ kWh} > 21700 \text{ kWh}$$

$$\Rightarrow C_{energy} = 1794\$/year > C_{energy} = 1736\$/year$$

However, the total length will be smaller:

$$L = 1043 \text{ m}$$

and the final life cycle cost:

$$LCC = \$141862$$

(e) The impact of pushing the pipes toward the end of the borehole is to reduce the borehole resistance. However, the use of spacers often bothers contractors who are often reluctant to use them.

(continued)

Example 12.18 (continued)

In our specific example, the impact is important. Even though a \$5000 extra cost was expected, the borehole resistance is now much lower:

$$R'_b = 0.113 \text{ K-m/W} \; < 0.197 \text{ K-m/W}$$

$$L = 870 \text{ m}$$

In this case, the 4×3 initial configuration will probably change and the final cost even lower. Keeping the same configuration, the result is:

$$LCC = \$135015$$

(f) and (g) The effect of changing the pipe diameter inside the loop is twofold: A smaller diameter will increase the pumping energy, increase the borehole resistance (less contact area), and decrease the cost. An increase in diameter will have an opposite effect. In this latter case, it is important to check that the flow remains turbulent, which is the case here.

Scenario	LCC
Base case	143133
(a)	134575
(b)	152233
(c)	142054
(d)	141862
(e)	135015
(f)	142374
(g)	144440

Of course, many other parameters can be analyzed: borehole radius, field configuration, using an hybrid configuration with backup system, etc. Once parameterized, it is easy to modify the different parameter values as needed. In this specific example, it was observed that the effect of grout and spacers was the most important. ■

Problems

Pr. 12.1 A building uses up 180 GJ of energy in the form of electricity. Two geothermal systems are possible to replace the system:

H(m)	% savings
310	60
495	72

We have the following conditions:

- Initial cost of electricity \$ 0.07/kWh.
- Installation fixed cost \$12500.
- Installation variable cost \$ 50/m.
- Discount rate 8%.
- Interest rate 4%.
- Inflation rate of energy 4%.
- Inflation rate (except energy) 3%.
- Down payment 20%
- Mortgage period 10 years.
- LCC period 20 years.

(a) Find the system with the best NPV.
(b) What would be the electricity price for the two NPV to be equal?

Answer: (a) (NPV case 1) = \$ 3688, (NPV case 2) = \$ 1280, (b) \$0.10

Pr. 12.2 Two systems are available to replace an old heating system in a building having an annual energy bill of \$ 8000.

- case 1: A first one having an installation cost of \$45,000 with an estimating energy bill of \$ 4000.
- case 2: A second one smaller having a cost of \$ 38,000 and an estimated future energy bill of \$ 4800.

The first system is assumed to work with water, whereas the second one will need a mixture of water-glycol due to the lower temperatures involved. Glycol is expected to be replaced every 4 years at a present cost of \$ 500 each time (the initial filling price is included in the installation cost). Compare the IRR for the two systems on a 15-year period assuming an inflation rate of 2.5% except for energy where a 4% inflation is assumed. The down payment is 10%. A 10-year mortgage at a 4% interest rate is taken for the remaining balance. A tax deduction of 20% is available for the interest payments and the resale value is expected to be \$ 2000 (case 1) and \$ 1000 (case 2).
Answer: IRR (case 1) = 15.2%, IRR (case 2) = 11.5%

Pr. 12.3 In the previous problem, find the MIRR with a financial rate of 4% and a reinvestment rate of 6%. *Answer: MIRR (case 1) = 13.3%, MIRR (case 2) = 10.0%*

References

1. Reidy, R. et al.: Guidelines for life cycle cost analysis. Stanf. Univ. Land Build **1**:1–30 (2005)
2. Park, C.S.: Fundamentals of Engineering Economics. 3rd edn. Prentice Hall, Upper Saddle River (2012). ISBN: 9780132775427
3. Duffie, J.A., Beckman, W.A.: Solar Engineering of Thermal Processes. Wiley, New York (2013)
4. Gabrielli, L., Bottarelli, M.: Financial and economic analysis for ground-coupled heat pumps using shallow ground heat exchangers. Sustain. Cities Soc. **20**, 71–80 (2016)
5. Wang, Z.: Heat pumps with district heating for the UKs domestic heating: individual versus district level. Energy Proc. **149**, 354–362 (2018)
6. Yogi Goswami, D.: Principles of Solar Engineering. CRC Press, Boca Raton (2022)
7. Lin, S.A.Y.: The modified internal rate of return and investment criterion. Eng. Econ. **21**(4), 237–247 (1976)
8. Kavanaugh, S., Rafferty, K.: Geothermal Heating and Cooling: Design of Ground-Source Heat Pump Systems. ASHRAE Atlanta (2014)
9. RS Means Company: Mechanical Costs With RSmeans Data 2021 (Means Mechanical Cost Data). Gordian (2021)
10. Caneta Research: Operating experiences with commercial ground-source heat pump systems. Technical report. ASHRAE, Atlanta, 1995
11. Retkowski, W., Thoming, J.: Thermoeconomic optimization of vertical ground-source heat pump systems through nonlinear integer programming. Appl. Energy **114**, 492–503 (2014). https://doi.org/10.1016/j.apenergy.2013.09.012
12. Vu, N.B. et al.: Life cycle cost analysis for ground-coupled heat pump systems including several types of heat exchangers. Int. J. Archit. Eng. Constr. **2**, 17–24 (2013)
13. Hakkaki-Fard, A. et al.: A techno-economic comparison of a direct expansion ground-source and an air-source heat pump system in Canadian cold climates. Energy **87**, 49–59 (2015). https://doi.org/10.1016/j.energy.2015.04.093
14. Henault, B., Pasquier, P., Kummert, M.: Financial optimization and design of hybrid ground-coupled heat pump systems. Appl. Thermal Eng. **93**, 72–82 (2016). https://doi.org/10.1016/j.applthermaleng.2015.09.088
15. Cui, Y. et al.: Techno-economic assessment of the horizontal geothermal heat pump systems: a comprehensive review. Energy Convers. Manag. **191**, 208–236 (2019)
16. Trovato, M.R., Nocera, F., Giuffrida, S.: Life-cycle assessment and monetary measurements for the carbon footprint reduction of public buildings. Sustainability **12**(8) (2020). https://doi.org/10.3390/su12083460
17. Dusseault, B., Pasquier, P.: Usage of the net present value-at-risk to design ground-coupled heat pump systems under uncertain scenarios. Renewable Energy **173**, 953–971 (2021). https://doi.org/10.1016/j.renene.2021.03.065
18. Robert, F., Gosselin, L.: New methodology to design ground coupled heat pump systems based on total cost minimization. Appl. Thermal Eng. **62**(2), 481–491 (2014)
19. Forrest, S., Rustum, R.: Design optimisation of marine wastewater outfalls. Int. J. Sustain. Water Environ. Syst. **8**(1), 7–12 (2016)
20. HDPE Supply. HDPE Pipe fittings and fusion equipment. https://hdpesupply.com/black-hdpe-polyethylene-pipe/. Accessed 12 Dec 2021
21. Buy Insulation Products. Fiberglass Pipe Insulation. https://www.buyinsulationproductstore.com/Fiberglass-Pipe-Insulation-SSL-ASJ/. Accessed 12 Dec 2021

13.1 Introduction

In the previous chapters, the basic concepts needed to solve most of the practical aspects associated with GSHP systems were addressed. The goal of this chapter is to introduce some more advanced issues that are associated with recent research trends.

13.2 Short-Time Thermal Response

In Chaps. 4, and 5, it was argued that the thermal response of a ground heat exchanger can be split into two parts: outside the borehole where the thermal mass of the ground plays a major role and inside the borehole where the thermal mass of the grout and the fluid is neglected. As already stated, this assumption is only valid for Fourier numbers larger than approximately 5, which corresponds with a typical borehole to the order of several hours to almost a day. This restriction is not too limiting since the time scales of heat transfer in the ground are often larger. However, in a typical ground heat exchanger design, the thermal response of pulses of several hours is needed (see Chap. 6), and the use of classical models like ICS or ILS can lead to overestimations [1, 2]. For this reason, several researches have been carried out recently to take into account the thermal capacity of the inside part of the borehole. Assuming that for a short time scale, the axial heat transfer can be neglected, the problem can be solved, having the solution of the following equations:

$$\frac{1}{\alpha_i}\frac{\partial T_i}{\partial t} = \frac{\partial^2 T_i}{\partial r^2} + \frac{1}{r}\frac{\partial T_i}{\partial r} + \frac{1}{r^2}\frac{\partial^2 T_i}{\partial \theta^2} \quad , \quad i = 1, 2, 3, 4$$

where

$i = 1 :$ water
$i = 2 :$ plastic pipe
$i = 3 :$ grout
$i = 4 :$ soil

$$(13.1)$$

The advection problem in the inner fluid region can also be added to the problem but neglected for now. This problem can be easily solved using numerical software. Some early work has been proposed by Yavusturk and Spitler [3] based on a numerical approach. A simplified approach based on an equivalent axisymmetric numerical model was proposed by Xu and Spitler [4] and used by Ahmadfard and Bernier [5]. Analytic tools to solve the real problem are not possible and to do so, some approximations have to be made. Some authors used the composite line-source approach [6], but the most simple approach is to replace the complex geometry by an equivalent one using an *equivalent radius* (Fig. 13.1b). Two further approximations can be easily accepted: The first one is to neglect the thermal capacity of the plastic pipe and to assume an almost uniform temperature for the water region. In such a case, the last problem can be simplified as:

$$\frac{1}{\alpha_g}\frac{\partial T_g}{\partial t} = \frac{\partial^2 T_g}{\partial r^2} + \frac{1}{r}\frac{\partial T_g}{\partial r} \qquad (13.2)$$

for the domain $r_e < r < r_b, t > 0$, where T_g represent the grout temperature:

Fig. 13.1 (**a**) Typical U-tube.
(**b**) Equivalent borehole

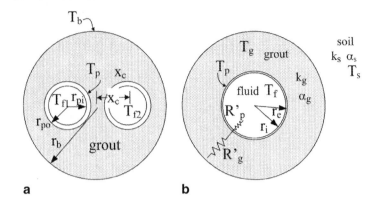

a   b

$$\frac{1}{\alpha_s}\frac{\partial T_s}{\partial t} = \frac{\partial^2 T_s}{\partial r^2} + \frac{1}{r}\frac{\partial T_s}{\partial r} \qquad (13.3)$$

for the domain $r_b < r, t > 0$, where T_s is the soil temperature. The boundary conditions for the grout-ground interface are:

$$T_g(r, 0) = T_s(r, 0) = T_o$$

$$T_g(r_b, t) = T_s(r_b, t)$$

$$-k_g\frac{\partial T_g}{\partial r}\bigg|_{r=r_b} = -k_s\frac{\partial T_s}{\partial r}\bigg|_{r=r_b} \qquad (13.4)$$

From the assumption of a well-stirred fluid in the inner part and assuming that the heat exchange is done in the fluid (neglecting the advection), the following boundary condition will be imposed at the inner radius:[1]

$$\pi r_i^2(\rho C_p)_f\frac{d\,T_f}{d\,t} = -q'{}_b + 2\pi r_e k_g\frac{\partial T_g}{\partial r}\bigg|_{r=r_e} \qquad (13.5)$$

$$\frac{T_f - T_g}{R'_p} = -2\pi r_e k_g\frac{\partial T_g}{\partial r}\bigg|_{r=r_e} \qquad (13.6)$$

R'_p is the unit length resistance between the fluid and the grout. It is often represented by a convection resistance and the plastic resistance associated with the real geometry (Fig. 13.1a):

$$R'_p = \frac{1}{4\pi r_{pi}\hbar} + \frac{\log\left(r_{po}/r_{pi}\right)}{4\pi k_p} \qquad (13.7)$$

Solutions of thermal problems are more easily solved using dimensionless variables. In a composite cylinder, several reference variables can be used, for example, the soil properties or the grout properties. The former was used by Beier and Smith [7] and the latter by Lamarche and Beauchamp [8, 9]. Even though those last results will be used in this section, they will be reformulated using the soil variables as reference

variables since it is a more logical choice. The last system is then rewritten as:

$$\gamma^2\frac{\partial \tilde{T}_g}{\partial \tilde{t}} = \frac{\partial^2 \tilde{T}_g}{\partial \tilde{r}^2} + \frac{1}{\tilde{r}}\frac{\partial \tilde{T}_g}{\partial \tilde{r}} \quad , \quad \tilde{r}_e < \tilde{r} < 1 \qquad (13.8)$$

$$\frac{\partial \tilde{T}_s}{\partial \tilde{t}} = \frac{\partial^2 \tilde{T}_s}{\partial \tilde{r}^2} + \frac{1}{\tilde{r}}\frac{\partial \tilde{T}_s}{\partial \tilde{r}} \quad , \quad \tilde{r} > 1 \qquad (13.9)$$

$$\tilde{T}_g(\tilde{r}, 0) = \tilde{T}_s(\tilde{r}, 0) = 0 \quad , \quad \tilde{T}_g(1, \tilde{t}) = \tilde{T}_s(1, \tilde{t}) \quad ,$$

$$\tilde{k}\frac{\partial \tilde{T}_g}{\partial \tilde{r}}\bigg|_{\tilde{r}=1} = -\frac{\partial \tilde{T}_s}{\partial \tilde{r}}\bigg|_{\tilde{r}=1} \qquad (13.10)$$

with:

$$\gamma = \sqrt{\frac{\alpha_s}{\alpha_g}} \quad , \quad \tilde{k} = \frac{k_g}{k_s} \quad , \quad \tilde{T} = k_s\frac{T - T_o}{q'_b} \quad ,$$

$$\tilde{t} = \frac{\alpha_s t}{r_b^2} \quad , \quad \tilde{r} = \frac{r}{r_b} \qquad (13.11)$$

The core boundary conditions are now:

$$\nu\frac{d\tilde{T}_f}{d\,\tilde{t}} = \frac{-1}{2\pi} + \tilde{r}_e\tilde{k}\frac{\partial \tilde{T}_g}{\partial \tilde{r}}\bigg|_{\tilde{r}=\tilde{r}_e} \qquad (13.12)$$

$$\tilde{T}_f - \tilde{T}_g = -\tilde{R}\,\tilde{r}_e\tilde{k}\frac{\partial \tilde{T}_g}{\partial \tilde{r}}\bigg|_{\tilde{r}=\tilde{r}_e} \qquad (13.13)$$

with:

$$\nu = \frac{\tilde{r}_i^2(\rho C_p)_f}{2(\rho C_p)_s} \quad , \quad \tilde{R} = 2\pi k_s R'_p \qquad (13.14)$$

Special simplified cases of the latter problem were solved in the literature. The case where the conductivity of the backfill material is very high (perfect conductor) is solved in Carslaw and Jaeger [10]. It is known as the *buried cable* solution. Although the approximation of a perfect conductor is not suitable to the grout material in which conductivity is relatively low, this solution has been successfully used by

---

[1]Remember that heat injection is negative in our convention.

Young [11] in the context of the short-time solution where he added the significant grout resistance to the pipe resistance to take into account the total resistance of the borehole. In [8], the solution of the previous problem with $R'_p = C_{p_f} = 0$ was given. Again, this assumption can be justified by adding the pipe resistance to the grout resistance, the same thing for the added mass of the fluid. The solution where the fluid capacity is kept but not the pipe resistance was given by Beier and Smith [7]. All analytical solutions found from several approximations of the previous problem share in common the initial solution in the Laplace domain. In Beier and Smith [7], the solution in the temporal domain was obtained using a numerical Laplace inversion. The first analytical solution of the whole problem was given by Javed and Claesson [12]. The solution is given by the following system:

$$T_f(\tilde{t}) = T_o + \frac{-q'_b}{k_s} \frac{2}{\pi} \int_0^\infty \frac{1 - e^{-u^2 \tilde{t}}}{u} L(u) d u \quad (13.15)$$

with:

$$L(u) = \frac{1}{2\pi} Im \cfrac{1}{vu^2 - \cfrac{1}{\tilde{R} + \cfrac{1}{K_p(u) + \cfrac{1}{R_t(u) + \cfrac{1}{K_b(u) + K_s(u)}}}}} \quad (13.16)$$

where:

$$K_s(u) = u \frac{J_1(u) - j \cdot Y_1(u)}{J_o(u) - j \cdot Y_o(u)} \quad (13.17)$$

$$K_p(u) = \frac{0.5\pi \tilde{r}_e \gamma u (J_1(\tilde{r}_e \gamma u) Y_o(\gamma u) - J_o(\gamma u) Y_1(\tilde{r}_e \gamma u)) - 1}{R_t(u)} \quad (13.18)$$

$$K_b(u) = \frac{0.5\pi \gamma u (J_1(\gamma u) Y_o(\tilde{r}_e \gamma u) - J_o(\tilde{r}_e \gamma u) Y_1(\gamma u)) - 1}{R_t(u)} \quad (13.19)$$

$$R_t(u) = \frac{\pi (J_o(\tilde{r}_e \gamma u) Y_o(\gamma u) - J_o(\gamma u) Y_o(\tilde{r}_e \gamma u))}{2\tilde{k}} \quad (13.20)$$

Lamarche [9] presented a closed-form analog of the previous results by applying the classical Laplace inversion theorem in the complex plane which was a generalization of their previous results for the case where $R'_p = C_{p_f} = 0$ [8]. The solution is given in the form of an improper integral:

$$T_f(\tilde{t}) = T_o + \frac{-q'_b}{k_s} \frac{8}{\pi^5 \gamma^2 \tilde{r}_e^2} \int_0^\infty \frac{\left(1 - e^{-\beta^2 \tilde{t}}\right)}{\beta^5 (\phi^2 + \psi^2)} d\beta \quad (13.21)$$

with:

$$\phi = \zeta_1 [\tilde{k}\gamma Y_o(\beta) J_1(\gamma\beta) - Y_1(\beta) J_o(\gamma\beta)] \\ - \zeta_2 [\tilde{k}\gamma Y_o(\beta) Y_1(\gamma\beta) - Y_1(\beta) Y_o(\gamma\beta)] \quad (13.22)$$

$$\psi = \zeta_2 [\tilde{k}\gamma J_o(\beta) Y_1(\gamma\beta) - J_1(\beta) Y_o(\gamma\beta)] \\ - \zeta_1 [\tilde{k}\gamma J_o(\beta) J_1(\gamma\beta) - J_1(\beta) J_o(\gamma\tilde{r}_e)] \quad (13.23)$$

$$\zeta_1 = (1 - v\tilde{R}\beta^2) Y_1(\tilde{r}_e\gamma\beta) - v\beta Y_o(\tilde{r}_e\gamma\beta)/(\tilde{r}_e\gamma\tilde{k}) \quad (13.24)$$

$$\zeta_2 = (1 - v\tilde{R}\beta^2) J_1(\tilde{r}_e\gamma\beta) - v\beta J_o(\tilde{r}_e\gamma\beta)/(\tilde{r}_e\gamma\tilde{k}) \quad (13.25)$$

Any of the last relations can be represented as *short-time response factors* as:

$$\Delta T = \frac{-q'_b}{k_s} G_{st}(Fo) \quad (13.26)$$

with:

$$G_{st}(Fo) = \frac{8}{\pi^5 \gamma^2 \tilde{r}_e^2} \int_o^\infty \frac{\left(1 - e^{-\beta^2 Fo}\right)}{\beta^5 (\phi^2 + \psi^2)} d\beta \quad (13.27)$$

or in Eskilson's formalism:

$$\Delta T = \frac{-q'_b}{2\pi k_s} g_{st}(\bar{t}, \bar{r}) \quad (13.28)$$

with:

$$g_{st}(\bar{t}, \bar{r}) = 2\pi G_{st}(Fo)|_{Fo=\bar{t}/(3\bar{r})^2} \quad (13.29)$$

A major difference between the last relation and the previously defined response factors (G or g) is that the definition given in (13.26) relates the mean fluid temperature to the load, whereas with the response factors previously discussed, the borehole temperature was first obtained and after, the fluid temperature was found by adding the effect of the borehole resistance which of course, generates an unphysical temperature jump at $t = 0$. In their work, Yavustruk and Spitler [3] and Xu and Spitler [4] used a finite-volume numerical code to generate more physical fluid and borehole temperatures at early times. However, they chose to keep the dimensionless temperature response (g-function) associated with the borehole temperature and still use the concept of borehole resistance to find the fluid temperature. This choice brought some singular behavior. A simple schematic can help to visualize. Let's assume that at $t = 0$, a heat pulse is generated in the fluid. Keeping our convention, this would represent a negative heat pulse. A possible temperature profile could look like Fig. 13.2. In the real case, a time lag can be observed for the borehole temperature due to the capacity of the borehole which would not be the case, for example, if the infinite cylinder source solution would be used (see Fig. 13.2).

Fig. 13.2 Influence of the
borehole capacity on the borehole
temperature

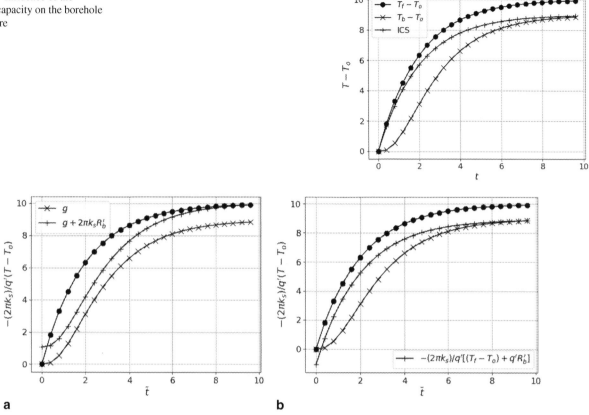

Fig. 13.3 (**a**) Classical definition of g-function influenced by the borehole capacity. (**b**) Yavustruk definition of short-time g-function

However, even if this temperature behavior is more realistic, defining the *g-function* as the reduced borehole temperature and adding the effect of the borehole resistance to find the fluid temperature would still lead to a temperature jump at t = 0 (Fig. 13.3a) given by:

$$T_f(t) - T_o = (T_b(t)_{st} - T_o) - q'_b R'_b$$

In order to eliminate the unphysical jump, they defined a corrected response factor by subtracting (Fig. 13.3b) the effect of that resistance by:[2]

$$T_f(t) - T_o = -\frac{q'_b}{2\pi k_s}\underbrace{\left[\frac{2\pi k_s}{-q'_b}\left((T_f(t)_{st} - T_o) + R'_b q'_b\right)\right]}_{g_{st,Yav-Spi}} - q'_b R'_b$$

It is easy to see that, with that definition, since the fluid temperature is almost zero for a short-time period, surprising negative values of thermal response factors and corresponding soil resistance will be obtained, which is not the case in the previous definition.

13.2.1 Equivalent Radius

In the previous section, the analytical expressions found were based on the concept of equivalent radius. This concept is not new as it was introduced in the early stage of GSHP systems by Bose et al. [13]. In their work, the equivalent radius was defined as:

$$r_e = \sqrt{2}r_{po} \qquad (13.30)$$

This radius was also chosen by Javed and Claesson [12] in their work. The reason is that the grout volume, and then its capacity, will be the same as the real geometry. However, the resulting thermal resistance of the equivalent annulus will be different than the real one, and the quasi-steady-state behavior will be different than the real borehole, which is very important. In order to solve this problem, the conductivity of the grout is modified in order that the annulus resistance will be the same as the real borehole:

$$\frac{\log(r_b/r_e)}{2\pi k_{eq}} = R'_g \qquad (13.31)$$

where the grout resistance can be found using classical methods described in Chap. 4. The thickness of the plastic pipe or its conductivity can also change in order that its resistance value will be equal to the real one:

---

[2] Again, there is a sign difference between [3] due to the sign convention.

$$\frac{\log\left(r_e/r_i\right)}{2\pi k_{p,eq}} = \frac{\log\left(r_{po}/r_{pi}\right)}{4\pi k_p} \qquad (13.32)$$

Another possibility for the equivalent radius, which was proposed by Sutton et al. [14], is to choose it such that the annulus resistance is the same as the borehole resistance. To do so, the equivalent radius is chosen as:

$$\frac{\log\left(r_b/r_e\right)}{2\pi k_g} = R'_g \qquad (13.33)$$

$$\Rightarrow r_e = r_b e^{-2\pi k_g R'_g} \qquad (13.34)$$

In such a case, the area occupied by the fluid and the grout is different than the real borehole and the volumetric capacity can be corrected as:

$$(\rho Cp)_{g,eq} = (\rho Cp)_g \frac{r_b^2 - 2r_{po}^2}{r_b^2 - r_e^2} \qquad (13.35)$$

Finally, depending on the equivalent inner radius chosen, the fluid capacity can be adjusted to meet the real fluid capacity:

$$\pi r_i^2 (\rho C_p)_{f,eq} = 2\pi r_{pi}^2 (\rho C_p)_f \qquad (13.36)$$

Actually the geometry of the inner tube is not important since the pipe resistance and convection resistance are replaced by a contact resistance whose value is adjusted to meet the real pipe resistance (Eq. (13.7)) and the fluid capacity is often artificially increased to take into account the added water in the whole loop. Impact of the choice of equivalent radius has been studied by Lamarche [9] where it was observed that small differences were observed but both can be used.

Example 13.1 In a well-known paper, Beir, Smith, and Spitler [15] provided a reference data set that was intended at the beginning to serve as reference data to compare different TRT methods. The experience was conducted in a sandbox whose properties were known and to which the TRT's results could be compared. Since the measured values were given for early times, they were also used to evaluate the accuracy of some short-time models [9, 12, 15]. They will be compared here to Eq. (13.26). The known reference measured properties were given as:

r_{po}	1.67 cm	k_g	0.73 W/mK	k_s	2.82 W/mK
r_b	6.3 cm	$(\rho C_p)_s$	1.92 MJ/m³ K	q'_b	57.7 W/m
x_c	2.65 cm	$(\rho C_p)_g$	3.84 MJ/m³ K	R'_p	? K-m/W
T_o	22°C	$(\rho C_p)_f$	4.16 MJ/m³ K	R'_g	? K-m/W

(continued)

The pipe resistance is given by (13.7). With the values given, a Reynolds number of 11000 was found and the pipe resistance found was in the order of $R'_p = 0.046$ m-K/W. Remember that Eq. (13.7) is the total pipe resistance which is half the pipe resistance associated with a single pipe. So the total borehole resistance can be found as:

$$R'_b = R'_g + 0.046 \text{ K-m/W}$$

The grout resistance can be found from the known analytical expressions given in Chap. 5, but here, it will be estimated from parameter estimation in order to avoid a bias associated with the uncertainty of the grout resistance. The parameter estimation was done in [9] and was found to be:

$$R'_g = 0.105 \text{ K-m/W} \Rightarrow R'_b = 0.151 \text{ K-m/W}$$

This value is somehow smaller than the one found by Beier et al. [15] (0.187 K-m/W) but will be used here. Using Eq. (13.34), as the equivalent radius, the following values are obtained:

$$r_e = 0.063 e^{-2\pi k_g R'_g} = 0.039 \text{ m}$$

$$(\rho Cp)_{g,eq} = 3.84 \frac{r_b^2 - 2r_{po}^2}{r_b^2 - r_e^2} = 5.34 \text{ MJ/m}^3$$

$$\alpha_s = \frac{2.82}{1.92e6} = 1.47e\text{-}06 \text{ m}^2/\text{s}$$

$$\alpha_{g,eq} = \frac{0.73}{5.34e6} = 1.37e\text{-}06 \text{ m}^2/\text{s}$$

Since, in the composite cylinder, the pipe resistance is associated with a contact resistance, we have $r_i = r_e$ and:

$$(\rho C_p)_{f,eq} = \frac{2r_{pi}^2}{r_e^2} (\rho C_p)_f = 1.02 \text{ MJ/m}^3$$

$$\gamma = \sqrt{\frac{\alpha_s}{\alpha_{g,eq}}} = 3.27 \quad , \quad \tilde{k} = \frac{k_g}{k_s} = 0.259 \quad ,$$

$$\tilde{r}_e = \frac{0.039}{0.063} = 0.618$$

$$v = \frac{\tilde{r}_i^2 (\rho C_p)_{f,eq}}{2(\rho C_p)_s} = 0.102 \quad , \quad \tilde{R} = 2\pi k_s R'_p = 0.81$$

(continued)

Example 13.1 (continued)

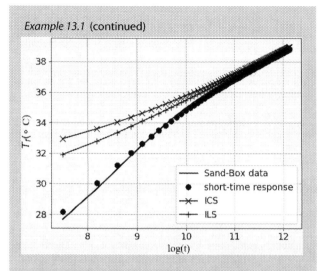

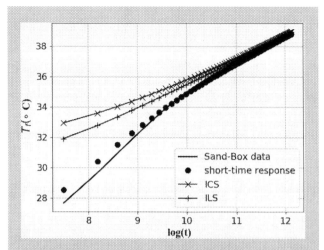

The result obtained is shown in the following figure where the classical solution obtained by the ICS and the ILS solutions are also shown. The result shows the typical impact of neglecting the borehole capacity for the first few hours and also, contrary to what is often believed, a better result obtained by the ILS than the ICS which was pointed out in [16].

Similar results can be obtained with a different choice of equivalent radius. Using Eq. (13.30) instead, one would obtain:

$$r_e = \sqrt{2} \cdot 0.0167 = 0.024 \text{ m}$$

$$(k)_{g,eq} = \frac{\log r_b/r_e}{2\pi R'_g} = 1.49 \text{ W/m-K}$$

$$\alpha_{g,eq} = \frac{1.49}{3.84e6} = 3.87\text{e-}07 \text{ m}^2/\text{s}$$

$$(\rho C_p)_{f,eq} = \frac{2r_{pi}^2}{r_e^2}(\rho C_p)_f = 2.78 \text{ MJ/m}^3$$

$$\gamma = \sqrt{\frac{\alpha_s}{\alpha_{g,eq}}} = 1.94 \quad, \quad \tilde{k} = \frac{k_{g,eq}}{k_s} = 0.527 \quad,$$

$$\tilde{r}_e = \frac{0.024}{0.063} = 0.374$$

$$v = \frac{\tilde{r}_i^2 (\rho C_p)_{f,eq}}{2(\rho C_p)_s} = 0.102 \quad, \quad \tilde{R} = 2\pi k_s R'_p = 0.81$$

It is seen that both approaches give different results but both are satisfactory to represent the short-time behavior of the borehole. In [9], the author presents similar results, but instead of estimating the unknown grout resistance using parameter estimation, both the grout resistance and the equivalent water capacitance were evaluated in order to minimize the RMS difference between the model and the experiment. The reason behind this approach is that the volume of water that is heated is larger than the amount of water in the borehole itself. This quantity is often unknown and is sometimes associated with an arbitrary multiplication factor [11]. The multiplication thus found was different depending on the equivalent radius chosen. Using Eq. (13.30), one finds:

$$v_{estimated} = 0.1643$$

which would imply:

$$(\rho_f C p_f)_{estimated} = 6.7 \text{ MJ/m}^3 \text{ K} = 1.61 \cdot (\rho_f C p_f)$$

Using Eq. (13.34):

$$v_{estimated} = 0.1353$$

$$(\rho_f C p_f)_{estimated} = 5.5 \text{ MJ/m}^3 \text{ K} = 1.32 \cdot (\rho_f C p_f)$$

The results obtained with these values are shown in the following figure:

(continued)

(continued)

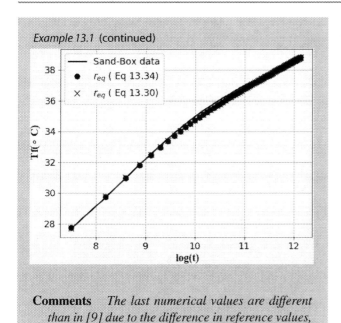

Example 13.1 (continued)

Comments   *The last numerical values are different than in [9] due to the difference in reference values, but the principle is the same.*

13.2.2 Modification of the Sizing Approach

It was shown in the previous section that the short-time response factor reaches the same value as the previous models described previously, but it relates now to the fluid temperature. For a single borehole, we have:

$$\lim_{t \to +\infty} (T_f - T_o) = -\frac{q'_b g_{st}}{2\pi k_s} \approx -\frac{q'_b}{2\pi k_s} \left({}_1g + 2\pi k_s R'_b \right)$$

$$(13.37)$$

This should also be true for bore fields if a short-time g-function would be available. The impact on the sizing method will be addressed using the ASHRAE method but it can be transposed to any others. Taking Eq. (6.34), for example:

$$L(T_o - T_f)$$

$$= \frac{1}{2\pi k_s} \left(q_a g(\bar{t}_f, \bar{r}, \bar{z}_o) + (q_m - q_a)g((\bar{t}_f - \bar{t}_y), \bar{r}, \bar{z}_o) \right.$$

$$+ (q_h - q_m)g((\bar{t}_f - \bar{t}_m), \bar{r}, \bar{z}_o) + q_h R'_b \right)$$

The logical step would be to replace all *g-functions* with their short-time equivalent and remove the borehole resistance term:

$$L(T_o - T_f) = \frac{1}{2\pi k_s} \left(q_a g_{st}(\bar{t}_f, \bar{r}, \bar{z}_o) + (q_m - q_a)g_{st} \right.$$

$$((\bar{t}_f - \bar{t}_y), \bar{r}, \bar{z}_o) + (q_h - q_m)g_{st}((\bar{t}_f - \bar{t}_m), \bar{r}, \bar{z}_o))$$

However, the short-time g-function was built for a single borehole. It is not impossible to generate new short-time functions for fields, but the simplest approach is to remember that for long-time periods (years and months), Eq. (13.37) holds and for short time periods (hours), the interference effects between boreholes are negligible, which implies that the field short-time response factor would be essentially the same as the single borehole expression already found. The previous expression would then become:

$$L(T_o - T_f) = \frac{1}{2\pi k_s} \left(q_a(g(\bar{t}_f, \bar{r}, \bar{z}_o) + 2\pi k_s R'_b) + \right.$$

$$(q_m - q_a)(g((\bar{t}_f - \bar{t}_y), \bar{r}, \bar{z}_o) + 2\pi k_s R'_b) +$$

$$(q_h - q_m)g_{st}((\bar{t}_f - \bar{t}_m), \bar{r}, \bar{z}_o))$$

And finally:

$$L_{st} = \frac{q_a R'_{g,st,a} + q_m R'_{g,st,m} + q_h R'_{g,st,h}}{T_o - T_f} \quad (13.38)$$

with:

$$R'_{g,st,a} = R'_{g,a} = \frac{g(\bar{t}_f, \bar{r}, \bar{z}_o) - g((\bar{t}_f - \bar{t}_y), \bar{r}, \bar{z}_o)}{2\pi k_s}$$

$$R'_{g,st,m}$$
$$= \frac{g((\bar{t}_f - \bar{t}_y), \bar{r}, \bar{z}_o) + 2\pi k_s R'_b - g_{st}((\bar{t}_f - \bar{t}_m), \bar{r}, \bar{z}_o)}{2\pi k_s}$$

$$R'_{g,st,h} = \frac{g_{st}((\bar{t}_f - \bar{t}_m), \bar{r}, \bar{z}_o)}{2\pi k_s}$$

$$(13.39)$$

13.3   Calculation of g-Function, Revisited

As discussed in Chaps. 4, and 5, the concept of *g-function* was introduced by Eskilson [17] as response factors of generic geothermal fields in order to evaluate their thermal response in time. They were generated with a numerical model and tabulated values were made available for design purposes in several sizing software. In Chap. 4, the superposition of FLS was used to generate analytically approximations of *g-function*. The solutions obtained are quite similar to Eskilson's results for small fields but larger difference are observed for larger bore fields. The reason for that is that, in (4.53), the same heat transfer is assumed in each borehole, whereas in Eskilson's work, the same borehole temperature was imposed. Although the real situation is probably different than either situations, Eskilson's g-function were used for many years and are still considered as reference values. As pointed out in Chap. 4, the work of Cimmino and Bernier [18] led to better accuracy imposing the same boundary conditions as Eskilson's using analytical tools. They first introduce Type-II g-function by imposing the same mean temperature and different heat transfer rates for each borehole while each individual borehole has a uniform heat flux distribution and Type-III boundary condition where the temperature along the borehole was assumed to be uniform.

13.3.1 Type-II, g-Function

In the case of several boreholes in parallel, the FLS can be used to evaluate the mutual thermal influence. Assuming a uniform heat transfer rate along the borehole, the mean temperature of one borehole in the field will be given by:[3]

$$\Delta T_i(\bar{t}) = \frac{1}{2\pi k_s} \sum_{k=0}^{t_k < t} \sum_{j=1}^{N_b} (q'_j(\bar{t}_k) - q'_j(\bar{t}_{k-1}))\,{}_1 g_{ij}(\bar{t} - \bar{t}_k, \bar{d}_{ij}) \tag{13.40}$$

or

$$\frac{2\pi k_s \Delta T_i(t)}{q'_m} = \sum_{k=0}^{t_k < t} \sum_{j=1}^{n_b} f_j(\bar{t}_k)\,{}_1 g_{ij}(\bar{t} - \bar{t}_k, \bar{d}_{ij}) \tag{13.41}$$

with:

$$q'_m = \frac{q}{L} \ , \ \ q'(\bar{t}_{-1}) = 0 \ , \ \ f_j(\bar{t}_k) = \frac{q'_j(\bar{t}_k) - q'_j(\bar{t}_{k-1})}{q'_m}$$

where, as in Chap. 4, ${}_1 g$ is the FLS response associated with a single borehole and $\bar{d}_{ii} = \bar{r}_b$. In the Type-I case, the heat transfer rate in each borehole is equal $q'_j = q'_m$. Since $f_j(0) = 1$, $f_j(k \neq 0) = 0$, and Eq. (13.41) reduces to:

$$\frac{2\pi k_s \Delta T_i(t)}{q'_m} = \sum_{j=1}^{n_b} {}_1 g_{ij}(\bar{t}, \bar{b}_{ij}) \tag{13.42}$$

Taking the mean value of the last expression gives Eq. (4.53), the Type-I *g-function* associated with the particular field. For Type-II, the heat transfer rate varies from one borehole to the other but also with time. So the time convolution has to be kept in the calculation. Moreover, each mean temperature is assumed to be equal $\Delta T_i(t) = \Delta T_j(t) = \Delta T(t)$, so at each borehole:

$$\frac{2\pi k_s \Delta T(t)}{q'_m} = \sum_{j=1}^{n_b} f_j \circledast {}_1 g_{ij} \tag{13.43}$$

This represents a $n_k \cdot n_b$ system of equations for $n_k \cdot (n_b + 1)$ unknowns: the $n_k \cdot n_b$ values of heat flux as well as the n_k values of reduced temperatures (Type-II *g-function*). The last

equations are found knowing that, at each time step, the total heat transfer rate per unit length is:

$$q'_m L = \sum_{j=1}^{n_b} q'_j(t_k) H_j \tag{13.44}$$

Assuming an equal height:

$$n_b = \sum_{j=1}^{n_b} q'_j(t_k) \tag{13.45}$$

Since this expression is true for all times:

$$\sum_{j=1}^{n_b} f_j(t_k) = n_b \quad , t_k = 0$$
$$\sum_{j=1}^{n_b} f_j(t_k) = 0 \quad , t_k > 0 \tag{13.46}$$

or more concisely:

$$\sum_{j=1}^{n_b} f_j(t_k) = n_b \delta(t) \tag{13.47}$$

with $\delta(t)$ the impulse Dirac function. In [19], Marcotte and Pasquier used the FFT method to evaluate the convolution product. In their work, Cimmino and Bernier [18] used the Laplace transform. Taking the Laplace transform. With $\Theta^* = \mathscr{L}(2\pi k_s \Delta T)$, $f^* = \mathscr{L}(f)$, $g^* = \mathscr{L}(g)$, the last system of equations can be put into the form:

$$\begin{bmatrix} g_{11}^* & g_{12}^* & \cdots & -1 \\ g_{21}^* & g_{22}^* & \cdots & -1 \\ \vdots & & & \vdots \\ 1 & 1 & \cdots & 0 \end{bmatrix} \begin{bmatrix} f_1^* \\ f_2^* \\ \vdots \\ g^* \end{bmatrix} = \begin{bmatrix} 0 \\ 0 \\ \vdots \\ n_b \end{bmatrix} \tag{13.48}$$

Several techniques can be used to evaluate the numerical Laplace transform [20]. In [18] the Fourier transform is used where they showed that:

$$H(s) = \mathscr{L}(h(t)) = \exp(-\sigma t)\mathscr{F}(h(t)) \tag{13.49}$$

and

$$h(t) = \mathscr{L}^{-1}(H(s)) = \exp(\sigma t)\mathscr{F}^{-1}(H(s)) \tag{13.50}$$

where:

$$\sigma = \frac{2\log(N)}{t_{max}} \tag{13.51}$$

---

[3]The $\bar{r}$, $\bar{z}_o$ dependence will be omitted for brevity.

with N, the number of terms in the time domain, and t_{max}, the value of time at the end of the last time step, and they used the fast Fourier transform (FFT) to numerically evaluate the Fourier transform. The whole algorithm will be as follows:

1. For all times (n_k), evaluate the values of g_{ij}.
2. Take the Laplace transform using the FFT algorithm.
3. For all frequencies, inverse the system (13.48), to obtain the Laplace transform of the unknown variables.
4. Evaluate the inverse Laplace transform to obtain the time values of the variables.
5. The Type-II *g-function* is just the inverse Laplace transform of Θ^*.
6. Of course, it is advantageous to use the symmetries to reduce the number of unknowns.

Example 13.2 A simple 4×1 field will be used to illustrate the method. Since the heat is more easily dissipated for borehole P_1 and P_4, it is expected that the heat transfer will be higher.

The usual Eskilson's default values of:

$$\bar{r}_b = 0.0005 \quad \text{and} \quad \bar{z}_o = 0.04$$

will be used unless specified. As in [18], the solution will be done for few values of time to illustrate the method. From the symmetry of the problem, one has:

$$f_1(t) = f_4(t)$$
$$f_2(t) = f_3(t)$$

So,

$$2\pi k_s \Delta T(t) = f_1 \circledast (\overbrace{{}_1g(\bar{r}_b) + {}_1g(3\bar{b})}^{G_{11}}) + f_2 \circledast (\overbrace{{}_1g(\bar{b}) + {}_1g(2\bar{b})}^{G_{12}})$$

$$2\pi k_s \Delta T(t) = f_1 \circledast (\overbrace{{}_1g(\bar{b}) + {}_1g(2\bar{b})}^{G_{21}}) + f_2 \circledast (\overbrace{{}_1g(\bar{r}_b) + {}_1g(\bar{b})}^{G_{22}})$$

Taking the Laplace transform (with $G^* = \mathscr{L}(G)$),

$$g^* = f_1^* G_{11}^* + f_2^* G_{12}^*$$
$$g^* = f_1^* G_{21}^* + f_2^* G_{22}^*$$

(continued)

The Laplace transform of (13.47) gives:

$$2f_1^* + 2f_2^* = 4$$

The simplified form of (13.48) is then:

$$\begin{bmatrix} G_{11}^* & G_{12}^* & -1 \\ G_{21}^* & G_{22}^* & -1 \\ 2 & 2 & 0 \end{bmatrix} \begin{bmatrix} f_1^* \\ f_2^* \\ g^* \end{bmatrix} = \begin{bmatrix} 0 \\ 0 \\ 4 \end{bmatrix}$$

As in the example given in [18], let's compute the *g-function* for seven times:

$$\bar{t} = [0, 0.05, 0.1, 0.15, 0.2, 0.25, 0.3]$$

and $\bar{b} = 0.05$ The time values of the FLS function ${}_1g$ are:

	Dimensionless time						
	0	.05	0.1	0.15	0.2	0.25	0.3
${}_1g(\bar{r})$	0	5.3209	5.6256	5.7948	5.9097	5.9953	6.0627
${}_1g(\bar{b})$	0	0.8115	1.0924	1.254	1.3652	1.4486	1.5147
${}_1g(2\bar{b})$	0	0.2912	0.5118	0.6524	0.7532	0.8306	0.8927
${}_1g(3\bar{b})$	0	0.0992	0.2472	0.3588	0.4445	0.5128	0.5688

and

	Dimensionless time						
	0	.05	0.1	0.15	0.2	0.25	0.3
G_{11}	0	5.4200	5.8728	6.1537	6.3542	6.5081	6.6315
$G_{12} =$	0	1.1027	1.6042	1.9063	2.1184	2.2792	2.4074
G_{21}							
G_{22}	0	6.1323	6.718	7.0488	7.2748	7.4439	7.5774

Note that, although (13.48) is symmetric for similar boreholes, the reduced matrix is not necessarily symmetric. Taking the Laplace transform gives:

	Dimensionless frequency					
G_{11}^*	6.18	0.22-3.60j	−1.49-1.73j	−1.82-0.51j	−1.82+0.51j	−1.49+1.73j	0.22+3.60j
G_{12}^*	1.58	−0.12-0.80j	−0.35-0.27j	−0.33-0.07j	−0.33+0.07j	−0.35+0.27j	−0.12+0.80j
G_{22}^*	7.04	0.23-4.09j	−1.70-1.94j	−2.05-0.57j	−2.05+0.57j	−1.70+1.94j	0.22+4.09j

The inversion in the Laplace domain gives:

(continued)

Example 13.2 (continued)

g^*	8.16	0.11- 4.63j	−1.94- 2.10j	−2.25- 0.60j	−2.25+ 0.60j	−1.93+ 2.10j	0.11+ 4.63j
f_1^*	1.08	1.08- 0.008j	1.07- 0.007j	1.07- 0.003j	1.07+ 0.003j	1.07+ 0.007j	1.08+ 0.008j
f_2^*	0.91	0.92+ 0.008j	0.93+ 0.007j	0.93+ 0.003j	0.93- 0.003j	0.93- 0.007j	0.92- 0.008j

And taking the inverse transform:

	Dimensionless time						
	0	0.05	0.1	0.15	0.2	0.25	0.3
g	0	6.8517	7.8624	8.4657	8.8884	9.2091	9.4646
f_1	1.0764	0.014	0.0053	0.0027	0.0016	0.0008	
f_2	0.9236	−0.014	−0.0053	−0.0027	−0.0016	−0.0008	

From the last results, the heat transfer history can be reconstructed from $q'(i) = q'(i-1) + f$:

	Dimensionless time					
	[0,0.05[	[.05,0.1[	[0.1,0.15[	[0.15,0.2[	[0.2,0.25[	[0.25,0.3[
q'_1	1.0764	1.0903	1.0956	1.0983	1.0999	1.1007
q'_2	0.9236	0.9097	0.9044	0.9017	0.9001	0.8993

Finally, the following table compares the Type-I, Type-II, and Eskilson's interpolated value of g-*function*:

	Dimensionless time						
	0	0.05	0.1	0.15	0.2	0.25	0.3
Type-I	0.0	6.8789	7.8996	8.5076	8.9329	9.2552	9.5118
Type-II	0.0	6.8517	7.8624	8.4657	8.8884	9.2091	9.4646
Eskilson	0.0	6.8890	7.8769	8.4583	8.8714	9.1927	9.4552

13.3.2 Type-III, g-Function

In the last example, a small bore field was used to better illustrate the method although it was pointed out that differences are more pronounced for large ones. In this latter case, Type-II is closer to Eskilson's functions but differences are still noticeable. Again the reason is that, although the mean temperature is constant for each borehole, the temperature varies along the borehole where as it is uniform for Eskilson's functions. To solve this problem, Cimmino and Bernier [21] proposed to split each borehole in segments having the same mean temperature but different heat transfer rates. The proposed line of solution was similar to the one described in the previous section except that the thermal response between

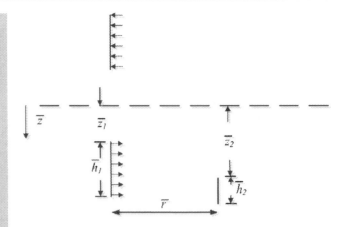

Fig. 13.4 Cimmino segment-to-segment response factor

two vertical segments being at different heights had to be found (see Fig. 13.4). The authors proved that this expression was given by:

$$\frac{\Delta T_{ij \to uv}(\bar{r}, \bar{t}) 2\pi k_s}{q'} = h_{ij \to uv}(\bar{r}, \bar{z}_1, \bar{z}_2, \bar{h}_1, \bar{h}_2)$$
$$= \frac{1}{2\bar{h}_2} \int_{3/\sqrt{4\bar{t}}}^{\infty} \frac{e^{-\bar{r}s^2} ds}{s^2}$$
$$\times [\, \text{ierf}((\bar{z}_2 - \bar{z}_1 + \bar{h}_2)s) - \text{ierf}((\bar{z}_2 - \bar{z}_1)s)$$
$$+ \text{ierf}((\bar{z}_2 - \bar{z}_1 - \bar{h}_1)s) - \text{ierf}((\bar{z}_2 - \bar{z}_1 + \bar{h}_2 - \bar{h}_1)s)$$
$$+ \text{ierf}((\bar{z}_2 + \bar{z}_1 + \bar{h}_2)s) - \text{ierf}((\bar{z}_2 + \bar{z}_1)s)$$
$$+ \text{ierf}((\bar{z}_2 + \bar{z}_1 + \bar{h}_1)s) - \text{ierf}((\bar{z}_2 + \bar{z}_1 + \bar{h}_2 + \bar{h}_1)s)]$$
$$(13.52)$$

which represents the average temperature increment on the v_{th} segment of the u_{th} borehole due to j_{th} segment on the i_{th} borehole. As before,

$$\bar{r} = \bar{r}_b \quad \text{if } i = u$$
$$\bar{r} = \bar{d}_{iu} \quad \text{if } i \neq u$$

Generalization of (13.43) applied at each v_{th} segment for every u_{th} borehole:

$$\frac{2\pi k_s \Delta T(t)}{q'_m} = \sum_{i=1}^{n_b} \sum_{j=1}^{n_s} f_{ij} \circledast h_{ij \to uv} \qquad (13.53)$$

Equation (13.47) becomes:

$$\sum_{i=1}^{n_b} \sum_{j=1}^{n_s} f_{ij}(t_k) = n_b \times n_s \, \delta(t) = n_t \delta(t) \qquad (13.54)$$

Equation (13.48) is now:

$$\begin{bmatrix} h^*_{11\to11} & h^*_{12\to11} & \cdots & h^*_{21\to11} & h^*_{22\to11} & \cdots & -1 \\ h^*_{11\to12} & h^*_{12\to12} & \cdots & h^*_{21\to12} & h^*_{22\to12} & \cdots & -1 \\ \vdots & & & & & & \vdots \\ h^*_{11\to21} & h^*_{12\to21} & \cdots & h^*_{21\to21} & h^*_{22\to21} & \cdots & -1 \\ h^*_{11\to22} & h^*_{12\to22} & \cdots & h^*_{21\to22} & h^*_{22\to22} & \cdots & -1 \\ \vdots & & & & & & \vdots \\ 1 & 1 & \cdots & 1 & 1 & \cdots & 0 \end{bmatrix} \begin{bmatrix} f^*_{11} \\ f^*_{12} \\ \vdots \\ f^*_{21} \\ f^*_{22} \\ \vdots \\ g^* \end{bmatrix} = \begin{bmatrix} 0 \\ 0 \\ \vdots \\ 0 \\ 0 \\ \vdots \\ n_t \end{bmatrix}$$

(13.55)

As before, the system is solved for every frequency and the inverse Laplace transform is used to obtain the solution in the time domain.

Example 13.3 Let's redo the last example with two segments on each borehole. The number of unknown heat flux f_{ij} is then eight. However, due to the symmetry, it is reduced to four:

$$f_{11} = f_{41}, \; f_{12} = f_{42}, \; f_{21} = f_{31}, \; f_{22} = f_{32}$$

So,

$$\frac{2\pi k_s \Delta T_{11}(t)}{q'_m}$$

$$= f_{11} \circledast \overbrace{(h(\bar{r}_b, \bar{z}_o, \bar{z}_o, \bar{h}_1, \bar{h}_1) + h(3\bar{b}, \bar{z}_o, \bar{z}_o, \bar{h}_1, \bar{h}_1))}^{G_{11}}$$

$$+ f_{12} \circledast \overbrace{(h(\bar{r}_b, \bar{z}_o + \bar{h}_1, \bar{z}_o, \bar{h}_2, \bar{h}_1) + h(3\bar{b}, \bar{z}_o + \bar{h}_1, \bar{z}_o, \bar{h}_2, \bar{h}_1))}^{G_{12}}$$

$$+ f_{21} \circledast \overbrace{(h(\bar{b}, \bar{z}_o, \bar{z}_o, \bar{h}_1, \bar{h}_1) + h(2\bar{b}, \bar{z}_o, \bar{z}_o, \bar{h}_1, \bar{h}_1))}^{G_{13}}$$

$$+ f_{22} \circledast \overbrace{(h(\bar{b}, \bar{z}_o + \bar{h}_1, \bar{z}_o, \bar{h}_2, \bar{h}_1) + h(2\bar{b}, \bar{z}_o + \bar{h}_1, \bar{z}_o, \bar{h}_2, \bar{h}_1))}^{G_{14}}$$

$$\frac{2\pi k_s \Delta T_{12}(t)}{q'_m}$$

$$= f_{11} \circledast \overbrace{(h(\bar{r}_b, \bar{z}_o, \bar{z}_o + \bar{h}_1, \bar{h}_1, \bar{h}_2) + h(3\bar{b}, \bar{z}_o, \bar{z}_o + \bar{h}_1, \bar{h}_1, \bar{h}_2))}^{G_{21}}$$

$$+ f_{12} \circledast \overbrace{(h(\bar{r}_b, \bar{z}_o + \bar{h}_1, \bar{z}_o + \bar{h}_1, \bar{h}_2, \bar{h}_2) + h(3\bar{b}, \bar{z}_o + \bar{h}_1, \bar{z}_o + \bar{h}_1, \bar{h}_2, \bar{h}_2))}^{G_{22}}$$

$$+ f_{21} \circledast \overbrace{(h(\bar{b}, \bar{z}_o, \bar{z}_o + \bar{h}_1, \bar{h}_1, \bar{h}_2) + h(2\bar{b}, \bar{z}_o, \bar{z}_o + \bar{h}_1, \bar{h}_1, \bar{h}_2))}^{G_{23}}$$

$$+ f_{22} \circledast \overbrace{(h(\bar{b}, \bar{z}_o + \bar{h}_1, \bar{z}_o + \bar{h}_1, \bar{h}_2, \bar{h}_2) + h(2\bar{b}, \bar{z}_o + \bar{h}_1, \bar{z}_o + \bar{h}_1, \bar{h}_2, \bar{h}_2))}^{G_{24}}$$

$$\frac{2\pi k_s \Delta T_{21}(t)}{q'_m}$$

$$= f_{11} \circledast \overbrace{(h(\bar{b}, \bar{z}_o, \bar{z}_o, \bar{h}_1, \bar{h}_1) + h(2\bar{b}, \bar{z}_o, \bar{z}_o, \bar{h}_1, \bar{h}_1))}^{G_{31}}$$

$$+ f_{12} \circledast \overbrace{(h(\bar{b}, \bar{z}_o + \bar{h}_1, \bar{z}_o, \bar{h}_2, \bar{h}_1) + h(2\bar{b}, \bar{z}_o + \bar{h}_1, \bar{z}_o, \bar{h}_2, \bar{h}_1))}^{G_{32}}$$

$$+ f_{21} \circledast \overbrace{(h(\bar{r}_b, \bar{z}_o, \bar{z}_o, \bar{h}_1, \bar{h}_1) + h(\bar{b}, \bar{z}_o, \bar{z}_o, \bar{h}_1, \bar{h}_1))}^{G_{33}}$$

(continued)

$$+ f_{22} \circledast \overbrace{(h(\bar{r}_b, \bar{z}_o + \bar{h}_1, \bar{z}_o, \bar{h}_2, \bar{h}_1) + h(\bar{b}, \bar{z}_o + \bar{h}_1, \bar{z}_o, \bar{h}_2, \bar{h}_1))}^{G_{34}}$$

$$\frac{2\pi k_s \Delta T_{22}(t)}{q'_m}$$

$$= f_{11} \circledast \overbrace{(h(\bar{b}, \bar{z}_o, \bar{z}_o + \bar{h}_1, \bar{h}_1, \bar{h}_2) + h(2\bar{b}, \bar{z}_o, \bar{z}_o + \bar{h}_1, \bar{h}_1, \bar{h}_2))}^{G_{41}}$$

$$+ f_{12} \circledast \overbrace{(h(\bar{b}, \bar{z}_o + \bar{h}_1, \bar{z}_o + \bar{h}_1, \bar{h}_2, \bar{h}_2) + h(2\bar{b}, \bar{z}_o + \bar{h}_1, \bar{z}_o + \bar{h}_1, \bar{h}_2, \bar{h}_2))}^{G_{42}}$$

$$+ f_{21} \circledast \overbrace{(h(\bar{r}_b, \bar{z}_o, \bar{z}_o + \bar{h}_1, \bar{h}_1, \bar{h}_2) + h(\bar{b}, \bar{z}_o, \bar{z}_o + \bar{h}_1, \bar{h}_1, \bar{h}_2))}^{G_{43}}$$

$$+ f_{22} \circledast \overbrace{(h(\bar{r}_b, \bar{z}_o + \bar{h}_1, \bar{z}_o + \bar{h}_1, \bar{h}_2, \bar{h}_2) + h(\bar{b}, \bar{z}_o + \bar{h}_1, \bar{z}_o + \bar{h}_1, \bar{h}_2, \bar{h}_2))}^{G_{44}}$$

As previously, we set:

$$\bar{r}_b = 0.0005, \bar{z}_o = 0.04, \bar{b} = 0.05 \text{ and } \bar{h}_1 = \bar{h}_2 = 0.5$$

and compute the *g-function* for seven times:

$$\bar{t} = [0, 0.05, 0.1, 0.15, 0.2, 0.25, 0.3]$$

The time values of the thermal response function are:

	Dimensionless time						
	0	.05	0.1	0.15	0.2	0.25	0.3
G_{11}	0	5.3237	5.7139	5.9392	6.0895	6.1978	6.2796
$G_{12} = G_{21}$	0	0.091	0.143	0.1873	0.2263	0.261	0.2923
$G_{13} = G_{31}$	0	1.0342	1.4666	1.7089	1.8679	1.9811	2.066
$G_{14} = G_{41}$	0	0.0627	0.1203	0.168	0.2092	0.2455	0.278
$G_{13} = G_{13}$	0	1.0342	1.4666	1.7089	1.8679	1.9811	2.066
G_{22}	0	5.3345	5.7457	5.9935	6.1663	6.2962	6.3987
$G_{23} = G_{32}$	0	0.0627	0.1203	0.168	0.2092	0.2455	0.278
$G_{24} = G_{42}$	0	1.0457	1.5012	1.7678	1.9505	2.0863	2.1927
G_{33}	0	5.9972	6.5024	6.7678	6.9373	7.0563	7.1448
$G_{34} = G_{43}$	0	0.127	0.194	0.2461	0.29	0.3282	0.3621
G_{44}	0	6.0135	6.5456	6.8375	7.0323	7.1751	7.2859

which would result in the following system in the Laplace domain:

$$\begin{bmatrix} G^*_{11} & G^*_{12} & G^*_{13} & G^*_{14} & -1 \\ G^*_{21} & G^*_{22} & G^*_{23} & G^*_{24} & -1 \\ G^*_{31} & G^*_{32} & G^*_{33} & G^*_{34} & -1 \\ G^*_{41} & G^*_{42} & G^*_{43} & G^*_{44} & -1 \\ 2 & 2 & 2 & 2 & 0 \end{bmatrix} \begin{bmatrix} f^*_{11} \\ f^*_{12} \\ f^*_{21} \\ f^*_{22} \\ g^* \end{bmatrix} = \begin{bmatrix} 0 \\ 0 \\ 0 \\ 0 \\ 8 \end{bmatrix}$$

The inversion in the Laplace domain gives the Type-III *g-function* given in the following table:

	Dimensionless time						
	0	0.05	0.1	0.15	0.2	0.25	0.3
Type-I	0.0	6.8789	7.8996	8.5076	8.9329	9.2552	9.5118
Type-II	0.0	6.8517	7.8624	8.4657	8.8884	9.2091	9.4646
Type-III	0.0	6.8517	7.8623	8.4655	8.8879	9.2083	9.4634
Eskilson	0.0	6.8890	7.8769	8.4583	8.8714	9.1927	9.4552

(continued)

Example 13.3 (continued)
It is observed that the difference between all calculations is very small. As it was already pointed out, the difference is larger for bigger fields and larger time values. The number of boreholes and time values was kept small in the example to make it easier to follow. Finally, the four different heat transfer rates can be calculated as usual and give:

	Dimensionless time					
	[0,0.05[	[.05,0.1[	[0.1,0.15[	[0.15,0.2[	[0.2,0.25[	[0.25,0.3[
q'_{11}	1.0782	1.0955	1.104	1.1098	1.1143	1.1175
q'_{12}	1.0746	1.0852	1.0872	1.0868	1.0855	1.0839
q'_{21}	0.9258	0.9152	0.9131	0.9133	0.9143	0.9157
q'_{22}	0.9215	0.9041	0.8957	0.8901	0.8859	0.8829

As a final comparison, Fig. 13.5 the three types of analytical g-functions are compared to the tabulated values given by Eskilson for a 4 × 4 field where $\bar{b} = 0.05$. The Type-III g-function was evaluated with the optimized g-function module *pygfunction* made available by Massimo Cimmino (https://github.com/MassimoCimmino/pygfunction) with eight segments per borehole.

The Fourier and Laplace transform are very powerful tools to analyze convoluted signals in the frequency domain. One of the disadvantages of the method is that it requires equally spaced time values. In the case of *g-function* generation, it is not practical since the time scale is very large. It is possible to generate several calculations with different time steps and interpolate the other values if necessary. The solution of the same problem in the time domain using nonuniform time steps were proposed by Cimmino [22] and Dusseault et al.

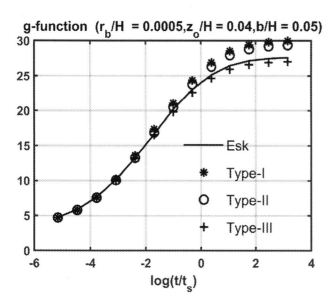

Fig. 13.5 Comparison of analytical g-functions

[23]. In Cimmino's work, each borehole was split into a finite number of segments where the heat transfer rate was assumed to be uniform along the segment. Lamarche [24] extended its approach assuming a linear variation of the heat transfer rate along the segments. Doing so, it was possible to achieve similar accuracy with a smaller number of segments.

The concept of *g-function* initiated by Eskilson and Claesson assumed that the borehole temperature was uniform along the borehole and equal for all boreholes (Type-III BC). Although this concept seemed logical and is still mostly accepted in the community, other choices can be even more logical like assuming a uniform inlet temperature along each borehole. This case was treated by Cimmino [25]. Another important element associated with the analytical *g-function* evaluations is the time required for each segment thermal response factors. The symmetry often observed in bore field makes it possible to identify identical pairs of line segments to reduce the number of evaluations. This problem was described also by Cimmino [26].

13.4   Moving Infinite Line Source

The moving infinite line source (MILS) is an analytical solution similar to the ILS that has been used for many years to represent a borehole subjected to groundwater advection [27–30]. The physical phenomenon can be modeled by the conduction-advection equation:

$$\rho C_p \frac{\partial T}{\partial t} + \rho_w C_{p,w} \vec{U} \cdot \nabla T = \nabla \cdot (k \nabla T) \qquad (13.56)$$

where $\rho_w, C_{p,w}$ are associated with the water properties, whereas ρ, C_p are properties of the porous medium with porosity ϕ:

$$\rho = \phi \rho_w + (1 - \phi)\rho_s$$
$$C_p = \phi C_{p,w} + (1 - \phi)C_{p,s}$$

Assuming constant properties and that the advection velocity moves in the x-direction, the last relation can be rewritten as:

$$\frac{\partial T}{\partial \tilde{t}} + Pe \frac{\partial T}{\partial \tilde{x}} = \frac{\partial^2 T}{\partial \tilde{x}^2} + \frac{\partial^2 T}{\partial \tilde{y}^2} + \frac{\partial^2 T}{\partial \tilde{z}^2} \qquad (13.57)$$

with:

$$\tilde{t} = \frac{\alpha t}{r_b^2} \equiv Fo \quad , \quad \tilde{x} = \frac{x}{r_b}, \tilde{y} = \frac{y}{r_b}, \tilde{z} = \frac{z}{r_b}$$

and the *Peclet number*, Pe, is defined as:

$$Pe = \frac{U_{eff} r_b}{\alpha} \quad , \quad U_{eff} = \frac{\rho_w C_{p,w} U}{\rho C_p} \qquad (13.58)$$

Several approaches can be used to solve this equation but the simplest way is to see that the moving infinite line source can be used to represent the solution. This solution comes from the source solution given in Chap. 4. The solution of the ILS came from Eq. (4.22):

$$T(x, y, t) - T_o = -\frac{1}{4\pi k} \int_o^t \frac{q'(\tau)e^{-d^2/(4\alpha(t-\tau))}}{t - \tau} d\tau \quad (4.22)$$

$$d^2 = (x - \xi)^2 + (y - \eta)^2$$

where the usual ILS solution was obtained assuming that $q'(\tau)$ was constant for $t > 0$ and that the line source was at the origin $\xi = \eta = 0$, i.e, $d^2 = x^2 + y^2 = r^2$. In the case of the MILS moving in the x-direction, the only change is that:

$$\xi = U_{eff}(t - \tau) \quad (13.59)$$

Replacing in Eq. (4.22):

$$T(x, y, t, U_{eff}) - T_o = -\frac{q'_o}{4\pi k} \int_0^t \frac{e^{-((x-U_{eff}(t-\tau))^2+y^2)/(4\alpha(t-\tau))}}{t - \tau} d\tau$$

$$(13.60)$$

With the usual change of variable:

$$\beta = \frac{r^2}{4\alpha(t - \tau)}$$

$$T(x, y, t, U_{eff}) - T_o = -\frac{q'_o}{4\pi k} e^{\frac{U_{eff}x}{2\alpha}} \int_{r^2/4\alpha t}^\infty \frac{e^{-\beta - \frac{(rU_{eff})^2}{(4\alpha)^2\beta}}}{\beta} d\beta$$

$$(13.61)$$

Again, it is worthwhile to rewrite it with dimensionless variables:

$$T(\tilde{r}, \theta, Fo, Pe) - T_o = -\frac{q'_o}{4\pi k} e^{\tilde{r}\cos(\theta)Pe/2} \int_{\tilde{r}^2/4Fo}^\infty \frac{e^{-\beta - \frac{(\tilde{r}Pe/4)^2}{\beta}}}{\beta} d\beta$$

$$(13.62)$$

$$T(\tilde{r}, \theta, Fo, Pe) - T_o = -\frac{q'_o}{k} G_{MILS}(Fo, \tilde{r}, \theta, Pe) \quad (13.63)$$

$$G_{MILS}(Fo, \tilde{r}, \theta, Pe) = \frac{e^{\tilde{r}\cos(\theta)Pe/2}}{4\pi} \int_{\tilde{r}^2/4Fo}^\infty \frac{e^{-\beta - \frac{(\tilde{r}Pe/4)^2}{\beta}}}{\beta} d\beta$$

$$(13.64)$$

It is interesting to remark that, although found in a different context, this is also the same solution as the leaky aquifer (Eq. (11.42)). It is straightforward to see that:

$$G_{MILS}(Fo, \tilde{r}, \theta, Pe) = \frac{e^{\tilde{r}\cos(\theta)Pe/2}}{4\pi} W(u, r/B) \quad (13.65)$$

with:

$$\frac{\tilde{r}}{4Fo} = u \quad , \quad \frac{\tilde{r}Pe}{4} = \frac{r}{B}$$

Several authors have suggested expressions to evaluate the Hantush well function [31–33] that can be used to estimate the MILS function as well. As in the case of the ILS, the finite moving finite line source (MFLS) can be determined in a similar way [30]. As for the leaky aquifer function, a steady state is reached whose solution is:

$$\lim_{t\to\infty} W(u, r/B) = W(0, r/B)) = 2K_o(r/B)$$

$$\lim_{Fo\to\infty} G_{MILS}(Fo, \tilde{r}, \theta, Pe) = \frac{e^{\tilde{r}\cos(\theta)Pe/2} K_o(\tilde{r}/2\,Pe)}{2\pi}$$

Again, as in the case of other response factors, the value at the borehole perimeter is the most pertinent value. However, contrary to other models, the solution is not axisymmetric. The temperature will vary along the periphery and the mean value is of interest:

$$\bar{T}_b(Fo, Pe) - T_o = -\frac{q'_o}{k} \bar{G}_{MILS}(Fo, 1, Pe)$$

$$\bar{G}_{MILS}(Fo, 1, Pe) = \frac{1}{\pi} \left(\int_0^\pi \frac{e^{\cos(\theta)Pe/2}}{4\pi} d\theta \right) \int_{1/4Fo}^\infty \frac{e^{-\beta - \frac{(Pe/4)^2}{\beta}}}{\beta} d\beta$$

which is given [29]

$$\bar{G}_{MILS}(Fo, 1, Pe) = \frac{1}{4\pi} I_o\left(\frac{Pe}{2}\right) \int_{1/4Fo}^\infty \frac{e^{-\beta - \frac{(Pe/4)^2}{\beta}}}{\beta} d\beta$$

$$(13.66)$$

13.5 Pile Heat Exchangers

Pile foundations are structural elements that were mostly used to mechanically support buildings that are built on soft soils. The concept of *energy piles* or thermo-active ground structures was initiated to take the opportunity of exchanging thermal energy with the soil without having to drill expensive boreholes but using instead the existing pile foundation. The concept of exchanging heat through existing structures could be extended to tunnels retaining walls and so on [34] but only structure piles will be described here. Their concept is quite similar to normal grouted boreholes. Plastic pipes are introduced in the concrete pile and exchange heat with the soil through a heat transfer fluid. In principle, it is then possible to use the same analysis tools described earlier to predict their thermal performance. Pile foundations have however specific characteristics that require some special attention:

(a) Their diameter is often large, which make their heat capacity non-negligible.
(b) Their height is normally small, which make the axial effect more important.
(c) The borehole resistance concept associated with slender boreholes needs to be adapted.
(d) The fluid passes often though ring-shaped pipes.
(e) etc.

13.5.1 Man's Model

Man et al. [35] suggested finding the thermal response of a superposition of continuous sources (see Chap. 4) that are distributed along a finite (or infinite) cylindrical surface. In the case of an infinite cylinder, this solution can be thought of as the same as the ICS model, the difference being that in the ICS case, the heat is originating from the cylindrical surface but no heat is stored inside the cylinder ("hollow cylinder"). In the case treated here, heat is stored inside the cylinder ("solid cylinder"), taking into account its heat capacity. The solution is strictly valid however if the thermal properties inside the cylinder are the same as outside.

Infinite Solid Cylinder

Superposition of sources solutions was discussed in Sect. 4.2.3; the case treated here can be analyzed by integrating the continuous source (4.19) along a finite cylindrical surface. Man et al. [35] used instead the superposition of point source. Since these sources are applied on a surface element located at a distance r_o from the origin, their heat content is:

$$\delta Q = q \, d\tau = q'' r_o d\phi' d\zeta d\tau$$

As for other models, Man et al. [35] analyzed the case of an infinite (1D) and finite (axisymmetric) models. In the former case, the superposition on sources is then:

$$\Delta T(x, y, z, t) = -\int_0^t d\tau \int_0^{2\pi} d\phi' \int_{-\infty}^{\infty} q''(\tau) F(t - \tau) r_o d\zeta$$

(13.67)

where F is the point source solution (4.17) and F_{imag} is its usual image. Its expression was given in Chap. 4 as:

$$F = \frac{1}{8\rho C_p (\pi \alpha t)^{3/2}} e^{-((x-\xi)^2 + (y-\eta)^2 + (z-\zeta)^2)/(4\alpha t)} \quad (4.17)$$

Since the sources are located at the periphery r_o of the cylinder:

$$(x - \xi)^2 + (y - \eta)^2 + (z - \zeta)^2$$
$$= x^2 + y^2 + \xi^2 + \eta^2 - 2x\xi - 2y\eta + (z - \zeta)^2$$
$$= r^2 + r_o^2 - 2rr_o \cos(\phi) \cos(\phi')$$
$$- 2rr_o \sin(\phi) \sin(\phi') + (z - \zeta)^2$$

In [35], the authors suggested that, since the final solution should not depend on z and ϕ, they put them equal to 0. Again, assuming that $q' = 2\pi r_o q''$ is constant:

$$\Delta T(r, t) = \frac{-q'}{16k\pi^2 \sqrt{\pi\alpha}} \int_0^t d\tau \int_0^{2\pi}$$
$$\frac{e^{-(r^2 + r_o^2 - 2rr_o \cos(\phi'))/(4\alpha(t-\tau))}}{(t-\tau)^{3/2}} d\phi' \int_{-\infty}^{\infty} e^{-\zeta^2/(4\alpha(t-\tau))} d\zeta$$

(13.68)

Integration with respect to ζ gives:

$$\int_{-\infty}^{\infty} e^{-\zeta^2/(4\alpha(t-\tau))} d\zeta = 2\sqrt{\alpha(t - \tau)}$$

$$\int_{-\infty}^{\infty} \exp(-\nu^2) d\nu = 2\sqrt{\pi\alpha(t - \tau)}$$

$$\Delta T(r, t) = \frac{-q'}{8k\pi^2} \int_0^t d\tau \int_0^{2\pi} \frac{e^{-(r^2 + r_o^2 - 2rr_o \cos(\phi'))/(4\alpha(t-\tau))}}{t - \tau} d\phi'$$

(13.69)

and integrating with respect to time with the usual change of variable:

$$u = \frac{r^2 + r_o^2 - 2rr_o \cos(\phi')}{4\alpha(t - \tau)}$$

$$\Delta T(r, t) = \frac{-q'}{4k\pi^2} \int_0^\pi d\phi' \int_{\frac{r^2 + r_o^2 - 2rr_o \cos(\phi')}{4\alpha t}}^{\infty} \frac{e^{-u}}{u} du$$

$$= \frac{-q'}{4k\pi^2} \int_0^\pi E_1\left(\frac{(r^2 + r_o^2 - 2rr_o \cos(\phi'))}{4\alpha t}\right) d\phi'$$

(13.70)

Using dimensionless variables gives:

$$\Delta T(\tilde{r}, Fo) = \frac{-q'}{4k\pi^2} \int_0^\pi E_1\left(\frac{(\tilde{r}^2 + 1 - 2\tilde{r}\cos(\phi'))}{4Fo}\right) d\phi' \tag{13.71}$$

This result can be easily interpreted as a superposition of infinite line source along the periphery of the cylinder. Another possibility is to integrate with respect to ϕ' first. Doing so would result in:

$$\Delta T(r, t) = \frac{-q'}{16k\pi^2\sqrt{\pi\alpha}} \int_0^t d\tau \int_{-\infty}^\infty d\zeta \frac{e^{-(r^2+r_o^2+\zeta^2)/(4\alpha(t-\tau))}}{(t-\tau)^{3/2}}$$

$$\int_0^{2\pi} e^{2rr_o\cos(\phi'))/(4\alpha(t-\tau))} d\phi' \tag{13.72}$$

The last integral gives:

$$\int_0^{2\pi} e^{\frac{2rr_o\cos(\phi')}{4\alpha(t-\tau)}} d\phi' = 2\pi I_o\left(\frac{rr_o}{2\alpha(t-\tau)}\right) \tag{13.73}$$

and integrating with respect to ζ leads to:

$$\Delta T(r, t) = \frac{-q'}{8k\pi\sqrt{\pi\alpha}} \int_0^t I_o\left(\frac{rr_o}{2\alpha(t-\tau)}\right) \frac{e^{-(r^2+r_o^2)/(4\alpha(t-\tau))}}{(t-\tau)^{3/2}} d\tau$$

$$\int_{-\infty}^\infty e^{-\zeta^2/(4\alpha(t-\tau))} d\zeta$$

$$= \frac{-q'}{4k\pi} \int_0^t I_o\left(\frac{rr_o}{2\alpha(t-\tau)}\right) \frac{e^{-(r^2+r_o^2)/(4\alpha(t-\tau))}}{t-\tau} d\tau \tag{13.74}$$

As usual, the solution can be reformulated in terms of dimensionless variables:

$$\Delta T(r, t) = \frac{-q'}{4k\pi} \int_0^{Fo} I_o\left(\frac{\tilde{r}}{2(Fo - Fo')}\right) \frac{e^{-(\tilde{r}^2+1)/(4(Fo-Fo'))}}{Fo - Fo'} dFo' \tag{13.75}$$

Either (13.71) or (13.75) can be formulated as the *infinite solid cylinder* (ISC) solution:

$$\Delta T(\tilde{r}, Fo) = \frac{-q'}{k} G_{ISC}(\tilde{r}, Fo) \tag{13.76}$$

Example 13.4 The last model can be compared with previous 1D models like the infinite line source or

the infinite cylindrical source. A significant difference between the new model and the others is that it is the only one that has a physical meaning inside the cylinder. Other solutions have mathematical values but without any physical significance inside the borehole; in each case, the heat transfer inside the cylinder is modeled with the concept of borehole resistance. In the case of the infinite solid cylinder, it represents the temperature change inside the cylinder when heat is exchanged at its periphery. In this example, the dimensionless temperature change (g-function) is compared only outside the borehole. In the figure at the left, the calculation is done for different radius values for a short-time step (Fo = 0.5) and in the right figure a larger time step (Fo = 5).

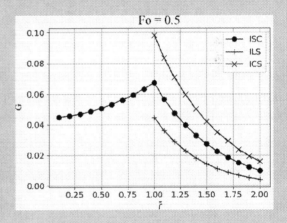

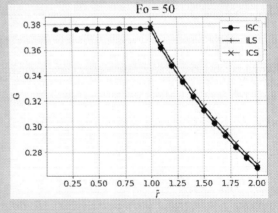

As expected, for large time values, all models behave similarly as the temperature variation inside the cylinder reaches the steady-flux regime. In the early times, the difference is important. As expected, the largest value of the temperature variation is at the periphery of the cylinder. It is interesting to note that, although the solution is a superposition of singular solutions (ILS is singular at the origin of the line), its summation is regular.

(continued)

Finite Solid Cylinder

In the previous section, the ICS solution was presented in two different equivalent mathematical expressions. The first one is more efficient to compute but the second is more straightforward to generalize for the finite cylinder. In this case, as it was done for the line source, it suffices to integrate along the finite distance of the borehole and to take into account its image. The solution is:

$$\Delta T(x, y, z, t) = -\int_0^t d\tau \int_0^{2\pi} d\phi' \int_{z_o}^{z_o+H} q''(\tau)(F - F_{imag})r_o d\zeta$$

(13.77)

The solution is now axisymmetric. From the previous development, Eq. (13.77) becomes:

$$\Delta T(r, z, t) = \frac{-q'}{8k\pi\sqrt{\pi\alpha}} \int_0^t I_o\left(\frac{rr_o}{2\alpha(t-\tau)}\right) \frac{e^{-(r^2+r_o^2)/(4\alpha(t-\tau))}}{(t-\tau)^{3/2}} d\tau$$
$$\int_{z_o}^{z_o+H} \left(e^{-(z-\zeta)^2/(4\alpha(t-\tau))} - e^{-(z+\zeta)^2/(4\alpha(t-\tau))}\right) d\zeta$$

(13.78)

The last integrals give:

$$\int_z^{H+z_o} e^{-(z-\zeta)^2/(4\alpha(t-\tau))} d\tau = \sqrt{\pi\alpha}\sqrt{(t-\tau)}\left[\text{erf}\left(\frac{z-z_o}{2\sqrt{\alpha(t-\tau)}}\right)\right.$$
$$\left. -\text{erf}\left(\frac{z-z_o-H}{2\sqrt{\alpha(t-\tau)}}\right)\right]$$

$$\int_z^{H+z_o} e^{-(z+\zeta)^2/(4\alpha(t-\tau))} d\tau = \sqrt{\pi\alpha}\sqrt{(t-\tau)}\left[-\text{erf}\left(\frac{z+z_o}{2\sqrt{\alpha(t-\tau)}}\right)\right.$$
$$\left. +\text{erf}\left(\frac{z+z_o+H}{2\sqrt{\alpha(t-\tau)}}\right)\right]$$

From which:

$$\Delta T(\tilde{r}, \tilde{z}, Fo) = \frac{-q'}{k}G_{FSC}(Fo, \tilde{r}, \tilde{z})$$

(13.79)

with:

$$G_{FSC}(Fo, \tilde{r}, \tilde{z})$$
$$= \frac{1}{8\pi}\int_0^{Fo} \frac{e^{-(\tilde{r}+1)^2/(4(Fo-Fo'))}I_o\left(\frac{\tilde{r}}{2(Fo-Fo')}\right)}{Fo-Fo'}$$

$$\left[\text{erf}\left(\frac{\tilde{z}-\tilde{z}_o}{2\sqrt{(Fo-Fo')}}\right) - \text{erf}\left(\frac{\tilde{z}-\tilde{z}_o-\tilde{H}}{2(Fo-Fo')}\right)\right.$$
$$\left. +\text{erf}\left(\frac{\tilde{z}+\tilde{z}_o}{2\sqrt{(Fo-Fo')}}\right) - \text{erf}\left(\frac{\tilde{z}+\tilde{z}_o+\tilde{H}}{2(Fo-Fo')}\right)\right] dFo'$$

(13.80)

13.5.2 Cui's Model

Man's model assumes that the heat is generated uniformly at the periphery of the pile represented by a "solid" cylinder. In practice, heat is transferred through pipes which form often a spiral coil. Cui et al. [36] first considered the spiral coil as a finite number of discrete rings (Fig. 13.6). The derivation of the thermal response follows the same steps as previously (Eq. (13.77)) where the integral (continuous summation) in the ζ direction is replaced by a *finite summation* of all the rings. The heat exchanged by each point source on a single ring is now given as:

$$\delta Q = \delta q_i \, d\tau = q\grave{\ }r_o d\phi' d\tau$$

with $q\grave{\ }$, the heat per unit length of ring. Assuming again that it is constant, it is given by:

$$\delta Q = \frac{q_i}{2\pi r_o}r_o d\phi' d\tau = \frac{q_i}{2\pi}d\phi' d\tau$$

The superposition of sources becomes simply:

$$\Delta T(r, z, t) = -\frac{q_i}{16k\pi^2\sqrt{\pi\alpha}}\int_0^t \frac{e^{-(r+r_o)^2/(4\alpha(t-\tau))}d\tau}{(t-\tau)^{3/2}}$$
$$\int_0^{2\pi} e^{\frac{2rr_o\cos(\phi')}{4\alpha(t-\tau)}} d\phi$$
$$\sum_{i=1}^N e^{-(z-\zeta_i)^2/(4\alpha(t-\tau))} - e^{-(z+\zeta_i)^2/(4\alpha(t-\tau))}$$

(13.81)

In previous models, the thermal response was given with respect to the heat per unit length of borehole height. This can be expressed as:

$$Nq_i = q = q'H \Rightarrow q_i = \frac{q'H}{N} = q'b$$

where b is the coil pitch. From (13.73), replacing as before the middle integral, it becomes:

$$\Delta T(\tilde{r}, \tilde{z}, Fo) = -\frac{q'}{k_s} G_{Ring}(Fo, \tilde{r}, \tilde{z}) \qquad (13.82)$$

with:

$$G_{Ring}(Fo, \tilde{r}, \tilde{z})$$

$$= \frac{\tilde{b}}{8\pi^{3/2}} \int_0^{Fo} \frac{e^{-(\tilde{r}+1)^2/(4(Fo-Fo'))} I_0\left(\frac{\tilde{r}}{2(Fo-Fo')}\right)}{(Fo-Fo')^{3/2}}$$

$$\left[\sum_{i=1}^{N} e^{-(\tilde{z}-\zeta_i)^2/(4(Fo-Fo'))} - e^{-(\tilde{z}+\zeta_i)^2/(4(Fo-Fo'))}\right] dFo'$$

$$(13.83)$$

where:

$$\zeta_i = \tilde{z}_o + (i - 1/2)\tilde{b}$$

13.5.3 Spiral Models

Extension of the previous source superposition principle to a real coil (Fig. 13.6b) has been reported by several authors [37, 38]. The principle is quite straightforward; the idea is to apply the source to a parameterized curve $s(x, y, z)$ representing the spiral coil, with:

$$\delta Q = q \, d\tau = q`ds d\tau$$

where $q`$ represents the heat per unit length of pipe, not the borehole. The superposition becomes:

$$\Delta T(x, y, z, t) = -\int_0^t d\tau \int_0^{s_f} q`(\tau)(F - F_{imag}) ds$$

$$(13.84)$$

The parameterization of the spiral is classical, and one has:

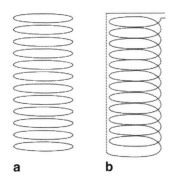

a   **b**

Fig. 13.6 (**a**) Ring-coil model. (**b**) Spiral-coil model

$$\begin{aligned}\xi &= r_o \cos(\phi') \\ \eta &= r_o \sin(\phi') \\ \zeta &= \frac{b\phi'}{2\pi}\end{aligned} \qquad (13.85)$$

The coil pitch which can also be related to the wave number:

$$\omega = \frac{2\pi}{b} = \frac{2\pi N_t}{H} \qquad (13.86)$$

with N_t the number of turns. From which one finds:

$$ds = \sqrt{d\xi^2 + d\eta^2 + d\zeta^2} = \sqrt{r_o^2 + \frac{b^2}{4\pi^2}} d\phi = \frac{\sqrt{1 + (\omega r_o)^2}}{\omega} d\phi' \qquad (13.87)$$

Replacing again the point source solution (4.17) and assuming constant heat transfer:

$$\begin{aligned}(x - \xi)^2 + (y - \eta)^2 + (z - \zeta)^2 \\ = x^2 + y^2 + \xi^2 + \eta^2 - 2x\xi - 2y\eta + (z - \zeta)^2 \\ = r^2 + r_o^2 - 2rr_o \cos(\phi - \phi') + (z - \zeta)^2\end{aligned}$$

$$\Delta T(r, z, \phi, t) = \frac{-q`\sqrt{1 + (\omega r_o)^2}}{8k\pi\omega\sqrt{\pi\alpha}} \int_0^t \frac{e^{-\frac{r^2+r_o^2}{4\alpha(t-\tau)}}}{(t-\tau)^{3/2}} d\tau$$

$$\int_{\phi_i}^{\phi_f} e^{\frac{2rr_o\cos(\phi-\phi')}{4\alpha(t-\tau)}} \left(e^{-\frac{(z-\zeta/\omega)^2}{4\alpha(t-\tau)}} - e^{-\frac{(z+\zeta/\omega)^2}{4\alpha(t-\tau)}}\right) d\phi'$$

$$(13.88)$$

As usual, we now relate the heat rate per unit pipe length to heat rate per unit borehole height:

$$q`L = q'H$$

Knowing that:

$$L = \int_0^{s_f} ds = \frac{\sqrt{1 + (\omega r_o)^2}}{\omega}(\phi_f - \phi_i)$$

$$= \frac{\sqrt{1 + (\omega r_o)^2}}{\omega} 2\pi N_t = \sqrt{1 + (\omega r_o)^2} H$$

$$\Delta T(r, z, \phi, t) = -\frac{q'b}{16k\pi^2\sqrt{\pi\alpha}} \int_0^t \frac{e^{-\frac{r^2+r_o^2}{4\alpha(t-\tau)}}}{(t-\tau)^{3/2}} d\tau$$

$$\int_{\phi_i}^{\phi_f} e^{\frac{2rr_o\cos(\phi-\phi')}{4\alpha(t-\tau)}} \left(e^{-\frac{(z-\zeta/\omega)^2}{4\alpha(t-\tau)}} - e^{-\frac{(z+\zeta/\omega)^2}{4\alpha(t-\tau)}}\right) d\phi'$$

$$(13.89)$$

This last expression was given by Man et al. [37] but was put in dimensionless form. Park et al. [38] recognize that the double integral can be simplified as usual. Defining:

$$A_-^2 = r^2 + r_o^2 - 2rr_o \cos(\phi - \phi') + (z - \zeta/\omega)^2$$

$$A_+^2 = r^2 + r_o^2 - 2rr_o \cos(\phi - \phi') + (z + \zeta/\omega)^2 \tag{13.90}$$

and will result in the usual error function:

$$\Delta T(r, z, \phi, t) = -\frac{q'b}{8k\pi^2} \int_{\phi_i}^{\phi_f} \frac{\mathrm{erfc}\left(\frac{A_-}{\sqrt{4\alpha t}}\right)}{A_-} - \frac{\mathrm{erfc}\left(\frac{A_+}{\sqrt{4\alpha t}}\right)}{A_+} d\phi' \tag{13.91}$$

In [38], the integration variable was changed as:

$$\phi' = \frac{2\pi}{b}\zeta$$

which lead to :

$$\Delta T(r, z, \phi, t) = -\frac{q'}{4k\pi} \int_{z_o}^{z_o+H} \frac{\mathrm{erfc}\left(\frac{A_-}{\sqrt{4\alpha t}}\right)}{A_-} - \frac{\mathrm{erfc}\left(\frac{A_+}{\sqrt{4\alpha t}}\right)}{A_+} d\zeta \tag{13.92}$$

$$A_-^2 = r^2 + r_o^2 - 2rr_o \cos(\phi - \omega\zeta) + (z - \zeta)^2$$

$$A_+^2 = r^2 + r_o^2 - 2rr_o \cos(\phi - \omega\zeta) + (z + \zeta)^2$$

Again, the use of dimensionless variables is encouraged and the last expression can be put in the form:

$$\Delta T(r, z, \phi, t) = -\frac{q'}{k} G_{spiral}(Fo, \tilde{r}, \tilde{z}, \phi) \tag{13.93}$$

with:

$$G_{spiral}(Fo, \tilde{r}, \tilde{z}, \phi) = \frac{1}{4\pi} \int_{\tilde{z}_o}^{\tilde{z}_o+\tilde{H}} \frac{\mathrm{erfc}\left(\frac{\tilde{A}_-}{\sqrt{4Fo}}\right)}{\tilde{A}_-} - \frac{\mathrm{erfc}\left(\frac{\tilde{A}_+}{\sqrt{4Fo}}\right)}{\tilde{A}_+} d\zeta \tag{13.94}$$

$$\tilde{A}_-^2 = \tilde{r}^2 + 1 - 2\tilde{r}\cos(\phi - \tilde{\omega}\zeta) + (\tilde{z} - \zeta)^2$$

$$\tilde{A}_+^2 = \tilde{r}^2 + 1 - 2\tilde{r}\cos(\phi - \tilde{\omega}\zeta) + (\tilde{z} + \zeta)^2$$

and $\tilde{\omega} = \frac{2\pi r_o}{b}$

Example 13.5 In [38], the authors compare their results with some measured values from a sandbox experiment. Without reproducing all of their results, a single comparison can be done as a first example. The parameters of one of the experiments were:

$$r_o = 12.5\,\mathrm{cm} \quad , \quad H = 4\,\mathrm{m} \quad , \quad b = 5\,\mathrm{cm} \quad , \quad z_o = 0$$

$$q = 400\,\mathrm{W} \quad , \quad k = 0.26\,\mathrm{W/m\text{-}K} \quad , \quad \alpha = 0.0201\,\mathrm{m^2/day}$$

The temperature increase was measured at a location of $r = 25$ cm, $\phi = 0$ and $z = 1.8$ m during 25 hours and compared with the spiral-coil model, the ICS, the ILS, and a 3D model. We can easily reproduce their analytical solutions and include the previously discussed models (FSC, ring).

Solution In their experiment, the heat was injected, so from our convention, we have:

$$q' = -400/4 = -100\,\mathrm{W/m}$$

The solution is shown in the following figures:

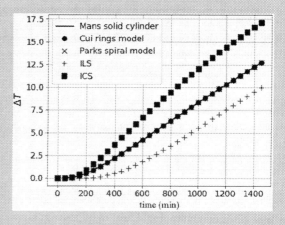

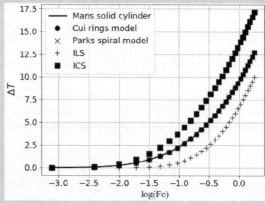

(continued)

Example 13.5 (continued)
In the figure at the left, the same results as those presented by Park et al. [38] are shown. The measured values are not shown but are very close to the spiral model results. In the figure at the right, the same results are shown in the more usual form, showing that with such a large coil radius, the Fourier number associated with the experiment is very small and explains that, in such a case, the ICS and ILS cannot be used. It is also observed that the finite solid cylinder, the ring model, and the spiral model all give, at least for this case, similar results. ■

Example 13.6 In the previous example, it was observed that the different coil models gave similar results, but, of course, the measured point was located at a distant point from the coil and the differences are expected to be higher near it. One point that has to be observed is that, although the FSC model is regular at the location of the source, it is not the case for the other models, and the temperature variation cannot be evaluated on the axis where the source line lies. Taking the same parameters as in the previous example, with a larger coil pitch (b = 50 cm) for clarity, the axial variation along the value of $r = 1.4 r_o$ is evaluated. This value will be calculated for the three models for $Fo = 1.5$ which represents approximately 28 hours.

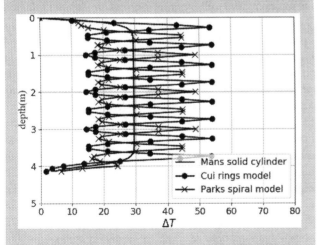

As expected, the FSC model does not take into account the spatial variation and the ring model over-estimates it. ■

Alternative to the Ring Model

The results shown so far in this section give the great versatility and power of superposition of source solutions to the construction of more complex applications. For example, Man's solid cylinder was interpreted as the temporal superposition of either instantaneous line source solution or instantaneous ring source, the latter being used to generate the FSC model, the ring model, and also the spiral model. In the latter case, inversion of the integration variable gave solution (13.94) which can now be interpreted as the spatial superposition of *continuous point source solution* (Eq. (4.19)).

$$F_c(t, x, y, z, \xi, \eta, \zeta) = \frac{1}{4\pi k_s d} \text{erfc} \left(\frac{d}{2\sqrt{\alpha t}} \right) \quad (4.19)$$

The same interpretation could have been used to generate the ring model and was actually used by Xiong et al. [39] in the context of horizontal slinky models (see Sect. 13.6. It can, of course, be used here in the context of pile models. The superposition of continuous rings can be generated as:

$$\Delta T(r, z, t) = -\frac{\grave{q}_i}{4\pi k} \sum_{i=1}^{N} \int_0^{2\pi}$$
$$\left(\frac{\text{erfc} \left(d_-/\sqrt{4\alpha t} \right)}{d_-} - \frac{\text{erfc} \left(d_+/\sqrt{4\alpha t} \right)}{d_+} \right) d\phi' \quad (13.95)$$

with again:

$$d_-^2 = r^2 + r_o^2 - 2rr_o \cos(\phi - \phi') + (z - \zeta_i))^2$$
$$d_+^2 = r^2 + r_o^2 - 2rr_o \cos(\phi - \phi') + (z + \zeta_i))^2$$

As before $\grave{q}_i$ is used for the heat transfer rate per unit ring length. It can be associated with the heat rate per unit borehole length as before:

$$q_i = 2\pi r_o \grave{q}_i = q'b$$

leading to the following alternative to the ring model:

$$\Delta T(\tilde{r}, \tilde{z}, Fo) = -\frac{q'}{k_s} G_{Ring}(Fo, \tilde{r}, \tilde{z})$$

with the new expression:

$$G_{Ring}(Fo, \tilde{r}, \tilde{z})$$

$$= \frac{\tilde{b}}{4\pi^2} \int_0^\pi \sum_{i=1}^N \left(\frac{\text{erfc}\left(\tilde{d}_-/\sqrt{4Fo}\right)}{\tilde{d}_-} - \frac{\text{erfc}\left(\tilde{d}_+/\sqrt{4Fo}\right)}{d_+} \right) d\phi'$$

$$(13.96)$$

$$\tilde{d}_-^2 = \tilde{r}^2 + 1 - 2\tilde{r}\cos(\phi') + (\tilde{z} - \zeta_i))^2$$
$$\tilde{d}_+^2 = \tilde{r}^2 + 1 - 2\tilde{r}\cos(\phi') + (\tilde{z} + \zeta_i))^2$$

$$\zeta_i = \tilde{z}_o + (i - 1/2)\tilde{b}$$

It can be verified that this solution is equivalent to (13.83) and, since it is more efficient, it will be used as our ring model.

13.5.4 Pipe and Fluid Temperature

In practice, it is not the ground temperature which is of importance but the fluid temperature. In the case of a spiral coil, it can be evaluated from the spiral model evaluating the thermal response on the pipe outer circumference. This was suggested by Man et al. [37] although we will apply Park's model here. Neglecting the slope of the spiral, the coordinates are given by:

$$r = r_o + r_p \cos(\theta)$$
$$\phi = \phi$$
$$z = \frac{\phi}{\omega} + r_p \sin(\theta)$$

The mean dimensionless temperature response can then be given as:

$$\Theta(Fo, \phi) = \frac{1}{2\pi} \int_0^{2\pi} G_{spiral}(Fo, \tilde{r}, \tilde{z}, \phi)d\theta \quad (13.97)$$

Man et al. [37] suggested evaluating the thermal response at four points, A, B, C, and D, associated with different angles, $\theta = 0, \pi/2, \pi, 3\pi/2$, instead of integrating over θ:

$$\Theta(Fo, \phi) = \frac{1}{4} \sum_{A,B,C,D} G_{spiral}(Fo, \tilde{r}, \tilde{z}, \phi) \quad (13.98)$$

Example 13.7

(a) Evaluate the mean pipe reduced temperature using the following parameters:

$$r_o = 12.5\,\text{cm}\,, r_p = 1\,\text{cm}\,, H = 5\,\text{m}\,, b = 50\,\text{cm}\,, z_o = 0$$

for $Fo = 5$ and $Fo = 500$.

(b) Redo the same for $H = 50$ m.

(continued)

The solution using (13.98) is shown in the figure:

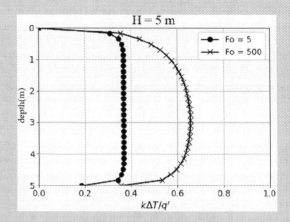

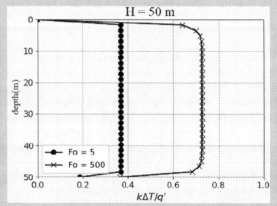

It is seen that for shorter coils, the temperature variation varies more along the coil for larger Fourier numbers. The coil pitch also influences the spatial variation of the temperature.

As it is the case for the vertical borehole, a mean temperature along the whole spiral coil can be evaluated. Man et al. [37] proposed instead using the value at the midpoint of the coil to reduce computing time. As expected from the previous example, it is expected that the difference between this value and the mean value will increase with Fourier number. As seen also from the previous example, the difference will be smaller if the temperature is evaluated at a different point than the middle one.

Example 13.8 In this example, we reproduce the results of [37] at least qualitatively since they didn't specify the height of their coil. In this example, the same parameters as before are used:

$$r_o = 12.5\,\text{cm}\,, r_p = 1\,\text{cm}\,, H = 5\,\text{m}\,, b = 50\,\text{cm}\,, z_o = 0$$

(continued)

Example 13.8 (continued)
and the mean pipe temperature along the coil and the
pipe temperature at $H/2$ and $H/4$ are shown:

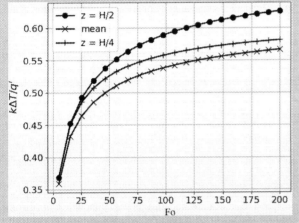

Of course, the difference is accentuated by the short
coil.

Finally, the fluid temperature can be calculated as it was
done for a classical borehole. One has:

$$T_f = T_p - q`R'_p \qquad (13.99)$$

with:

$$R'_p = \frac{\log\left(\frac{r_{po}}{r_{pi}}\right)}{2\pi k_{pipe}} + \frac{1}{2\pi r_{pi}\hbar_f} \qquad (13.100)$$

Again replacing the heat rate per coil length with the heat rate
per borehole height:

$$T_f = T_p - \frac{q'}{\sqrt{1+(\omega r_o)^2}} R'_p \qquad (13.101)$$

Assuming a linear temperature variation, the following results are obtained:

$$T_{fo} - T_{fi} = \frac{q'H}{\dot{m}C_p}$$

$$\frac{T_{fo} + T_{fi}}{2} = T_f = T_p - \frac{q'}{\sqrt{1+(\omega r_o)^2}} R'_p$$

$$T_{fo} = T_p + q'\left(\frac{H}{2\dot{m}C_p} - \frac{R'_p}{\sqrt{1+(\omega r_o)^2}}\right)$$

$$\qquad (13.102)$$

$$T_{fi} = T_p - q'\left(\frac{H}{2\dot{m}C_p} + \frac{R'_p}{\sqrt{1+(\omega r_o)^2}}\right)$$

In all the previous models, the heat transfer rate was assumed
to be uniform along the coil. This approximation is not
always fulfilled. It is possible to split the coil into section with
unknown heat distribution with the total heat transfer being
known. This was the approach used by Leroy and Bernier
[40].

13.6 Slinky Model for Horizontal Systems, Xiong's Model

The principle of superposition of source's solutions was also
used by Li et al. [41] and by Xiong et al. [39] in relation
with slinky models in horizontal trenches. The model will be
referred as Xiong's model since it is mostly known and used
in Energy+ but they are basically the same. A slinky configuration can be regarded as a flattened coil laid horizontally or
vertically in a trench but doing so, the parameterization of the
path is less trivial. Instead, Li et al. [41] and Xiong et al. [39]
took the same approach as Cui et al. and modeled the Slynky
as a superposition of rings, neglecting then the contribution
of the tube along the side. A schematic of a slinky shape is
shown in Fig. 13.7.

The parameters of the geometry are associated with two
main parameters: the diameter of the ring D and the pitch p.
The total length of trench and tubing for one trench are given
by [42]:

$$\begin{aligned} H_{trench} &= N_t p + D \\ L_{tube} &= N_t(\pi D + 2p) + D(\pi/2 + 1) \end{aligned}$$

with N_t the number of rings per trench. For a large number
of rings, the end contribution given by the second term of the
last equations can be neglected:

$$\begin{aligned} L_{trench} &\approx N_t p \\ L_{tube} &\approx N_t(\pi D + 2p) \end{aligned} \qquad (13.103)$$

The $2p$ term inside the parenthesis represents the tube length
along the forward and return path shown in the figure. If n_r
trenches are present (see Fig. 13.8):

$$N_r = n_r N_t$$

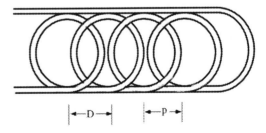

Fig. 13.7 Schematic of a slinky (from [39])

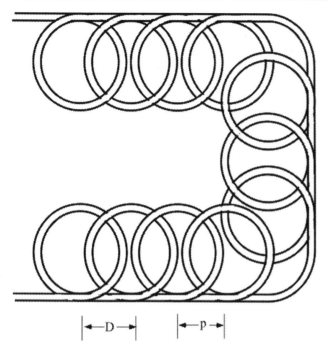

Fig. 13.8   Schematic of a slinky in two trenches

Contrary to the other models, Li and Xiong did not start with a superposition of *instantaneous sources* but *continuous sources* along the centerline of the pipe and evaluate its contribution at two points of the pipe. From (4.19), the effect of one ring to a target point, taking into account the effect of the image part, is then given, for the horizontal slinky, by:[4]

$$\Delta T(P_i, P_j, t) = -\frac{q`_i R}{4\pi k_s} \int_0^{2\pi} \frac{\mathrm{erfc}\left(\frac{d(P_i, P_j)}{2\sqrt{\alpha t}}\right)}{d(P_i, P_j)}$$

$$-\frac{\mathrm{erfc}\left(\frac{d(P_{imag,i}, P_j)}{2\sqrt{\alpha t}}\right)}{d(P_{imag,i}, P_j)} d\omega \qquad (13.104)$$

$$d(P_i, P_j) = \sqrt{(x_i - x_j)^2 + (y_i - y_j)^2 + (z_i - z_j)^2}$$
$$d(P_{imag,i}, P_j) = \sqrt{(x_i - x_j)^2 + (y_i - y_j)^2 + (-z_i - z_j)^2}$$
$$(13.105)$$

13.6.1   Mean Pipe Temperature

Equation (13.104) can be used to evaluate the temperature variation anywhere in the soil. Again, in practice, the

temperature at the pipe outside radius (r_p) is of particular importance. Let's start by looking at the temperature increase along the pipe of a ring i due to the source line in the center of the same ring i, and we then have ($R = D/2$, the ring radius) (see Fig. 13.9):

$$x_i = R\cos(\omega), \, y_i = R\sin(\omega), \, z_i = -h$$

$$x_j = \cos(\phi)(R + r_p \cos(\gamma)), \, y_j = \sin(\phi)(R + r_p \cos(\gamma)),$$

$$z_j = -h + r_p \sin(\gamma))$$

Since in the case of ΔT_{ii}, the response does not depend on ϕ, this angle can be put equal to zero and one can evaluate as suggested by Li et al. [41]:

$$\Delta \bar{T} = \frac{1}{2\pi} \int_0^{2\pi} \Delta T(P_i, P_j, t) d\gamma \qquad (13.106)$$

As pointed out by Man et al. [37], the integral can be replaced by evaluating the temperature at several key points, for example, $\gamma = [0, \pi/2, \pi, 3\pi/2]$.

Xiong et al. [39] suggested almost the same thing using only two points $\gamma = [0, \pi]$, but instead of taking the mean of the temperature variation at those two points, which would have been maybe more logical, they chose to evaluate the temperature variation at the mean distance (see Fig. 13.10):

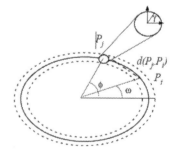

Fig. 13.9   Pipe geometry

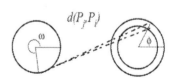

Fig. 13.10   Distance from ring j to i

---

[4]In [39], heat is positive when injected; the sign convention is then different here.

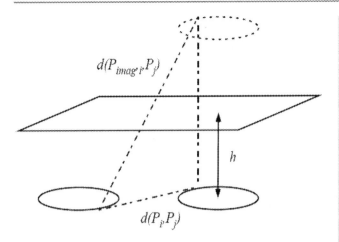

Fig. 13.11 Image contribution

$$d(P_i, P_j) = \frac{d(P_i, P_{ji}) + d(P_i, P_{jo})}{2}$$

$$d(P_i, P_{ji}) = \left[(x_{oi} + (R - r_p)\cos\phi - x_{oj} - R\cos\omega)^2 + (y_{oi} + (R - r_p)\sin\phi - y_{oj} - R\sin\omega)^2\right]^{1/2}$$

$$d(P_i, P_{jo}) = \left[(x_{oi} + (R + r_p)\cos\phi - x_{oj} - R\cos\omega)^2 + (y_{oi} + (R + r_p)\sin\phi - y_{oj} - R\sin\omega)^2\right]^{1/2}$$
(13.107)

where $(x_{oi}, y_{oi}, 0)$ is the location of the center of the ring i. In the case of a single ring, those values are zero. Since the target point lies in the $x - y$ plane, the image point is now (see Fig. 13.11):

$$d(P_{imag,i}, P_j) = \sqrt{d(P_i, P_j)^2 + 4h^2}$$
(13.108)

Example 13.9 Evaluate the real mean pipe reduced temperature variation $(\Theta = \Delta T k_s/q')$ of the single ring model and Xiong's approximation for the case where:

$$\frac{r_p}{R} = 0.05, \frac{h}{R} = 3.5$$

Solution For only one ring, the real mean is evaluated using Eq. (13.106). Xiong's approximation is found evaluating Eq. (13.104) with the distance given (13.107). The solution can also be compared to the evaluation of Eq. (13.104) at different points on the perimeter as in Man's suggestion for the coil model. The solution is shown in the following figure. As for the other variables, the Fourier number is associated with the ring radius:

$$Fo = \frac{\alpha t}{R^2}$$

(continued)

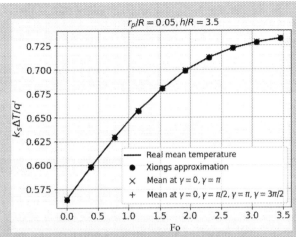

which shows that all approximations give almost the same value as the real mean value. Of course, it is expected that the difference might be larger for a pipe with a larger diameter and when the soil distance is shorter. The same calculation is done for the case where:

$$\frac{r_p}{R} = 0.25, \frac{h}{R} = 0.5$$

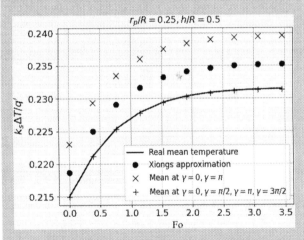

Comments *Of course, the values given in the last case are not typical. For example, this would give a 20 cm diameter pipe for a 80 cm ring diameter which does not make sense and in practice, for typical values, Xiong's approximation can be used.*

For multiple rings, the pipe temperature is dependent on the angle location ϕ. The mean pipe temperature increase from ring i to ring j is then found integrating as:

$$\Delta T_{i,j}(t) = -\frac{q\grave{}_i R}{8\pi^2 k_s} \int_0^{2\pi} \int_0^{2\pi} \frac{\text{erfc}\left(\frac{d(P_i,P_j)}{2\sqrt{\alpha t}}\right)}{d(P_i,P_j)}$$

$$-\frac{\text{erfc}\left(\frac{d(P_{imag,i},P_j)}{2\sqrt{\alpha t}}\right)}{d(P_{imag,i},P_j)} d\omega d\phi \qquad (13.109)$$

The superposition of all ring contributions will give the temperature variation for a single ring:

$$\Delta T_j(t) = \sum_{i=1}^{N_r} \Delta T_{i,j} \qquad (13.110)$$

Finally, the mean value of all those temperatures is found from:

$$\Delta \bar{T}(t) = \text{mean} \left(\Delta T_j(t)\right) = \frac{1}{N_r} \sum_{j=1}^{N_r} \sum_{i=1}^{N_r} \Delta T_{i,j} \qquad (13.111)$$

Xiong et al. [39] define a dimensionless temperature response factor (g-function) assuming that the heat transfer per unit ring length was uniform along slinky as:

$$g_{slinky}(t) = \frac{2\pi k_s}{q\grave{}} \Delta \bar{T}$$

$$= \sum_{i=1}^{N_r} \sum_{j=1}^{N_r} \frac{1}{4\pi N_r} \int_0^{2\pi} \int_0^{2\pi} \frac{\text{erfc}\left(\frac{d(P_i,P_j)}{2\sqrt{\alpha t}}\right)}{d(P_i,P_j)}$$

$$-\frac{\text{erfc}\left(\frac{d(P_{imag,i},P_j)}{2\sqrt{\alpha t}}\right)}{d(P_{imag,i},P_j)} d\omega d\phi$$

$$(13.112)$$

The 2π was used to be coherent with the formalism of *g-function* proposed by Eskilson and Claesson, but, of course, it can be omitted to be associated with the *g-function* formalism used so far in this chapter. In either case, it is preferable to use dimensionless variables. Finally, it is also more relevant to express the response with respect to heat transfer per trench length instead of pipe length. As previously stated, the tubing length is not uniquely defined. Using Eq. (13.103), for example, one would get:

$$N_r(2\pi R + 2p)q\grave{} = N_r p q'$$

$$q\grave{} = \frac{\tilde{p}}{2\pi + 2\tilde{p}} q' \qquad (13.113)$$

So a slightly different form as the one suggested by Xiong would be:

$$\Delta \bar{T}(Fo) = -\frac{q'}{k_s} G_{slinky}(Fo) \qquad (13.114)$$

with:

$$G_{slinky}(Fo)$$

$$= \sum_{i=1}^{N_r} \sum_{j=1}^{N_r} \frac{\tilde{p}}{8\pi^2(2\pi + 2\tilde{p})N_r}$$

$$\int_0^{2\pi} \int_0^{2\pi} \frac{\text{erfc}\left(\frac{\tilde{d}(P_i,P_j)}{2\sqrt{Fo}}\right)}{\tilde{d}(P_i,P_j)} - \frac{\text{erfc}\left(\frac{\tilde{d}(P_{imag,i},P_j)}{2\sqrt{Fo}}\right)}{\tilde{d}(P_{imag,i},P_j)} d\omega d\phi$$

$$(13.115)$$

with:

$$\tilde{d}(P_i,P_{ji}) = \left[\tilde{x}_{oi} + (1-\tilde{r}_p)\cos\phi - \tilde{x}_{oj} - \cos\omega)^2 + (\tilde{y}_{oi} + (1-\tilde{r}_p)\sin\phi - \tilde{y}_{oj} - \sin\omega)^2\right]^{1/2}$$

$$\tilde{d}(P_i,P_{jo}) = \left[(\tilde{x}_{oi} + (1+\tilde{r}_p)\cos\phi - \tilde{x}_{oj} - \cos\omega)^2 + (\tilde{y}_{oi} + (1+\tilde{r}_p)\sin\phi - \tilde{y}_{oj} - \sin\omega)^2\right]^{1/2}$$

$$(13.116)$$

where:

$$\tilde{r}_p = \frac{r_p}{R} \quad , \quad \tilde{h} = \frac{h}{R} \quad , \quad \tilde{d} = \frac{d}{R} \quad , \quad Fo = \frac{\alpha t}{R^2}$$

Several simplifications to accelerate the calculation of these response factors were proposed in [39] but will not be discussed here. Besides those, an interesting one was suggested by Li et al. [41]. From Fig. 13.12, it can be seen that the temperature variation is symmetric with respect to a plane joining the two centers. From this, the last relation can be simplified as:

$$G_{slinky}(Fo)$$

$$= \sum_{i=1}^{N_r} \sum_{j=1}^{N_r} \frac{\tilde{p}}{4\pi^2(2\pi + 2\tilde{p})N_r}$$

$$\int_\zeta^{\zeta+\pi} \int_0^{2\pi} \frac{\text{erfc}\left(\frac{\tilde{d}(P_i,P_j)}{2\sqrt{Fo}}\right)}{\tilde{d}(P_i,P_j)} - \frac{\text{erfc}\left(\frac{\tilde{d}(P_{imag,i},P_j)}{2\sqrt{Fo}}\right)}{\tilde{d}(P_{imag,i},P_j)} d\omega d\phi$$

$$(13.117)$$

Fluid Temperature

As for all other configurations, the fluid temperature is deduced from the pipe temperature assuming a temperature variation profile inside the pipe. Again, the simplest case is

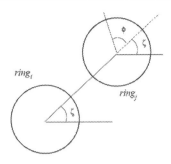

Fig. 13.12 Symmetry in the temperature variation

to assume a linear profile in which case:

$$T_f = \frac{T_{fi} + T_{fo}}{2} = T_p - q`R'_p$$

$$\frac{T_{fi} - T_{fo}}{2} = -\frac{q}{2\dot{m}C_p}$$

(13.118)

with R'_p given as usual by Eq. (13.100):

$$T_{fo} = T_p + q'\left(\frac{H_{tr}}{2\dot{m}C_p} - \frac{\tilde{p}R'_p}{2\pi + 2\tilde{p}}\right)$$

$$T_{fi} = T_p - q'\left(\frac{H_{tr}}{2\dot{m}C_p} + \frac{\tilde{p}R'_p}{2\pi + 2\tilde{p}}\right)$$

(13.119)

Example 13.10 As a first example, the temperature response factors as defined by Xiong et al. [39] will be evaluated for the slinky configurations used in Fujii et al. experiments [43]. The different parameters are given in the following table:

Parameter	Slinky1	Slinky 2	Slinky 3
Diameter of the loop	0.8 m	0.8 m	0.8 m
Buried depth	1.5 m	1.5 m	1.5 m
Distance between trenches	2 m	2 m	2 m
Outside pipe diameter	34 mm	34 mm	34 mm
Inside pipe diameter	24 mm	24 mm	24 mm
Pitch	0.4 m	0.6 m	0.8 m

In each case, there were two trenches of 35 m in length separated by 2 m given a total of 72 m trench. Xiong et al. simplified the geometry by representing the slinky as two 36 m trenches. Neglecting end effects, the total number of rings was then:

$$72/p = 180, 120 \text{ and } 90 \text{ rings}$$

(continued)

Xiong et al. presented their results in function of time, so the results would depend on the soil diffusivity which was not given but the results are qualitatively similar:

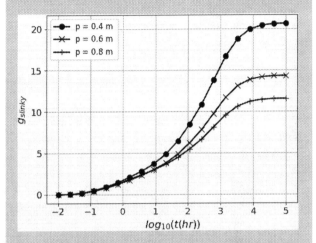

Since the thermal response factors are associated with the soil resistance, it was argued that a larger pitch is better due to the less interferences between coils. However, it is sometimes interesting to evaluate the performance with respect to heat transfer per trench length. To do so, using Eq. (13.113), we can multiply the previous results by:

$$\frac{\tilde{p}}{2\pi + 2\tilde{p}}$$

and get:

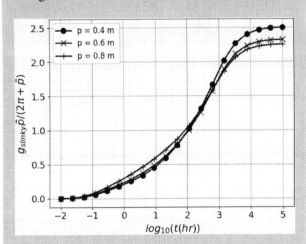

The results are quite different.

Example 13.11 As a second example, we will compare the experimental results given by Fujii et al. [43] during three 5-hour thermal response tests (TRT), which they published using the same three slinkys described in the previous example. The results give the mean fluid temperature with respect to time and were not given in dimensionless form but instead as:[a]

$$\frac{T_f - T_o}{q'} = \frac{G_{slinky}}{k_s} + \frac{\tilde{p}}{2\pi + 2\tilde{p}} R'_p$$

The results are then dependent on the soil conductivity which wa given as 1.09 W/m-K but also on the diffusivity (or volumetric capacity) which was not given. The value was assumed to be $\alpha_s \approx 0.08$ m²/d. The pipe conductivity was reported as 3.5 W/m-k, so its resistance was:

$$R'_{cond} = \frac{\log(34/24)}{2\pi 3.5} = 0.158 \text{ W/m-K}$$

The flow rate changed from test to test but the effect on the convection resistance was negligible. In each case, the Reynolds number was higher than 10,000 and neglecting possible centrifugal effect, the convection coefficient was in the order of 2000 W/m²-K, given a small contribution to the total resistance:

$$R'_p \approx 0.158 + \frac{1}{2000\pi 0.024} \approx 0.165 \text{ W/m-K}$$

Using the model previously described, the following results are obtained:

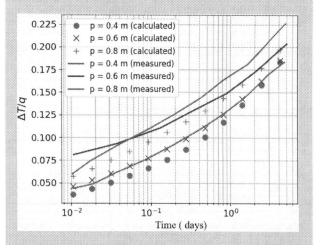

___
[a]Again, the sign convention is different since in Fujjii's paper injection of heat was positive.

Several reasons might explain the discrepancies between the model and the measured values, some of which can be attributed to the accuracy of the measurements themselves like the strange behavior between loop 2 and 3 in the early times of the experiment. Others can be associated with the specificities associated with the porous media associated with the overburden, discussed in [43], whereas the theoretical model assumes as usual pure conduction heat transport in a homogeneous ground. Nevertheless, the qualitative thermal behavior is reproduced, and, as reported in Fujii et al., the tight slinky seems to be a better choice when one looks for heat per trench length and not per pipe length. From these results, that was the configuration that they kept for a long-term measurement. However, as usual, the cost of the pipe will be higher and compromise has to be made. From our previous example, it was observed that in the long term, the advantage of the short pitch slinky seems to be lost, but unfortunately, Fujii's experiment did not compare the different configurations in the long run. As a final remark, one has to notice that the results are highly dependent on the trench length to pipe length ratio.

From the previous examples, it was shown that the advantages and the drawbacks of the slinky configuration are highly dependent whether one looks at the heat exchange per pipe length and heat transfer per trench length. Does the user has a specific trench space available? Does the increase in cost due to the increase in pipe length eliminate a possible slight advantage of one configuration on the other one? There does not seem to have a clear answer on that subject yet.

13.6.2 Vertical Slinky

Finally, from the same analysis, Xiong et al. [39] propose a similar model for the vertical slinky. The only difference is in the evaluation of the source to target distances:

$$\tilde{d}(P_i, P_j) = \frac{\tilde{d}(P_i, P_{ji}) + \tilde{d}(P_i, P_{jo})}{2} \tag{13.120}$$

$$\tilde{d}(P_i, P_{ji}) = \Big[(\tilde{x}_{oi} + (1 - \tilde{r}_p)\cos\phi - \tilde{x}_{oj} - \cos\omega)^2 + (\tilde{y}_{oi} - \tilde{y}_{oj})^2$$

$$+ (\tilde{z}_{oi} + (1 - \tilde{r}_p)\sin\phi - \tilde{z}_{oj} - \sin\omega)^2 \Big]^{1/2} \tag{13.121}$$

(continued)

$\tilde{d}(P_i, P_{jo})$

$$= \left[(\tilde{x}_{oi} + (1 + \tilde{r}_p) \cos\phi - \tilde{x}_{oj} - \cos\omega)^2 + (\tilde{y}_{oi} - \tilde{y}_{oj})^2 \right.$$

$$\left. + (\tilde{z}_{oi} + (1 + \tilde{r}_p) \sin\phi - \tilde{z}_{oj} - \sin\omega)^2 \right]^{1/2} \quad (13.122)$$

$$\tilde{d}(P_{imag,i}, P_j) = \frac{\tilde{d}(P_{imag,i}, P_{ji}) + \tilde{d}(P_{imag,i}, P_{jo})}{2} \quad (13.123)$$

$\tilde{d}(P_{imag,i}, P_{ji})$

$$= \left[(\tilde{x}_{oi} + (1 - \tilde{r}_p) \cos\phi - \tilde{x}_{oj} - \cos\omega)^2 + (\tilde{y}_{oi} - \tilde{y}_{oj})^2 \right.$$

$$\left. + (\tilde{z}_{oi} - 2\tilde{h} + (1 - \tilde{r}_p) \sin\phi - \tilde{z}_{oj} - \sin\omega)^2 \right]^{1/2}$$

$$(13.124)$$

$\tilde{d}(P_{imag,i}, P_{ji})$

$$= \left[(\tilde{x}_{oi} + (1 + \tilde{r}_p) \cos\phi - \tilde{x}_{oj} - \cos\omega)^2 + (\tilde{y}_{oi} - \tilde{y}_{oj})^2 \right.$$

$$\left. + (\tilde{z}_{oi} - 2\tilde{h} + (1 + \tilde{r}_p) \sin\phi - \tilde{z}_{oj} - \sin\omega)^2 \right]^{1/2}$$

$$(13.125)$$

Example 13.12 The monthly loads of a typical building are given in the following table. It is assumed that these loads will remain the same for all subsequent years.

Month	Q(MWh)	Q(MWh)
January	15	0.0
February	10	0.0
March	10	0.0
April	6	0.0
May	2	−5
June	0.0	−8
July	0.0	−10
August	0.0	−10
September	0.0	−3
October	1.5	0.0
November	10	0.0
December	10	0.0

(continued)

The peak load is 40 kW in heating mode and 20 kW in cooling mode. A GSHP is used with a COP = 4 (heating) and 5 (cooling). The peak load block is chosen to be 4 hours and a 10-year lifespan is used. We have the following data:

Ground:

$$k_s = 1.5 \text{ W/m-K}, \alpha_s = 0.0648 \text{ m}^2/\text{day}, T_o = 10°C$$

$$A_s = 8K, t_{shift} = 15 \text{ d}$$

$$d_{po} = 35 \text{ mm}, d_{pi} = 30 \text{ mm}, k_p = 0.4 \text{ W/m-K},$$

$$\hbar_f = 5000 \text{W/m}^2\text{-K}$$

Fluid:

$$\dot{V} = 2 \text{ l/s}, \rho = 1030 \text{ kg/m}^3, Cp = 3.90 \text{ kJ/kg-K}$$

The minimum temperature at the heat pump entrance is −4°C and 30°C in cooling mode. Using the ASHRAE method, find the total trench and pipe length using a two-pipe (in series) horizontal system shown in Fig. 13.13 and using a horizontal slinky with a pitch of 0.5 m and a ring diameter of 1 m. Assume only one trench.

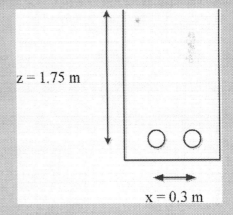

Fig. 13.13 Two-pipes trench

Solution The classical ASHRAE method will be applied. For the horizontal layout, the soil resistance will be evaluated with the method of images:

$$R_{hori} = \frac{1}{2\pi k_s} \left(I(X_1) + I(X_2) - (I(X_{im,1}) \right.$$

$$\left. + I(X_{im,2})) \right)$$

where:

(continued)

Example 13.12 (continued)

$$X_1 = \frac{r_p}{2\sqrt{\alpha t}} \ , \quad X_2 = \frac{x}{2\sqrt{\alpha t}} \ , \quad X_{im,1} = \frac{z}{\sqrt{\alpha t}} \ ,$$

$$X_{im,2} = \frac{\sqrt{(2z)^2 + x^2}}{\sqrt{2\alpha t}}$$

The results are:

$R'_{hori,a} = 0.116$ K-m/W , $R'_{hori,m} = 0.471$ K-m/W ,

$R'_{hori,h} = 0.235$ K-m/W

$$R'_p = \frac{\log(35/30)}{2\pi\,0.4} + \frac{1}{6000\pi\,0.03} = 0.063 \text{ K-m/W}$$

$$q_a = \frac{\sum Q_h(COP_h - 1)/COP_h + \sum Q_c(COP_c + 1)/COP_c}{8760}$$

$$= 1.0 \text{ kW}$$

In heating mode, the worst month is January:

$$q_m = \frac{15000(COP_h - 1)}{744\,COP_h} = 15.1 \text{ kW}$$

$$q_h = \frac{40(COP_h - 1)}{COP_h} = 30 \text{ kW}$$

During the coldest day of the year, the temperature at 1.75 m is:

$$T_o + \Delta T = 10 - 3.4 = 6.6\,°C$$

$$T_{fo} = -2, T_{fi} = -2 - \frac{30}{1030 \cdot 3.9} = -5.83\,°C \Rightarrow$$

$$T_{f,h} = -3.9\,°C$$

$$L_{he} = 1000\frac{1.0 \cdot 0.116 + 15.2 \cdot 0.471 + 30(0.235 + 0.063)}{6.6 - (-3.92)}$$

$$= 1538 \text{ m}$$

In cooling mode, at the warmest day (July 15):

$$q_m = -\frac{10000(COP_c + 1)}{744\,COP_c} = -16.1 \text{ kW}$$

$$q_h = -\frac{20(COP_c + 1)}{COP_c} = 24 \text{ kW}$$

$$T_o + \Delta T = 10 + 3.4 = 13.4\,°C$$

$$T_{fo} = 30, T_{fi} = 30 - \frac{-24}{1030 \cdot 3.9} = 33\,°C \Rightarrow T_{f,h} = 31.5\,°C$$

$$L_{co} = 1000\frac{-16.1 \cdot 0.471 - 24(0.235 + 0.063)}{13.4 - (31.5)} = 812 \text{ m}$$

(continued)

$$\Rightarrow L_{horiz,pipe} = 1538 \text{ m} \ , L_{horiz,trench} = 769 \text{ m}$$

For the slinky case, everything remains the same except for the soil resistances. The calculation is iterative since the trench length has to be known. As a first guess, the same trench length can be used:

$$L_{tr} \approx 769 \text{ m} \Rightarrow N_{rings} = \frac{769 - 1}{0.5} = 1536$$

Using Eq. (13.115) will give:

$$G_f = 0.335, G_1 = 0.27, G_2 = 0.06$$

$$R'_{hori,a} = \frac{0.335 - 0.27}{1.5} = 0.043 \text{ K-m/W}$$

$$R'_{hori,m} = \frac{0.27 - 0.06}{1.5} = 0.139 \text{ K-m/W} \ ,$$

$$R'_{hori,h} = \frac{0.06}{1.5} = 0.04 \text{ K-m/W}$$

$$R'^*_p = 0.063\frac{1}{2\pi + 2} = 0.0087 \text{ K-m/W}$$

$$L_{he,tr} = 1000\frac{1.0 \cdot 0.043 + 15.2 \cdot 0.139 + 30(0.04 + 0.0087)}{6.6 - (-3.92)}$$

$$= 346 \text{ m}$$

which, as seen in the previous calculation, will be higher than in cooling mode. After convergence, the result will be:

$$L_{slinky,tr} = 343.5 \text{ m} \ , L_{slinky,pipe} = 2839 \text{ m} \ , N_{rings} = 685$$

A result indicates that the trench will be smaller but the pipe length will be higher, so the final decision will depend on the cost, the area available, the willingness of the contractor to install a slinky configuration, etc.

Problems

Pr. 13.1 Redo Problem 6.2 where the short-time response factor is used to estimate the effect of the last pulse. Use Eq. (13.30) for the equivalent radius.

Answer: 150 m

Pr. 13.2 Following the steps given in Sect. 4.2.6, derive Eq. (13.52).

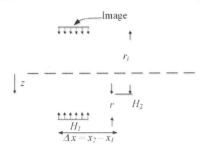

Fig. Pr_13.1 Horizontal segment-to-segment response factor

Pr. 13.3 With the same approach, derive the expression for the dimensionless mean temperature increment a horizontal pipe segment of length H_2 due to another horizontal pipe segment of length H_1.

Answer: see [44]

$$\frac{\Delta T_{1\to 2}(\bar{r},\bar{t})2\pi k_s}{q'}$$

$$= h_{ij\to uv}(\bar{r}, \bar{z}_1, \bar{z}_2, \bar{h}_1, \bar{h}_2)$$

$$= \frac{1}{2\bar{h}_2} \int_{3/\sqrt{4t}}^{\infty} \frac{e^{-\bar{r}s^2} - e^{-\bar{r}_i s^2}}{s^2}$$

$$\times [\ \mathrm{ierf}((\bar{x}_2 - \bar{x}_1 + \bar{h}_2)s) - \mathrm{ierf}((\bar{x}_2 - \bar{x}_1)s)$$

$$+ \mathrm{ierf}((\bar{x}_2 - \bar{x}_1 - \bar{h}_1)s) - \mathrm{ierf}((\bar{x}_2 - \bar{x}_1 + \bar{h}_2 - \bar{h}_1)s)]ds$$

References

1. Ahmadfard, M., Bernier, M.: A review of vertical ground heat exchanger sizing tools including an inter-model comparison. Renew. Sust. Energ. Rev. **110**, 247–265 (2019)
2. Lamarche, L.: Short-time modelling of geothermal systems. In: Proceedings of ECOS 2016, pp. 19–23 (2016)
3. Yavusturk, C., Splitter, J.D.: A short time step response factor model for vertical ground loop heat exchangers. ASHRAE Trans. **105**(2), 475 (1999)
4. Xu, X., Spitler, J.D.: Modeling of vertical ground loop heat exchangers with variable convective resistance and thermal mass of the uid. In: Proceedings of Ecostock (2006)
5. Ahmadfard, M., Bernier, M.: Modifications to ASHRAE's sizing method for vertical ground heat exchangers. Science and Technology for the Built Environment **24**(7), 803–817 (2018)
6. Li, M., Lai, A.C.K.: New temperature response functions (G functions) for pile and borehole ground heat exchangers based on compositemedium line-source theory. Energy **38**(1), 255–263 (2012)
7. Beier, R.A., Smith, M.D.: Minimum duration of in-situ tests on vertical boreholes. ASHRAE Trans. **109**(2), 475–486 (2003)
8. Lamarche, L., Beauchamp, B.: New solutions for the short-time analysis of geothermal vertical boreholes. Int. J. Heat Mass Transf. **50**(7), 1408–1419 (2007). http://dx.doi.org/10.1016/j.ijheatmasstransfer.2006.09.007
9. Lamarche, L.: Short-time analysis of vertical boreholes, new analytic solutions and choice of equivalent radius. Int. J. Heat Mass Transf. **91**(12), 800–807 (2015). http://dx.doi.org/10.1016/j.ijheatmasstransfer.2006.09.007
10. Carslaw, H.S., Jaeger, J.C.: Conduction of Heat in Solids. 2nd edn. (1959)
11. Young, T.R.: Development, verification, and design analysis of the borehole fluid thermal mass model for approximating short term borehole thermal response. Master's thesis. Oklahoma State University, New York (2004)
12. Javed, S., Claesson, J.: New analytical and numerical solutions for the short-term analysis of vertical ground exchangers. ASHRAE Trans. **117**(1), 3–12 (2011)
13. Bose, J.E., Parker, J.D., McQuiston, F.C.: Design data manual for closed-loop ground-coupled heat pump systems. American Society of Heating, Refrigerating and Air-Conditioning Engineers, Atlanta, Ga (1985). isbn: 0910110417
14. Sutton, M.G., et al.: An algorithm for approximating the performance of vertical bore heat exchangers installed in a stratified geological regime. ASHRAE Trans. **108**(2), 177–184 (2002)
15. Beier, R.A., Smith, M.D., Spitler, J.D.: Reference data sets for vertical borehole ground heat exchanger models and thermal response test analysis. Geothermics **40**(1), 79–85 (2011). https://doi.org/10.1016/j.geothermics.2010.12.007
16. Lamarche, L.: Short-term behavior of classical analytic solutions for the design of ground-source heat pumps. Renew. Energy **57**(0), 171–180 (2013)
17. Eskilson, P.: Thermal analysis of heat extraction systems. PhD thesis. Lund University, Sweden (1987)
18. Cimmino, M., Bernier, M.: A semi-analytical method to generate gfunctions for geothermal bore fields. Int. J. Heat Mass Transf. **70**(0), 641–650 (2014)
19. Marcotte, D., Pasquier, P.: On the estimation of thermal resistance in borehole thermal conductivity test. Renew. Energy **33**(11), 2407–2415 (2008)
20. Lamarche, L., Kajl, S., Beauchamp, B.: A review of methods to evaluate borehole thermal resistances in geothermal heat-pump systems. Geothermics **39**(2), 187–200 (2010)
21. Cimmino, M., Bernier, M.: Experimental determination of the gfunctions of a small-scale geothermal borehole. Geothermics **56**(0), 60–71 (2015)
22. Cimmino, M.: Fast calculation of the g-functions of geothermal borehole fields using similarities in the evaluation of the finite line source solution. J. Build. Perform. Simul. **11**(6), 655–668 (2018)
23. Dusseault, B., Pasquier, P., Marcotte, D.: A block matrix formulation for eficient g-function construction. Renew. Energy **121**, 249–260 (2018). https://doi.org/10.1016/j.renene.2017.12.092
24. Lamarche, L.: g-function generation using a piecewise-inear profile applied to ground heat exchangers. Int. J. Heat Mass Transf. **115**, 354–360 (2017)
25. Cimmino, M.: The effects of borehole thermal resistances and fluid flow rate on the g-functions of geothermal bore fields. Int. J. Heat Mass Transf. **91**, 1119–1127 (2015)
26. Cimmino, M.: Fast calculation of the g-functions of geothermal borehole fields using similarities in the evaluation of the finite line source solution. J. Build. Perform. Simul. **11**(6), 655–668 (2018). https://doi.org/10.1080/19401493.2017.1423390
27. Eskilson, P.: Thermal analysis of heat extraction systems. PhD thesis. Lund University, Sweden (1987)
28. Sutton, M.G., Nutter, D.W., Couvillion, R.J.: A ground resistance for vertical bore heat exchangers with groundwater flow. J. Energy Resour. Technol. **125**(3), 183–189 (2003)
29. Diao, N., Li, Q., Fang, Z.: Heat transfer in ground heat exchangers with groundwater advection. Int. J. Therm. Sci. **43**(12), 1203–1211 (2004)
30. Molina-Giraldo, N., et al.: A moving finite line source model to simulate borehole heat exchangers with groundwater advection. Int. J. Therm. Sci. **50**(12), 2506–2513 (2011)
31. Hunt, B.: Calculation of the leaky aquifer function. J. Hydrol. **33**(1), 179–183 (1977) https://doi.org/10.1016/0022-1694(77)90107-X

32. Veling, E.J.M., Maas, C.: Hantush Well Function revisited. J. Hydrol. **393**(3), 381–388 (2010). https://doi.org/10.1016/j.jhydrol. 2010.08.033

33. Pasquier, P., Lamarche, L.: Analytic expressions for the moving infinite line source model. Geothermics **103**, 102413 (2022)

34. Brandl, H.: Energy foundations and other thermo-active ground structures. Géotechnique **56**(2), 81–122 (2006)

35. Man, Y., et al.: A new model and analytical solutions for borehole and pile ground heat exchangers. Int. J. Heat Mass Transf. **53**(13-14), 2593–2601 (2010)

36. Cui, P., et al.: Heat transfer analysis of pile geothermal heat exchangers with spiral coils. Appl. Energy **88**(11), 4113–4119 (2011). https://doi.org/10.1016/j.apenergy.2011.03.045

37. Man, Y., et al.: Development of spiral heat source model for novel pile ground heat exchangers. Hvac&R Research **17**(6), 1075–1088 (2011)

38. Park, S., et al.: Characteristics of an analytical solution for a spiral coil type ground heat exchanger. Comput. Geotech. **49**, 18–24 (2013)

39. Xiong, Z., Fisher, D.E., Spitler, J.D.: Development and validation of a Slinky? ground heat exchanger model. Appl. Energy **141**, 57–69 (2015)

40. Leroy, A., Bernier, M.: Development of a novel spiral coil ground heat exchanger model considering axial effects. Appl. Therm. Eng. **84**, 409–419 (2015)

41. Li, H., Nagano, K., Lai, Y.: A new model and solutions for a spiral heat exchanger and its experimental validation. Int. J. Heat Mass Transf. **55**(15–16), 4404–4414 (2012)

42. Jones, F.R.: Closed-loop geothermal systems : slinky installation guide. eng. Stillwater, Okl.: Distributed by International Ground Source Heat Pump Association (1994). isbn: 0929974042

43. Fujii, H., et al.: Numerical modeling of slinky-coil horizontal ground heat exchangers. Geothermics **41**, 55–62 (2012)

44. Lamarche, L.: Horizontal ground heat exchangers modelling. Appl. Therm. Eng. **155**, 534–545 (2019)

See Figs. A.1, A.2, A.3, A.4, A.5, A.6, A.7, A.8, A.9, A.10, A.11, A.12, A.13, A.14, A.15, A.16, A.17, A.18, A.19, and A.20.

Variable Speed ECM Motor

AHRI/ASHRAE/ISO 13256-1
English (IP) Units

Model	Capacity Modulation	Airflow Clg/Htg	Water Loop Heat Pump				Ground Water Heat Pump				Ground Loop Heat Pump			
			Cooling EWT 86°F		Heating EWT 68°F		Cooling EWT 59°F		Heating EWT 50°F		Cooling Full Load 77°F Part Load 68°F		Heating Full Load 32°F Part Load 41°F	
		cfm	Capacity Btu/h	EER Btu/h per W	Capacity Btu/h	COP	Capacity Btu/h	EER Btu/h per W	Capacity Btu/h	COP	Capacity Btu/h	EER Btu/h per W	Capacity Btu/h	COP
036	Full	1300/1500	32,000	18.0	50,000	5.3	38,000	31.5	41,000	4.6	36,000	22.0	32,000	3.5
	Part		11,000	21.0	17,000	7.5	13,000	47.2	14,000	5.9	14,000	37.0	13,000	5.3
048	Full	1500/1800	41,000	17.6	67,000	5.0	49,000	31.7	55,000	4.3	46,000	21.7	43,000	3.6
	Part		16,000	22.5	24,000	7.6	19,200	53.2	19,000	5.9	19,000	41.0	16,000	5.3
060	Full	1800/2200	50,000	16.3	78,000	4.8	60,000	28.6	65,000	4.3	56,000	19.4	51,000	3.5
	Part		20,000	21.7	29,000	7.5	23,200	45.8	23,000	6.0	23,000	36.0	20,000	5.1

Fig. A.1 AHRI/ISO 13256-1 performance rating example [1]

ASHRAE/AHRI/ISO 13256-1. English (I-P) Units

Model	Fan Motor	Water Loop Heat Pump				Ground Water Heat Pump				Ground Loop Heat Pump			
		Cooling 86°F		Heating 68°F		Cooling 59°F		Heating 50°F		Cooling 77°F		Heating 32°F	
		Capacity Btuh	EER Btuh/W	Capacity Btuh	COP	Capacity Btuh	EER Btuh/W	Capacity Btuh	COP	Capacity Btuh	EER Btuh/W	Capacity Btuh	COP
TSH/V006	PSC	6,300	15.7	8,000	5.4	7,400	25.5	6,300	4.4	6,700	18.5	4,800	3.4
TSH/V009	PSC	9,300	15.3	11,100	4.8	11,100	25.2	9,400	4.3	10,000	18.1	7,100	3.4
TSH/V012	PSC	11,700	15.4	13,800	4.5	13,300	24.6	11,800	4.0	12,300	18.1	9,500	3.5
TSH/V/D 018	PSC	18,600	15.0	23,000	5.2	21,300	24.8	18,600	4.5	19,500	18.4	14,500	3.6
	ECM	19,200	16.5	23,300	5.9	22,100	26.3	18,900	4.9	20,200	19.4	14,500	3.9
TSH/V/D 024	PSC	23,800	16.9	30,800	5.9	26,500	26.4	24,100	5.0	25,300	18.9	19,200	4.0
	ECM	23,900	17.9	30,400	6.1	26,900	28.5	23,800	5.2	25,500	20.9	19,100	4.2
TSH/V/D 030	PSC	28,000	16.3	35,500	5.5	31,000	24.9	28,400	4.7	29,200	18.5	22,200	3.9
	ECM	28,000	17.3	35,100	5.8	30,800	26.7	28,000	4.9	29,200	19.4	22,000	4.1
TSH/V/D 036	PSC	33,400	17.0	40,400	5.6	35,400	23.2	33,200	4.7	34,300	19.2	25,900	4.1
	ECM	33,500	18.1	39,900	5.9	35,400	24.9	32,600	4.9	34,600	20.4	25,600	4.3
TSH/V/D 042	PSC	38,500	17.2	46,300	5.6	43,700	25.4	36,200	4.7	40,100	19.4	28,700	3.8
	ECM	39,400	19.6	45,100	6.0	44,400	29.5	35,200	5.2	40,700	21.9	27,400	4.1
TSH/V/D 048	PSC	47,100	14.8	58,000	4.7	53,600	21.3	47,500	4.0	49,600	16.8	36,600	3.4
	ECM	48,900	17.2	57,700	5.2	53,700	23.9	45,700	4.4	50,600	18.8	36,100	3.7
TSH/V/D 060	PSC	62,400	15.9	73,900	5.0	68,500	23.1	58,200	4.2	63,900	17.2	46,900	3.7
	ECM	63,200	17.2	73,200	5.4	68,900	24.9	58,200	4.6	64,400	18.4	46,400	3.9
TSH/V/D 070	PSC	71,000	14.6	82,100	4.6	78,000	21.1	66,100	4.0	72,900	16.2	53,800	3.4
	ECM	71,100	15.7	82,000	4.8	78,100	23.0	65,200	4.1	73,000	17.2	53,000	3.6

Cooling capacities based upon 80.6°F DB, 66.2°F WB entering air temperature
Heating capacities based upon 68°F DB, 59°F WB entering air temperature
All ratings based upon operation at lower voltage of dual voltage rated models

Fig. A.2 AHRI/ISO 13256-1 performance rating example [2]

Air Flow Corrections (Compressor Speeds 4-12)

Airflow		Cooling				Heating		
CFM Per Ton of Clg	% of Nominal	Total Cap	Sens Cap	Power	Heat of Rej	Htg Cap	Power	Heat of Ext
240	60	0.928	0.747	0.936	0.929	0.961	1.097	0.938
275	69	0.940	0.789	0.946	0.941	0.967	1.081	0.948
300	75	0.952	0.831	0.957	0.953	0.974	1.064	0.959
325	81	0.964	0.873	0.968	0.965	0.980	1.048	0.969
350	88	0.976	0.916	0.979	0.976	0.987	1.032	0.979
375	94	0.988	0.958	0.989	0.988	0.993	1.016	0.990
400	**100**	**1.000**	**1.000**	**1.000**	**1.000**	**1.000**	**1.000**	**1.000**
425	106	1.020	1.023	1.004	1.018	1.010	0.966	1.018
450	113	1.056	1.042	1.008	1.035	1.020	0.932	1.036
475	119	1.072	1.079	1.011	1.053	1.029	0.898	1.054
500	125	1.087	1.095	1.015	1.070	1.039	0.865	1.071
520	130	1.099	1.113	1.019	1.088	1.049	0.831	1.089

Fig. A.3 Air flow correction factors [1]

TS024-070 with ECM Fan Motor

Airflow	Cooling					Heating		
% of Rated	Total Capacity	Sensible Capacity	S/T	Power	Heat of Rejection	Heating Capacity	Power	Heat of Extraction
72	0.925	0.850	0.919	0.951	0.950	0.957	1.124	0.942
80	0.954	0.903	0.946	0.966	0.968	0.973	1.072	0.963
88	0.974	0.941	0.966	0.977	0.982	0.984	1.037	0.979
96	0.992	0.981	0.989	0.992	0.995	0.995	1.010	0.994
100	1.000	1.000	1.000	1.000	1.000	1.000	1.000	1.000
104	1.007	1.018	1.011	1.009	1.005	1.005	0.993	1.006
112	1.017	1.052	1.035	1.027	1.013	1.012	0.986	1.015
120	1.023	1.082	1.058	1.047	1.019	1.019	0.990	1.022

Fig. A.4 Air flow correction factors [2]

Heating Capacity Corrections

Ent Air DB °F	Heating Corrections		
	Htg Cap	Power	Heat of Ext
45	1.062	0.739	1.158
50	1.050	0.790	1.130
55	1.037	0.842	1.096
60	1.025	0.893	1.064
65	1.012	0.945	1.030
68	1.005	0.976	1.012
70	**1.000**	**1.000**	**1.000**
75	0.987	1.048	0.970
80	0.975	1.099	0.930

Fig. A.5 Heating capacity correction factors [1]

Fig. A.6 Heating capacity
correction factors [2]

Unit Sizes 024-070

Heating			
Entering Air DB °F	Heating Capacity	Power	Heat of Extraction
45	1.032	0.777	1.089
50	1.029	0.817	1.077
55	1.025	0.859	1.062
60	1.018	0.903	1.044
65	1.010	0.950	1.024
70	1.000	1.000	1.000
75	0.988	1.052	0.974
80	0.974	1.107	0.944

Cooling Capacity Corrections

Entering Air WB °F	Total Clg Cap	Sensible Cooling Capacity Multipliers - Entering DB °F										Power Input	Heat of Rejection
		60	65	70	75	80	80.6	85	90	95	100		
55	0.898	0.723	0.866	1.048	1.185	*	*	*	*	*	*	0.985	0.913
60	0.912		0.632	0.880	1.078	1.244	1.260	*	*	*	*	0.994	0.927
63	0.945			0.768	0.960	1.150	1.175	*	*	*	*	0.996	0.954
65	0.976			0.694	0.881	1.079	1.085	1.270	*	*	*	0.997	0.972
66.2	0.983			0.655	0.842	1.040	1.060	1.232	*	*	*	0.999	0.986
67	**1.000**			0.616	0.806	**1.000**	1.023	1.193	1.330	1.480	*	**1.000**	**1.000**
70	1.053				0.693	0.879	0.900	1.075	1.205	1.404	*	1.003	1.044
75	1.168					0.687	0.715	0.875	1.040	1.261	1.476	1.007	1.141

Fig. A.7 Cooling capacity correction factors [1]

Unit Sizes 024-070

Cooling												
Entering Air WB°F	Total Capacity	Sensible Cooling Capacity Multiplier - Entering DB °F									Power	Heat of Rejection
		60	65	70	75	80	80.6	85	90	95		
50	0.7491	0.7663	*	*	*	*	*	*	*	*	0.9894	0.8389
55	0.8265	0.5937	0.8724	1.0816	*	*	*	*	*	*	0.9927	0.8886
60	0.9040		0.6709	0.8826	1.1211	*	*	*	*	*	0.9959	0.9383
65	0.9814			0.6624	0.8850	1.0986	1.1140	*	*	*	0.9992	0.9881
66.2	1.0000			0.6065	0.8268	1.0394	1.0536	1.2294	*	*	1.0000	1.0000
67	1.0124			0.5685	0.7879	1.0000	1.0133	1.1891	1.3838	*	1.0005	1.0080
70	1.0589	Operation not recommended			0.6391	0.8521	0.8599	1.0361	1.2347	1.4461	1.0025	1.0378
75	1.1363				0.6056	0.5981	0.7783	0.9861	1.2256		1.0058	1.0875

* = Sensible capacity equals total capacity
AHRI/ISO/ASHRAE 13256-1 uses entering air conditions of Cooling - 80.6°F DB/66.2°F WB, 1
and Heating - 68°F DB/59°F WB entering air temperature

Fig. A.8 Cooling capacity correction factors [2]

Antifreeze Type	Antifreeze % by wt	Heating	Cooling	Pressure Drop
EWT - °F [°C]		30 [-1.1]	90 [32.2]	30 [-1.1]
Water	0	1.000	1.000	1.000
Ethylene Glycol	10	0.973	0.991	1.075
	20	0.943	0.979	1.163
	30	0.917	0.965	1.225
	40	0.890	0.955	1.324
	50	0.865	0.943	1.419
Propylene Glycol	10	0.958	0.981	1.130
	20	0.913	0.969	1.270
	30	0.854	0.950	1.433
	40	0.813	0.937	1.614
	50	0.770	0.922	1.816
Ethanol	10	0.927	0.991	1.242
	20	0.887	0.972	1.343
	30	0.856	0.947	1.383
	40	0.815	0.930	1.523
	50	0.779	0.911	1.639
Methanol	10	0.957	0.986	1.127
	20	0.924	0.970	1.197
	30	0.895	0.951	1.235
	40	0.863	0.936	1.323
	50	0.833	0.920	1.399

WARNING: Gray area represents antifreeze concentrations greater than 35% by weight and should be avoided due to the extreme performance penalty they represent.

Fig. A.9 Antifreeze capacity correction factors [1]

Antifreeze Type	Antifreeze %	Cooling EWT 90°F			Heating EWT 30°F		WPD Corr. Fct. EWT 30°F
		Total Cap	Sens Cap	Power	Htg Cap	Power	
Water	0	1.000	1.000	1.000	1.000	1.000	1.000
Propylene Glycol	5	0.995	0.995	1.003	0.989	0.997	1.070
	15	0.986	0.986	1.009	0.968	0.990	1.210
	25	0.978	0.978	1.014	0.947	0.983	1.360
Methanol	5	0.997	0.997	1.002	0.989	0.997	1.070
	15	0.990	0.990	1.007	0.968	0.990	1.160
	25	0.982	0.982	1.012	0.949	0.984	1.220
Ethanol	5	0.998	0.998	1.002	0.981	0.994	1.140
	15	0.994	0.994	1.005	0.944	0.983	1.300
	25	0.986	0.986	1.009	0.917	0.974	1.360
Ethylene Glycol	5	0.998	0.998	1.002	0.993	0.998	1.040
	15	0.994	0.994	1.004	0.980	0.994	1.120
	25	0.988	0.988	1.008	0.966	0.990	1.200

Fig. A.10 Antifreeze capacity correction factors [2]

NV036 - 100% Full Load

HEATING - EAT 70°F

EWT °F	Flow gpm	WPD psi	WPD ft	Airflow cfm	HC MBtu/h	Power kW	HE MBtu/h	LAT °F	COP	HWC MBtu/h
20	5.5	1.09	2.5	Operation not recommended						
	8.0	2.69	6.2							
	11.5	3.51	8.1	1150	31.6	2.84	21.9	95.4	3.26	4.7
				1500	32.6	2.87	22.8	90.1	3.32	4.3
30	5.5	1.06	2.5	1150	35.3	2.74	25.9	98.4	3.77	5.2
				1500	36.0	2.79	26.4	92.2	3.78	4.8
	8.0	2.61	6.0	1150	35.6	2.77	26.1	98.6	3.77	5.3
				1500	36.3	2.82	26.7	92.4	3.77	4.8
	11.5	3.41	7.9	1150	36.1	2.92	26.1	99.1	3.62	5.4
				1500	37.4	2.89	27.5	93.1	3.79	5.0
40	5.5	1.02	2.4	1150	40.9	2.73	31.6	103.0	4.40	5.6
				1500	41.5	2.77	32.1	95.6	4.39	5.0
	8.0	2.53	5.9	1150	41.3	2.74	32.0	103.3	4.42	5.8
				1500	42.3	2.80	32.8	96.1	4.43	5.1
	11.5	3.31	7.6	1150	42.5	2.80	32.9	104.2	4.44	5.9
				1500	43.3	2.85	33.6	96.7	4.46	5.2
50	5.5	0.99	2.3	1150	46.0	2.85	36.2	107.0	4.73	6.0
				1500	46.4	2.88	36.6	98.7	4.72	5.6
	8.0	2.46	5.7	1150	46.4	2.84	36.7	107.4	4.79	6.5
				1500	47.6	2.91	37.7	99.4	4.80	5.7
	11.5	3.20	7.4	1150	47.5	2.93	37.5	108.2	4.74	6.6
				1500	48.5	2.94	38.5	99.9	4.83	6.8
60	5.5	0.95	2.2	1150	50.8	2.87	41.0	110.9	5.18	6.1
				1500	51.6	2.89	41.7	101.9	5.23	5.7
	8.0	2.38	5.5	1150	52.0	2.88	42.2	111.9	5.29	6.7
				1500	52.8	2.92	42.8	102.6	5.29	5.8
	11.5	3.10	7.2	1150	53.2	2.93	43.2	112.8	5.31	6.9
				1500	54.3	2.97	44.1	103.5	5.36	6.0
70	5.5	0.90	2.1	1150	54.9	2.95	44.8	114.2	5.45	6.6
				1500	56.4	3.00	46.2	104.8	5.51	6.2
	8.0	2.30	5.3	1150	56.3	3.02	46.6	115.8	5.52	6.9
				1500	57.9	2.95	47.8	105.7	5.75	6.4
	11.5	3.00	6.9	1150	58.7	3.08	48.2	117.3	5.58	7.6
				1500	60.0	3.01	49.7	107.0	5.84	6.6
80	5.5	0.87	0.1	1150	60.2	3.15	49.4	118.4	5.59	7.2
				1500	61.3	3.16	50.5	107.8	5.68	6.1
	8.0	2.22	5.1	1150	63.0	3.18	52.1	120.7	5.80	7.5
				1500	63.4	3.17	52.6	109.1	5.85	6.3
	11.5	2.90	6.7	1150	63.9	3.23	52.8	121.4	5.79	7.7
				1500	65.1	3.25	54.0	110.2	5.87	7.2
90	5.5	0.84	1.9	1150	64.7	3.35	51.3	122.1	5.66	7.4
				1500	65.8	3.29	54.5	110.6	5.86	6.7
	8.0	2.14	5.0	1150	68.2	3.41	56.5	124.9	5.86	7.8
				1500	69.0	3.33	57.6	112.6	6.07	7.2
	11.5	2.79	6.5	1150	68.5	3.48	56.6	125.1	5.76	8.6
				1500	70.3	3.42	58.6	113.4	6.01	7.9
100	5.5	0.80	1.8	Operation not recommended						
	8.0	2.07	4.8							
	11.5	2.69	6.2							
110	5.5	0.77	1.8							
	8.0	1.99	4.6							
	11.5	2.59	6.0							
120	5.5	0.73	1.7							
	8.0	1.91	4.4							
	11.5	2.49	5.8							

COOLING - EAT 80/67°F

EWT °F	Flow gpm	WPD psi	WPD ft	Airflow cfm	TC MBtu/h	SC MBtu/h	S/T Ratio	Power kW	HR MBtu/h	EER	HWC MBtu/h
20	4.5	0.70	1.6	Operation not recommended							
	7.0	1.70	3.9								
	9.0	2.81	6.5								
30	4.5	0.68	1.6	1000	39.2	27.3	0.70	0.98	42.6	40.1	---
				1300	38.5	27.4	0.71	1.09	42.2	35.2	---
	7.0	1.65	3.8	1000	39.2	27.3	0.70	0.95	42.5	41.5	---
				1300	38.0	28.1	0.74	1.14	41.9	33.2	---
	9.0	2.73	6.3	1000	39.3	27.5	0.70	0.95	42.6	41.5	---
				1300	38.6	27.8	0.72	1.07	42.3	36.1	---
40	4.5	0.66	1.5	1000	41.1	29.6	0.72	1.10	44.9	37.4	---
				1300	41.0	30.3	0.74	1.23	45.2	33.3	---
	7.0	1.60	3.7	1000	41.2	29.6	0.72	1.06	44.8	38.8	---
				1300	40.8	30.9	0.76	1.24	45.0	32.8	---
	9.0	2.65	6.1	1000	41.5	29.8	0.72	1.06	45.1	39.1	---
				1300	41.3	30.7	0.74	1.18	45.3	35.1	---
50	4.5	0.64	1.5	1000	44.0	32.3	0.73	1.25	48.3	35.1	---
				1300	44.2	33.6	0.76	1.40	48.9	31.5	---
	7.0	1.55	3.6	1000	44.1	32.4	0.73	1.21	48.2	36.5	---
				1300	44.2	33.9	0.77	1.38	48.8	32.1	---
	9.0	2.56	5.9	1000	44.5	32.6	0.73	1.20	48.6	37.1	---
				1300	44.6	33.9	0.76	1.31	49.3	34.0	---
60	4.5	0.62	1.4	1000	40.3	30.9	0.77	1.45	45.3	27.8	1.9
				1300	41.3	32.8	0.79	1.59	46.7	25.9	2.0
	7.0	1.50	3.5	1000	40.5	31.0	0.77	1.40	45.2	28.9	2.1
				1300	41.5	33.0	0.79	1.55	46.7	26.8	2.1
	9.0	2.48	5.7	1000	40.9	31.2	0.76	1.39	45.6	29.5	2.0
				1300	41.7	33.1	0.79	1.49	46.7	28.0	2.2
70	4.5	0.60	1.4	1000	38.9	30.3	0.78	1.68	44.6	23.2	2.7
				1300	36.8	29.7	0.81	1.84	43.1	20.0	2.9
	7.0	1.45	3.4	1000	39.1	30.5	0.78	1.62	44.6	24.2	3.1
				1300	41.0	33.0	0.80	1.74	47.0	23.6	3.1
	9.0	2.40	5.5	1000	36.2	26.0	0.72	1.62	41.7	22.3	3.2
				1300	41.0	33.3	0.81	1.69	46.8	24.3	3.3
80	4.5	0.57	1.3	1000	37.8	30.7	0.81	1.95	44.5	19.4	4.0
				1300	38.8	32.8	0.85	2.08	45.9	18.7	4.0
	7.0	1.40	3.2	1000	38.1	30.9	0.81	1.88	44.5	20.3	4.2
				1300	39.2	32.8	0.84	1.98	46.0	19.8	4.2
	9.0	2.32	5.4	1000	38.5	31.2	0.81	1.85	44.8	20.8	4.4
				1300	39.4	32.2	0.84	1.94	46.0	20.5	4.4
90	4.5	0.55	1.3	1000	34.2	29.4	0.86	2.25	41.9	15.2	5.3
				1300	34.5	30.9	0.90	2.39	42.6	14.5	5.3
	7.0	1.35	3.1	1000	34.5	29.7	0.86	2.16	41.9	15.9	5.6
				1300	34.9	32.2	0.89	2.26	42.5	15.4	5.6
	9.0	2.24	5.2	1000	34.8	30.0	0.86	2.12	42.1	16.4	5.9
				1300	35.2	31.2	0.89	2.19	42.7	16.1	5.6
100	4.5	0.53	1.2	1000	33.6	28.9	0.86	2.61	42.5	12.9	6.8
				1300	33.9	30.3	0.90	2.69	43.0	12.6	6.7
	7.0	1.30	3.0	1000	33.9	29.3	0.87	2.50	42.4	13.5	7.1
				1300	34.2	30.5	0.89	2.56	42.9	13.4	7.0
	9.0	2.16	5.0	1000	34.2	29.6	0.87	2.44	42.5	14.0	7.5
				1300	34.5	30.9	0.89	2.49	43.0	13.9	6.2
110	4.5	0.51	1.2	1000	29.7	26.7	0.90	3.01	40.0	9.9	8.1
				1300	30.0	27.8	0.93	3.04	40.4	9.9	8.0
	7.0	1.25	2.9	1000	30.0	27.1	0.90	2.88	39.8	10.4	8.5
				1300	30.3	28.2	0.93	2.91	40.2	10.4	8.5
	9.0	2.07	4.8	1000	30.6	27.5	0.99	2.81	37.4	9.9	8.5
				1300	30.6	28.6	0.93	2.84	40.3	10.8	8.1
120	4.5	0.49	0.8	1000	28.0	25.7	0.92	3.40	39.6	8.2	9.9
				1300	27.6	26.0	0.95	3.31	38.9	8.3	10.2
	7.0	1.20	2.8	1000	28.3	26.1	0.92	3.26	39.4	8.7	10.3
				1300	27.7	26.5	0.96	3.27	38.9	8.5	10.5
	9.0	1.99	4.6	1000	28.4	26.6	0.94	3.17	39.2	8.9	10.7
				1300	28.2	26.8	0.95	3.17	39.0	8.9	10.7

Fig. A.11 Water-to-air real performance data [1]

NV048 - 100% Full Load

HEATING - EAT 70°F

EWT °F	Flow gpm	WPD psi	WPD ft	Airflow cfm	HC MBtu/h	Power kW	HE MBtu/h	LAT °F	COP	HWC MBtu/h
20	6.5	1.40	3.2	Operation not recommended						
	10.0	2.85	6.6							
	13.5	4.79	11.1	1500	40.6	3.67	28.1	95.1	3.24	6.05
				1800	41.6	3.72	28.9	91.4	3.28	5.6
30	6.5	1.36	3.2	1500	44.9	3.86	31.7	97.7	3.41	6.6
				1800	45.8	3.93	32.4	93.5	3.41	6.1
	10.0	2.77	6.4	1500	45.3	3.90	32.0	98.0	3.40	6.8
				1800	46.2	3.97	32.6	93.8	3.41	6.1
	13.5	4.65	10.7	1500	45.9	4.11	31.9	98.4	3.27	6.9
				1800	47.6	4.07	33.7	94.5	3.43	6.3
40	6.5	1.32	3.1	1500	49.7	3.94	36.3	100.7	3.70	7.0
				1800	50.4	4.00	36.8	95.9	3.70	6.2
	10.0	2.69	6.2	1500	50.2	3.95	36.7	101.0	3.72	7.3
				1800	51.4	4.04	37.6	96.4	3.73	6.4
	13.5	4.51	10.4	1500	51.6	4.04	37.8	101.8	3.74	7.5
				1800	52.6	4.11	38.6	97.1	3.75	6.5
50	6.5	1.28	3.0	1500	58.0	4.04	44.2	105.8	4.20	7.5
				1800	58.6	4.09	44.6	100.1	4.20	7.0
	10.0	2.60	6.0	1500	58.6	4.03	44.8	106.2	4.26	8.2
				1800	60.1	4.13	46.0	100.9	4.27	7.2
	13.5	4.37	10.1	1500	59.9	4.16	45.7	107.0	4.22	8.4
				1800	61.2	4.17	47.0	101.5	4.30	8.6
60	6.5	1.24	2.9	1500	63.2	3.99	49.6	109.0	4.64	7.6
				1800	64.2	4.03	50.5	103.0	4.67	7.1
	10.0	2.52	5.8	1500	64.7	4.01	51.0	110.0	4.73	8.4
				1800	65.7	4.07	51.8	103.8	4.73	7.2
	13.5	4.23	9.8	1500	66.2	4.08	52.2	110.8	4.75	8.6
				1800	67.5	4.13	53.4	104.7	4.79	7.4
70	6.5	1.20	2.8	1500	67.6	4.02	53.9	111.7	4.93	8.1
				1800	68.3	3.94	54.9	105.1	5.08	7.5
	10.0	2.44	5.6	1500	71.3	4.00	57.7	114.0	5.23	8.6
				1800	71.6	4.02	57.9	106.9	5.23	7.9
	13.5	4.09	9.5	1500	72.6	4.61	56.9	114.8	4.62	9.4
				1800	74.3	4.10	60.3	108.2	5.31	8.2
80	6.5	1.16	2.7	1500	74.9	4.01	61.3	116.3	5.48	9.0
				1800	76.3	4.01	62.6	109.3	5.57	7.7
	10.0	2.35	5.4	1500	78.3	4.08	64.4	118.3	5.63	9.5
				1800	79.0	4.03	65.2	110.6	5.74	7.9
	13.5	3.95	9.1	1500	79.6	4.13	65.5	119.1	5.65	9.6
				1800	81.1	4.16	67.0	111.7	5.72	9.0
90	6.5	1.12	2.6	1500	81.0	4.04	67.3	120.0	5.88	9.7
				1800	82.3	4.03	68.6	112.4	5.99	8.6
	10.0	2.27	5.2	1500	86.3	4.13	71.2	122.7	6.05	9.5
				1800	86.3	4.03	72.6	114.4	6.28	8.9
	13.5	3.81	8.8	1500	85.7	4.35	70.9	122.9	5.77	10.4
				1800	88.0	4.20	73.7	115.3	6.14	9.7
100	6.5	1.08	2.5	Operation not recommended						
	10.0	2.19	5.1							
	13.5	3.67	8.5							
110	6.5	1.04	2.4							
	10.0	2.11	4.9							
	13.5	3.53	8.2							
120	6.5	1.00	2.3							
	10.0	2.02	4.7							
	13.5	3.39	7.8							

COOLING - EAT 80/67°F

EWT °F	Flow gpm	WPD psi	WPD ft	Airflow cfm	TC MBtu/h	SC MBtu/h	S/T Ratio	Power kW	HR MBtu/h	EER	HWC MBtu/h
20	5.5	1.05	2.4	Operation not recommended							
	8.0	2.00	4.6								
	10.5	2.94	6.8								
30	5.5	1.02	2.4	1000	52.6	36.3	0.69	1.21	56.7	43.4	—
				1400	51.5	36.3	0.70	1.35	56.1	38.2	—
	8.0	1.94	4.5	1000	52.5	36.2	0.69	1.17	56.5	45.0	—
				1400	50.9	37.3	0.73	1.41	55.7	36.0	—
	10.5	2.85	6.6	1000	52.7	36.5	0.69	1.17	56.7	45.0	—
				1400	51.7	36.9	0.71	1.32	56.2	39.2	—
40	5.5	0.99	2.3	1000	53.8	37.2	0.69	1.40	58.6	38.5	—
				1400	53.7	38.1	0.71	1.56	59.0	34.3	—
	8.0	1.88	4.3	1000	53.9	37.2	0.69	1.35	58.5	39.9	—
				1400	53.4	38.8	0.73	1.58	58.8	33.8	—
	10.5	2.77	6.4	1000	54.3	37.4	0.69	1.35	58.9	40.3	—
				1400	54.0	38.6	0.71	1.49	59.1	36.2	—
50	5.5	0.96	2.2	1000	56.7	38.9	0.69	1.64	62.3	34.7	—
				1400	56.9	40.4	0.71	1.83	63.2	31.1	—
	8.0	1.82	4.2	1000	56.9	39.0	0.69	1.58	62.2	36.0	—
				1400	56.9	40.8	0.72	1.80	63.1	31.7	—
	10.5	2.68	6.2	1000	57.4	39.2	0.68	1.57	62.8	36.6	—
				1400	57.5	40.8	0.71	1.71	63.3	33.6	—
60	5.5	0.93	2.2	1000	51.2	36.5	0.71	1.88	57.7	27.2	2.6
				1400	52.4	38.8	0.74	2.06	59.4	25.4	2.8
	8.0	1.76	4.1	1000	51.4	36.7	0.71	1.81	57.6	28.4	2.9
				1400	52.7	39.0	0.74	2.00	59.5	26.3	2.9
	10.5	2.60	6.0	1000	51.9	36.9	0.71	1.80	58.0	28.9	2.8
				1400	52.9	39.2	0.74	1.93	59.5	27.5	3.0
70	5.5	0.90	2.1	1000	48.8	35.0	0.72	2.22	56.4	22.0	3.4
				1400	47.6	34.4	0.72	2.52	56.2	18.9	3.7
	8.0	1.71	3.9	1000	49.0	35.2	0.72	2.13	56.3	23.0	3.9
				1400	51.5	38.0	0.74	2.30	59.3	22.4	3.9
	10.5	2.51	5.8	1000	44.0	29.9	0.68	2.17	51.4	20.3	3.9
				1400	51.5	38.4	0.75	2.23	59.1	23.1	4.1
80	5.5	0.87	2.0	1000	47.5	35.2	0.74	2.56	56.2	18.5	5.4
				1400	48.7	37.6	0.77	2.73	58.1	17.8	5.4
	8.0	1.65	3.8	1000	47.8	35.5	0.74	2.46	56.2	19.4	5.7
				1400	49.2	37.6	0.76	2.60	58.1	18.9	5.7
	10.5	2.42	5.6	1000	48.3	35.8	0.74	2.42	56.5	19.9	5.9
				1400	49.5	38.0	0.77	2.53	58.1	19.6	5.9
90	5.5	0.84	1.9	1000	42.7	33.4	0.78	2.94	52.8	14.5	6.7
				1400	43.0	35.1	0.82	3.12	53.7	13.8	6.6
	8.0	1.59	3.7	1000	43.0	33.7	0.78	2.83	52.7	15.2	7.0
				1400	43.5	35.1	0.81	2.95	53.5	14.8	7.0
	10.5	2.34	5.4	1000	43.5	34.1	0.78	2.77	52.9	15.7	7.3
				1400	43.9	35.5	0.81	2.86	53.7	15.3	7.4
100	5.5	0.81	1.9	1000	42.4	32.9	0.78	3.45	54.2	12.3	8.9
				1400	42.8	34.5	0.81	3.55	54.9	12.0	8.8
	8.0	1.53	3.5	1000	42.8	33.3	0.78	3.30	54.1	12.9	9.3
				1400	43.2	34.7	0.80	3.39	54.8	12.8	9.2
	10.5	2.25	5.2	1000	43.2	33.7	0.78	3.23	54.2	13.4	9.8
				1400	43.6	35.1	0.80	3.29	54.9	13.2	8.2
110	5.5	0.78	1.8	1000	38.8	31.7	0.82	3.93	52.2	9.9	10.6
				1400	39.2	33.0	0.84	3.97	52.7	9.9	10.5
	8.0	1.47	3.4	1000	39.2	32.1	0.82	3.77	52.1	10.4	11.1
				1400	39.6	33.5	0.85	3.80	52.6	10.4	11.1
	10.5	2.17	5.0	1000	39.6	32.5	0.82	3.67	52.1	10.8	11.6
				1400	40.0	33.9	0.85	3.71	52.7	10.8	10.5
120	5.5	0.75	1.7	1000	36.1	29.1	0.80	4.57	51.7	7.9	12.8
				1400	35.6	29.5	0.83	4.45	50.8	8.0	13.1
	8.0	1.42	3.3	1000	36.5	29.5	0.81	4.38	51.5	8.3	13.2
				1400	35.8	30.1	0.84	4.39	50.8	8.2	13.6
	10.5	2.08	4.8	1000	36.6	30.1	0.82	4.26	51.2	8.6	13.8
				1400	36.4	30.4	0.83	4.26	50.9	8.5	14.0

Performance capacities shown in thousands of Btu/h

6/28/12

Fig. A.12 Water-to-air real performance data [1]

NV060 - 100% Full Load

HEATING - EAT 70°F

EWT °F	Flow gpm	WPD psi	WPD ft	Airflow cfm	HC MBtu/h	Power kW	HE MBtu/h	LAT °F	COP	HWC MBtu/h
20	8.5	2.00	4.6	Operation not recommended						
	13.0	3.52	8.1							
	17.0	6.55	15.1	1800	51.4	5.23	33.6	96.5	2.88	8.1
				2200	52.4	5.27	34.4	92.3	2.91	7.5
30	8.5	1.95	4.5	1800	56.8	5.07	39.5	99.2	3.28	8.4
				2200	57.9	5.16	40.3	94.4	3.29	7.8
	13.0	3.42	7.9	1800	57.3	5.12	39.8	99.5	3.28	8.6
				2200	58.4	5.21	40.6	94.6	3.28	7.9
	17.0	6.36	14.7	1800	58.1	5.40	39.7	99.9	3.15	8.8
				2200	60.2	5.35	42.0	95.3	3.30	8.0
40	8.5	1.90	4.4	1800	64.8	5.16	47.2	103.3	3.68	8.9
				2200	65.7	5.24	47.8	97.7	3.68	8.2
	13.0	3.32	7.7	1800	65.4	5.18	47.7	103.7	3.70	9.2
				2200	66.9	5.29	48.9	98.2	3.71	8.4
	17.0	6.17	14.3	1800	67.2	5.30	49.1	104.6	3.72	9.5
				2200	68.5	5.38	50.2	98.8	3.73	8.6
50	8.5	1.85	4.3	1800	72.2	5.28	54.2	107.1	4.01	9.5
				2200	72.9	5.34	54.7	100.7	4.00	8.6
	13.0	3.21	7.4	1800	73.0	5.27	55.0	107.5	4.06	9.9
				2200	74.9	5.39	56.4	101.5	4.07	8.9
	17.0	5.98	13.8	1800	74.6	5.44	56.0	108.4	4.02	10.3
				2200	76.2	5.45	57.6	102.1	4.10	9.3
60	8.5	1.80	4.2	1800	80.8	5.43	62.3	111.6	4.37	10.1
				2200	82.1	5.47	63.5	104.6	4.40	9.0
	13.0	3.11	7.2	1800	82.8	5.45	64.2	112.6	4.45	10.6
				2200	84.0	5.53	65.1	105.3	4.45	9.5
	17.0	5.79	13.4	1800	84.6	5.55	65.7	113.5	4.47	11.1
				2200	86.4	5.61	67.2	106.3	4.51	9.9
70	8.5	1.75	4.0	1800	88.4	5.68	69.0	115.5	4.56	10.6
				2200	89.9	5.63	70.7	107.8	4.68	9.5
	13.0	3.01	7.0	1800	93.2	5.65	73.9	117.9	4.83	11.2
				2200	93.6	5.68	74.2	109.4	4.83	10.0
	17.0	5.60	12.9	1800	95.0	5.91	74.8	118.8	4.71	11.9
				2200	97.1	5.80	77.3	110.9	4.91	10.6
80	8.5	1.68	3.9	1800	97.6	5.74	78.0	120.2	4.98	11.8
				2200	99.4	5.75	79.8	111.8	5.06	9.9
	13.0	2.91	6.7	1800	102.0	5.84	82.1	122.5	5.12	12.2
				2200	102.9	5.78	83.2	113.3	5.21	10.6
	17.0	5.41	12.5	1800	103.7	6.02	83.1	123.3	5.04	12.6
				2200	105.7	5.96	85.4	114.5	5.19	11.2
90	8.5	1.60	3.7	1800	106.4	5.99	85.9	124.7	5.20	11.7
				2200	108.1	5.93	87.8	115.5	5.34	10.3
	13.0	2.80	6.5	1800	112.0	6.08	91.2	127.6	5.40	12.6
				2200	113.3	5.93	93.1	117.7	5.60	11.1
	17.0	5.22	12.1	1800	112.5	6.40	90.6	127.9	5.15	13.4
				2200	115.5	6.18	94.4	118.6	5.48	11.9
100	8.5	1.55	3.6	Operation not recommended						
	13.0	2.70	6.2							
	17.0	5.03	11.6							
110	8.5	1.50	3.5							
	13.0	2.60	6.0							
	17.0	4.84	11.2							
120	8.5	1.40	3.2							
	13.0	2.50	5.8							
	17.0	4.65	10.7							

COOLING - EAT 80/67°F

EWT °F	Flow gpm	WPD psi	WPD ft	Airflow cfm	TC MBtu/h	SC MBtu/h	S/T Ratio	Power kW	HR MBtu/h	EER	HWC MBtu/h
20	6.5	1.21	2.8	Operation not recommended							
	10.0	2.70	6.2								
	13.5	4.20	9.7								
30	6.5	1.17	2.7	1500	69.5	46.4	0.67	1.86	75.9	37.4	---
				1800	68.1	46.4	0.68	2.08	75.2	32.8	---
	10.0	2.62	6.1	1500	69.5	46.3	0.67	1.80	75.7	38.6	---
				1800	67.3	47.7	0.71	2.18	74.7	30.9	---
	13.5	4.08	9.4	1500	69.7	46.7	0.67	1.81	75.9	38.6	---
				1800	68.4	47.2	0.69	2.04	75.3	33.6	---
40	6.5	1.14	2.6	1500	68.6	46.5	0.68	2.08	75.7	32.9	---
				1800	68.4	47.6	0.70	2.33	76.3	29.3	---
	10.0	2.55	5.9	1500	68.6	46.5	0.68	2.01	75.5	34.1	---
				1800	68.0	48.5	0.71	2.35	76.1	28.9	---
	13.5	3.96	9.1	1500	69.2	46.8	0.68	2.01	76.0	34.4	---
				1800	68.8	48.3	0.70	2.23	76.4	30.9	---
50	6.5	1.10	2.5	1500	68.6	47.0	0.69	2.36	76.6	29.1	---
				1800	68.8	48.8	0.71	2.64	77.8	26.1	---
	10.0	2.47	5.7	1500	68.7	47.1	0.69	2.27	76.5	30.2	---
				1800	68.8	49.3	0.72	2.59	77.6	26.6	---
	13.5	3.83	8.9	1500	69.4	47.4	0.68	2.26	77.1	30.7	---
				1800	69.5	49.3	0.71	2.46	77.9	28.2	---
60	6.5	1.07	2.5	1500	63.8	45.2	0.71	2.68	72.9	23.8	3.1
				1800	65.2	48.0	0.74	2.93	75.2	22.3	3.1
	10.0	2.39	5.5	1500	64.0	45.4	0.71	2.58	72.8	24.8	2.9
				1800	65.5	48.2	0.74	2.85	75.3	23.0	2.9
	13.5	3.71	8.6	1500	64.6	45.7	0.71	2.55	73.3	25.3	2.7
				1800	65.9	48.4	0.74	2.74	75.2	24.1	2.7
70	6.5	1.03	2.4	1500	60.2	44.1	0.73	3.03	70.5	19.9	4.6
				1800	62.7	47.2	0.75	3.41	74.4	18.4	4.6
	10.0	2.31	5.3	1500	60.5	44.4	0.73	2.91	70.4	20.8	4.3
				1800	63.5	47.9	0.75	3.34	74.2	20.3	4.3
	13.5	3.59	8.3	1500	62.2	44.0	0.71	2.86	71.9	21.7	4.2
				1800	63.5	48.4	0.76	3.04	73.9	20.9	4.1
80	6.5	0.99	2.3	1500	58.1	43.6	0.75	3.48	70.0	16.7	6.1
				1800	59.6	46.6	0.78	3.71	72.3	16.1	6.1
	10.0	2.23	5.2	1500	58.5	43.9	0.75	3.34	69.9	17.5	5.8
				1800	60.2	46.6	0.77	3.53	72.3	17.1	5.8
	13.5	3.47	8.0	1500	59.1	44.3	0.75	3.29	70.3	18.0	5.5
				1800	60.5	47.1	0.78	3.43	72.2	17.6	5.5
90	6.5	0.96	2.2	1500	54.1	42.2	0.78	3.98	67.7	13.6	8.2
				1800	54.5	44.5	0.82	4.22	68.9	12.9	8.2
	10.0	2.15	5.0	1500	54.5	42.7	0.78	3.83	67.6	14.2	7.8
				1800	55.0	44.5	0.81	3.99	68.7	13.8	7.8
	13.5	3.34	7.7	1500	55.0	43.1	0.78	3.75	67.9	14.7	7.3
				1800	55.6	44.9	0.81	3.87	68.8	14.4	7.4
100	6.5	0.92	2.1	1500	51.3	41.4	0.81	4.50	66.6	11.4	10.3
				1800	51.7	43.3	0.84	4.64	67.5	11.1	10.3
	10.0	2.07	4.8	1500	51.7	41.9	0.81	4.31	66.4	12.0	9.7
				1800	52.2	43.6	0.84	4.42	67.3	11.8	9.8
	13.5	3.22	7.5	1500	52.2	42.4	0.81	4.22	66.6	12.4	9.2
				1800	52.8	44.1	0.84	4.30	67.4	12.3	9.2
110	6.5	0.89	2.1	1500	46.4	39.0	0.84	5.06	63.6	9.2	13.0
				1800	46.8	40.6	0.87	5.11	64.3	9.2	13.1
	10.0	2.00	4.6	1500	46.8	39.5	0.84	4.85	63.4	9.7	12.3
				1800	47.3	41.2	0.87	4.89	64.0	9.7	12.4
	13.5	3.10	7.2	1500	47.3	40.0	0.85	4.73	63.4	10.0	11.6
				1800	47.8	41.7	0.87	4.77	64.1	10.0	11.7
120	6.5	0.85	2.0	1500	42.8	38.6	0.90	5.75	62.5	7.4	16.1
				1800	42.2	39.2	0.93	5.60	61.3	7.5	16.2
	10.0	1.92	4.4	1500	43.3	39.2	0.90	5.51	62.1	7.9	15.2
				1800	42.5	39.9	0.94	5.52	61.3	7.7	15.3
	13.5	2.98	6.9	1500	43.5	40.0	0.92	5.36	61.8	8.1	14.3
				1800	43.2	40.3	0.93	5.36	61.5	8.1	14.5

Performance capacities shown in thousands of Btu/h   6/28/12

Fig. A.13 Water-to-air real performance data [1]

Performance Data – TS H/V/D 024 (ECM Blower)

950 CFM Nominal (Rated) Airflow Cooling, 950 CFM Nominal (Rated) Airflow Heating

Performance capacities shown in thousands of Btuh

EWT °F	GPM	WPD PSI	WPD FT	Cooling - EAT 80/67°F Airflow CFM	TC	SC	Sens/ Tot Ratio	kW	HR	EER	HWC	Heating - EAT 70°F Airflow CFM	HC	kW	HE	LAT	COP	HWC
20	6.0	1.9	4.4	Operation not recommended								680	15.4	1.64	10.4	90.9	2.7	1.7
	6.0	1.9	4.4									950	16.1	1.46	11.1	85.7	3.2	1.5
30	3.0	0.9	2.1	680	28.1	16.5	0.59	0.87	31.8	32.5	0.8	680	17.1	1.65	12.1	93.3	3.0	2.0
	3.0	0.9	2.1	950	30.4	19.4	0.64	0.91	33.5	33.4	0.8	950	17.8	1.47	12.8	87.4	3.6	1.8
	4.5	1.2	2.7	680	27.5	15.6	0.57	0.80	30.9	34.4	0.7	680	17.9	1.65	12.9	94.4	3.2	2.1
	4.5	1.2	2.7	950	29.7	18.3	0.62	0.84	32.5	35.3	0.7	950	18.7	1.47	13.7	88.2	3.7	1.9
	6.0	1.7	4.0	680	26.9	14.9	0.56	0.77	30.2	34.9	0.6	680	18.4	1.65	13.3	95.0	3.3	2.1
	6.0	1.7	4.0	950	29.1	17.6	0.60	0.81	31.8	35.9	0.6	950	19.2	1.47	14.2	88.7	3.8	1.9
40	3.0	0.7	1.5	680	28.1	17.1	0.61	0.96	32.2	29.3	1.0	680	19.8	1.66	14.7	96.9	3.5	2.4
	3.0	0.7	1.5	950	30.4	20.2	0.66	1.01	33.9	30.1	1.1	950	20.7	1.48	15.6	90.1	4.1	2.1
	4.5	1.0	2.4	680	28.2	16.7	0.59	0.88	32.0	31.9	0.9	680	20.8	1.66	15.7	98.3	3.7	2.6
	4.5	1.0	2.4	950	30.5	19.7	0.65	0.93	33.7	32.8	0.9	950	21.7	1.48	16.7	91.2	4.3	2.3
	6.0	1.6	3.6	680	28.1	16.4	0.58	0.86	31.7	32.8	0.8	680	21.3	1.67	16.2	99.0	3.7	2.6
	6.0	1.6	3.6	950	30.4	19.3	0.64	0.90	33.4	33.7	0.8	950	22.3	1.49	17.2	91.7	4.4	2.3
50	3.0	0.5	1.2	680	27.4	17.2	0.63	1.07	31.8	25.5	1.4	680	22.5	1.67	17.3	100.6	3.9	2.8
	3.0	0.5	1.2	950	29.6	20.2	0.68	1.13	33.5	26.2	1.5	950	23.5	1.49	18.4	92.9	4.6	2.5
	4.5	0.9	2.1	680	28.0	17.2	0.61	0.99	32.2	28.3	1.1	680	23.7	1.69	18.5	102.2	4.1	2.9
	4.5	0.9	2.1	950	30.3	20.2	0.67	1.04	33.8	29.1	1.2	950	24.7	1.50	19.6	94.1	4.8	2.6
	6.0	1.5	3.3	680	28.2	17.1	0.61	0.95	32.2	29.6	1.0	680	24.3	1.69	19.1	103.1	4.2	3.0
	6.0	1.5	3.3	950	30.5	20.1	0.66	1.00	33.9	30.5	1.1	950	25.4	1.50	20.3	94.8	5.0	2.7
60	3.0	0.4	0.9	680	26.2	16.8	0.64	1.20	30.9	21.8	1.8	680	25.2	1.70	20.0	104.3	4.4	3.1
	3.0	0.4	0.9	950	28.3	19.7	0.70	1.26	32.6	22.5	1.9	950	26.4	1.51	21.2	95.7	5.1	2.8
	4.5	0.8	1.8	680	27.1	17.1	0.63	1.10	31.6	24.6	1.5	680	26.6	1.71	21.3	106.2	4.6	3.4
	4.5	0.8	1.8	950	29.3	20.1	0.69	1.16	33.3	25.3	1.6	950	27.8	1.52	22.6	97.1	5.4	3.0
	6.0	1.4	3.1	680	27.5	17.2	0.62	1.06	31.9	26.1	1.3	680	27.4	1.72	22.0	107.3	4.7	3.5
	6.0	1.4	3.1	950	29.8	20.2	0.68	1.11	33.6	26.8	1.4	950	28.6	1.53	23.4	97.9	5.5	3.1
70	3.0	0.3	0.6	680	24.6	16.1	0.66	1.33	29.8	18.5	2.3	680	28.0	1.72	22.6	108.1	4.8	3.6
	3.0	0.3	0.6	950	26.6	19.0	0.71	1.40	31.4	19.0	2.4	950	29.3	1.53	24.0	98.5	5.6	3.2
	4.5	0.7	1.7	680	25.8	16.6	0.65	1.23	30.6	21.0	1.9	680	29.6	1.74	24.2	110.3	5.0	3.8
	4.5	0.7	1.7	950	27.8	19.5	0.70	1.29	32.3	21.6	2.0	950	31.0	1.55	25.7	100.2	5.9	3.4
	6.0	1.3	2.9	680	26.3	16.8	0.64	1.18	31.0	22.3	1.8	680	30.5	1.75	25.0	111.6	5.1	3.9
	6.0	1.3	2.9	950	28.4	19.8	0.70	1.24	32.7	22.9	1.9	950	31.9	1.56	26.6	101.1	6.0	3.5
80	3.0	0.2	0.5	680	22.8	15.4	0.67	1.48	28.5	15.4	2.9	680	30.9	1.76	25.3	112.0	5.1	3.9
	3.0	0.2	0.5	950	24.7	18.1	0.73	1.56	30.0	15.8	3.0	950	32.2	1.57	26.9	101.4	6.0	3.5
	4.5	0.7	1.5	680	24.1	15.9	0.66	1.38	29.4	17.4	2.5	680	32.7	1.80	27.1	114.6	5.3	4.2
	4.5	0.7	1.5	950	26.0	18.7	0.72	1.45	30.9	17.9	2.6	950	34.2	1.60	28.8	103.3	6.3	3.7
	6.0	1.2	2.7	680	24.7	16.2	0.66	1.32	29.8	18.7	2.3	680	33.8	1.81	28.1	116.0	5.5	4.3
	6.0	1.2	2.7	950	26.7	19.0	0.71	1.39	31.4	19.2	2.4	950	35.3	1.61	29.8	104.4	6.4	3.8
85	3.0	0.2	0.4	680	21.9	14.9	0.68	1.6	27.8	14.0	3.2	680	32.3	1.79	26.7	114.0	5.3	4.1
	3.0	0.2	0.4	950	23.7	17.6	0.74	1.65	29.3	14.4	3.4	950	33.8	1.6	28.4	102.9	6.2	3.7
	4.5	0.6	1.5	680	23.1	15.5	0.67	1.46	28.7	15.9	2.8	680	34.4	1.8	28.6	116.8	5.5	4.3
	4.5	0.6	1.5	950	25.0	18.2	0.73	1.54	30.2	16.4	2.9	950	35.9	1.6	30.4	105.0	6.5	3.9
	6.0	1.1	2.6	680	23.7	15.8	0.66	1.40	29.2	17.0	2.6	680	35.5	1.8	29.6	118.3	5.6	4.4
	6.0	1.1	2.6	950	25.7	18.5	0.72	1.48	30.7	17.5	2.7	950	37.1	1.6	31.5	106.1	6.6	4.0
90	3.0	0.1	0.3	680	21.0	14.5	0.69	1.65	27.2	12.7	3.5	680	33.8	1.81	28.1	116.0	5.5	4.3
	3.0	0.1	0.3	950	22.7	17.1	0.75	1.74	28.6	13.0	3.7	950	35.3	1.61	29.8	104.4	6.4	3.8
	4.5	0.6	1.4	680	22.2	15.1	0.68	1.54	28.0	14.4	3.0	680	36.0	1.85	30.1	119.0	5.7	4.5
	4.5	0.6	1.4	950	24.0	17.7	0.74	1.62	29.5	14.8	3.2	950	37.6	1.65	31.9	106.6	6.7	4.0
	6.0	1.1	2.5	680	22.8	15.4	0.67	1.48	28.5	15.4	2.9	680	37.2	1.89	31.2	120.7	5.8	4.6
	6.0	1.1	2.5	950	24.7	18.1	0.73	1.56	30.0	15.8	3.0	950	38.9	1.68	33.1	107.9	6.8	4.1
100	3.0	0.1	0.2	680	19.2	13.8	0.72	1.84	26.0	10.4	4.2	Operation not recommended						
	3.0	0.1	0.2	950	20.7	16.2	0.78	1.94	27.3	10.7	4.4							
	4.5	0.5	1.2	680	20.3	14.2	0.70	1.72	26.7	11.8	3.7							
	4.5	0.5	1.2	950	21.9	16.7	0.76	1.81	28.1	12.1	3.9							
	6.0	1.0	2.2	680	20.9	14.5	0.69	1.66	27.1	12.5	3.5							
	6.0	1.0	2.2	950	22.6	17.1	0.76	1.75	28.5	12.9	3.7							
110	3.0	0.0	0.1	680	17.5	13.2	0.75	2.07	25.0	8.5	5.0							
	3.0	0.0	0.1	950	18.9	15.5	0.82	2.18	26.4	8.7	5.3							
	4.5	0.5	1.0	680	18.5	13.5	0.73	1.93	25.6	9.6	4.6							
	4.5	0.5	1.0	950	20.0	15.9	0.80	2.03	26.9	9.8	4.8	Operation not recommended						
	6.0	0.8	1.9	680	19.0	13.7	0.72	1.86	25.9	10.2	4.3							
	6.0	0.8	1.9	950	20.5	16.1	0.78	1.96	27.2	10.5	4.5							
120	3.0	0.0	0.0	680	16.2	12.9	0.79	2.32	24.6	7.0	6.0							
	3.0	0.0	0.0	950	17.5	15.1	0.86	2.44	25.8	7.2	6.3							
	4.5	0.3	0.8	680	16.9	13.0	0.77	2.17	24.8	7.8	5.4							
	4.5	0.3	0.8	950	18.3	15.3	0.84	2.28	26.1	8.0	5.7							
	6.0	0.6	1.5	680	17.3	13.1	0.76	2.10	25.0	8.2	5.1							
	6.0	0.6	1.5	950	18.7	15.4	0.82	2.21	25.3	8.5	5.4							

Fig. A.14 Water-to-air real performance data [2]

Performance Data – TS H/V/D 048 (ECM Blower)

1,550 CFM Nominal (Rated) Airflow Cooling, 1,650 CFM Nominal (Rated) Airflow Heating

Performance capacities shown in thousands of Btuh

EWT °F	GPM	WPD PSI	WPD FT	Cooling Airflow CFM	TC	SC	Sens/Tot Ratio	kW	HR	EER	HWC	Heating Airflow CFM	HC	kW	HE	LAT	COP	HWC
20	12.0	4.2	9.6	Operation not recommended								1200	31.7	3.45	21.3	94.5	2.7	3.7
	12.0	4.2	9.6									1650	33.1	3.07	22.7	88.6	3.2	3.3
30	6.0	1.1	2.6	1100	47.7	26.3	0.55	1.71	54.8	27.9	1.7	1200	34.4	3.45	24.0	96.6	2.9	3.9
	6.0	1.1	2.6	1550	51.5	30.9	0.60	1.80	57.7	28.6	1.8	1650	36.0	3.07	25.5	90.2	3.4	3.5
	9.0	2.3	5.3	1100	44.3	23.4	0.53	1.57	50.8	28.2	1.8	1200	35.8	3.46	25.4	97.6	3.0	3.9
	9.0	2.3	5.3	1550	47.9	27.6	0.58	1.65	53.5	29.0	1.9	1650	37.4	3.08	26.9	91.0	3.6	3.5
	12.0	3.8	8.7	1100	42.3	21.9	0.52	1.50	48.6	28.2	1.8	1200	36.5	3.46	26.1	98.2	3.1	4.0
	12.0	3.8	8.7	1550	45.7	25.8	0.56	1.58	51.1	28.9	1.9	1650	38.2	3.08	27.7	91.4	3.5	3.6
40	6.0	0.9	2.1	1100	50.6	29.2	0.58	1.95	58.6	26.0	1.9	1200	39.1	3.50	28.5	100.2	3.3	4.2
	6.0	0.9	2.1	1550	54.7	34.4	0.63	2.05	61.7	26.7	2.0	1650	40.9	3.11	30.3	92.9	3.9	3.7
	9.0	2.1	4.8	1100	49.0	27.5	0.56	1.80	56.9	27.3	1.7	1200	40.8	3.52	30.2	101.5	3.4	4.3
	9.0	2.1	4.8	1550	53.0	32.3	0.61	1.89	59.4	28.0	1.8	1650	42.7	3.13	32.0	94.0	4.0	3.8
	12.0	3.5	8.0	1100	47.8	26.4	0.55	1.72	55.0	27.8	1.7	1200	41.8	3.53	31.1	102.3	3.5	4.4
	12.0	3.5	8.0	1550	51.7	31.0	0.60	1.81	57.9	28.6	1.8	1650	43.7	3.14	33.0	94.5	4.1	3.9
50	6.0	0.8	1.9	1100	51.3	30.6	0.60	2.18	60.1	23.5	2.3	1200	44.1	3.56	33.2	104.0	3.6	4.5
	6.0	0.8	1.9	1550	55.4	36.0	0.65	2.29	63.2	24.2	2.4	1650	46.0	3.17	35.2	95.8	4.3	4.0
	9.0	1.9	4.4	1100	51.0	29.8	0.58	2.02	59.3	25.3	2.0	1200	46.2	3.61	35.1	105.6	3.7	4.6
	9.0	1.9	4.4	1550	55.2	35.1	0.64	2.12	62.4	26.0	2.1	1650	48.2	3.21	37.3	97.1	4.4	4.1
	12.0	3.3	7.6	1100	50.6	29.2	0.58	1.94	58.6	26.1	1.9	1200	47.3	3.63	36.2	106.5	3.8	4.7
	12.0	3.3	7.6	1550	54.7	34.3	0.63	2.04	61.6	26.8	2.0	1650	49.4	3.23	38.4	97.7	4.5	4.2
60	6.0	0.8	1.8	1100	50.4	30.8	0.61	2.43	60.0	20.8	2.8	1200	49.2	3.68	37.9	107.9	3.9	4.8
	6.0	0.8	1.8	1550	54.5	36.3	0.67	2.55	63.2	21.4	2.9	1650	51.4	3.27	40.2	98.8	4.6	4.3
	9.0	1.8	4.2	1100	51.1	30.8	0.60	2.25	60.2	22.7	2.4	1200	51.6	3.73	40.1	109.8	4.1	5.1
	9.0	1.8	4.2	1550	55.3	36.2	0.65	2.37	63.4	23.3	2.5	1650	53.9	3.32	42.6	100.3	4.8	4.5
	12.0	3.1	7.3	1100	51.3	30.6	0.60	2.17	60.0	23.6	2.2	1200	52.9	3.77	41.4	110.9	4.1	5.2
	12.0	3.1	7.3	1550	55.4	36.0	0.65	2.28	63.2	24.3	2.3	1650	55.3	3.35	43.9	101.0	4.8	4.6
70	6.0	0.8	1.7	1100	48.4	30.3	0.62	2.69	58.9	18.0	3.5	1200	54.3	3.80	42.6	111.9	4.2	5.3
	6.0	0.8	1.7	1550	52.4	35.6	0.68	2.83	62.0	18.5	3.7	1650	56.8	3.38	45.3	101.9	4.9	4.7
	9.0	1.8	4.1	1100	49.9	30.7	0.62	2.50	59.8	19.9	2.9	1200	57.1	3.87	45.1	114.1	4.3	5.6
	9.0	1.8	4.1	1550	53.9	36.1	0.67	2.63	62.9	20.5	3.1	1650	59.7	3.44	47.9	103.5	5.1	5.0
	12.0	3.1	7.1	1100	50.4	30.8	0.61	2.41	60.0	21.0	2.8	1200	58.6	3.90	46.5	115.2	4.4	5.7
	12.0	3.1	7.1	1550	54.5	36.3	0.67	2.53	63.2	21.6	2.9	1650	61.2	3.47	49.4	104.4	5.2	5.1
80	6.0	0.8	1.7	1100	45.8	29.2	0.64	2.98	57.2	15.4	4.4	1200	59.5	3.92	47.3	115.9	4.4	5.8
	6.0	0.8	1.7	1550	49.6	34.3	0.69	3.13	60.2	15.8	4.6	1650	62.2	3.49	50.2	104.9	5.2	5.2
	9.0	1.7	4.0	1100	47.7	30.0	0.63	2.78	58.4	17.2	3.7	1200	62.5	4.00	50.1	118.2	4.6	6.2
	9.0	1.7	4.0	1550	51.5	35.3	0.68	2.92	61.5	17.7	3.9	1650	65.3	3.56	53.1	106.6	5.4	5.5
	12.0	3.0	7.0	1100	48.5	30.3	0.62	2.68	59.0	18.1	3.4	1200	64.1	4.05	51.5	119.5	4.6	6.3
	12.0	3.0	7.0	1550	52.5	35.6	0.68	2.82	62.1	18.6	3.6	1650	67.0	3.60	54.7	107.6	5.5	5.6
85	6.0	0.8	1.7	1100	44.3	28.5	0.64	3.1	56.3	14.2	4.9	1200	62.0	3.99	49.6	117.9	4.6	6.1
	6.0	0.8	1.7	1550	47.9	33.5	0.70	3.31	59.2	14.6	5.2	1650	64.8	3.6	52.7	106.4	5.3	5.5
	9.0	1.7	4.0	1100	46.3	29.4	0.63	2.93	57.5	15.8	4.2	1200	65.1	4.1	52.4	120.2	4.7	6.5
	9.0	1.7	4.0	1550	50.0	34.5	0.69	3.09	60.5	16.3	4.4	1650	68.0	3.6	55.6	108.2	5.5	5.8
	12.0	3.0	6.9	1100	47.2	29.7	0.63	2.83	58.1	16.7	3.9	1200	66.7	4.1	53.9	121.5	4.8	6.6
	12.0	3.0	6.9	1550	51.0	35.0	0.69	2.98	61.2	17.2	4.1	1650	69.7	3.7	57.2	109.1	5.6	5.9
90	6.0	0.8	1.8	1100	42.8	27.8	0.65	3.31	55.3	12.9	5.4	1200	64.5	4.06	51.9	119.8	4.7	6.4
	6.0	0.8	1.8	1550	46.3	32.8	0.71	3.48	58.2	13.3	5.7	1650	67.4	3.61	55.1	107.8	5.5	5.7
	9.0	1.7	4.0	1100	44.9	28.8	0.64	3.09	56.6	14.5	4.7	1200	67.7	4.15	54.8	122.2	4.8	6.7
	9.0	1.7	4.0	1550	48.5	33.8	0.70	3.25	59.6	14.9	4.9	1650	70.7	3.69	58.2	109.7	5.6	6.0
	12.0	3.0	6.9	1100	45.8	29.2	0.64	2.98	57.2	15.4	4.4	1200	69.3	4.18	56.3	123.5	4.9	7.0
	12.0	3.0	6.9	1550	49.6	34.4	0.69	3.13	60.3	15.8	4.6	1650	72.5	3.72	59.8	110.7	5.7	6.2
100	6.0	0.8	1.8	1100	39.7	26.4	0.66	3.68	53.3	10.8	6.6							
	6.0	0.8	1.8	1550	42.9	31.0	0.72	3.87	56.1	11.1	6.9							
	9.0	1.7	4.0	1100	41.7	27.3	0.65	3.43	54.5	12.1	5.8							
	9.0	1.7	4.0	1550	45.1	32.1	0.71	3.61	57.4	12.5	6.1							
	12.0	3.0	6.8	1100	42.7	27.8	0.65	3.32	55.2	12.9	5.4							
	12.0	3.0	6.8	1550	46.2	32.7	0.71	3.49	58.1	13.2	5.7							
110	6.0	0.7	1.7	1100	36.6	25.0	0.68	4.11	51.6	8.9	8.0	Operation not recommended						
	6.0	0.7	1.7	1550	39.6	29.4	0.74	4.32	54.4	9.2	8.4							
	9.0	1.7	3.9	1100	38.5	25.8	0.67	3.83	52.6	10.0	7.1							
	9.0	1.7	3.9	1550	41.6	30.4	0.73	4.03	55.4	10.3	7.5							
	12.0	2.9	6.7	1100	39.5	26.3	0.67	3.71	53.2	10.6	6.7							
	12.0	2.9	6.7	1550	42.7	30.9	0.72	3.90	56.0	10.9	7.0							
120	6.0	0.7	1.5	1100	34.0	24.0	0.71	4.60	50.6	7.4	9.5							
	6.0	0.7	1.5	1550	36.7	28.2	0.77	4.84	53.2	7.6	10.0							
	9.0	1.6	3.7	1100	35.5	24.6	0.69	4.30	51.1	8.3	8.6							
	9.0	1.6	3.7	1550	38.4	28.9	0.75	4.52	53.8	8.5	9.0							
	12.0	2.8	6.5	1100	36.4	24.9	0.69	4.15	51.5	8.8	8.1							
	12.0	2.8	6.5	1550	39.3	29.3	0.75	4.36	54.2	9.0	8.5							

Interpolation is permissible; extrapolation is not

Fig. A.15 Water-to-air real performance data [2]

Performance Data – TS H/V/D 060 (ECM Blower)

1,950 CFM Nominal (Rated) Airflow Cooling, 2,050 CFM Nominal (Rated) Airflow Heating

Performance capacities shown in thousands of Btuh

EWT °F	GPM	WPD PSI	WPD FT	Cooling - EAT 80/67°F Airflow CFM	TC	SC	Sens/Tot Ratio	kW	HR	EER	HWC	Heating - EAT 70°F Airflow CFM	HC	kW	HE	LAT	COP	HWC
20	15.0	7.2	16.5	Operation not recommended								1475	38.5	4.07	26.2	94.2	2.8	4.3
	15.0	7.2	16.5									2050	40.2	3.62	27.8	88.2	3.3	3.8
30	7.5	1.3	3.0	1400	68.4	43.6	0.64	2.41	78.5	28.4	1.9	1475	42.0	4.14	29.5	96.3	3.0	4.5
	7.5	1.3	3.0	1950	74.0	51.3	0.69	2.53	82.6	29.2	2.0	2050	43.9	3.68	31.3	89.8	3.5	4.0
	11.3	3.5	8.1	1400	66.0	41.7	0.63	2.26	75.5	29.2	1.7	1475	43.7	4.17	31.1	97.5	3.1	4.6
	11.3	3.5	8.1	1950	71.4	49.0	0.69	2.38	79.5	30.0	1.8	2050	45.7	3.71	33.0	90.6	3.6	4.1
	15.0	6.1	14.1	1400	64.2	40.4	0.63	2.20	73.5	29.2	1.7	1475	44.7	4.19	32.0	98.1	3.1	4.6
	15.0	6.1	14.1	1950	69.5	47.6	0.69	2.31	77.4	30.1	1.8	2050	46.7	3.73	34.0	91.1	3.7	4.1
40	7.5	0.9	2.0	1400	69.5	44.8	0.65	2.63	80.3	26.4	2.2	1475	48.2	4.25	35.3	100.3	3.3	4.8
	7.5	0.9	2.0	1950	75.1	52.8	0.70	2.77	84.6	27.1	2.3	2050	50.4	3.78	37.5	92.8	3.9	4.3
	11.3	2.9	6.7	1400	68.9	44.0	0.64	2.46	79.2	28.0	1.9	1475	50.6	4.29	37.5	101.8	3.5	4.9
	11.3	2.9	6.7	1950	74.5	51.8	0.70	2.59	83.4	28.8	2.0	2050	52.9	3.82	39.8	93.9	4.1	4.4
	15.0	5.3	12.2	1400	68.2	43.4	0.64	2.39	78.2	28.6	1.8	1475	51.9	4.33	38.7	102.6	3.5	5.1
	15.0	5.3	12.2	1950	73.7	51.0	0.69	2.51	82.3	29.4	1.9	2050	54.2	3.85	41.1	94.5	4.1	4.5
50	7.5	0.6	1.4	1400	68.6	45.0	0.66	2.89	80.3	23.7	2.8	1475	55.1	4.38	41.7	104.6	3.7	5.2
	7.5	0.6	1.4	1950	74.1	52.9	0.71	3.04	84.5	24.4	2.9	2050	57.5	3.90	44.2	96.0	4.3	4.6
	11.3	2.5	5.7	1400	69.4	45.0	0.65	2.69	80.5	25.8	2.3	1475	58.0	4.44	44.4	106.4	3.8	5.4
	11.3	2.5	5.7	1950	75.0	52.9	0.71	2.83	84.7	26.5	2.4	2050	60.6	3.95	47.2	97.4	4.5	4.8
	15.0	4.7	10.9	1400	69.5	44.8	0.64	2.61	80.2	26.7	2.2	1475	59.7	4.47	46.0	107.5	3.9	5.5
	15.0	4.7	10.9	1950	75.1	52.7	0.70	2.74	84.4	27.4	2.3	2050	62.4	3.98	48.8	98.2	4.6	4.9
60	7.5	0.5	1.1	1400	66.3	44.3	0.67	3.18	78.9	20.9	3.4	1475	62.2	4.52	48.3	109.1	4.0	5.6
	7.5	0.5	1.1	1950	71.7	52.2	0.73	3.34	83.1	21.5	3.6	2050	65.0	4.02	51.3	99.4	4.7	5.0
	11.3	2.2	5.1	1400	68.1	44.9	0.66	2.96	80.0	23.0	2.9	1475	65.8	4.60	51.6	111.3	4.2	5.8
	11.3	2.2	5.1	1950	73.6	52.8	0.72	3.11	84.2	23.7	3.1	2050	68.7	4.09	54.8	101.1	4.9	5.2
	15.0	4.3	10.0	1400	68.8	45.0	0.65	2.85	80.3	24.1	2.7	1475	67.8	4.63	53.4	112.5	4.3	6.0
	15.0	4.3	10.0	1950	74.3	53.0	0.71	3.00	84.6	24.8	2.8	2050	70.8	4.12	56.7	102.0	5.0	5.3
70	7.5	0.5	1.1	1400	63.2	43.2	0.68	3.52	76.9	18.0	4.3	1475	69.5	4.68	55.1	113.7	4.4	6.1
	7.5	0.5	1.1	1950	68.3	50.8	0.74	3.70	80.9	18.5	4.5	2050	72.7	4.16	58.5	102.8	5.1	5.4
	11.3	2.1	4.8	1400	65.5	44.1	0.67	3.26	78.4	20.1	3.6	1475	73.6	4.77	58.8	116.2	4.5	6.4
	11.3	2.1	4.8	1950	70.9	51.9	0.73	3.43	82.6	20.7	3.8	2050	76.9	4.24	62.4	104.7	5.3	5.7
	15.0	4.1	9.5	1400	66.6	44.4	0.67	3.15	79.1	21.2	3.3	1475	75.8	4.81	60.8	117.6	4.6	6.6
	15.0	4.1	9.5	1950	72.0	52.3	0.73	3.31	83.3	21.8	3.5	2050	79.2	4.28	64.6	105.8	5.4	5.9
80	7.5	0.5	1.1	1400	59.5	41.8	0.70	3.90	74.4	15.3	5.3	1475	76.8	4.83	61.7	118.2	4.7	6.6
	7.5	0.5	1.1	1950	64.4	49.1	0.76	4.10	78.3	15.7	5.6	2050	80.2	4.30	65.5	106.2	5.5	5.9
	11.3	2.0	4.6	1400	62.2	42.8	0.69	3.62	76.2	17.2	4.6	1475	81.1	4.95	65.7	120.9	4.8	7.1
	11.3	2.0	4.6	1950	67.2	50.4	0.75	3.81	80.2	17.6	4.8	2050	84.7	4.40	69.7	108.3	5.6	6.3
	15.0	4.0	9.2	1400	63.5	43.3	0.68	3.49	77.1	18.2	4.2	1475	83.4	5.01	67.8	122.4	4.9	7.3
	15.0	4.0	9.2	1950	68.6	51.0	0.74	3.67	81.1	18.7	4.4	2050	87.2	4.46	72.0	109.4	5.7	6.5
85	7.5	0.5	1.1	1400	57.6	41.0	0.71	4.1	73.2	14.1	5.9	1475	80.2	4.92	64.9	120.4	4.8	7.0
	7.5	0.5	1.1	1950	62.3	48.2	0.77	4.33	77.1	14.4	6.2	2050	83.8	4.4	68.9	107.9	5.6	6.2
	11.3	2.0	4.6	1400	60.3	42.0	0.70	3.83	74.9	15.8	5.1	1475	84.6	5.0	68.8	123.1	4.9	7.4
	11.3	2.0	4.6	1950	65.1	49.5	0.76	4.03	78.9	16.3	5.4	2050	88.4	4.5	73.1	109.9	5.8	6.6
	15.0	4.0	9.1	1400	61.6	42.6	0.69	3.69	75.8	16.8	4.7	1475	86.9	5.1	70.9	124.5	5.0	7.6
	15.0	4.0	9.1	1950	66.6	50.1	0.75	3.88	79.8	17.3	5.0	2050	90.8	4.6	75.3	111.0	5.8	6.8
90	7.5	0.5	1.2	1400	55.7	40.2	0.72	4.34	72.0	12.8	6.5	1475	83.7	5.01	68.0	122.5	4.9	7.3
	7.5	0.5	1.2	1950	60.2	47.3	0.79	4.56	75.8	13.2	6.8	2050	87.4	4.46	72.2	109.5	5.7	6.5
	11.3	2.0	4.5	1400	58.3	41.3	0.71	4.03	73.7	14.5	5.7	1475	88.1	5.15	72.0	125.3	5.0	7.8
	11.3	2.0	4.5	1950	63.1	48.6	0.77	4.24	77.5	14.9	6.0	2050	92.1	4.58	76.4	111.5	5.9	6.9
	15.0	3.9	9.1	1400	59.7	41.8	0.70	3.88	74.5	15.4	5.2	1475	90.4	5.22	74.0	126.7	5.1	8.0
	15.0	3.9	9.1	1950	64.5	49.2	0.76	4.08	78.5	15.8	5.5	2050	94.4	4.64	78.6	112.6	6.0	7.1
100	7.5	0.5	1.1	1400	51.9	38.7	0.75	4.85	69.9	10.7	7.9	Operation not recommended						
	7.5	0.5	1.1	1950	56.1	45.5	0.81	5.10	73.5	11.0	8.3							
	11.3	1.9	4.5	1400	54.4	39.7	0.73	4.51	71.2	12.1	6.9							
	11.3	1.9	4.5	1950	58.8	46.7	0.79	4.74	75.0	12.4	7.3							
	15.0	3.9	9.0	1400	55.7	40.2	0.72	4.34	72.0	12.8	6.5							
	15.0	3.9	9.0	1950	60.2	47.3	0.79	4.56	75.8	13.2	6.8							
110	7.5	0.4	0.8	1400	48.6	37.6	0.77	5.45	68.5	8.9	9.4							
	7.5	0.4	0.8	1950	52.5	44.2	0.84	5.73	72.1	9.2	9.9							
	11.3	1.8	4.2	1400	50.7	38.3	0.76	5.05	69.3	10.0	8.4	Operation not recommended						
	11.3	1.8	4.2	1950	54.8	45.0	0.82	5.31	72.9	10.3	8.8							
	15.0	3.8	8.8	1400	51.8	38.7	0.75	4.86	69.8	10.7	7.9							
	15.0	3.8	8.8	1950	56.1	45.5	0.81	5.11	73.5	11.0	8.3							
120	7.5	0.1	0.3	1400	46.1	37.2	0.81	6.15	68.3	7.5	11.3							
	7.5	0.1	0.3	1950	49.8	43.7	0.88	6.47	71.9	7.7	11.9							
	11.3	1.6	3.8	1400	47.6	37.3	0.78	5.69	68.2	8.4	10.1							
	11.3	1.6	3.8	1950	51.4	43.9	0.85	5.98	71.8	8.5	10.6							
	15.0	3.7	8.5	1400	48.5	37.6	0.77	5.48	68.4	8.8	9.5							
	15.0	3.7	8.5	1950	52.4	44.2	0.84	5.76	72.0	9.1	10.0							

Fig. A.16 Water-to-air real performance data [2]

O25 - Performance Data

Cooling Capacity

EST °F	Flow GPM	ELT °F	LLT °F	TC MBTUH	Power kW	HR MBTUH	EER	LST °F	LLT °F	TC MBTUH	Power kW	HR MBTUH	EER	LST °F	LLT °F	TC MBTUH	Power kW	HR MBTUH	EER	LST °F
			Load Flow-4 GPM						Load Flow-5.5 GPM						Load Flow-7 GPM					
30	4	50	36.6	25.9	0.96	29.2	27.0	45.0	39.3	26.7	0.96	30.0	27.8	45.5	41.9	27.5	0.96	30.8	28.6	45.9
		70	55.5	28.0	0.96	31.3	29.2	46.1	58.5	28.6	0.96	31.8	29.7	46.4	61.4	29.1	0.96	32.3	30.3	46.7
		90	74.5	30.2	0.96	33.4	31.4	47.2	77.7	30.4	0.96	33.7	31.7	47.4	81.0	30.6	0.96	33.9	31.9	47.5
		110	93.4	32.3	0.96	35.6	33.6	48.3	96.9	32.3	0.96	35.5	33.6	48.3	100.5	32.2	0.96	35.5	33.5	48.3
	5.5	50	36.9	25.4	0.93	28.6	27.3	41.6	39.5	26.2	0.93	29.3	28.1	41.9	42.1	26.9	0.93	30.1	28.9	42.3
		70	56.0	27.1	0.93	30.3	29.2	42.4	58.9	27.6	0.93	30.7	29.7	42.6	61.7	28.1	0.93	31.2	30.2	42.8
		90	75.2	28.8	0.93	31.9	31.0	43.1	78.3	29.0	0.93	32.2	31.3	43.2	81.4	29.2	0.93	32.4	31.5	43.3
		110	94.3	30.5	0.93	33.6	32.9	43.8	97.7	30.4	0.93	33.6	32.9	43.8	101.0	30.4	0.93	33.6	32.9	43.8
	7	50	37.2	24.9	0.90	28.0	27.7	38.2	39.7	25.6	0.90	28.7	28.4	38.4	42.3	26.3	0.90	29.4	29.2	38.7
		70	56.5	26.1	0.90	29.2	29.1	38.6	59.3	26.6	0.90	29.7	29.7	38.7	62.0	27.1	0.90	30.1	30.2	38.9
		90	75.9	27.4	0.89	30.4	30.6	39.0	78.8	27.6	0.89	30.6	30.9	39.0	81.8	27.8	0.89	30.9	31.2	39.1
		110	95.3	28.6	0.89	31.6	32.1	39.3	98.4	28.6	0.89	31.6	32.1	39.3	101.6	28.6	0.89	31.6	32.1	39.3
50	4	50	37.3	24.6	1.24	29.8	21.2	64.8	39.8	25.4	1.24	29.6	21.8	65.2	42.3	26.2	1.24	30.4	22.5	65.7
		70	55.4	28.3	1.25	32.6	24.0	66.8	58.3	29.0	1.25	33.3	24.5	67.2	61.2	29.7	1.25	34.0	25.0	67.5
		90	73.4	32.1	1.26	36.4	26.7	68.8	76.8	32.7	1.26	37.0	27.1	69.1	80.2	33.3	1.26	37.6	27.5	69.4
		110	91.5	35.9	1.27	40.2	29.4	70.7	95.3	36.4	1.27	40.7	29.7	71.0	99.1	36.9	1.27	41.2	29.9	71.3
	5.5	50	37.5	24.3	1.20	28.4	20.3	61.5	39.9	25.1	1.20	29.2	21.0	61.9	42.4	25.9	1.20	30.0	21.6	62.2
		70	55.7	27.8	1.20	31.9	23.1	63.0	58.5	28.5	1.20	32.6	23.7	63.3	61.4	29.2	1.20	33.3	24.3	63.6
		90	73.9	31.3	1.21	35.4	25.9	64.4	77.2	31.8	1.21	35.9	26.4	64.7	80.5	32.4	1.21	36.5	26.8	64.9
		110	92.1	34.7	1.21	38.9	28.7	65.9	95.8	35.2	1.21	39.3	29.1	66.1	99.5	35.7	1.21	39.8	29.4	66.3
	7	50	37.6	24.1	1.16	28.0	22.0	58.3	40.0	24.9	1.16	28.8	22.8	58.5	42.4	25.7	1.16	29.6	23.5	58.7
		70	56.0	27.2	1.16	31.2	24.5	59.2	58.8	27.9	1.16	31.9	25.1	59.4	61.6	28.6	1.16	32.5	25.8	59.6
		90	74.3	30.4	1.16	34.3	27.1	60.1	77.5	30.9	1.16	34.9	27.5	60.3	80.7	31.5	1.16	35.4	28.0	60.4
		110	92.7	33.6	1.16	37.5	29.6	61.0	96.3	34.0	1.16	37.9	29.9	61.2	99.9	34.4	1.16	38.3	30.2	61.3
70	4	50	38.0	23.2	1.51	28.4	15.4	84.6	40.4	24.0	1.52	29.2	15.8	85.0	42.7	24.8	1.52	30.0	16.3	85.5
		70	55.2	28.6	1.53	33.9	18.7	87.5	58.1	29.5	1.54	34.8	19.2	87.9	61.0	30.4	1.54	35.7	19.7	88.4
		90	72.4	34.1	1.55	39.4	22.0	90.3	75.9	35.0	1.56	40.3	22.5	90.8	79.4	36.0	1.56	41.3	23.1	91.3
		110	Operation not recommended																	
	5.5	50	38.0	23.2	1.47	28.2	15.8	81.4	40.3	24.1	1.47	29.1	16.4	81.8	42.7	25.0	1.47	30.0	17.0	82.1
		70	55.3	28.5	1.48	33.5	19.3	83.6	58.2	29.4	1.48	34.4	19.9	84.0	61.1	30.3	1.48	35.3	20.5	84.3
		90	72.6	33.7	1.49	38.8	22.7	85.8	76.1	34.7	1.49	39.7	23.3	86.2	79.5	35.6	1.49	40.7	23.9	86.5
		110	Operation not recommended																	
	7	50	38.0	23.2	1.42	28.0	16.3	78.3	40.3	24.2	1.42	29.0	17.1	78.5	42.6	25.1	1.41	29.9	17.8	78.8
		70	55.4	28.3	1.42	33.1	19.9	79.8	58.3	29.2	1.42	34.1	20.6	80.0	61.1	30.1	1.41	35.0	21.3	80.3
		90	72.8	33.4	1.42	38.2	23.5	81.3	76.2	34.3	1.42	39.1	24.2	81.5	79.6	35.2	1.42	40.0	24.8	81.8
		110	90.2	38.5	1.42	43.3	27.1	82.8	94.2	39.4	1.42	44.2	27.7	83.0	98.2	40.2	1.42	45.0	28.3	83.3
90	4	50	39.3	20.9	1.93	27.4	11.6	104.1	41.4	21.5	1.94	28.1	12.0	104.5	43.5	22.2	1.94	28.8	12.3	104.9
		70	56.3	26.6	1.96	33.3	14.5	107.2	59.0	27.4	1.97	34.1	14.9	107.6	61.7	28.3	1.97	35.0	15.3	108.0
		90	Operation not recommended																	
		110																			
	5.5	50	39.2	20.9	1.88	27.3	11.1	101.1	41.3	21.6	1.88	28.0	11.5	101.4	43.4	22.4	1.88	28.8	11.9	101.7
		70	56.3	26.7	1.89	33.1	14.1	103.4	59.0	27.5	1.90	34.0	14.5	103.8	61.7	28.3	1.90	34.8	14.9	104.1
		90	Operation not recommended																	
		110																			
	7	50	39.2	21.0	1.83	27.2	12.3	98.0	41.3	21.7	1.83	28.0	12.8	98.2	43.4	22.5	1.83	28.7	13.3	98.5
		70	56.2	26.8	1.83	33.0	15.6	99.7	58.9	27.6	1.83	33.8	16.1	100.0	61.6	28.4	1.83	34.6	16.6	100.2
		90	Operation not recommended																	
		110																			
110	4	50	40.5	18.5	2.35	26.5	7.9	123.7	42.3	19.1	2.36	27.1	8.1	124.0	44.2	19.6	2.36	27.7	8.3	124.3
		70	57.3	24.6	2.39	32.7	10.3	126.9	59.8	25.4	2.40	33.5	10.6	127.3	62.3	26.2	2.40	34.4	10.9	127.7
		90	Operation not recommended																	
		110																			
	5.5	50	40.4	18.6	2.30	26.4	8.1	120.7	42.3	19.2	2.30	27.0	8.3	120.9	44.2	19.8	2.30	27.6	8.6	121.2
		70	57.2	24.9	2.31	32.8	10.8	123.3	59.7	25.6	2.32	33.5	11.1	123.6	62.2	26.4	2.32	34.3	11.4	123.9
		90	Operation not recommended																	
		110																			
	7	50	40.4	18.7	2.24	26.3	8.3	117.8	42.2	19.3	2.24	26.9	8.6	117.9	44.1	19.9	2.24	27.5	8.9	118.1
		70	57.0	25.2	2.24	32.8	11.3	119.7	59.6	25.9	2.24	33.5	11.6	119.9	62.2	26.6	2.24	34.2	11.9	120.1
		90	Operation not recommended																	
		110																			

Fig. A.17 Water-to water cooling performance [3]

Heating Capacity

Source			Load Flow-4 GPM							Load Flow-5.5 GPM						Load Flow-7 GPM					
EST °F	Flow GPM	ELT °F	LLT °F	HC MBTUH	Power kW	HE MBTUH	COP	LST °F	LLT °F	HC MBTUH	Power kW	HE MBTUH	COP	LST °F	LLT °F	HC MBTUH	Power kW	HE MBTUH	COP	LST °F
25	5.5	60	Operation not recommended																	
		80																		
		100																		
		120																		
	7	60	71.0	21.4	1.28	17.0	4.90	20.0	68.1	21.5	1.26	17.2	5.02	19.9	66.4	21.6	1.23	17.4	5.15	19.9
		80	90.6	20.5	1.70	14.7	3.54	20.7	87.7	20.7	1.67	14.9	3.62	20.6	86.1	20.8	1.65	15.1	3.70	20.5
		100	110.1	19.7	2.12	12.4	2.72	21.3	107.4	19.8	2.09	12.7	2.77	21.3	105.9	19.9	2.06	12.9	2.83	21.2
		120	129.7	18.8	2.54	10.1	2.17	22.0	127.1	19.0	2.51	10.4	2.21	21.9	125.6	19.1	2.48	10.6	2.26	21.9
30	4	60	71.4	22.2	1.29	17.8	5.04	20.8	69.1	22.4	1.26	18.1	5.21	20.7	66.7	22.6	1.23	18.4	5.38	20.5
		80	91.0	21.4	1.71	15.6	3.67	22.0	88.7	21.6	1.68	15.9	3.78	21.8	86.4	21.8	1.64	16.2	3.88	21.7
		100	110.6	20.6	2.12	13.4	2.84	23.1	108.4	20.8	2.09	13.6	2.91	23.0	106.2	20.9	2.06	13.9	2.98	22.8
		120	130.2	19.8	2.54	11.1	2.28	24.3	128.1	20.0	2.51	11.4	2.33	24.1	125.9	20.1	2.47	11.7	2.38	24.0
	5.5	60	71.8	22.9	1.29	18.4	5.19	22.6	69.3	23.0	1.26	18.7	5.35	22.5	66.8	23.2	1.23	19.0	5.53	22.4
		80	91.3	21.9	1.71	16.1	3.76	23.5	88.9	22.1	1.68	16.4	3.87	23.4	86.6	22.3	1.64	16.7	3.98	23.3
		100	110.8	21.0	2.12	13.7	2.90	24.5	108.6	21.2	2.09	14.1	2.97	24.4	106.3	21.4	2.06	14.4	3.05	24.2
		120	130.3	20.1	2.54	11.4	2.31	25.4	128.2	20.3	2.51	11.7	2.37	25.3	126.0	20.5	2.47	12.1	2.43	25.2
	7	60	72.1	23.5	1.29	19.1	5.34	24.4	69.6	23.7	1.26	19.3	5.50	24.3	67.0	23.8	1.23	19.6	5.67	24.2
		80	91.6	22.4	1.71	16.6	3.85	25.1	89.1	22.6	1.68	16.9	3.96	25.0	86.7	22.8	1.64	17.2	4.07	24.9
		100	111.0	21.4	2.12	14.1	2.95	25.8	108.7	21.6	2.09	14.5	3.03	25.7	106.4	21.9	2.06	14.8	3.12	25.6
		120	130.5	20.3	2.54	11.6	2.34	26.6	128.3	20.6	2.51	12.1	2.41	26.5	126.2	20.9	2.47	12.5	2.48	26.3
50	4	60	75.2	29.4	1.31	24.9	6.55	37.2	71.9	29.5	1.27	25.1	6.79	37.1	68.7	29.5	1.23	25.3	7.03	37.0
		80	94.5	28.2	1.74	22.3	4.74	38.5	91.4	28.3	1.69	22.5	4.89	38.4	88.3	28.3	1.65	22.7	5.04	38.3
		100	113.9	27.0	2.16	19.6	3.65	39.9	111.0	27.1	2.11	19.9	3.75	39.7	108.0	27.2	2.06	20.1	3.85	39.6
		120	133.3	25.9	2.59	17.0	2.91	41.2	130.5	25.9	2.54	17.3	2.99	41.1	127.7	26.0	2.48	17.5	3.07	41.0
	5.5	60	75.7	30.4	1.31	25.9	6.79	39.6	72.3	30.4	1.27	26.1	7.02	39.5	69.0	30.5	1.23	26.3	7.25	39.5
		80	95.0	29.1	1.74	23.2	4.90	40.7	91.8	29.1	1.69	23.4	5.04	40.6	88.6	29.2	1.65	23.6	5.19	40.6
		100	114.3	27.7	2.16	20.4	3.76	41.8	111.3	27.8	2.11	20.6	3.86	41.7	108.2	27.9	2.06	20.9	3.96	41.6
		120	133.6	26.4	2.59	17.6	2.99	42.9	130.7	26.5	2.54	17.9	3.06	42.8	127.8	26.6	2.48	18.2	3.15	42.7
	7	60	76.2	31.5	1.32	27.0	6.98	42.1	72.7	31.4	1.27	27.1	7.23	42.0	69.2	31.4	1.23	27.2	7.48	42.0
		80	95.4	30.0	1.74	24.0	5.02	42.9	92.1	30.0	1.69	24.2	5.18	42.9	88.8	30.0	1.65	24.4	5.34	42.8
		100	114.7	28.5	2.17	21.1	3.83	43.8	111.5	28.5	2.11	21.3	3.95	43.7	108.4	28.6	2.06	21.6	4.06	43.6
		120	133.9	27.0	2.59	18.1	3.04	44.7	131.0	27.1	2.54	18.4	3.13	44.6	128.0	27.3	2.48	18.8	3.22	44.5
70	4	60	78.9	36.6	1.33	32.1	8.06	53.5	74.8	36.5	1.28	32.1	8.37	53.4	70.7	36.4	1.23	32.2	8.67	53.4
		80	98.1	35.0	1.77	29.0	5.81	55.0	94.2	35.0	1.71	29.1	6.00	55.0	90.3	34.9	1.65	29.3	6.20	54.9
		100	117.3	33.5	2.20	25.9	4.45	56.6	113.5	33.4	2.14	26.1	4.59	56.5	109.8	33.4	2.07	26.3	4.73	56.4
		120	136.4	31.9	2.64	22.9	3.54	58.2	132.9	31.9	2.57	23.1	3.65	58.1	129.4	31.9	2.49	23.4	3.75	57.9
	5.5	60	79.6	38.0	1.34	33.4	8.34	56.6	75.3	37.9	1.28	33.5	8.65	56.6	71.1	37.7	1.23	33.5	8.98	56.6
		80	98.7	36.3	1.77	30.2	6.00	57.9	94.7	36.2	1.71	30.3	6.19	57.9	90.6	36.1	1.65	30.4	6.40	57.8
		100	117.8	34.5	2.21	27.0	4.58	59.2	114.0	34.5	2.14	27.2	4.72	59.1	110.1	34.4	2.07	27.3	4.87	59.0
		120	136.9	32.8	2.64	23.7	3.63	60.5	133.3	32.8	2.57	24.0	3.74	60.4	129.6	32.8	2.49	24.3	3.85	60.3
	7	60	80.3	39.4	1.34	34.8	8.61	59.7	75.9	39.2	1.29	34.8	8.95	59.7	71.5	39.0	1.23	34.8	9.29	59.7
		80	99.3	37.5	1.77	31.4	6.19	60.7	95.3	37.3	1.71	31.5	6.40	60.7	91.0	37.2	1.65	31.6	6.61	60.7
		100	118.3	35.5	2.21	28.0	4.72	61.8	114.4	35.5	2.14	28.2	4.86	61.7	110.4	35.4	2.07	28.3	5.01	61.7
		120	137.3	33.6	2.64	24.6	3.73	62.8	133.6	33.6	2.57	24.8	3.84	62.7	129.9	33.6	2.49	25.1	3.95	62.6
90	4	60	82.7	44.0	1.37	39.3	9.41	69.7	77.7	43.7	1.31	39.2	9.79	69.8	72.8	43.4	1.25	39.1	10.17	69.8
		80	101.6	41.9	1.80	35.7	6.80	71.6	96.9	41.7	1.73	35.8	7.07	71.6	92.2	41.5	1.66	35.9	7.33	71.5
		100	Operation not recommended																		
		120																		
	5.5	60	83.5	45.5	1.38	40.8	9.66	73.6	78.2	44.8	1.32	40.3	9.96	73.8	73.0	44.1	1.26	39.8	10.30	73.9
		80	102.3	43.2	1.81	37.0	6.99	75.2	97.4	42.8	1.74	36.8	7.22	75.2	92.5	42.4	1.66	36.7	7.46	75.2
		100	Operation not recommended																		
		120																		
	7	60	84.2	47.0	1.39	42.3	9.91	77.6	78.7	45.9	1.33	41.4	10.16	77.8	73.2	44.8	1.26	40.5	10.42	78.1
		80	102.9	44.5	1.82	38.3	7.18	78.7	97.8	43.8	1.74	37.9	7.38	78.8	92.7	43.2	1.67	37.5	7.59	79.0
		100	Operation not recommended																		
		120																		

Fig. A.18 Water-to water heating performance [3]

040 - Performance Data

Cooling Capacity

Source EST °F	Flow GPM	ELT °F	Load Flow-5 GPM LLT °F	TC MBTUH	Power kW	HR MBTUH	EER	LST °F	Load Flow-7.5 GPM LLT °F	TC MBTUH	Power kW	HR MBTUH	EER	LST °F	Load Flow-10 GPM LLT °F	TC MBTUH	Power kW	HR MBTUH	EER	LST °F
30	5	50	34.0	38.7	1.38	43.4	28.0	47.9	37.8	39.8	1.39	44.5	28.7	48.4	41.6	40.9	1.39	45.8	29.4	48.8
		70	54.5	37.5	1.27	41.8	29.4	47.2	58.3	38.0	1.25	42.3	30.4	47.4	62.0	38.6	1.23	42.8	31.4	47.8
		90	75.1	36.2	1.17	40.2	31.1	46.6	78.8	36.2	1.12	40.0	32.6	46.5	82.5	36.2	1.06	39.9	34.1	46.4
		110	95.6	35.0	1.06	38.6	33.0	45.9	99.3	34.5	0.98	37.8	35.3	45.6	103.0	33.9	0.90	37.0	37.7	45.2
	7.5	50	34.5	37.7	1.32	42.2	28.6	43.2	38.2	38.6	1.33	43.1	29.1	43.5	41.9	39.4	1.33	43.9	29.6	43.8
		70	55.1	36.2	1.23	40.4	29.6	42.6	58.7	36.7	1.21	40.8	30.3	42.8	62.3	37.1	1.19	41.2	31.2	42.9
		90	75.7	34.7	1.13	38.6	30.7	42.1	79.3	34.8	1.09	38.5	31.9	42.1	82.8	34.9	1.05	38.5	33.1	42.0
		110	96.3	33.2	1.04	36.7	32.1	41.6	99.8	32.9	0.98	36.2	33.7	41.4	103.3	32.6	0.92	35.7	35.6	41.2
	10	50	34.9	36.7	1.26	41.0	29.1	38.5	38.5	37.3	1.27	41.6	29.5	38.6	42.2	37.9	1.27	42.2	29.8	38.7
		70	55.6	34.9	1.18	38.9	29.7	38.0	59.1	35.3	1.17	39.3	30.3	38.1	62.6	35.7	1.16	39.6	30.9	38.2
		90	76.3	33.2	1.09	36.9	30.3	37.6	79.7	33.3	1.07	37.0	31.2	37.6	83.1	33.5	1.04	37.1	32.1	37.6
		110	97.1	31.4	1.01	34.8	31.1	37.2	100.3	31.4	0.97	34.7	32.4	37.1	103.5	31.3	0.93	34.5	33.7	37.1
50	5	50	38.3	35.7	1.76	41.7	21.6	67.2	38.6	37.4	1.76	43.4	22.6	67.9	41.9	39.2	1.76	45.2	23.5	68.6
		70	53.4	40.2	1.73	46.1	24.5	69.0	57.3	41.5	1.72	47.4	25.5	69.5	61.2	42.8	1.71	48.6	26.4	70.1
		90	71.5	44.8	1.71	50.6	27.4	70.9	76.0	45.6	1.69	51.4	28.5	71.2	80.4	46.4	1.66	52.1	29.5	71.5
		110	89.6	49.4	1.68	55.1	30.4	72.7	94.7	49.7	1.65	55.3	31.7	72.8	99.7	50.1	1.62	55.6	33.0	72.9
	7.5	50	39.1	35.3	1.69	41.0	20.9	62.8	38.8	36.9	1.68	42.6	21.9	63.3	42.1	38.5	1.68	44.2	22.9	63.8
		70	57.8	39.2	1.65	44.8	23.7	64.0	58.7	40.3	1.64	45.9	24.5	64.4	61.4	41.5	1.63	47.1	25.4	64.7
		90	76.5	43.0	1.62	48.6	26.5	65.2	77.8	43.8	1.60	49.2	27.3	65.4	80.8	44.5	1.59	49.9	28.1	65.7
		110	95.2	46.9	1.59	52.3	29.5	66.5	95.4	47.2	1.56	52.5	30.2	66.5	100.2	47.5	1.54	52.7	30.9	66.6
	10	50	42.8	34.9	1.61	40.4	23.0	59.0	38.9	36.4	1.61	41.9	23.8	58.6	42.2	37.9	1.61	43.3	24.7	58.8
		70	62.1	38.1	1.57	43.5	25.3	59.0	60.1	39.1	1.57	44.5	26.1	59.2	61.7	40.2	1.56	45.5	26.9	59.4
		90	81.5	41.3	1.54	46.5	27.6	59.6	79.6	41.9	1.52	47.1	28.4	59.7	81.2	42.6	1.51	47.7	29.1	59.8
		110	100.8	44.5	1.50	49.6	30.0	60.2	96.2	44.7	1.48	49.7	30.8	60.3	100.7	44.9	1.46	49.9	31.5	60.3
70	5	50	36.6	32.6	2.14	39.9	15.2	86.5	39.4	35.0	2.14	42.3	16.4	87.4	42.3	37.4	2.13	44.7	17.6	88.4
		70	52.3	43.0	2.19	50.5	19.6	90.8	56.3	45.0	2.20	52.5	20.5	91.6	60.3	47.0	2.20	54.5	21.4	92.5
		90	68.0	53.3	2.25	61.0	23.7	95.2	73.2	55.0	2.26	62.7	24.4	95.8	78.3	56.6	2.26	64.3	25.0	96.5
		110	Operation not recommended																		
	7.5	50	36.5	32.9	2.05	39.8	16.0	82.3	39.4	35.2	2.04	42.2	17.2	83.1	42.2	37.6	2.04	44.5	18.5	83.8
		70	56.9	42.1	2.08	49.2	20.2	85.3	58.7	44.0	2.08	51.1	21.2	85.9	60.5	45.9	2.08	52.9	22.1	86.5
		90	73.9	51.4	2.11	58.6	24.3	89.4	76.4	52.7	2.12	59.9	24.9	88.8	78.8	54.1	2.12	61.3	25.5	89.3
		110	Operation not recommended																		
	10	50	36.4	33.1	1.96	39.8	16.9	78.2	39.3	35.5	1.95	42.1	18.2	78.7	42.2	37.8	1.94	44.4	19.5	79.2
		70	61.5	41.2	1.97	48.0	20.9	79.9	61.1	43.0	1.96	49.7	21.9	80.2	60.8	44.7	1.96	51.4	22.8	80.6
		90	79.8	49.4	1.98	56.1	24.9	81.6	79.6	50.5	1.98	57.2	25.5	81.8	79.4	51.6	1.97	58.3	26.1	82.0
		110	86.3	57.5	1.99	64.3	28.9	83.3	92.1	58.0	1.99	64.8	29.1	83.4	97.9	58.5	1.99	65.3	29.4	83.5
90	5	50	38.0	29.1	2.74	38.4	11.4	105.8	40.6	30.9	2.74	40.3	12.2	106.6	43.2	32.8	2.74	42.2	13.0	107.4
		70	53.8	39.2	2.80	48.8	15.0	110.1	57.5	41.1	2.81	50.7	15.7	110.9	61.1	43.0	2.81	52.6	16.4	111.7
		90	69.6	49.4	2.86	59.1	18.4	114.4	74.3	51.3	2.87	61.1	19.0	115.2	79.0	53.3	2.88	63.1	19.6	116.0
		110	Operation not recommended																		
	7.5	50	41.0	29.2	2.65	38.3	11.0	101.8	40.6	31.1	2.64	40.1	11.8	102.4	43.2	33.0	2.64	42.0	12.5	103.0
		70	57.9	38.9	2.68	48.1	14.5	104.9	59.6	40.8	2.68	49.9	15.2	105.5	61.2	42.7	2.68	51.8	15.9	106.1
		90	74.9	48.6	2.71	57.9	17.9	108.0	77.0	50.5	2.71	59.7	18.6	108.6	79.2	52.3	2.71	61.6	19.3	109.2
		110	Operation not recommended																		
	10	50	43.9	29.4	2.55	38.1	12.5	97.9	40.5	31.3	2.54	40.0	13.4	98.2	43.1	33.3	2.54	41.9	14.3	98.6
		70	62.0	38.7	2.56	47.4	16.2	99.8	61.7	40.5	2.55	49.2	17.0	100.1	61.3	42.3	2.54	51.0	17.8	100.5
		90	80.1	47.9	2.56	56.6	19.8	101.7	79.8	49.6	2.56	58.4	20.6	102.0	79.4	51.4	2.55	60.1	21.3	102.4
		110	Operation not recommended																		
110	5	50	39.5	25.5	3.34	36.9	7.6	125.2	41.8	26.9	3.35	38.3	8.0	125.8	44.2	28.2	3.35	39.6	8.4	126.3
		70	55.4	35.5	3.41	47.1	10.4	129.4	58.7	37.3	3.42	48.9	10.9	130.2	61.9	39.1	3.42	50.8	11.4	130.9
		90	Operation not recommended																		
		110																			
	7.5	50	39.4	25.6	3.24	36.7	7.9	121.4	41.8	27.0	3.24	38.1	8.3	121.8	44.1	28.5	3.24	39.5	8.8	122.2
		70	59.0	35.8	3.28	46.9	10.9	124.5	60.4	37.6	3.28	48.8	11.5	125.1	61.9	39.5	3.28	50.7	12.1	125.7
		90	Operation not recommended																		
		110																			
	10	50	39.4	25.7	3.14	36.4	8.2	117.5	41.7	27.2	3.14	37.9	8.7	117.8	44.1	28.7	3.13	39.4	9.2	118.1
		70	62.6	36.1	3.14	46.8	11.5	119.6	62.2	38.0	3.14	48.7	12.1	120.0	61.8	39.9	3.13	50.6	12.8	120.4
		90	Operation not recommended																		
		110																			

8/20/09

Fig. A.19 Water-to-water cooling performance data [3]

040 - Performance Data cont.

Heating Capacity

Source EST °F	Flow GPM	ELT °F	Load Flow-5 GPM LLT °F	HC MBTUH	Power kW	HE MBTUH	COP	LST °F	Load Flow-7.5 GPM LLT °F	HC MBTUH	Power kW	HE MBTUH	COP	LST °F	Load Flow-10 GPM LLT °F	HC MBTUH	Power kW	HE MBTUH	COP	LST °F
25	7.5	60							Operation not recommended											
		80																		
		100																		
		120																		
	10	60	72.2	29.7	1.83	23.5	4.76	20.2	68.2	29.7	1.78	23.6	4.89	20.1	66.1	29.6	1.72	23.7	5.04	20.1
		80	91.9	28.8	2.42	20.6	3.50	20.8	87.9	28.8	2.36	20.7	3.58	20.7	85.9	28.7	2.29	20.9	3.67	20.7
		100	111.5	28.0	3.00	17.7	2.73	21.3	107.7	27.9	2.94	17.9	2.78	21.3	105.7	27.8	2.87	18.0	2.84	21.3
		120	131.2	27.1	3.59	14.8	2.21	21.9	127.4	27.0	3.52	15.0	2.25	21.9	125.5	26.9	3.44	15.2	2.29	21.9
30	5	60	72.7	30.9	1.84	24.6	4.92	19.8	69.5	30.8	1.79	24.7	5.06	19.8	66.3	30.7	1.73	24.8	5.20	19.8
		80	92.4	30.0	2.42	21.8	3.64	21.0	89.3	30.0	2.36	21.9	3.73	21.0	86.2	29.9	2.30	22.0	3.81	20.9
		100	112.0	29.2	2.99	19.0	2.85	22.2	109.0	29.1	2.93	19.1	2.91	22.1	106.0	29.0	2.86	19.3	2.97	22.1
		120	131.7	28.3	3.57	16.1	2.32	23.4	128.7	28.3	3.50	16.3	2.37	23.3	125.8	28.2	3.43	16.5	2.41	23.2
	7.5	60	73.3	32.3	1.84	26.0	5.14	22.1	70.0	32.2	1.78	26.1	5.29	22.1	66.6	32.2	1.73	26.3	5.46	22.0
		80	92.8	31.0	2.43	22.7	3.74	23.1	89.6	31.0	2.36	23.0	3.85	23.0	86.4	31.1	2.29	23.2	3.97	22.9
		100	112.3	29.8	3.02	19.5	2.90	24.0	109.2	29.9	2.94	19.9	2.98	23.9	106.2	30.0	2.86	20.2	3.07	23.9
		120	131.8	28.6	3.61	16.3	2.32	25.0	128.9	28.7	3.52	16.7	2.39	24.9	125.9	28.9	3.43	17.1	2.46	24.8
	10	60	73.9	33.6	1.84	27.3	5.35	24.4	70.4	33.6	1.78	27.5	5.54	24.3	66.9	33.6	1.72	27.7	5.72	24.3
		80	93.2	32.0	2.44	23.7	3.85	25.1	89.9	32.1	2.37	24.1	3.99	25.0	86.6	32.2	2.29	24.4	4.12	25.0
		100	112.6	30.5	3.04	20.1	2.94	25.9	109.5	30.7	2.95	20.6	3.05	25.8	106.4	30.9	2.86	21.1	3.16	25.6
		120	131.9	28.9	3.64	16.5	2.33	26.6	129.0	29.2	3.54	17.1	2.42	26.5	126.1	29.5	3.43	17.8	2.52	26.3
50	5	60	76.9	41.1	1.88	34.7	6.37	35.7	72.7	40.8	1.81	34.7	6.61	35.7	68.4	40.6	1.74	34.6	6.84	35.7
		80	96.3	39.5	2.47	31.1	4.67	37.2	92.2	39.3	2.39	31.1	4.82	37.2	88.1	39.1	2.30	31.2	4.97	37.1
		100	115.6	37.9	3.06	27.4	3.61	38.7	111.7	37.7	2.96	27.6	3.72	38.6	107.8	37.6	2.87	27.8	3.84	38.5
		120	134.9	36.3	3.65	23.8	2.90	40.2	131.2	36.2	3.54	24.1	2.99	40.1	127.5	36.2	3.44	24.4	3.08	39.9
	7.5	60	77.7	43.0	1.86	36.6	6.77	38.9	73.2	42.7	1.80	36.5	6.95	38.9	68.7	42.3	1.74	36.4	7.15	38.9
		80	96.9	41.1	2.46	32.7	4.89	40.1	92.7	40.9	2.38	32.7	5.03	40.0	88.4	40.7	2.30	32.8	5.18	40.0
		100	116.1	39.1	3.06	28.7	3.75	41.3	112.3	39.1	2.97	29.0	3.86	41.2	108.0	39.0	2.87	29.2	3.98	41.1
		120	135.4	37.2	3.66	24.7	2.98	42.4	131.5	37.3	3.55	25.2	3.08	42.3	127.7	37.4	3.44	25.6	3.19	42.2
	10	60	78.5	44.9	1.84	38.6	7.14	42.0	73.8	44.5	1.79	38.4	7.29	42.1	69.1	44.1	1.74	38.2	7.43	42.1
		80	97.6	42.6	2.45	34.3	5.08	42.9	93.1	42.5	2.38	34.3	5.23	42.9	88.7	42.3	2.30	34.4	5.37	42.9
		100	116.7	40.4	3.07	30.0	3.85	43.8	112.5	40.4	2.97	30.3	3.99	43.8	108.3	40.4	2.87	30.6	4.13	43.7
		120	135.8	38.2	3.68	25.6	3.03	44.7	131.9	38.4	3.56	26.3	3.16	44.6	128.0	38.6	3.44	26.9	3.29	44.5
70	5	60	81.2	51.3	1.92	44.7	7.83	51.5	75.8	50.9	1.83	44.6	8.16	51.6	70.4	50.4	1.74	44.5	8.49	51.7
		80	100.2	48.9	2.52	40.3	5.69	53.4	95.1	48.6	2.42	40.4	5.91	53.4	90.0	48.3	2.31	40.4	6.13	53.3
		100	119.2	46.6	3.12	35.9	4.37	55.2	114.4	46.4	3.00	36.1	4.54	55.1	109.5	46.2	2.88	36.4	4.70	55.0
		120	138.2	44.2	3.72	31.5	3.48	57.0	133.7	44.2	3.59	31.9	3.61	56.8	129.1	44.1	3.45	32.3	3.75	56.7
	7.5	60	82.1	53.7	1.88	47.3	8.37	55.6	76.5	53.1	1.81	46.9	8.58	55.7	70.8	52.5	1.75	46.5	8.82	55.8
		80	101.1	51.1	2.49	42.6	6.00	57.1	95.7	50.7	2.40	42.5	6.18	57.1	90.4	50.3	2.31	42.4	6.38	57.1
		100	120.0	48.5	3.11	37.9	4.57	58.5	115.0	48.3	2.99	38.1	4.73	58.4	109.9	48.1	2.88	38.3	4.90	58.4
		120	138.9	45.9	3.72	33.2	3.61	59.9	134.2	45.9	3.58	33.6	3.75	59.8	129.5	45.9	3.45	34.1	3.90	59.6
	10	60	83.1	56.1	1.84	49.8	8.93	59.7	77.2	55.4	1.80	49.2	9.04	59.9	71.3	54.6	1.75	48.6	9.14	60.0
		80	102.0	53.2	2.47	44.8	6.32	60.8	96.4	52.8	2.39	44.6	6.47	60.8	90.8	52.3	2.31	44.4	6.62	60.8
		100	120.8	50.4	3.09	39.8	4.77	61.8	115.5	50.2	2.99	40.0	4.93	61.8	110.3	50.0	2.88	40.2	5.09	61.7
		120	139.6	47.5	3.72	34.8	3.74	62.8	134.7	47.6	3.58	35.4	3.90	62.7	129.8	47.7	3.44	36.0	4.06	62.6
90	5	60	85.4	61.5	1.99	54.7	9.05	67.4	78.5	59.0	1.88	52.5	9.20	68.3	71.6	56.4	1.77	50.4	9.34	69.2
		80	104.5	59.4	2.56	50.7	6.79	69.1	97.8	56.7	2.44	48.3	6.80	70.1	91.1	53.9	2.32	46.0	6.80	71.0
		100							Operation not recommended											
		120																		
	7.5	60	86.2	63.7	2.01	56.8	9.30	72.6	79.0	60.4	1.89	53.9	9.37	73.5	71.8	57.1	1.77	51.0	9.44	74.3
		80	105.5	61.9	2.58	53.1	7.03	73.8	98.4	58.3	2.45	49.9	6.97	74.7	91.3	54.7	2.33	46.8	6.90	75.6
		100	124.8	60.1	3.15	49.3	5.59	75.0	117.8	56.2	3.01	45.9	5.47	76.0	110.8	52.4	2.88	42.6	5.33	76.9
		120							Operation not recommended											
	10	60	87.1	65.8	2.02	58.9	9.54	77.9	79.5	61.8	1.90	55.3	9.55	78.6	71.9	57.7	1.77	51.7	9.55	79.3
		80	106.5	64.3	2.59	55.4	7.27	78.6	99.0	59.9	2.46	51.5	7.13	79.4	91.5	55.6	2.33	47.6	7.00	80.2
		100	125.9	62.7	3.16	51.9	5.82	79.3	118.4	58.1	3.02	47.8	5.62	80.2	111.0	53.4	2.88	43.6	5.43	81.0
		120							Operation not recommended											

Fig. A.20 Water-to-water heating performance data [3]

Nominal size	SDR-9			SDR-11		
	Internal diameter	Outside diameter	Thickness	Internal diameter	Outside diameter	Thickness
(in)	(cm)	(cm)	(mm)	(cm)	(cm)	(mm)
3/4	2.07	2.67	2.96	2.18	2.67	2.42
1	2.60	3.34	3.71	2.73	3.34	3.04
1 1/4	3.28	4.22	4.68	3.45	4.22	3.83
1 1/2	3.75	4.83	5.36	3.95	4.83	4.39
2	4.69	6.03	6.70	4.94	6.03	5.48
3	6.91	8.89	9.88	7.27	8.89	8.08
4	8.89	11.43	12.70	9.35	11.43	10.39
5	10.99	14.13	15.70	11.56	14.13	12.84
6	13.09	16.83	18.70	13.77	16.83	15.30

Nominal size	SDR-13.5			SDR-15.5		
	Internal diameter	Outside diameter	Thickness	Internal diameter	Outside diameter	Thickness
(in)	(cm)	(cm)	(mm)	(cm)	(cm)	(mm)
3/4	2.27	2.67	1.98	2.32	2.67	1.72
1	2.85	3.34	2.47	2.91	3.34	2.15
1 1/4	3.59	4.22	3.12	3.67	4.22	2.72
1 1/2	4.11	4.83	3.57	4.20	4.83	3.11
2	5.14	6.03	4.47	5.25	6.03	3.89
3	7.57	8.89	6.59	7.74	8.89	5.74
4	9.74	11.43	8.47	9.96	11.43	7.37
5	12.03	14.13	10.46	12.30	14.13	9.11
6	14.33	16.83	12.46	14.66	16.83	10.86

© The Author(s), under exclusive license to Springer Nature Switzerland AG 2023
L. Lamarche, *Fundamentals of Geothermal Heat Pump Systems*,
https://doi.org/10.1007/978-3-031-32176-4

The Hardy Cross method is the equivalent of Kirchhoff's loop method applied to piping networks where the nonlinear relationship between head and flow rates lead to an iterative procedure. Looking at Fig. C.1, if the flow rates are known, then the head at one point is the same, so the head variations along a closed loop (i) should be zero:

$$\sum_{j=1}^{N_i}(gh)_{j,i} = 0 \qquad (C.1)$$

where N_i is the number of pipes in the loop i.[1] For example, in loop L_1, one has:

$$(gh)_{pump} - (gh)_{p1} - (gh)_{p3} - (gh)_{p5} = 0$$

For loop L_2:

$$-(gh)_{p2} + (gh)_{p4} + (gh)_{p3} = 0$$

The + sign is associated with the loop path going in reverse direction with the assumed flow path and:

$$-(gh)_{p4} - (gh)_{p6} + (gh)_{p5} = 0$$

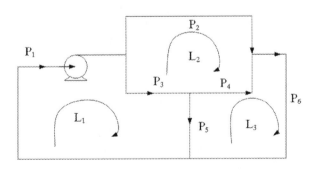

Fig. C.1 Network 1

for the last loop. Since the flow rates are unknowns, using initial guess for the flow rates, we will have:

$$(gh)_{pump} - (gh)_{p1} - (gh)_{p3} - (gh)_{p5} = y_1 \neq 0$$

$$-(gh)_{p2} + (gh)_{p4} + (gh)_{p3} = y_2 \neq 0$$

$$-(gh)_{p4} - (gh)_{p6} + (gh)_{p5} = y_3 \neq 0$$

The new flow rates are evaluated loop by loop in the Hardy Cross method. In this chapter, the pipe flow rates will be associated with letter q and the loop flow rates the letter[2] Q. In the Hardy Cross method, the unknowns are the loop flow rate, so that the pipe flow rates are given with respect to the unknowns. In the example given, that will be given as:

$$q_1 = Q_1 \; , \; q_2 = Q_2 \; , \; q_3 = Q_1 - Q_2$$
$$\qquad (C.2)$$
$$q_4 = Q_3 - Q_2 \; , \; q_5 = Q_1 - Q_3 \; , \; q_6 = Q_3$$

In loop L_2, for example, one has:

$$-(gh)_1(q_1 + \Delta Q_2) + (gh)_4(q_4 - \Delta Q_2) + (gh)_3(q_3 - \Delta Q_2)$$

From the Darcy-Weisbach equation, we have that:

$$(gh) = \left(f\frac{L + \sum L_e}{D} + \sum K\right)\frac{8q^2}{\pi^2 D^4} = Rq^2 \quad (C.3)$$

So if the pipe flow is in the same direction as the loop flow:

$$(gh)(q + \Delta Q) = Rq^2 + 2Rq\,\Delta Q + (\Delta Q)^2$$
$$\approx Rq^2 + 2Rq\,\Delta Q = (qh)_o + 2(gh)_o/q\,\Delta Q$$

And if the pipe flow is in the opposite direction:

$$(gh)(q - \Delta Q) = Rq^2 - 2Rq\,\Delta Q + (\Delta Q)^2$$

---

[1] The analysis is done here using specific energy (J/kg) but can be done of course using head (m) or pressure (Pa).

[2] Although the symbol q was not used for flow rates before, it will be used in this Appendix for brevity and should not bring too much confusion.

© The Author(s), under exclusive license to Springer Nature Switzerland AG 2023
L. Lamarche, *Fundamentals of Geothermal Heat Pump Systems*,
https://doi.org/10.1007/978-3-031-32176-4

$$\approx Rq^2 - 2Rq\,\Delta Q = (qh)_o - 2(gh)_o/q\,\Delta Q$$

where $(gh)_o$ is associated with the losses from the previous iteration. Forcing the new loop losses to zero will give the loop flow rate increment ΔQ. So for loop 2, one has:

$$-(gh)_{o,2} - \frac{2(gh)_{o,2}}{q_2}\Delta Q_2 + (gh)_{o,4}$$

$$-\frac{2(gh)_{o,4}}{q_4}\Delta Q_2 - (gh)_{o,3} - \frac{2(gh)_{o,3}}{q_3}\Delta Q_2 = 0$$

$$\Delta Q_2 = \frac{-(gh)_{o,2} + (gh)_{o,4} + (gh)_{o,3}}{2((gh)_{o,2}/q_2 + (gh)_{o,4}/q_4 + (gh)_{o,3}/q_3)}$$

$$= \frac{y_2}{2\sum |(gh)/q|}$$

In the case where a pump is present in the loop (L_1), the head is normally given by a polynomial. If, for example, it is given by a third-degree polynomial, it will be given by:

$$(gh)_{pump} = a_o + a_1 q + a_2 q^2 + a_3 q^3 \qquad (C.4)$$

Applying a similar analysis will give:

$$(gh)_{pump}(q + \Delta Q) = (gh)_{pump,o} + (a_1 + 2a_2 q + 3a_3 q^2)\Delta Q$$

$$= (gh)_{pump,o} + \frac{\delta(gh)_{pump}}{\delta Q}\Delta Q$$

The loop flow rate increment for loop 1 is then:

$$y_1 + \left(\frac{\delta(gh)_{pump}}{\delta Q} - \frac{2(gh)_{o,1}}{q_1} - \frac{2(gh)_{o,3}}{q_3} - \frac{2(gh)_{o,5}}{q_5}\right)\Delta Q_1 = 0$$

$$\Delta Q_1 = \frac{y_1}{2\sum |(gh)/q| - \frac{\delta(gh)_{pump}}{\delta Q}}$$

So in general, the increment in each loop will be given by:

$$\Delta Q = \frac{(gh)_{pump} \pm (gh)_i}{2\sum |(gh)_i/q| - \frac{\delta(gh)_{pump}}{\delta Q}} \qquad (C.5)$$

with the + sign when the pipe flow rate is in the opposite direction of the loop flow rate and the - sign when they are in the same direction. Once all the loop flow rates are known, the pipe flow rates are given by (C.2).

Example C.1 The network shown in Fig. C.1 will be used as a first example. The parameters of each pipe are given in the following table:

(continued)

Pipe	L(m)	D(m)	K	Le/D	ϵ(m)
1	80	0.025	2	0	4.5e-5
2	15	0.025	16	0	4.5e-5
3	10	0.025	8	0	4.5e-5
4	20	0.025	0	0	4.5e-5
5	12	0.025	8	0	4.5e-5
6	25	0.025	0	0	4.5e-5

Only one pump is present in pipe 1, and its head curve (in J/kg) is given by:

$$(gh)_{pump} = 180 + 1900q - 1e6q^2 - 6e7q^3$$

where the flow rate is in m^3/s. A "network" instance from the "network" class defined in the hydraulic_md library should be defined. Before a list of all "pipes" in the "network" have to be instantiated. An example is:

```
npipes = 6          # number of pipes in the network
# vectors
Dpipe = 0.025*np.ones(npipes)
# diameter in meters
eppipe = 45e-6*np.ones(npipes)
# pipe rugosity in meters
Lepipe = np.zeros(npipes)
# pipe total equival length in meters
Lpipe = np.array([80.0,15.0,10.0,20.0,12.0,25.0])
# pipe length in meters
Kpipe = np.array([2.0,16.0,8.0,0.0,8.0,0.0])
# pipe total singular coefficents (dimensionless)
# Pump characterics
Pompe = [np.array([-6e7,-1e6,1.9e3,1.8e2]),
         np.array([]),np.array([]),\
         np.array([]),np.array([]),np.array([])]
my_pipes = []
for i in range(0,npipes):
        my_pipes.append(pipes(Lpipe = Lpipe[i],
        Dpipe = Dpipe[i],eppipe = eppipe[i],\
        Lepipe = Lepipe[i],Kpipe = Kpipe[i],
        Pompe = Pompe[i]))
```

The attribute ("Pompe") is an array given the polynomial pump characteristics where the flow rate is in m^3/s and the head in J/kg. If a pipe has no pump, an empty array is given. After that, the topology of the network is given. The network topology consists of

(continued)

Example C.1 (continued)

defining which pipe is in each loop. A minus sign is assigned if the flow rate is assumed to be in the opposite sign than the loop flow rate. In this example, we would have:

```
# network topology
nloop = 3
nloopi = 0
# number of loops where the flow rate is imposed
mloop = [np.array([1,3,5]),np.array([2,-3,-4]),
    np.array([4,-5,6])]
```

The attribute *nloopi* is the number of loop where the flow rate is imposed (see Example 3). Here it is zero. Pay attention that in Python, arrays and lists start at index 0, so in a list of pipes, the first pipe would be pipe[0], but in the case of the topology nomenclature, the first pipe is labeled 1 and so on; this is to avoid having sign(0) = 0. If the final pipe flow rate is not in the direction assumed, it will have a minus sign, so, for example, if the flow rate in pipe 4 is negative, it means that the flow goes from right to left and not in the direction assumed at first. Once the pipes and the topology are defined, an instance of network is created:

```
Trefk = 300
fluide = 'Water'
rho =PropsSI('D','T',Trefk,'P',patm,fluide)
mu = PropsSI('viscosity','T',Trefk,'P',patm,fluide)
nu = mu/rho
my_network = network(nloopi = nloopi,
    mloop = mloop, my_pipes = my_pipes,nu = nu)
```

The attributes are the number of loops where the flow is imposed, the network topology, the pipes, and the kinematic viscosity of the fluid used. Once the network is defined, we can solve it using the Hardy Cross method:

```
# initial values
qbi = np.array([0.001,0.0005,0.00075])
qb = my_network.hardy_cross(qbi)
```

The final results are:

$$Q_{L1} = 0.00094997 \, \text{m}^3/\text{s}$$
$$Q_{L2} = 0.00040983 \, \text{m}^3/\text{s}$$
$$Q_{L3} = 0.0004409 \, \text{m}^3/\text{s}$$

(continued)

The pipe flow rate and head losses can be retrieved with the following methods:

```
qpipes = my_network.get_qpipes()
gh,ghp = my_network.get_gh()
```

The results are:

	q(l/s)	$(gh)_{loss}$ (J/kg)	$(gh)_{pump}$ (J/kg)
1	0.95	157.922	180.851
2	0.41	11.543	0.000
3	0.54	11.488	0.000
4	0.03	0.056	0.000
5	0.51	11.441	0.000
6	0.44	11.385	0.000

It is easy to verify that Eq. (C.1) holds for each loop and that the continuity equation holds for each node. The pump fluid work will then be:

$$\dot{W} = 0.95 \cdot 180.851 = 171.80 \, \text{W}$$

∎

Example C.2 Let's redo the last example but with a similar pump as the previous one in pipe 2 (Fig. C.2).

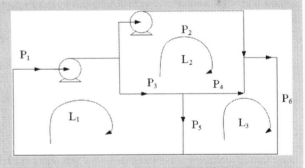

Fig. C.2 Network 2

Exactly the same input has to be given except that the pump list will look like:

```
# Pump characterics
Pompe = [np.array([-6e7,-1e6,1.9e3,1.8e2]),
    np.array([-6e7,-1e6,1.9e3,1.8e2]), \
np.array([]),np.array([]),np.array([]),np.array([])]
```

(continued)

Example C.2 (continued)

The final results are:

Pipe	q(l/s)	$(gh)_{loss}$(J/kg)	$(gh)_{pump}$(J/kg)
1	1.03	185.699	180.829
2	1.51	145.613	180.385
3	−0.47	8.936	0.000
4	−0.77	25.837	0.000
5	0.30	4.066	0.000
6	0.74	29.902	0.000

The minus sign for the flow rates in pipes 3 and 4 means that the final flow rate goes in the opposite direction as assumed by the arrows. Again, it is easy to verify that the continuity laws are consistent. For example, in loop 3, the head loss will now be given by:

$$(gh)_4 - (gh)_6 + (gh)_7 = 25.837 - 29.902 + 4.066 = 0$$

The energy is subtracted (losses) when the loop direction is the same as the pipe flow direction; otherwise, it is added.   ∎

Example C.3 As a final example, we will redo Example 1 where the inlet flow rate is imposed at 0.5 l/s (Fig. C.3).

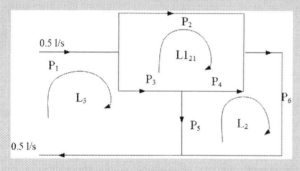

Fig. C.3   Network 3

(continued)

The only thing to pay attention is to assign all the loops where the flow is imposed at the end of the loop list in the topology. So the only difference between this example and Example 1 will be:

```
nloopo = 1 # number of loops where the flow rate
is imposed
# network topology
mloop = [np.array([2,-3,-4]),np.array([4,-5,6]),
    np.array([1,3,5])]
qbi = np.array([0.0001,0.0002,0.0005])
```

The correct flow rate for loop 3 (0.5 l/s) should be given in the initial value array for loop 3; the others are, in theory, unimportant, but too far values can lead to divergence. The final results are:

Pipe	q(l/s)	$(gh)_{loss}$(J/kg)	$(gh)_{pump}$(J/kg)
1	0.50	47.064	53.889
2	0.22	3.395	0.000
3	0.28	3.371	0.000
4	0.01	0.024	0.000
5	0.27	3.453	0.000
6	0.23	3.430	0.000

The mechanical device that supplies the flow rate in pipe 1 should overcome losses of:

$$\begin{aligned}
(gh)_{tot} &= (gh)_1 + (gh)_2 + (gh)_6 \\
&= (gh)_1 + (gh)_3 + (gh)_5 \\
&= (gh)_1 + (gh)_3 + (gh)_4 + (gh)_6 \\
&= 53.889 \ (J/kg)
\end{aligned}$$

                                                                                            ∎

D.1 Geothermal Library (geothermal_md.py)

G_function_ ics(Fo,rt=1)
    # Solve Eq. (4.8), Infinite cylindrical source
    # $Fo = \alpha t/r_b^2, rt = r/r_b$
Example:
> > > from geothermal_md import *
> > > G_function_ ics(10,2)
0.15518264194318915
> > > G_function_ ics(10,1)
0.26274805279766855
> > > G_function_ ics(10)
0.26274805279766855

G_function(Fo)
    # Solve Eq. (4.11)
    # $Fo = \alpha t/r_b^2$
Example:
> > > G_function(10)
0.2626987095045498

G_function_ ils(Fo,rt=1)
    # Solve Eq. (4.12), Infinite line source
    # $Fo = \alpha t/r_b^2, rt = r/r_b$
Example:
> > > G_function_ ils(10,2)
0.14506367943154536
> > > G_function_ ils(10)
0.24959540821048082

I_function(X)
    # Solve Eq. (4.16)
    # $X = r/2\sqrt{\alpha t}$
Example:
> > > X = np.sqrt(1/10)/2
> > > I = I_function(X)
> > > print(I/(2*np.pi))
0.24959540821048082

G_function_ fls(Fo, Ht = 2000,zot = 80)
    # Solve Eq. (4.45)
    # $Fo = \alpha t/r_b^2, Ht = H/r_b, zot = zo/r_b$
Example:
> > > G_function_ fls(10,2000,80)
0.24938396409753524
> > > G_function_ fls(10)
0.24938396409753524
> > > G_function_ fls(10,1000)
0.24917252005492121

g_function_ fls(tt, rbb = 0.0005,zob = 0.04)
    # Solve Eq. (4.45)
    # $tt = 9\alpha t/H^2, rbb = r_b/H, zob = zo/H$
Example:
> > > g_function_ fls(.001,0.0005,0.04)
3.4418322393304797
> > > g_function_ fls(.001)
3.4418322393304797
> > > g_function_ fls(.001,0.001)
2.7500081538278676
> > > g_function_ fls(.001) - log(2)
2.7486850587705343

compute_g_function(zb,tt,rbb=0.0005,zob =0.04)
    # Evaluate type-I g-function (Eq. (4.53))
    # $tt = 9\alpha t/H^2, zb = r_b/H, zob = zo/H$
    # zb : 2D-array (Nb x 2) giving $[\bar{x} - \bar{y}]$ positions of Nb boreholes
Example:
> > > zb = np.array([[0.0, 0.0], [0.0, 0.1], [0.1, 0.0], [0.1, 0.1]])
0.1 = b/H
> > > compute_g_function(zb,1)
9.87571047075779

Rb_Paul(kg,rb,rp,cas)
    # Solve Eq. (5.8)
    # kg : grout conductivity, rb : borehole radius
    # rp : pipe external radius, cas : 'a' 'b' or 'c'
> > > Rb_Paul(1.7,0.075,0.0165,'b')
0.08432772646349938

Rb_Sharqawi(kg,rb,rp,xc)
    # Solve Eq. (5.10)
    # kg : grout conductivity, rb : borehole radius
    # rp : pipe external radius, xc : center to center half distance
Example:
> > > Rb_Sharqawi(1.7,0.075,0.0165,0.025)
0.08731063478010241

Rb_linesource(kg,ks,rb,rp,xc)
    # Solve Eq. (5.11)
    # kg : grout conductivity, ks : soil conductivity, rb : borehole radius
    # rp : pipe external radius, xc : center to center half distance
> > > Rb_linesource(1.7,2.5,0.075,0.0165,0.025)
(0.08974591112070972, 0.19962881111192546)
> > > Rg,Rga = Rb_linesource(1.7,2.5,0.075,0.0165,0.025)
> > > print(Rg)
0.08974591112070972,
> > > print(Rga)
0.19962881111192546

Rb_linesource_2UT(kg,ks,rb,rp,xc,cas = '13-24')
    # Solve Eq. (5.48)

kg : grout conductivity, ks : soil conductivity, rb : borehole radius
rp : pipe external radius, xc : center to center half distance
Example:
\>>> Rb_linesource_2UT(1.7,2.5,0.075,0.0165,0.025)
(0.08013079719786986, 0.09981440555596273)
\>>> Rb_linesource_2UT(1.7,2.5,0.075,0.021,0.025,'12-34')
(0.08013079719786986, 0.03846045569135939)

Rb_multipole(kg,ks,rb,rp,Rp,J,zb)
# Evaluate borehole resistance with multipole method
# kg : grout conductivity, ks : soil conductivity, rb : borehole radius
# rp : pipe external radius, Rp : pipe resistance, J : order of multipole
# zb : array of complex numbers giving pipes locations
Example:
\>>> xc = 0.025
\>>> z = np.array([0 + xc*1j,xc,0-xc*1j,-xc])
\>>> Rb_multipole(1.7,2.5,0.075,0.0165,0,10,z)
(0.06505874736070819, 0.03853298865908427)
\>>> Rb_multipole(1.7,2.5,0.075,0.0165,0,0,z)
(0.08013079719786986, 0.03846045569135939)

Tp_ashrae(nx,ny,d,qp,ald,ks,n_years,nrings=3,Hb = -999)
# Evaluaute the temperature penalty as proposed by Kavanaugh Rafferty
# nx : number boreholes in x, nx : number boreholes in y
# d : distance between boreholes (m), ks : soil conductivity,
# qp : heat rate per unit lenght (W/m),ald : soil diffusivity (m^2/day)
# n_years : number of years, nrings : number of rings taken (3 : default)
# H : height of borehole, to compute C_f, if not given $C_f = 1$
Example:
\>>> Tp_ashrae(2,2,5,10,0.1,2.5,10)
3.0296825710085487
\>>> Tp_ashrae(2,2,5,10,0.1,2.5,10,Hb = 100)
2.992279082477579

Tp_ils(nx,ny,d,qp,ald,ks,n_years)
# Evaluate the temperature penalty with the ILS
# nx : number boreholes in x, nx : number boreholes in y
# d : distance between boreholes (m), ks : soil conductivity,
# qp : heat rate per unit length (W/m),ald : soil diffusivity (m^2/day)
# n_years : number of years
Example:
\>>> Tp_ils(2,2,5,10,0.1,2.5,10)
3.1338275496746655

Tp_fls(nx,ny,d,qp,ald,ks,n_years,H=100.0,zob = 0.04)
# Evaluate the temperature penalty with the FLS
# nx : number boreholes in x, nx : number boreholes in y
# d : distance between boreholes (m), ks : soil conductivity,
# qp : heat rate per unit lenght (W/m),ald : soil diffusivity (m^2/day)
# n_years : number of years
Example:
\>>> Tp_fls(2,2,5,10,0.1,2.5,10)
2.7508742550301144

Tp_Bernier(nx,ny,d,qp,ald,ks,n_years,H=100.0)
# Evaluate the temperature penalty with the relation of Philippe - Bernier
# nx : number boreholes in x, nx : number boreholes in y
# d : distance between boreholes (m), ks : soil conductivity,
# qp : heat rate per unit length (W/m),ald : soil diffusivity (m^2/day)

n_years : number of years
Example:
\>>> Tp_Bernier(2,2,5,10,0.1,2.5,10)
2.917597754985589

sdr_pipe(pipe,sdr)
# Return the inner and outer diameters of SDR pipind in meters
# pipe : nominal value in inch (0.5,.75,1,1.25,1.5,2,3,4,5,6)
# sdr : sdr-value nominal value in inch
    (0.5,.75,1,1.25,1.5,2,3,4,5,6)
Example:
\>>> sdr_pipe(1,11)
2.917597754985589
(0.02732809090909091, 0.033401)
\>>> din,dout = sdr_pipe(1,11)
\>>> print(din)
0.02732809090909091
\>>> print(dout)
0.033401

Pulses_ashrae(q_ground,nbloc = 4,flag = 1)
# Return the 5 pulses for the ASHRAE method from the hourly ground loads
# q_ground : 8760 hourly ground loads (+ heating, − Cooling)
# nbloc : Number of peak hours (default = 4)
# flag : 1 real monthly hours (default), 2 all months = 730 hours
Example:
\>>> loads = np.loadtxt("ground_loads.txt") # ground loads in kWatts
\>>> heating_loads = loads[:,0]
\>>> cooling_loads = loads[:,1]
\>>> ground_loads = heating_loads - cooling_loads
\>>> qa,qm_heating,qm_cooling,qh_heating,qh_cooling =
\>>>   Pulses_ashrae(ground_loads)
\>>> print(qa,qm_heating,qm_cooling,qh_heating,qh_cooling) # kW
1.300262414383562   4.58054435483871   -1.3862298387096774
8.9015625 -4.35625

leaky_function(u,rB)
# Solve Hantush integral (Eq. (11.42))
# $u = r^2/4\alpha t, rB = r/B$
Example:
\>>> leaky_function(.8,1)
0.2582813389770071

Delta_T_earth(z,ts,tshift,ald,amp)
# Return the earth temperature variation at day ts
# and maximum variation at depth z
# z (m) : depth, ts (day) : day when temperature is evaluated,
# tshift (day) : coldest day of the year
# ald (m^2/day) diffusivity, amp : Amplitude variation
Example:
\>>> DT,DTmax = Delta_T_earth(1.25,10,15,0.08,10)
\>>> print(DT)
-5.836452622369801
\>>> print(DTmax)
6.636441935294081

G_function_st(Fo,ret,gamma,kt,nu,Rt)
# Solve Eq. (13.27)
# $Fo = \alpha t/r_b^2$, ret $= r_{eq}/r_b$, gamma $= \sqrt{\alpha_{soil}/\alpha_{grout,eq}}$
# kt $= k_{grout}/k_{soil}$, nu $= \frac{\bar{r}_i^2(\rho C_p)_{f,eq}}{2(\rho C_p)_{soil}}$
# Rt $= 2\pi k_{soil} R'_p$
Example:
\>>> G_function_st(0.2,0.45,1.5,.4,0.15,0.5)
0.13932413977976044

G_function_ isc(Fo,rt=1)
Solve Eq. (13.71), infinite solid cylinder

Fo $= \alpha t / r_b^2$, rt $= r/r_b$

Example:

\> \> \> G_function_isc(15)

0.2825040547695048

G_function_ fsc(Fo,rt,zt,ht,zot)
    # Solve Eq. (13.80), finite solid cylinder
    # Fo $= \alpha t / r_b^2$, rt $= r/r_b$, zt $= z/r_b$
    # ht $= H/r_b$, zot $= z_o/r_b$

Example:

\> \> \> G_function_fsc(15,1,500,1000,60)

0.28250403513887706

G_function_ring(Fo,rt,zt,Ht,zot,bt)
    # Solve Eq. (13.83), vertical ring's model
    # Fo $= \alpha t / r_b^2$, rt $= r/r_b$, zt $= z/r_b$
    # ht $= H/r_b$, zot $= z_o/r_b$, bt $= b/r_b$

Example:

\> \> \> G_function_ring(15,1,500,1000,60,0.4)

0.27371609694338006

G_function_spiral(Fo,rt,zt,phi,Ht,zot,bt)
    # Solve Eq. (13.94), vertical coil model
    # Fo $= \alpha t / r_b^2$, rt $= r/r_b$, zt $= z/r_b$
    # phi : azimututh angle, ht $= H/r_b$, zot $= z_o/r_b$, bt $= b/r_b$

Example:

\> \> \> G_function_spiral(15,1,500,0,1000,60,0.5)

0.2900087039269048

G_function_slinky(Rb,Fo,rpt,zt,pt,ntr = 1,r1=7,r2=25)
    # Solve Eq. (13.115), Horizontal Xiong's slinky model
    # Rb : array of position for the centreas of rings
    # Fo $= \alpha t / Rring^2$, rpt $= r_p / Rring$, zt $= z_o / Rring$
    # pt : $p/Rring$, ntr : number of trench,
    # r1 : if r > r1, approximate function of Xiong
    # r2 : if r > r2, ring-to ring influence = 0

\> \> \>p = 0.5 # pitch

\> \> \>Dring = 1.0

\> \> \>Rring = Dring/2

\> \> \>do = 0.06

\> \> \>rp = do/2

\> \> \>z = 1.5

\> \> \>ald = 0.1

\> \> \>t = 2

\> \> \>Fo = ald*t/Rring**2

\> \> \>zt = z/Rring

\> \> \>pt = p/Rring

\> \> \>rpt = rp/Rring

\> \> \>Rb = array([[0, 0.],[pt, 0.],[2*pt, 0.]])

\> \> \>G = G_function_slinky(Rb,Fo,rpt,zt,pt,1)

\> \> \>print(G)

0.1020869336704691

D.2   Heat Exchangers Library (heat_exchanger_md.py)

shell_tube_eff(NTU,Cr,nshell=1)
    # Efficiency of shell and tube heat exchnager nshell shell passes
    # NTU : number of thermal units, Cr = Cmin/Cmax <=1

Example:

\> \> \> shell_tube_eff(2,.2)

0.7921458051796771

shell_tube_NTU(ep,Cr,nshell=1)
    # Number of thermal units of shell and tube heat exchanger nshell shell passes

    # ep : efficiency <= 1, Cr = Cmin/Cmax <=1

Example:

\> \> shell_tube_NTU(0.7,.2)

1.4072612655267316

counter_flow_eff(NTU,Cr)
    # Efficiency of counter flow heat exchanger
    # NTU : number of thermal units, Cr = Cmin/Cmax <=1

Example:

\> \> counter_flow_eff(2,.2)

0.8316863996353212

counter_flow_NTU(ep,Cr)
    # Number of thermal units of counter-flow heat exchnager
    # ep : efficiency <= 1, Cr = Cmin/Cmax <=1

Example:

\> \> counter_flow_NTU(.7,.2)

1.3164373932391902

D.3   Heat Pump Library (heat_pump_md.py)

D.3.1   ISO 13256 Functions

air_flow_corr(x_air,df,mode = 'cooling'):
        # air flow correction capacity (total,sensible(cooling), total(heating) and power
        # x_air : fraction of nominal air flow
        # df : dataframe associated with HP family,6 columns :
        # 'DEB' (% flow), 'TC_CORR' (total capacity correction)
        # 'SC_CORR' (sensible capacity correction)
        # 'C_POW_CORR' (cooling power correction)
        # 'HC_CORR' (heating capacity correction)
        # 'H_POW_CORR' (heating power correction)

Example:

\> \> \>import pandas as pd

\> \> \>from heat_pump_md import *

\> \> \>file = '..\\data\\wf_air_flow_corr.xls' # File associated with Fig. A.3

\> \> \>df = pd.read_excel(file)

\> \> \>x = 0.5   # 50% of nominal air flow

\> \> \>tc_corr,sc_corr,pow_corr = air_flow_corr(x,df) # Cooling mode by default

\> \> \>print(tc_corr,sc_corr,pow_corr)

0.928 0.747 0.936

\> \> \>hc_corr,heating_pow_corr = air_flow_corr(x,df,'heating')

\> \> \>print(hc_corr,heating_pow_corr)

0.961 1.097

air_temp_heating_corr(dry_bulb,df):
        # dry_bulb temp correction on capacity and power in heating mode
        # dry_bulb : dry bulb temperature (F)
        # df : dataframe associated with HP family,3 columns :
        # 'EA_DBT' air dry bulb temperatures (F)
        # 'HC_CORR' (heating capacity correction)
        # 'POW_CORR' (heating power correction)

Example:

\> \> \>file = '..\\data\\wf_hc_temp_corr.xls' # File associated with Fig. A.5

\> \> \>df = pd.read_excel(file)

\> \> \>T = 60 # F

\> \> \>hc_corr,heating_pow_corr = air_heating_temp_corr(T,df)

\> \> \>print(hc_corr,heating_pow_corr)

1.025 0.893

```
> > >hc_corr,heating_pow_corr = air_flow_corr(x,df,'heating')
> > >print(hc_corr,heating_pow_corr)
0.961 1.097
```

air_temp_cooling_corr(dry_bulb,wet_bulb,df):
       # temp correction on capacity (total + sensible) and power in cooling mode
       # dry_bulb : dry bulb temperature (F)
       # wet_bulb : dry bulb temperature (F)
       # df : dataframe associated with HP family,3 columns :
       # 'EA_WBT' air wet bulb temperatures (F)
       # 'EA_DBT' air dry bulb temperatures (F)
       # 'TC_CORR' (cooling total capacity correction)
       # 'SC_CORR' (cooling sensible capacity correction)
       # 'POW_CORR' (cooling power correction)

Example:
```
> > >file = '..\\data\\wf_cc_temp_corr.xls' # File associated with
Fig. A.7
> > >df = pd.read_excel(file)
> > >Tdb = 80 # F
> > >Twb = 70 # F
> > >tc_corr,sc_corr,cooling_pow_corr = air_cooling_temp_corr
(Tdb,Twb,df)
> > >print(tc_corr,sc_corr,cooling_pow_corr)
1.053 0.879 1.003
```

WA_heat_pump(EWT_S,DEB_S,CFM_L,mode,df):
       # Table look-up for water-air performances
       # EWT_S : entering water temperature source side (F)
       # DEB_S : flow rate source side (gpm)
       # CFM_L : flow rate load side (cfm)
       # mode, 'heating' or 'cooling'
       # df : dataframe associated with HP family,14 columns :
       # 'EWT_SH' entering water temperature source side, heating mode (F)
       # 'EWT_SC' entering water temperature source side, cooling mode (F)
       # 'CFM_C' air flow rate cooling mode (cfm)
       # 'DEB_SH' water source flow rate, heating mode (gpm)
       # 'DEB_SC' water source flow rate, cooling mode (gpm)
       # 'CFM_H' air flow rate heating mode mode (cfm)
       # 'TCC' (cooling total capacity (kBTU/hr))
       # 'SCC' (cooling sensible capacity (kBTU/hr))
       # 'POWC' (power cooling mode (kW)), 'EER' (cooling mode)
       # 'HCH' (heating total capacity (kBTU/hr)),'POWH' (power heating mode (kW))
       # 'COP' (heating mode), 'PSI' nominal pressure losses (psi)

Example:
```
> > >heat_pump = '..\\data\\nv036.xls'
> > >df = pd.read_excel(heat_pump)
> > >EWT = 50 # F
> > >gpm = 7.5
> > >cfm = 1200
> > >TC,SC,Pow,EER = WA_heat_pump(EWT,gpm,cfm,'cooling',df)
> > >print ('nominal total capacity = ',TC, ' kbtu/hr' )
nominal total capacity = 44.266666666666666 kbtu/hr
> > >print ('nominal power = ',Pow, ' kW')
nominal power = 1.31083333333333 kW
> > >print ('nominal EER = ',EER)
nominal EER = 33.93333333333334

> > >TC,Pow,COP = WA_heat_pump(EWT,gpm,cfm,'heating',df)
> > >print ('nominal total capacity = ',TC, ' kbtu/hr' )
nominal total capacity = 46.46857142857143 kbtu/hr
```

```
> > >print ('nominal power = ',Pow, ' kW')
nominal power = 2.8508571428571434 kW
> > >print ('nominal COP = ',COP)
nominal COP = 4.778857142857143
```

WA_hp_psi(EWT_S,DEB_S,df):
       # Pressure loss (psi) for water-air heat pump
       # EWT_S : entering water temperature source side (F)
       # DEB_S : flow rate source side (gpm)
```
> > >heat_pump = '..\\data\\nv036.xls'
> > >df = pd.read_excel(heat_pump)
> > >EWT = 50 # F
> > >gpm = 8.0
> > >Dp = WA_hp_psi(EWT,gpm,df)
> > >print(Dp)
2.4599999999999875
```

WW_heat_pump(EWT_S,DEB_S,EWT_L,DEB_L,mode,df):
       # Table look-up for water-water performances
       # EWT_S : entering water temperature source side (F)
       # DEB_S : flow rate source side (gpm)
       # EWT_L : entering water temperature load side (F)
       # DEB_L : flow rate load side (gpm)
       # mode, 'heating' or 'cooling'
       # df : dataframe associated with HP family,14 columns :
       # 'EWT_SH' entering water temperature source side, heating mode (F)
       # 'EWT_SC' entering water temperature source side, cooling mode (F)
       # 'DEB_LC' water flow rate load side cooling mode (gfm)
       # 'DEB_SH' water flow rate source side, heating mode (gpm)
       # 'DEB_SC' water flow rate, source side cooling mode (gpm)
       # 'DEB_LH' water flow rate load side heating mode mode (gfm)
       # 'TC' (cooling total capacity (kBTU/hr))
       # 'POWC' (power cooling mode (kW)), 'EER' (cooling mode)
       # 'HCH' (heating total capacity (kBTU/hr)),'POWH' (power heating mode (kW))
       # 'COP' (heating mode)

Example:
```
> > >heat_pump = '..\\data\\nsw025.xls'
> > >df = pd.read_excel(heat_pump)
> > >EWTS = 30 # F
> > >gpms = 5.5
> > >EWTL = 80
> > >gpml = 5.5
> > >TC,Pow,COP = WW_heat_pump(EWTS,gpms,EWTL,gpml,
'heating',df)
> > >print ('nominal total capacity = ',TC, ' kbtu/hr' )
nominal total capacity = 22.1 kbtu/hr
> > >print ('nominal power = ',Pow, ' kW')
nominal power = 1.68 kW
> > >print ('nominal COP = ',COP)
nominal COP = 3.87

> > >EWTS = 70 # F
> > >EWTL = 70
> > >TC,Pow,EER = WW_heat_pump(EWTS,gpms,EWTL,gpml,
'cooling',df)
> > >print ('nominal total capacity = ',TC, ' kbtu/hr' )
nominal total capacity = 29.4 kbtu/hr
> > >print ('nominal power = ',Pow, ' kW')
nominal power = 1.48 kW
> > >print ('nominal EER = ',EER)
nominal EER = 19.9
```

D.4   Hydraulic Library (hydraulic_md.py)

Colebrook(Re,eD=0)
>	# Evaluate the friction factor f using the Colebrook relation
>	# Re : Reynolds number, eD = epsilon/D

Example:
```
>>> from hydraulic_md import *
>>> Colebrook(15000,0.001)
0.02961128491536815
>>> Colebrook(15000)
0.02818510786784595
```

motor_efficiency(HP,x_mot,case = 'H')
>	# gives the motor electrical efficiency (8.5)
>	# HP Rated power (HP)
>	# x_mot : part load < 1
>	# case : 'H', 'M' or 'L'

Example:
```
>>> motor_efficiency(2,0.7)
0.852672863601433
>>> motor_efficiency(2,0.7,'L')
0.7314306204955595
```

VFD_efficiency(HP,x_vfd)
>	# gives the VFD electrical efficiency (8.6)
>	# HP Rated power (HP)
>	# x_vfd : part load < 1
>	# case : 'H', 'M' or 'L'

Example:
```
>>>VFD_efficiency(2,0.7)
0.927696463261273
```

sing_head_loss(type,size)
>	# Equivalent length (in m) of some piping defined in:
>	# '..\\data\\SI_equivalent_lengths.csv'
>	# type : 'Socket U-bend','Socket 90','Socket tee-branch',
>	# 'Socket tee-straight','Socket reducer','Butt U-bend',
>	# 'Butt 90','Butt tee-branch','Butt tee-straight',
>	# 'Butt reducer','Unicoil'
>	# size (nominal inch) : 0.75,1,1.25,1,2,3,4,6
>	# Other can be added if needed

Example:
```
>>>Leq = sing_head_loss('Butt 90',1.25)
>>>print(Leq) 3.0 >>>Leq = sing_head_loss('Butt 90',1.75)
No value given for this size
```

Cv_valve(type,siz)
>	# Cv coefficient (in $m^3/hr/\sqrt{bar}$) of some piping defined in:
>	# '..\\data\\SI_Cv_valves.csv'
>	# type : 'Ball valves', 'Swing check valves',
>	# 'Hoze kit', 'Y-strainer'
>	# size (nominal inch) : 0.75,1,1.25,1,2,3,4,6,8
>	# Other can be added if needed

Example:
```
>>>Cv1 = Cv_valve('Ball valves',2)
>>>print(Cv1)
91
```

class network:

See Appendix C

D.5   Finance Library (finance_md.py)

pwf(N,i,d,cas = 'end')
>	# gives the present worth factor
>	# N : number of terms
>	# i : inflation rate (fraction)
>	# d : discount rate (fraction)
>	# cas : 'end' inflation at the end of term
>	# cas : 'deb' inflation at the start of the term

Example:
```
>>>from finance_md import *
>>>pwf(20,0,0.08)
9.818147407449297
```

fwf(N,i,d,cas = 'end')
>	# gives the future worth factor
>	# N : number of terms
>	# i : inflation rate (fraction)
>	# d : discount rate (fraction)
>	# cas : 'end' inflation at the end of term
>	# cas : 'deb' inflation at the start of the term

Example:
```
>>>fwf(20,0,0.08)
45.76196429811637
```

present_value(A,N,i,d,cas = 'end')
>	# gives the present value of a cash flow
>	# that will occur after N periods
>	# A : value of future cash flow
>	# i : inflation rate (fraction)
>	# d : discount rate (fraction)
>	# cas : 'end' inflation at the end of term
>	# cas : 'deb' inflation at the start of the term

Example:
```
>>>present_value(1000,11,0,0.08)
428.88285933767054
```

mirr(dpa,gains,rr,fr)
>	# gives the modified internal rate return
>	# dpa : down paiment
>	# rr : reinvestment rate
>	# fr : financial rate
```
>>>dpa = 100000
>>>rr = 0.12
>>>fr = 0.1
>>>gains = array([39000,30000,-21000,10000,46000])
>>>miir = mirr(dpa,gains,rr,fr)
>>>print ('miir is = ',miir)
miir is = 0.06779139635193943
```

D.6   Design Library (design_md.py)

class ground:
>	# Attributes:
>	# ksoil : soil conductivity (W/m-K)
>	# alsoil : soil diffusivity (m^2/s)
>	# To : Undisturbed ground temperature
>	# def __init__(self, ksoil=1, alsoil = 1e-6, To = 10):

```
class borehole:
      # Attributes:
      # nx : number of boreholes in x-direction
      # ny : number of boreholes in y-direction
      # Rb : borehole resistance (m-K/W)
      # Rint : borehole internal resistance (m-K/W)
      # rb : borehole radius (m)
      # dist : distance between boreholes (m)
      # CCf : product flow rate times specific heat (W/K)
      # def __init__(self, nx=1, ny = 1, dist = 6.0, rb = 0.07, Rb =
0.1, CCf=2000, Rint = 0.4):

class ashrae_params:
      # parameters for the ashrae method:
      # Attributes:
      # qa : annual pulse (W)
      # qm_heating : monthly pulse heating (W)
      # qm_cooling : monthly pulse cooling (W)
      # qh_heating : hourly pulse heating (W)
      # qh_cooling : hourly pulse cooling (W)
      # Tfo_heating : min temp heating (°C)
      # Tfo_cooling : max temp cooling (°C)
      # n_years : number of years
      # n_bloc : number of hours (peak load)
      # n_rings : number of rings for penalty temperature if flag_Tp
= 'ASHRAE'
      # flag_Tp = 'ASHRAE', 'ILS', 'FLS', 'Bernier'
      # flag_inter = pipe-to-pipe heat transfe if 'TRUE'
      # def __init__(self, qa=0, qm_heating=0, qm_cooling=0,
qh_heating=0, qh_cooling=0,Tfo_heating = 0,
      # Tfo_cooling = 25,n_years = 20, n_bloc = 4,n_rings =
3,flag_Tp = 'ASHRAE',flag_inter = False):

class eed_params:
      # parameters for the swedish method:
      # Attributes:
      # q_months : monthly ground loads (W)
      # qh_heating : hourly pulse heating (W)
      # qh_cooling : hourly pulse cooling (W)
      # Tfo_heating : min temp heating (°C)
      # Tfo_cooling : max temp cooling (°C)
      # init_month : first month of the simulation
      # n_years : number of years
      # n_bloc : number of hours (peak load)
      # n_rings : number of rings for penalty temperature if flag_Tp
= 'ASHRAE'
      # zo : borehole depth (m)
      # flag_inter = pipe-to-pipe heat transfe if 'TRUE'
      # def __init__(self, init_month = 0, q_months =[],
qh_heating=0, qh_cooling=0,
      # Tfo_heating = 0,Tfo_cooling = 25,n_years = 20, n_bloc =
4,zo = 4,flag_inter = False):

class borefield:        # Attributes:
      # params : ashrae_params or eed_params instance
      # ground : ground instance
      # borehole : borehole instance
      # Methods:
      # Compute_L_ashrae() : calculate length with the ashrae
method
      # Compute_L_eed() : calculate length with the swedish method
```
Example:
```
> > >from design_md import *
> > >my_ground = ground(ksoil=2,alsoil=1e-6,To = 10)
> > >my_borehole = borehole(nx=1,ny=1,rb = 0.07,Rb = 0.1,Rint =
0.2,CCf = 3000)
> > >parama = ashrae_params(qa= 200,qm_heating = 1500,
qm_cooling = 0,
```

```
qh_heating= 8000,qh_cooling=0,Tfo_heating = 0,
Tfo_cooling = 25,n_years = 10,n_bloc = 4)
> > >first_design = borefield(params = parama,ground = my_ground,
borehole = my_borehole)
> > >L1 = first_design.Compute_L_ashrae()
> > >print('L = ',L1)
L = 164.25562190380612
> > >print (first_design.Rbs)
0.1 > > >parama.flag_inter = True
> > >L3 = first_design.Compute_L_ashrae()
> > >print('L = ',L3)
L = 167.90511622477027
> > >print (first_design.Rbs)

0.10517011695469923
```

D.7   Conversion (conversion_md.py)

```
def m3s_cfm (cfm=1):
def m3s_gpm (gpm=1):
def cfm_m3s (mcs=1):
def gpm_m3s (mcs=1):
def gpm_ls (ls=1):
def ls_gpm (gpm=1):
def ls_cfm (cfm=1):
def cfm_ls (ls=1):
def F_C (C=1):
def C_F (F=1):
def ls_kW_gpm_ton(gpm_ton):
def W_m2K_BTU_hrft2F (BTU_hrft2F=1):
def W_mK_BTU_hrftF (BTU_hrftF=1):
def W_HP (HP=1):
def HP_W (W=1):
def W_BTUhr (BTU_hr=1):
def BTUhr_W (W = 1):
def W_kW_hp_ton(hp_ton):
def J_BTU (btu =1):
def MJ_m2_kBTU_f2 (kbtu_f2 =1):
def W_hp (hp = 1):
def kWh_MTOE(mtoe = 1):
def MJ_MTOE(mtoe = 1):
def Pa_ftw (ftw = 1):
def Pa_inw (inw = 1):
def inw_Pa (Pa = 1):
def mw_psi (psi = 1):
def kPa_psi (psi = 1):
def Pa_bar(bar = 1):
def psi_bar(bar = 1):
def psi_kPa(kpa = 1):
def bar_psi(psi = 1):
def kPa_psi(psi = 1):
def m_ft (ft =1):
def m2_ft2(ft =1):
def m3_ft3(ft =1):
def m3_gal(gal =1):
def gal_m3(m = 1):
def ft_m (m =1):
def ft2_m2 (m =1):
def ft3_m3 (m =1):
def gal_m3 (m =1):
def m3_gal (g =1):
def kg_lb(lb = 1):
def lb_kg(kg = 1):
def gpm_psi_m3h_bar(kv = 1):
def gpm_psi_ls_kPa(kv = 1):
```

```
def m3h_bar_gpm_psi(cv = 1):
def ls_kPa_gpm_psi(cv = 1):
def m2hr_bar_gpm_psi(cv = 1):
```

References

1. Water Furnace, Specification Catalog. https://www.waterfurnace. com/literature/5series/sc2500an.pdf. Accessed 17 July 2019

2. Climate Master, Tranquility Series. https://www.climatemaster. com/download/18.274be999165850ccd5b5b73/1535543867815/ lc377-climatemaster-commercial-tranquility-20-single-stage-ts-series-water-source-heat-pump-submittal-set.pdf.   Accessed   17 July 2021

3. Water Furnace, Specification Catalog. https://www.waterfurnace. com/literature/5series/sc2506wn.pdf. Accessed 17 July 2019

Index

© The Editor(s) (if applicable) and The Author(s), under exclusive license to Springer Nature Switzerland AG 2023
L. Lamarche, *Fundamentals of Geothermal Heat Pump Systems*,
https://doi.org/10.1007/978-3-031-32176-4

Printed in the United States
by Baker & Taylor Publisher Services